This book retells the history of Western industrialization revealing possibilities unexplored in the nineteenth century, variants of which have come to transform the economies of the late twentieth. It shows that economic actors have historically been more aware of the great strategic choices they faced than standard theory credits them for being, and this surprising acuity allows them to imagine and put into practice solutions which current theories of industrial organization have scarcely anticipated. The book is therefore at one and the same time a contribution to a substantive revision of the history of mechanized production and a propaedeutic in a form of explanation that approximates the knowledge of the actor to the knowledge of the theorist. The volume groups essays presented by a multinational team of historians and social scientists drawing on intensive primary research on a wide range of firms, regions, sectors and national economies in Western Europe and the United States from the eighteenth century to the present day.

Studies in Modern Captalism*Etudes sur le Capitalisme Moderne

World of Possibilities

Studies in modern captalism . Etudes sur le capitalisme moderne

This series is devoted to an attempt to comprehend capitalism as a world-system. It will include monographs, collections of essays and colloquia around specific themes, written by historians and social scientists united by a common concern for the study of large-scale long-term social structure and social change.

The series is a joint enterprise of the Maison des Sciences de l'Homme in Paris and the Fernand Braudel Center for the Study of Economies, Historical Systems, and Civilizations at the State University of New York at Binghamton.

Other books in the series

MARIE-CLAIRE BERGERE: *The Golden Age of the Chinese bourgeoisie 1911–1937*

IVAN T. BEREND AND GYORGY RANKI: *The European periphery and industrialization, 1780–1914*

FERENC FEHER: *The frozen revolution: An essay on Jacobinism*

GEORGES HAUPT: *Aspects of international socialism 1871–1914*

HURI ISLAMOGLU-INAN (ed.): *The Ottoman empire and the world-economy*

CAGLAR KEYDER: *The definition of a peripheral economy: Turkey 1923–1929*

LUISA PASSERINI: *Fascism and popular memory: the cultural experience of the Turin working class*

LUCETTE VALENSI: *Tunisian peasants in the eighteenth and nineteenth centuries*

IMMANUEL WALLERSTEIN: *The capitalist world-economy: essays*

IMMANUEL WALLERSTEIN: *The politics of the world-economy: the states, the movements and the civilisations*

IMMANUEL WALLERSTEIN: *Geopolitics and geoculture: essays on the changing world-system*

STUART WOOLF (ed.): *Domestic strategies: work and family in France and Italy 1600–1800*

This book is published as part of the joint publishing agreement established in 1977 between the Fondation de la Maison des Sciences de l'Homme and the Press Syndicate of the University of Cambridge. Titles published under this arrangement may appear in any European language or, in the case of volumes of collected essays, in several languages.

New books will appear either as individual titles or in one of the series which the Maison des Sciences de l'Homme and the Cambridge University Press have jointly agreed to publish. All books published jointly by the Maison des Sciences de l'Homme and the Cambridge University Press will be distributed by the Press throughout the world.

World of Possibilities

Flexibility and Mass Production in Western Industrialization

Edited by

Charles F. Sabel

Columbia University School of Law

and

Jonathan Zeitlin

University of Wisconsin-Madison

MAISON DES SCIENCES DE L'HOMME

CAMBRIDGE
UNIVERSITY PRESS

PUBLISHED BY THE PRESS SYNDICATE OF THE UNIVERSITY OF CAMBRIDGE
The Pitt Building, Trumpington Street, Cambridge CB2 1RP, United Kingdom

CAMBRIDGE UNIVERSITY PRESS
40 West 20th Street, New York, NY 10011–4211, USA
10 Stamford Road, Oakleigh, Melbourne 3166, Australia

First published 1997

Printed in the United Kingdom at the University Press, Cambridge

Typeset in Plantin 9.5/11.5

A catalogue record for this book is available from the British Library

Library of Congress cataloguing in publication data

World of possibilities: flexibility and mass production in western industrialization / edited by Charles F. Sabel and Jonathan Zeitlin.
p. cm. – (Studies in modern capitalism = Etudes sur le capitalisme moderne)
Papers from an international seminar on "Historical alternatives to mass production."
ISBN 0 521 49555 5
1. Industrialization–History–19th century–Congresses.
2. Economic history–1750–1918–Congresses.
I. Sabel, Charles F. II. Zeitlin, Jonathan.
III. Maison des sciences de l'homme (Paris, France)
IV. Series: Studies in modern capitalism.
HC53.W63 1997
338.09–dc20 96–13016 CIP

ISBN 0 521 49555 5 hardback
ISBN 2 7351 0751 5 (hardback France only)

VN

Contents

Contributors

HÅKON WITH ANDERSEN is Professor of Modern History at the University of Trondheim.

RUDOLF BOCH is Professor of Economic History at the University of Chemnitz-Zwickau.

VITTORIO CAPECCHI is Professor of Sociology at the University of Bologna.

ALAIN COTTEREAU is a CNRS researcher at the Centre d'Etudes des Mouvements Sociaux, Ecole des Hautes Etudes en Sciences Sociales, Paris.

ALAIN DEWERPE is Director of Studies in History at the Ecole des Hautes Etudes en Sciences Sociales, Paris.

PEER HULL KRISTENSEN is Associate Professor of Sociology at Copenhagen Business School.

CARLO PONI is Professor of Economic History at the University of Bologna.

CHARLES F. SABEL is Professor at the Columbia University School of Law.

JEAN SAGLIO is Director of Research at the CNRS workshop Economie des Changements Techniques (Université Lumière Lyon 2 and Maison Rhône-Alpes des Sciences de l'Homme) and the Ministère du Travail, de l'Emploi et de la Formation Professionnelle.

PHILIP SCRANTON is Professor of History at Rutgers University in Camden, New Jersey, and Director of the Center for the History of Business, Technology, and Society at the Hagley Museum and Library, Wilmington, Delaware.

BEATRICE VEYRASSAT teaches economic history at the University of Geneva.

JONATHAN ZEITLIN is Professor of History, Sociology and Industrial Relations at the University of Wisconsin-Madison.

Acknowledgments

This volume, as the introduction explains more fully, is the fruit of an international seminar on "Historical Alternatives to Mass Production" sponsored by the Maison des Sciences de l'Homme. Our first and greatest debt is thus to the other participants in the seminar, who made it an extraordinary model of rigorous yet generous intellectual exchange, especially those whose contributions could not be included here: Yves Schwartz, Michael Sonenscher, John Styles and Steven Tolliday. We would also like to thank Maurice Aymard of the MSH for his indispensable encouragement and support; Eric Hobsbawm for originally suggesting the formation of a seminar on this theme; Jean Saglio and Yves Lequin of the Maison Rhône-Alpes des Sciences del l'Homme for hosting two productive and enjoyable meetings in Lyons; and the anonymous referees of Cambridge University Press for their helpful comments and criticisms.

Stories, strategies, structures: rethinking historical alternatives to mass production

Charles F. Sabel and Jonathan Zeitlin

This book grows out of an effort to rethink the history of industry and the writing of history more generally in the light of the vast transformation of the advanced economies that began nearly twenty years ago and may continue for at least as long into the future. Its proximate cause was a Parisian seminar in which economic historians and sociologists, predominantly from Western Europe, responded in studies of their own to our earlier work on artisanal, flexible alternatives to mass production,[1] and we responded in many joint discussions and now in this Introduction to them.

There is nothing novel in such an undertaking, for the most common of historical commonplaces is that every epoch rewrites history according to its own preoccupations. Yet predictable as they are, such efforts often produce results that disconcert by calling into question connections to the past that were among the few fixed points left in a turbulent age.

The essays in this volume may well disconcert in a narrower intellectual sense as well. Most current economic history is meant to show that methods and models for understanding the economy in the present can illuminate the rationality of previously obscure realms of business activity in the past. But our aim is to show that the strategic reflections and deliberate institutional innovations of nineteenth- and early twentieth-century economies are no less instructive than any analysis based on current theory. The participants in the seminar that gave rise to this volume are united by the conviction that these historical economies, be they composed of large firms or small, must be taken figuratively as worthy interlocutors, not, at best, as bright natives who have anticipated one or another result of modern economics and organizational theory. Indeed, because of certain affinities between our age and the formative period of mechanization, it is arguable that we have as much to learn from such a hypothetical exchange as they.

Two experiences, we believe, characterize the economic life of our age and distinguish it from the immediately preceding history that formed the horizon

[1] See Charles F. Sabel and Jonathan Zeitlin. "Historical Alternatives to Mass Production: Politics, Markets and Technology in Nineteenth-Century Industrialization," *Past and Present*, 108 (August 1985), 133–176.

of our deepest expectations. The first is the sense of fragility, and especially of institutional fragility as a continuing, perhaps permanent feature of economic life. It is not so much the idea of progress itself which is at stake here. Like it or not, it is hard to escape the conclusion that our entire economic life is being continually transformed by the introduction of new technologies and organizational forms; and in that minimal sense it is indubitably progressing. Rather the sense of fragility goes to the once commonsensical idea that progress would lead to the gradual consolidation of particular forms of economic organization, and hence to an ever more certain sense of how best to deploy technology, allocate labor and capital, and link supply of particular products to demand. Today, on the contrary, it is commonsensical to believe that the way many of these things are done depends on constantly shifting background conditions whose almost insensible mutation can produce abrupt redefinitions of the appropriate way to organize economic activity.[2] For the economic agents themselves, this perception is expressed in the resigned and monitory truism that nothing is so constant as change itself.[3] For historians, as well as for other social scientists, the study of the economy has become the study of adjustment to ever changing contexts: the principles which explain at once the stability and vulnerability to change of these contexts, and the turmoil characteristic of the transition from one to the next.[4]

The second experience is one of the recombinability and interpenetration of different forms of economic organization: the rigid and the flexible, the putatively archaic and the certifiably modern, the hierarchical and the market-conforming, the trusting and the mistrustful. During much of the post-war period, the self-conception and the strategic choices of the economic actors in the advanced countries were firmly rooted in the belief that these distinctions meant something. Indeed, it was precisely by drawing these distinctions that they and the economists, historians and sociologists closest to them came to define their identity in the first place and to establish a vision of historical progress in which they were the heirs and culmination of all that had gone before.[5] Today these

[2] For one of many statements of this perspective, see Michael J. Piore and Charles F. Sabel, *The Second Industrial Divide: Possibilities for Prosperity* (New York: Basic Books, 1984).

[3] See for example Tom Peters, *Thriving on Chaos: Handbook for a Management Revolution* (New York: Knopf, 1987); and Rosabeth Moss Kanter, *When Giants Learn to Dance* (New York: Simon and Schuster, 1989).

[4] For examples drawn respectively from the sociology of organizations, accounting and economic history, see Neil Fligstein, *The Transformation of Corporate Control* (Cambridge, MA: Harvard University Press, 1990); H. Thomas Johnson and Robert S. Kaplan, *Relevance Lost: The Rise and Fall of Management Accounting* (Boston: Harvard Business School Press, 1987); and Edward H. Lorenz, *Economic Decline in Britain: The Shipbuilding Industry, 1890–1970* (Oxford: Clarendon Press, 1991).

[5] Important statements of this view include Clark Kerr, John T. Dunlop, Frederick H. Harbison, and Charles Meyer, *Industrialism and Industrial Man* (Cambridge MA: Harvard University Press, 1960); Walt W. Rostow, *The Stages of Economic Growth: A Non-Communist Manifesto* (Cambridge: Cambridge University Press, 1960); Daniel Lerner, *The Passing of Traditional Society* (Glencoe IL: The Free Press, 1958); and Alfred D. Chandler, Jr., *The Visible Hand: The Managerial Revolution in American Business* (Cambridge MA: Harvard University Press, 1977).

distinctions, to the extent that they are honored at all, are honored in the breach. It is not that the economic actors have set out deliberately to destroy old certitudes and transgress established boundaries, but suddenly the repertoire of economic forms deemed appropriate to current conditions contains types such as the small firm which twenty years ago were viewed as close to extinction and combinations of types – such as the small contractor collaborating as an equal with a much larger customer in the design of a new product – which were quite literally unthinkable. It is as though the prehistoric and imaginary creatures in the industrial bestiary had suddenly come to life.

The central theme of this book is that the experience of fragility and mutability which seemed so novel and disorienting today has been, in fact, the definitive experience of the economic actors in many sectors, countries and epochs in the history of industrial capitalism. Precisely because they have been aware of the complex dependence of every form of economic organization on multiple and shifting background conditions, they have constantly experimented with institutional designs that until recently would have been judged economic solecisms. For the same reasons, they have rarely interpreted economic and technological progress as continual and ineluctable progression towards a single set of practices that in their self-perfection would ultimately pass into a sphere of transhistorical permanency. But this double perception of mutability and fragility, we will argue further, has not led them to exalt catch-as-catch-can muddling through as the organizing principle of reflection and action. What we find instead is an extraordinarily judicious, well-informed and continuing debate within firms, and between them and public authorities, as to the appropriate responses to an economy whose future is uncertain, but whose boundary conditions at least in the middle term are taken to be clear. To anticipate, where many observers in the post-war period saw the economy as steadily increasing in efficiency through the ever more specialized use of resources, and therefore paying an acceptable price in increased rigidity for previously unimaginable increases in well-being, throughout most of the history of industrial capitalism, and again today, the economic actors have tried with considerable success to increase efficiency without jeopardizing and indeed sometimes even increasing flexibility.

To reconnect this experience of the present to a new apprehension of the past is a three-part undertaking. The first step (part I) is to give an analytically disciplined account of the economic actors' sense of the fragility of their world, and the implications of this experience for their strategic choices. Our purpose here is to show that most firms in nineteenth- and early twentieth-century Europe and the United States, neither mired in tradition nor blinded by the prospect of a radiant future, carefully weighed the choices between mass production and what we would now call flexible specialization. Where possible, they developed sophisticated hedges for reducing their risks by avoiding a definitive choice in favor of either alternative. Where choice was obligatory, they hedged again by anticipating in the design of one organization the need to reconstruct it for alternative purposes. Understanding how the actors could engage in this kind of strategic reflection and how their choices influenced the structure of the economy will require us to rethink the place of the firm in economic history. It will also shed light on a wide range of organizational forms which have either escaped historians' attention or been dismissed as cases of stunted growth, failed transitions or institutional degeneration.

The second step (part II) is to look for some institutional regularities that underpin the frenzied recombination of economic forms both today and in the industrial past. As firms hedge against volatility by adopting new organizational forms, we shall see, they encounter anew problems of economic governance – especially, how to create and police trust without by those very efforts destroying it – that were familiar to their nineteenth-century counterparts. Then as now, the favored solution to these problems was an apparently syncretic combination of organizational features that moots conventional distinctions between the traditional and the modern and between the public or collective and the private. By extending our discussion of hedging responses and the agents' relation to their context upon which these responses depend, we aim here as well to discover the compositional logic which renders orderly and meaningful an otherwise ramshackle assembly of monitoring systems, collective bargaining arrangements and public provision of services which firms normally provide themselves.

Finally, by way of summary and elaboration, we want to indicate how this new way of speaking about economic history changes substantively the way the history of mechanization is told. Most accounts of that history, our own included, have emphasized epochal change. Typically the emphasis has fallen on a transition from a world of decentralized handicraft production to a world of concentrated factories in which specialized machines turn out standard products.[6] Our earlier account placed the emphasis on a seesaw battle between two forms of mechanization, one familiar from the standard accounts and the other employing flexible machines to manufacture a much more diversified range of products. The lesson we draw from this seminar is that the epochs are less epochal and the choices less stark than it once appeared. In the light of the actors' sense of their own vulnerability to a changing environment and their capacity to hedge against that vulnerability by developing complex hybrid forms, the fauna of economic life appear at once more differentiated and more restricted in their range of variation than we previously supposed.

Because firms are good at hedging their organizational bets about future changes in the environment, and because these hedges often require the revitalization of earlier features of economic organization, there are fewer outward manifestations of revolutions or breaks in economic history than analysis of the actors' own perceptions of changing standards of efficiency and their institutional concomitants suggests. Precisely because the agents know how radically views of productive efficiency can change, we found, they tended to build organizations which minimized the disruption of acting on their new

[6] The classic synthesis of this view of the history of mechanization is David S. Landes, *The Unbound Prometheus: Technological Change and Industrial Development in Western Europe from 1750 to the Present* (Cambridge: Cambridge University Press, 1972).

convictions. Seen this way, each period of economic history anticipates many of the strategies of its successors and ultimately comes to pursue them with means that it has inherited from its predecessors. Hence the very idea of distinguishing radically different epochs of development according to their dominant economic strategy becomes as debatable as the idea of trying unambiguously to distinguish firms within any one period according to their differing strategic orientations. The notions of contrasting strategies and distinct practices do, as we shall see, continue to serve an important analytic role: it is the actors' perception of the advantages and disadvantages of polar possibilities which leads them to hedge their strategies in the first place; and hence it is the appreciation of the full range of possible diversity which provokes the search for ever more various ways of avoiding risky bets on the extreme positions.

To avoid confusion among those readers who, like ourselves, are not fully habituated to the postmodern notion that all texts, being commentaries on preceding ones, are in the end produced by them, and that given such inter-textuality the distinctions between reading and writing or commenting and composing, or even different texts, become negligible, we need to clarify at the outset who we are. Henceforth, as authors and editors, we will speak for ourselves of what we learned in contributing to this volume. As we would like to believe that what we learned is what the other contributors were teaching us and themselves, we will associate our conclusions with theirs. But being ourselves sceptical of the notion that in conversation all ideas are one, we invite scepticism on your part as well, and advise even more cautious comparison than usual of our introductory readings and the essays that follow.

Narratives, models and strategies

Economic agents, we found again and again in the course of the seminar's work, do not maximize so much as they strategize. By this cryptic locution, we mean that they are at least as much concerned with determining, in all senses, the context they are in as they are in pursuing what they take to be their advantage within any context. Self-interested adjustment to conditions taken as given thus proceeds paradoxically yet as a matter of course together with efforts to find or create a more advantageous set of constraints. In short, the decisions of the agents in the cases under consideration here very often disregard Schumpeter's classic distinction between adaptive reactions which accept given constraints as binding and creative ones which do not.[7] This section argues that dichotomies of this sort, whatever their utility in extreme cases, make it difficult to grasp the calculating deliberations of economic agents in a wide range of circumstances. They consequently obstruct interpretation of the organizational designs which

[7] For this distinction see Joseph Schumpeter, *The Theory of Economic Development* (Oxford: Oxford University Press, 1961), especially p. 60; and the discussion of Schumpter's ideas in William Lazonick, "Industrial Organization and Technology Change: The Decline of the British Cotton Industry," *Business History Review*, 57, 2 (1983), 491–516.

result from their considerations. To formulate a more adequate view, we begin by criticizing the rigid distinction between maximizing actor and constraining context which underlies this and related dichotomies that ramify throughout most of economic history. Our aim is to clear the way for a recharacterization of the actors that takes account both of their capacity for self-reflection and the resulting interplay between the constitution of their interests and identity on the one hand and the context within which they operate on the other.

Economic historians are fond of saying that their work enriches the understanding of the firm as a maximizing agent presented in standard economic theory. Most such claims are wishful thinking, as some recent work in economic history itself acknowledges.[8] As a rule, in fact, economic history merely illustrates how firms or groups of firms take advantage or not of the opportunities for profit presented by particular circumstances. The underlying notion is that one set of circumstances, call them accidental or historical as you will, offer these opportunities and a second, perhaps distantly related set, determine the use to which they are put. Contexts constrain, actors react; and economic history traces the circumstances under which actors come to appreciate and act in accordance with the logic of their situation.

There are, to be sure, important variations on this theme. One trope is simply to demonstrate that however ruinous the outcome, profit-seeking individuals or firms subject to the relevant constraints would have been ill-advised to do otherwise than the historical actors in fact did. The point of the analysis is to show precisely that economic actors are rational in the sense of pursuing their best alternatives however impoverished or disastrous these may be. Like the lawyers who cry *fiat justitia, pereat mundus*, economic historians of this persuasion seem to find it almost ennobling that real persons could pursue the logic of short-term maximization even at the cost of their long-term decline.[9]

A second trope of analysis, often favored by those scornful of the first, looks beyond instances of adjustment to local circumstance to those rarer but more spectacular instances where firms or individuals extend the productive powers of nations and civilizations by grasping and turning to their immediate benefit some essential feature of an alleged logic of material progress. Accounts of this sort have a long and familiar pedigree extending back to such founders of classical political economy as Smith and Marx. Recast as business history, post-war versions of this story conflated the discovery by the large American corporation of special-purpose machines and the administrative structures required to manage high-throughput mass-distribution manufacturing with the perfection of human

[8] See for example Peter Temin (ed.), *Inside the Business Enterprise: Historical Perspectives on the Use of Information* (Chicago: University of Chicago Press, 1991).

[9] The most extreme statements of this position are associated with the attempt to rehabilitate the rationality of Victorian enterpreneurs whose attachment to gentlemanly values had been held to have inhibited their pursuit of business advantage: for an overview of this research, see Donald N. McCloskey and Lars G. Sandberg, "From Damnation to Redemption: Judgments on the Late Victorian Entrepreneur," in Donald N. McCloskey, *Enterprise and Trade in Victorian Britain* (London: Allen and Unwin, 1981), pp. 55–72.

productive powers.[10] More recent versions give pride of place to the large Japanese corporations, whose success in besting the American competition is attributed to the resolution of the classical conflict between gains in efficiency and losses in flexibility arising from specialization of resources by soliciting the participation of shop-floor workers in the continuous adjustment of the firm to changing circumstances.[11] But in all their variants, these accounts bear the marks of Greek tragedy, for they bind the heroic exercise of inventive freedom to a chastening, even humiliating recognition of the economic constraints imposed by the historical logic of material progress. The demiurge becomes in the end a humble problem-solver, a common enough problem as we will see later.[12]

Note furthermore that both managers and entrepreneurs can be the protagonists of economic histories about adjustment to local circumstance as well as those that recount the alleged *telos* of mechanization. In some accounts of the failure of Lancashire cotton firms to adopt new spinning and weaving technologies, it is individual entrepreneurs who are caught in the trap of short-term maximization within given constraints.[13] In some accounts of the failure of American corporations to avail themselves of the innovations exploited by the Japanese, on the other hand, it is the incrementalism of professional managers bound by accounting rules of their own making who are held responsible.[14] Alternatively, in many accounts of the nineteenth-century rise of these same corporations, such managers acting as a collective body are credited with the decisive administrative and technical breakthroughs.[15] But in several of the most recent writings on the creation of electric utilities, conversely, the shape of the industry as a whole is determined by the particular interventions and alliances of individual manager-

[10] See Alfred G. Chandler, Jr., *Strategy and Structure: Chapters in the History of American Enterprise* (Cambridge MA: MIT Press, 1962); *The Visible Hand; Scale and Scope: The Dynamics of Industrial Capitalism* (Cambridge MA: Harvard University Press, 1990).

[11] See James P. Womack, Daniel T. Jones, and Daniel Roos, *The Machine that Changed the World* (New York: Rawson Associates, 1990).

[12] The clearest example of a writer who appreciates these tragic possibilities is David Landes, who believes firmly that economic and technological progress culminates in something like the modern American corporation, and yet writes almost exclusively about the ultimately futile struggles of entrepreneurs and family firms to create a world in which domesticity and industry are united under their sovereign control. See particularly his "Religion and Enterprise: The Case of the French Textile Industry," in Edward G. Carter II, Robert Forster and John N. Moody (eds.), *Enterprise and Entrepreneurs in Nineteenth- and Twentieth-Century France* (Baltimore: Johns Hopkins University Press, 1976), pp. 41–86; *Revolution in Time: Clocks and the Making of the Modern World* (Cambridge MA: Harvard University Press, 1983); and "Small is Beautiful. Small is Beautiful?," in Fondazione ASSI di Storia e Studi sull'Impresa-Instituto per la Storia dell'Umbria Contemporanea, *Piccola e grande impresa: un problema storico* (Milan: Franco Angeli, 1987), pp. 15–34.

[13] See William Lazonick, "The Cotton Industry," in Bernard Elbaum and William Lazonick (eds.), *The Decline of the British Economy* (Oxford: Clarendon Press, 1986), pp. 18–50; and *Competitive Advantage on the Shop Floor* (Cambridge MA: Harvard University Press, 1990), chs. 3–5; and Lazonick and William Mass, "The British Cotton Industry and International Comparative Advantage: The State of the Debates," *Business History*, 32, 4 (1990), 9–65.

[14] See Johnson and Kaplan, *Relevance Lost*.

[15] The classic statement of this view is Chandler, *The Visible Hand*.

entrepreneurs able to mobilize through their social and business connections the resources needed to impose the one solution among the many possible outcomes which they personally favored.[16] Other recent writing on the same theme, however, adopts a point of view much less tied to classical presuppositions and closer to our own; and we will return to it later.

In important ways, and despite efforts to the contrary, our own earlier work on historical alternatives to mass production remains ensnared in the rigid distinction between context and actor that we are criticizing. A central theme of that work was the claim that mechanization could have proceeded down a track characterized by the use of flexible equipment operated by skilled workers to produce specialized goods instead of down the path of mass production of standardized goods which in the event proved victorious. The argument turned on the demonstration that in those regions where skilled labor was ample and tied by guild institutions or agricultural inheritance patterns to a particular locale, where demand remained differentiated because of the continuing influence of aristocratic tastes or the aggregation through extensive trading networks of the diverse wants of far-flung customers, mechanization accommodated itself to the need for flexibility. Where, on the other hand, as in the United States, immigrant masses provided both an abundant supply of unskilled labor and an extensive market for standardized consumption goods, mass produc[illegible]nd having thus established its dominance in a key econo[illegible]nt continued, manufacturers elsewhere were moved by co[illegible]f it to emulate American success rather than press forwar[illegible]own promising but as yet unproven alternative. The whole process was compared to the neo-Darwinian idea of a branching point in which closely-related species arising in different environments come into competition and victory in the struggle depends not on the inherent superiority of one or the other but rather on the accidental advantages which they respectively draw from the historical conditions of the contest.[17] Alternatively the process was compared to the shift from

[16] These arguments come in two variants, both of which assume that technology is more malleable and hence less determinative of outcomes than in the view of classical political economy. In the first variant, associated with the work of McGuire, Granovetter and Schwartz, the outcome reflects the balance of power between rival networks that link business associates to sources of capital, whereas in the second variant, associated with that of Thomas Hughes, the outcomes reflect the particular relation between political authorities and system-building manager-entrepreneurs. Both variants aim to demonstrate that purely economic accounts cannot explain the outcome rather than to criticize the idea that the actors' decisions are completely determined by their context. For the first variant, see Patrick McGuire, Mark Granovetter, and Michael Schwartz, *The Social Construction of Industry: Human Agency in the Development, Diffusion, and Institutionalization of the Electric Utility Industry* (Cambridge: Cambridge University Press, forthcoming); for the second, see Thomas P. Hughes, *Networks of Power: Electrification in Western Society, 1880–1930* (Baltimore: Johns Hopkins University Press, 1983).

[17] The neo-Darwinian analogy was taken from the work of Stephen J. Gould: see his "Bushes and Ladders in Human Evolution," in *Ever Since Darwin* (New York: Norton, 1977), pp. 56–63. Evolutionary analogies are frequently applied in modeling technological development: see Richard R. Nelson and Sidney G. Winter, *An Evolutionary Theory of Economic Change* (Cambridge MA: Harvard University Press, 1982).

one paradigm, or set of organizing ideas about technology, to another, where the outcome depended in part on the sequence of breakthroughs in the competing schools and in part on the alignment of interests in the community of those forced to make choices among competing technologies.[18]

Read this way, as many readers plainly did, our story can be understood as an extension of the standard accounts, and even as a step towards reconciling the localist and teleological variants. Instead of one logic of development, we envisioned two mirror-image outcomes and a whole class of mixed cases between them. Accordingly, we attempted to show that mass production was a special case of a more general set of possible outcomes; but the same argument implied that some local responses to particular environments represented viable alternatives to historically dominant forms even if they were later marginalized by the course of events. Thus the distinction between breakthroughs that advanced productive capacities along the trajectory traced by the logic of economic development and profit-maximizing adjustments which imperiled the actors' future by serving their immediate interests gave way to a world in which a distinction between "creative" and "adaptive" responses, if meaningful at all, could only be determined in long historical retrospect. No matter that we also argued that consumption patterns and producers' expectations about the technologically possible and the commercially feasible, all of which helped shape the choice of mechanization strategy, were themselves the outcome of complex political struggles. Our extensions and reconciliations nonetheless played so much on conventional ideas of the relation between actors and environments that they could not help but reinforce them.

Nothing that we have learned in this seminar leads us to believe that technology or the economy is less malleable than we originally thought. What we have learned is that the economic agents are more continuously and subtly aware of that malleability than ever we imagined. In order to do justice to this awareness and its consequences for industrial organization, it is necessary to attack directly the distinction between actor and context and the neo-Darwinian analogies as they appeared in our rough-hewn presentation. Three overlapping criticisms of the evolutionary analogy have been adumbrated in current social-scientific debate, and it is useful to underscore them here by way of introduction to a new characterization of economic agency consistent with the seminar's findings.

Firstly, in evolutionary models strictly speaking, adaptation occurs through the extinction of populations unfit for their environment and the expansion of populations that are. In human societies and in economies more specifically, adaptation occurs through the reconstruction of existing institutions to meet new demands rather than the wholesale replacement of one set of institutions by another.[19] It is plain that whole societies very rarely die out whereas it is equally

[18] The reference here was to Thomas Kuhn, *The Structure of Scientific Revolutions*, 2nd edn (Chicago: University of Chicago Press, 1970); for the application of this notion to technological development, see Giovanni Dosi, "Technological Paradigms and Technological Trajectories," *Research Policy*, 11 (1982), 147–162.

[19] See Phillippe van Parijs, *Evolutionary Explanation in the Social Sciences* (Totowa NJ: Rowman and Littlefield, 1981).

[illegible]vious that the institutions of which any society is composed are transformed to [illegible] new purposes radically and repeatedly in historical time. It may appear that [illegible]nic adjustment, by contrast, proceeds more often by extinction than by [illegible]ction, because the rates of natality and mortality of firms in modern [illegible]re high enough to suggest the possibility of rapid regeneration by this m[illegible]s the experience of diverse economies in the early phases of mec[illegible]n, during the rise of mass production in the late nineteenth century, and in the current period of restructuring shows, the most profound changes in the organization of production also require extensive redefinition of social hierarchies, patterns of education and socialization, and the boundaries of the household. Adjustment through natality and mortality, in other words, carries the burden of change when the changes are small. Adaptation of existing institutions does the real work when there is real work to be done.[20]

Secondly, adjustment proceeds in society and the economy by adaptation rather than natural selection because humans are sentient and more particularly strategic in the sense that using their wits they can find indirect means to their ends. It may seem superfluous to raise this point because it is hard to see how any account of human activity as maximizing behavior of the sort typical of economic history could not at least tacitly assume that the agents are aware enough of the logic of their situation to act purposefully in response to it. But this idea of sentience and strategy as the virtually automatic response to an unambiguous situation strips self-consciousness of one of its defining features: the ability to consider alternatives, meaning alternative responses to the same situation and speculations about the relative possibilities of creating alternative situations.[21] Unless this were so, it would be hard to understand how individuals acting in society could reconstruct the formative institutions upon which economic activity depends as they evidently do from time to desperate time. Another way to make this point is to say that economic agents make their environment even as they are formed by it, whereas in most accounts of natural selection organisms adapt or otherwise to environments which are given by nature. In a moment we will see that the ability to imagine these possibilities is constitutive of the economic agents' identity and indispensable to explaining the hedging strategies by which they both adapt to change and reduce the need for adaptation.

Thirdly, social and economic adaptation proceeds more by the recombination and re-elaboration of existing institutions than by the introduction of wholly novel features as suggested in the cruder formulations of natural selection for more competitive mutations.[22] In some ways, this point mirrors the preceding

[20] See for example the reconstruction of Japanese society under the Meiji as a response to the threat of Western economic and military superiority in Eleanor Westney, *Imitation and Innovation: The Transfer of Western Organization Patterns to Meiji Japan* (Cambridge MA: Harvard University Press, 1987).

[21] See Pierre Bourdieu, *Outline of a Theory of Practice*, trans. Richard Nice (Cambridge: Cambridge University Press, 1977).

[22] For the importance of recombination and re-elaboration of ancestral features in evolutionary biology, see Stephen J. Gould, *Ontogeny and Phylogeny* (Cambridge MA: Harvard University Press, 1977).

one. The present is connected to the future by the possibility of imagining alternatives; the present is connected to the past because of the necessity of imagining the future as a re-elaboration, however fanciful, of what has gone before.

The characterization of economic agents that emerges against the backdrop of these distinctions is more familiar to those forms of sociology and philosophy which take the act of choosing in time among alternative courses of action as constitutive of individual and collective identities than it is in the economic histories of maximizing agents in their constraining contexts. A good example is the cognate schools of sociology in which individuals constitute themselves as such precisely by deciding just how – which is very often to say when – they will bring honor to themselves by honoring their social obligations; and society is reconstituted – and sometimes transformed – by the intended and unintended consequences of these decisions. Philosophical treatments begin with the intuition that human life is intelligible only as narration – a story with a beginning, middle and meaningful end. From this it follows that to be perceived as human all experience must be emplotted or embedded in a story whose outcome is uncertain yet constrained by prior events in the sense that whatever happens, however surprising its occurrence, must be immediately intelligible in the light of what has gone before once it has in fact occurred. Action in this philosophic world is strategic too, for to act at all is to contemplate a choice among possible future selves, each shaped and connected to the others by its derivation from a common past. Notice that generalizations, theories and models, as these terms are conventionally understood in scientific debate, can all be ingredients in the self-understanding of individuals or groups formed through such strategic choices. Who one is thus depends on and can be shaped by what one thinks as well as what one does and particular views are no less identifying merely for being exposed to scientific debate.[23]

[23] This idea of the connection of past and future in the construction of identity through self-definition is developed in a line of work on phenomenology, hermeneutics, and narration. Within this current broad of thought, one school holds that the most elemental human perception of, say, the undifferentiated passage of time is already emplotted in the sense of always presupposing a beginning and anticipating a conclusion. This is roughly speaking the position of Husserl. The contrary school holds that humans can have a pretheoretic experience of time or life as a series of disconnected events, but that almost all attempts to reflect on these experiences result in stories or models with narrative form. This is, again, roughly speaking, the position of Augustine and later writers on narrative such as Hayden White and Louis Mink. The first school believes, consequently, that narratives capture reality because human life is really narrative, whereas the second school believes that humans are condemned to falsify in narration what they experience as chronicle or chaos. Given that the only consequence of belief in the intelligibility of a pre-narrative world is to discredit narratives without showing how to increase their verisimilitude, we side by conviction and calculation with the first school. For an excellent analysis of these debates, see David Carr, *Time, Narrative and History* (Bloomington IN: Indiana University Press, 1986), and "Narrative and the Real World: An Argument For Continuity," *History and Theory*, 25, 2 (1986), pp. 117–131. Note particularly his discussion of Paul Ricoeur, who straddles both schools and whose *Time and Narrative*, trans. Kathleen McLaughlin and David Pellauer, 3 vols. (Chicago: University of Chicago Press, 1984–88) strongly influenced our own conclusions.

The seminar's findings both in their substance and the form in which they are presented redouble our rejection of the standard distinction between actors and context and corroborate instead the alternative view of agency as the strategic, simultaneous constitution of actors and contexts. Firstly and most directly, both in large firms and industrial districts of small ones, the actors were astonishingly well aware of the choices they faced. Whereas intellectuals from Marx to the present have commonly confused some distant possibility of automation with an imminent reality, the economic actors of the nineteenth and early twentieth centuries took a much more judicious view of the claims of machine vendors, successful competitors and intellectual enthusiasts for one or another scheme of rationalization. For one thing, the actors were extraordinarily well-informed about the operation and diversity of their foreign competitors. They could typically see, even when academic observers could not, connections between the use of machines and the organization of the firm on the one hand and the structure of the latter's markets and institutional environment on the other. This meant, paradoxically, that precisely because they could see the kinds of connections between context and strategic choice so dear to economic historians that their own choices were more nuanced and their own relation to their context – not least in its overseas periphery – more complex than the academic theories which served as the building blocks for their own practical deliberations suggest. What stands out is the repeated well-considered rejection of successful foreign models, not because they were new or foreign, but because they embodied a variant of modernity inappropriate to local circumstances broadly conceived.

During the eighteenth and early nineteenth centuries, for example, the silk merchants and weavers of Lyons carefully monitored but did not imitate the policies of design routinization, subdivision of labor and price competition pursued by their Spitalfields counterparts, preferring alternative strategies based on rapid style change, increasingly flexible machinery and the skillful exploitation of fashionable markets for high-value products. During the late nineteenth and early twentieth centuries, similarly, Swiss watchmakers, Solingen cutlers and British metalworking firms were equally attentive to the progress of the American system of mass production. Much as they admired the efficiency of American methods, detailed accounts of the American system in trade journals and technical society proceedings typically emphasized that this efficiency depended on standardization of the product which was wholly incompatible with the current or expected organization of their respective markets. As a rule, therefore, such writings advocated rationalization of traditional batch production rather than a change in strategy premised on a rapid homogenization of world demand. During the 1980s, finally, the Bolognese automatic machinery makers chose to respond to increasing Japanese competition by further specializing their products and rationalizing production accordingly rather than imitating the Japanese practice of offering fewer variants at lower prices.

But the best indication that the refusal to adopt successful foreign models wholesale was not based on insularity was the concomitant aggressiveness with which firms did incorporate those aspects of foreign experience which served their constantly evolving definition of locally appropriate strategies. The most striking and recurrent examples of this was the alacrity with which firms serving specialized markets encouraged the standardization and automated production of those semi-finished goods and components which served in effect as the chassis upon which an almost unlimited variety of final products could be constructed. Examples range from the introduction of yarns spun or thrown automatically to standard counts, cutlery steels forged by steam-powered hammers, and watch movements produced by an extraordinary ensemble of dedicated machine tools. At bottom, this selective rejection and acceptance of particular elements from foreign innovations reflected a constant, permanently provisional re-evaluation of local strategy and, more specifically, an anxious effort to avoid entrapment in any given organization of production and its associated markets. The firms under consideration here rarely expected cataclysmic once-and-for-all changes in their context of operation. Yet they wanted to be well-situated to react to abrupt shifts in the conditions of competition, however unlikely; and, more to the point, they wanted to drift along with development insofar as there was any clear direction of drift.

Intentions of this sort might seem aimlessness or, at best, passivity masquerading as prudence. But it is clear from the archival record that the determination that things were drifting in a particular direction and the decision to drift along with them were themselves the products of the same kind of well-informed but passionate debates which underlay the apparently bolder decisions to affirm or modify some aspects of local practice in the face of a provocative foreign success. Indeed it was often the case that incremental changes in industrial organization conceived of as adjustments to relatively insignificant changes in business practice could in the aggregate amount to programs of renovation as profound in their consequences as those directly proclaimed as such.

Consider, for example, a few of the piecemeal reactions of the decentralized weaving and knitting districts of Philadelphia at the turn of the twentieth century to the growing dominance of mass retailers over their final markets and increased competition from vertically-integrated manufacturers based in other regions. To counter the first threat, they sought more direct contacts with their consumers and reduced risks by gearing production more closely to incoming orders; to counter the second, they looked for market niches where production runs were typically longer than their customary average but shorter than those which attracted the mass producers. This diversification in the direction of standardization went hand in hand with an intensified search for new specialty items that would be invulnerable to imitation by the integrated producers and new technologies which reduced the cost of producing them.

British engineering during the early twentieth century is an analogous case. The success of American exports in British and Imperial markets at the turn of the century provoked a shocked investigation into the sources of their advantage. The upshot was the adaptation during the course of the following decades of American machine tools, factory layout, planning methods and incentive payment systems to the requirements of shops producing much smaller batches than those of their trans-Atlantic counterparts. Even in historical retrospect, these apparently modest adjustments – which did not appear such to contemporaries – seemed adequate to the demands and possibilities of the situation, allowing British engineering firms to hold their own or even regain markets in many sectors.

Sophistication, however, is obviously no guarantee of success. In characterizing the agents as self-reflectively and judiciously strategic we do not mean to suggest that their strategies always won the day. As every observer of financial markets knows, on any given day there are in circulation sophisticated prognoses of everything and the opposite of everything. Consequently, the next day regardless of what happens, there is always a sophisticated prescient account of why that and only that occurrence could happen; and it is easy but clearly incorrect to assume that the account was prescient because sophisticated. So too with the larger movements of economic affairs in historical time. In reconstructing cases of successful adjustment to local circumstance or pioneering apprehensions of putative logics of production, economic historians typically suggest even if they rarely declare that analytic clarity was a precondition for material advancement. In arguing that analytic sophistication means something more to the agents than to the economic historians, and that analytic sophistication is much more common among the former than the latter suppose, we want to undermine this imputation and underscore what should have been obvious in the first place: strategic reflection is a necessary but not sufficient condition for competitive success.

In extreme cases, in fact, self-reflective, well-informed contemplation of alternatives leads to paralysis by well-known paradoxes. One of these is Hegel's vortex of bad infinity where nothing can be chosen because the price of choosing any one thing is the abandonment of the freedom to choose among all possibilities.[24] Another is the situation of Buridan's ass who starved to death unable to decide between the competing attractions of two equidistant morsels. Developments at the Italian shipyard and metalworking complex Ansaldo during the first two decades of this century illustrate the dilemmas of irresolution captured in these paradoxes and the relation between them. As a manufacturer of artillery shells they saw the possibilities of rationalization in the example of Henry Ford's assembly line; as a builder of naval warships they prized their capacity to respond to the extraordinarily specific requirements of

[24] For the vortex of bad infinity, see Georg Wilhelm Friedrich Hegel, *The Phenomenology of Mind* (Translated by J.V. Bailey, New York: Harper and Row, 1967), pp. 207–213; and for the origins of Hegel's concept, see Laurence W. Dickey, *Hegel: Religion, Economics and the Politics of Spirit, 1770–1807* (Cambridge: Cambridge University Press, 1987), pp. 246–252 and 270–272.

technically demanding customers. Efforts by competing factions to impose their own single model on the firm unleashed a war of analysis. Successive engagements revealed with greater and greater clarity the advantages and disadvantages of each model and the importance in choosing between them of knowing precisely what was most imponderable: the future composition of demand for the firm's products. Like Buridan's ass, the firm was inclined in one direction and then another, investing substantially with every turn of its corporate head, but fear of choosing once and for all and thereby binding itself in a world that especially after World War I was becoming less rather than more predictable deterred a decisive commitment to either line of action. The result was that depending on one's angle of vision Ansaldo went bankrupt either because it was paralyzed by the oscillation between two incompatible strategies or because it sacrificed its actual possibilities for some modest, compromised success that combined both to the hypothetical possibility of finding an ideal solution.

The conception of the actors as defining themselves strategically in the very act of constituting their context shapes the scope and form of the essays in this volume as well as their substance. In most standard writing about economic history, the actors' motives and thence their self-conception is inferred from their behavior. Because, as we saw, contexts in these accounts are held to be highly constraining, successful behavior is motivated by an understanding of the world consistent with the constraints, failure is a misapprehension of them. The actors' own occasional comprehensive reports of their situation are treated as illustrations of an exceptional psychological awareness of higher-order connections whose power to shape outcomes was in no way dependent on such accidents of understanding.

But the relation between the actors' awareness and their actions is plainly different when the actors' choices depend on their articulation of stories about possible developments and these stories may contain models of those various possibilities and assessment of their probability. In that case, there is no clear distinction between choosing a strategy and thinking about the actors' strategic choices. Where many outcomes are possible, it is simply impossible to deduce the actors' motives from the outcome, for among those motives there were intentions to construct worlds which did not succeed. If you believe that the failed intentions had to fail because they did not capture some essential feature of the local or global logic of progress, then it is natural to consider these strivings as confusions without consequence except insofar as their origins pique sociological curiosity. But if you believe as we do that such logics are only loosely defined at any given moment and are constantly being redefined by the intended and unintended consequences of pursuing them, then analysis of the actors' intentions provides an indispensable independent source of information about the range and robustness of the constraints that they faced and created.

Paul David's recent analysis of Thomas Edison's retreat from the electric-power generating industry is an elegant demonstration of this point.[25] Edison was, among many other things, a pioneer in the direct-current (DC) power-generation industry which he hoped to organize systematically and control in both the United States and Europe during the 1880s and early 90s. In standard accounts, he was driven from the field by a rival consortium favoring the use of alternating-current (AC) power which was easier to transmit over long distances; and in these accounts Edison's stubborn advocacy of DC and above all his denunciation of AC as uncontrollably dangerous are regarded as signs of either a constitutional rigidity or the dulling consequences of age and success. In David's story, by contrast, Edison knew what he was doing every step of the way, and given his own preferences and possibilities had good reasons for his doings, including the denunciation of AC.

AC, as David shows, was never as unequivocably superior to DC during the "battle of the systems" as commonly supposed. For one thing, DC could be used for a much wider range of industrial purposes at the time than AC whose principal use was in illumination. The average rate of utilization of DC networks during a 24-hour period was therefore substantially higher than the utilization rate of an AC network which was taxed much more by night than by day. Edison was thus right in believing that the lower cost of transmitting AC current was therefore offset by the higher cost of building such under-utilized networks, and hence in concluding that DC power-generation could hold its own in densely populated urban markets where short distances made transmission costs less important than the load factor or utilization rate.

In David's view, furthermore, Edison also understood the possibility of combining DC and AC networks into a single system by means of various power conversion devices invented during these years. His retreat from the industry reflected, therefore, not an inability to assess the possibilities of the moment, but on the contrary an extraordinary ability to do so, in combination with a ruthless assessment of his own preferences and capacity for action. Edison was and knew himself to be a system-builder. If he could not shape and control the fundamental features of an industry, creating it at least in his eyes *ex nihilo*, he wanted out. There was, moreover, a coalition of foreign financial interests led by the Deutsche Bank which was willing to buy Edison out and thus allow him to recover his initial investment and go on to other ventures. Seeing this possibility, Edison began a public-relations campaign attacking the safety of AC as a way of increasing the value of his holdings. The case of Edison is extreme because by abandoning the field in this way, he put the DC coalition in a much weaker position than it would otherwise have been. His personal hedging strategy thus shifted the balance of power within the industry. But what is exceptional about the case is surely the power of an individual to enact personal preferences in such a consequential way. The larger theme of the continuity between the actors'

[25] See Paul David, "The Hero and the Herd in Technological History: Reflections on Thomas Edison and the Battle of the Systems," in Patrice Higonnet, David S. Landes, and Henry Rosovsky (eds.), *Favorites of Fortune: Technology, Growth, and Economic Development Since the Industrial Revolution* (Cambridge MA: Harvard University Press, 1991), pp. 72–119.

strategy and the categories of analysis, we believe, is characteristic of the far more normal situation where firms must engage in extensive hedging and the compromise among different technological systems which that entails, precisely because individuals cannot realize their choices alone.

Plainly, however, a reconstruction of the actors' self-definition, even if it considers the differences among economic competitors and political adversaries, is not by itself a comprehensive analysis of the actors and their world. There is no all-encompassing point of view, simple or composite, within an economic situation from which to reconstruct its comprehensive meaning, any more than there is an exhaustive perspective within a novel or political controversy. Observers and readers know things which parochial actors, master strategists, fictional characters and authors do not: for example, which expectations about the development of an economic context, political compromise or new technology will prove correct; what happens when actors get what they want, and what actors in other context make of their success; how characters develop in the course of novels; how authors' own intentions about their works may be influenced by such development, and how this is revealed and recast by the responses of readers. Accounts which consider such things are different from the stories that they recount, but the difference is one of contemplative level or degree not of analytic kind.

A first conspicuous homology between the actors' experience and the accounts of that experience presented here is that the accounts, like the experience, are emplotted. They have a story line, which is to say a beginning, a middle and an end. Most economic history explains events even if they are extended in time: the diffusion of hybrid seeds, the rate of investment in steel-making technology in two countries, the emergence of a new type of financial institution.[26] If writings in this tradition embrace long time periods, then the intent is precisely to prove that time did not matter. The analysis of centuries of price data, for example, is meant to demonstrate the existence of a cyclical pattern of variation, a secular change owing to the operation of some fundamental and ineluctable logic of production, or the disruptive effects of an exogenous shock.[27] Historical time is just the

[26] On the diffusion of hybrid seeds, see Zvi Griliches, "Hybrid Corn: An Exploration in the Economics of Technological Change," *Econometrica*, 25, 4 (October 1957), pp. 501–522; "Research Costs and Social Returns: Hybrid Corn and Related Innovations," *Journal of Political Economy*, 66, 5 (October 1958), pp. 419–431. On rates of investment in British and American steelmaking, see Peter Temin, "The Relative Decline of the British Steel Industry," in Henry Rosovsky (ed.), *Industrialization in Two Systems: Essays in Honor of Alexander Gerschenkron* (New York: John Wiley and Sons, 1966), pp. 140–155. On the creation of investment banks, compare Alexander Gershenkron, *Economic Backwardness in Historical Perspective* (Cambridge MA: Harvard University Press, 1966); and J. Bradford De Long, "Did J. P. Morgan's Men Add Value? An Economist's Perspective on Financial Capitalism," in Temin, *Inside the Business Enterprise*, pp. 205–236.

[27] See for example, B. H. Slicher van Bath, *The Agrarian History of Western Europe, A.D. 500–1850* (New York: St. Martins, 1963); Wilhelm Abel, *Agrarian Fluctuations in Europe from the Thirteenth to the Twentieth Centuries*, trans. Olive Ordish (London: Methuen, 1990); and Emmanuel Le Roy Ladurie, *The Peasants of Languedoc*, trans. John Day (Urbana IL: University of Illinois Press, 1976).

medium in which the underlying lawfulness of economic life is revealed. In contrast, the essays in this volume often take the form of the *Bildungsroman*. In the beginning of these novels, naive characters experiment with their energies and resources. In time, they come to know their conflicting desires, lose their illusions through adventures that test their powers and make them prey to cynicism, and in the end, victorious or beaten, they are left reconciled to a world which is different for their successes and disappointments.

Thus the protagonists of the studies in this volume are regional economies, geographically concentrated industrial sectors or individual firms; and their theme is the life course of these economic entities. Often these studies took as their starting point the discovery of the surprising vitality of small-scale, decentralized production in the face of large-scale industrial competition; and the authors' original intent in extending the work was correspondingly to grasp better the sources of vitality and vulnerability of these protagonists. But as investigations along these lines proceeded within the setting of the seminar, the guiding questions were reformulated. The robustness of these areas as evidenced in an extraordinary capacity for self-transformation and reconstruction in the aftermath of wars, revolutions and economic shocks made mere demonstration of the vitality of a type of industrial organization underexposed in previous discussion seem an exercise in the obvious. Yet simultaneously, paradoxically and provocatively, these same investigations revealed complexities of structure and strategic motivation which blurred the line between the novel industrial organizations and the more familiar ones, and thus raised from a new vantage point the original question of the significance of "alternative" forms of mechanization. But to tell the story of this organizational and motivational complexity, it proved in case after case necessary to expand the scope of the study to stretch from the origins of, say, a regional economy virtually to the present day. Simultaneously it became necessary to broaden the study from the economy narrowly conceived to spheres of local, national and international politics and political economy that had originally appeared inconsequentially remote from the topic at hand. By these almost insensible extensions, arguments or briefs in favor of a surprising economic possibility became narratives about the vicissitudes of whole communities whose survival depended on their ability to avoid precisely the kinds of choices which first the standard account of industrialization and then our earlier natural-selection account of possible worlds had suggested were inevitable.

There is a second and complementary interpretation of the narrative turn in the studies collected here which depends more on the spirit of our times and less on the reciprocal influence of the participants in the seminar. In our time as in the late eighteenth century, a belief in the rational intelligibility of the world and a corresponding faith in humankind's ability to progress by turning reason to this purpose is colliding with a historicist sense of the almost ineffably complex and contingent preconditions of all human achievements and hence of their fragility and irreproducibility. To read Gibbon's *Decline and Fall* (six volumes written between 1776 and 1788) is to observe this collision in the eighteenth century in intellectual slow motion.[28] In their incomparably more modest way, these studies too reason about the place and power of reason in history by narrating the stories of groups whose existence depended on their capacity to recognize that the world was much more malleable than it seemed and yet more unpredictably refractory than this discovery could suggest.

Given that the studies here are narratives, it should not be surprising that they use narrative techniques more common in works of fiction than in historical monographs. Flashbacks, for example, are used in Boch's account of Solingen cutlery and Anderson's account of Norwegian shipping/shipbuilding. These flashbacks both replay before the reader's eyes extensions of the narrative required by a reformulation of the argument and dramatize the story by suggesting that the outcome remained always uncertain until it became known.

A second technique common to most of the essays is polyphony, in which different voices or points of view, including the narrator's are presented without establishing a hierarchical order amongst them. Here the authors meet the actors half-way in recognizing both the latters' capacity for theoretical comprehension of their world and the limits of their own capacity even as retrospective observers to evaluate fully and disentangle competing contemporaneous views. Sometimes in an effort to regain control of the narrative in full recognition of the impossibility of doing so, polyphony gives way to multiple retellings of the tale, each presenting a particular viewpoint.[29]

Thus Scranton's account of the Philadelphia textile industry first presents a narrator's survey of the competing logics of batch and bulk production, each with its strengths and weaknesses; it then retraces its steps to tell a story of the collision between these two models in which the ultimate victory of bulk production depended on historical contingencies rather than any intrinsic superiority. In Saglio's account of the plastics district of Oyonnax, the story is told once from the point of view of a sociologist who asserts the competence to understand the system as such, and then again from the point of view of a sociologist who, eliciting first-person accounts of events crucial to the formation and operation of the system, is unsure of what the actors understand that he does not and vice versa.

But to return this part of the argument to its starting point: all the doubts that surface in the use of these narrative techniques are the expression of a common certitude regarding the actors' character as strategic agents and hence the affinities between their choices and the reflection upon which all commentary on

[28] For an account of this process, see David Womersley, *The Transformation of "The Decline and Fall of the Roman Empire"* (Cambridge: Cambridge University Press, 1988).

[29] On polyphony, see Mikkhail Bakhtin "Discourse in the Novel," in *The Dialogic Imagination*, ed. Michael Holquist, trans Caryl Emerson and Michael Holquist (Austin: University of Texas Press, 1981); and *Problems of Dostoevsky's Poetics*, ed. and trans. by Caryl Emerson (Minneapolis: University of Minnesota Press, 1984); and Tzvetan Todorov, *Mikhail Bakhtin: The Dialogic Principle*, trans. Wlad Godzich (Minneapolis: University of Minnesota Press, 1984).

those choices depends. Just as the actors join an interpretation of their past and an adumbration of their future in a present that unfolds with the tensions of a plot, so the retelling of the resulting story, if it is to give a sense of what really happened and what could have happened, must maintain a sense of the tension of that junction despite knowledge of how everything turned out. It is easier to do this than it may at first seem because the actors' choices, as we have already suggested, so often resemble our own and because by calling attention to the contingency of the historical outcomes we find an expression, however indirect, for the sense of economic vulnerability which is today as close to a universal experience as anything in an age of increasing diversity and differentiation can be. But beyond this and perhaps of greater practical importance, there is yet another connection between the then of these historical accounts and our now: the attempt, sometimes successful and sometimes not, to create by design economies flexible enough to survive in an increasingly turbulent environment by reconstructing institutions formed to serve other purposes. The next part of this essay discusses the kinds of institutions which have performed well under such circumstances, explains why in the light of much current theory any attempt to create such institutions willfully appears a fool's errand, and how faced with equivalent problems the economies considered here have addressed them in a way that illuminates again the reciprocal relation between the constitution of actors and the formation of the context in which they make their choices.

Trust, tradition and politics: the governance of flexible economies

Most of the economies discussed in this volume competed by continuously renewing and expanding their product lines and introducing new production technologies which lowered the incremental cost of this diversification. For the nineteenth- and early twentieth-century firms studied here and again for an increasing number of large and small firms today, the central problem of industrial organization was and is the coordination of decentralization. By looking briefly at how this problem arises and why it appears so intractable from the point of view of most current economic theory, we will understand more clearly how the agents' self-conception led them to blur the lines between the economy and the society and relatedly between business strategies often held to be antithetic.

The more stable the economy, the more it pays to organize production in large, vertically-integrated firms in which output from each department or station becomes the input for the next. Economies of scale reduce production costs and thus extend the market sufficiently to justify the risks of investing huge sums in rigid equipment that can be turned to no other use. The more volatile the economy, conversely, the more it pays to organize each step in the production process as an independent business. Under these conditions, each business is encouraged to protect itself against shifts in demand by learning to use its resources in the most flexible possible way, and the risk of anyone holding assets too specialized to be adapted to a changing environment is reduced to a minimum. The owners of these specialized businesses with generalizing capacities can then contract with one another to produce whatever the market requires.[30]

The organizational problem is that the decentralization of authority required for a permanent-innovation economy creates the temptations of opportunism and free-riding which are the core issues in such otherwise distinct discussions as principal-agent and property rights theory and transaction-cost economics.[31] Parties to all agreements in such an economy enter those agreements expecting to encounter unforeseen problems. If they thought otherwise they would not seek out partners whose distinctive characteristic was the capacity to solve novel difficulties. But if parties enter into an exchange which cannot be specified accurately in advance, then there is the danger that one will induce the other to make the investments necessary to adapt general-purpose resources to the common task only to refuse then to do the same unless the original agreement is renegotiated to reward this abuse of power. Fearing such victimization, no one will enter agreements in the first place.

Even supposing, however, that the fear of opportunism is banished, there remains a second and longer-term problem in the coordination of decentralization. The more specialized each unit, the less willing and able it is to engage in such activities as training young people, collecting information on potential export markets, controlling the quality of raw materials and final products, or conducting research in areas not of immediate relevance to their daily business concerns. The temptation here is notably to free-ride, for example, by hiring workers who have been trained at others' expense or by appropriating others' research, perhaps by hiring the persons who did it. Anticipating such raids, however, those who might otherwise be inclined to invest in the training or the research for their own purposes will be deterred from doing so by the fear that their investments will become subsidies for their competitors. Again the result is paralysis through anticipation of a situation which no one wants but no one appears able to change.[32]

So long as actors are conceived, as they are in neo-classical economics, as having a fixed set of interests whose satisfaction they act to maximize, there is in principle no way out of this dilemma. Agents must always fear that their partners' compliance with agreements is a ploy to sucker them into a ruinous treachery. The best that can be hoped for is what John Rawls calls a *modus vivendi* in which

30 For a more extended analysis of this polarity, see Piore and Sabel, *The Second Industrial Divide*.

31 For a good introduction to these issues, see Oliver Hart, "An Economist's Perspective on the Theory of the Firm," *Columbia Law Review*, 89, 7 (November 1989), pp. 1,757–1,774.

32 For an analysis of this dilemma as market failure in a modern context, see David Finegold and David Soskice, "The Failure of Training in Britain: Analysis and Prescription," *Oxford Review of Economic Policy*, 4, 3 (1988), 21–53.

each party calculates that the others' calculations make it profitable to comply with the terms of the agreement.[33] But the slightest revision of expectations, including reappraisal of the other party's assessment of the situation, can cause even this fragile approximation of cooperation to break down. Where more durable forms of cooperation exist, they are explained reasonably enough as exceptions to the rule that calculating agents cannot sustain such relations. What is exceptional in these cases, the argument goes, is precisely that because of some accident of socialization the agents' interests are so synchronized or their sense of solidarity and mutual responsibility so strong that they do not calculate the costs and benefits of exchanges amongst themselves in the normal way. In particular, each forbears from exploiting the vulnerabilities of the other parties to an exchange and knows that this forbearance will be reciprocated. This forbearance is called trust and hence trust is considered as a historically given cement of society that makes possible a contractual regime which could not arise spontaneously from an antecedent chain of contracts.[34] Paradoxically, then, a flexible production system that simulates a competitive economy proves possible according to the theory of that economy only on the condition that the agents not calculate their competitive advantage.[35]

As the essays show, the actors in the economies studied here understood the problems of opportunism and free-riding. But they no more saw their choices as those depicted in neo-classical theory than saw themselves as depicted in that theory. Indeed, it is precisely because the actors conceived of themselves as having choices in the special sense of having choices about who they are that they could envisage, and sometimes realize, solutions to the coordination problems excluded by the postulates of the neo-classical view.

For the actors, to begin with, it is clear that the world of the opportunistic *modus vivendi* and the world of genuine forbearing cooperation shaded and could be transformed into one another. The actors, we said earlier, defined their individuality by creating particular places for themselves in a community whose larger destiny was the frame for the plot of their working lives. Hence it is not surprising that they saw mistrust and the paralyzing mutual suspicion which it presumed not as an expression of human action in the state of nature, but as a degenerate condition: a reasonable but regrettable individual response to the collective failure to maintain a moral order upon which all individual identity ultimately rested. Conversely, it is unsurprising that they rejected the view of cooperation as the result of some primordial solidarity or identity of interests.

33 See John Rawls, "Justice as Fairness: Political not Metaphysical," *Philosophy and Public Affairs*, 14, 3 (Summer 1985), especially p. 247.

34 For trust as the product of unexamined social norms, see Jon Elster, *The Cement of Society: A Study of Social Order* (Cambridge: Cambridge University Press, 1989).

35 For a discussion of trust as knowing reciprocal forbearance from opportunism and the idea of reflexive, sociable selves on which it depends, see Charles F. Sabel, "Studied Trust: Building New Forms of Co-operation in a Volatile Economy," in Frank Pyke and Werner Sengenberger (eds.), *Industrial Districts and Local Economic Regeneration* (Geneva: International Institute for Labor Studies, 1992), pp. 215–250.

Rather they saw it as the expression of a morally well-ordered world in which individuals who respected the generally acknowledged standards of fair dealing could succeed according to their talents and fortunes.

One consequence is that the actors did not distinguish as sharply between states of trust and mistrust as the conventional understanding would suggest. Rather they acknowledged within their economies moral borderlands where normal fair dealing was relaxed and activity was governed by the sparer rules of decency that apply, among other places, in love and war. Poni's Lyonese merchants often spied on each other; but as countless spy novels attest, there is a deontology of treachery, and a spy remains a peculiar sort of honorable person by complying with it. Sometimes, moreover, the borderlands were not so much behaviors at the edge of the acceptable, but clearly designated settings in which sub-standard behavior was tolerated, on condition that it be marked off from less dubious activity. Both Poni and Cotterau, for example, demonstrate that in Lyons knock-offs, rejected designs and seconds – a mixture of the honest failures and the failed con jobs of the flexible, artisanal economy – were sold in special secondary lower-priced markets.

A second and closely related consequence was that for the actors the problem of creating trust from mistrust or protecting trust when it was endangered by opportunism was for them not a matter of transforming human nature, but rather and more modestly a question of how to design institutions that would enable each to act with that confidence in the probity of the others which all were presumed to acknowledge as beneficial and just. In this sense, the problem of creating the institutions required to sustain cooperation was like the process by which a people in subjecting itself to a constitution both chooses and expresses what kind of a people it wants to be.[36] It is only a slight exaggeration to say that the economic history of the regional economies studied here is primarily the history of their constitutional deliberations, looking forward as well as backward, and no more rejecting particular institutions or practices because they were modern than because they were traditional.

A central institutional problem in the regional economies of Lyons, Solingen, Oyonnax, the Swiss Jura, the Norwegian shipping and shipbuilding industry and the Danish economy as a whole, to choose some notable examples, was to create a system of collective tutelage which monitored the fluid exchanges among private parties without intruding into them. Without monitoring, clearly, isolated acts of opportunism could lead to defensive reactions that undermined all cooperation. Intrusive monitoring, as, for instance, by detailed elaboration of commercial law, would fail because of the same circumstance that made it necessary to decentralize authority over production in the first place: the impossibility of writing contracts covering all the contingencies relevant to an exchange.

[36] For an extended treatment of this connection between deliberation as exercised in the institutionalization in the economy and political constitutionalism, see Charles F. Sabel, "Constitutional Ordering in Historical Context," in Fritz W. Scharpf (ed.), *Games in Hierarchies and Networks* (Boulder CO: Westview Press 1993), pp. 65–124.

One of the solutions to this dilemma, especially in regional economies, was the creation of joint boards of arbitration or conciliation whose members represented variously weighted combinations of the groups cooperating in production. In Lyons, for instance, the *Conseil de Prud'hommes* became the key institution for adjudicating individual disputes over product quality and remuneration between merchant-manufacturers and master weavers on the one hand and between the latter and their journeymen on the other. By determining when compensation was required to break existing contracts, such adjudication helped to prevent poaching of workers or subcontractors during booms and imposition of an undue share of the costs of adjustment on weaker parties during slumps. In Solingen, similarly, the *Vergleichskammer* was responsible for supervising the individual application of collective agreements attaching minimum piece rates to each type of cutlery. A cognate solution emerged in the Swiss Jura between 1876 and 1914. It combined collective bargaining agreements between trade unions and employers with a system of adjudicating disputes arising under them based on the model of the French *Prud'hommes* but also providing for the possibility of conciliation and arbitration by cantonal authorities.

A closely related mechanism of dispute resolution and market regulation was the *tarif* or piece price list which played a key role in determining wages and product quality in centers such as Solingen, Oyonnax and the Swiss Jura. Like the *Conseils de Prud'hommes*, the tariffs depended on the self-regulatory powers of the parties. In this they resembled nothing so much as a modern system of collective bargaining in which the parties create a private-law regime which will govern their behavior for a period stipulated by the agreement itself.[37]

Other systems depended on the provision of rules by "outside" organizations accepted as neutral (and hence in some senses insiders) because they had a manifest interest of their own in providing accurate and fair assessments which the parties could use to guide their affairs. In order to determine the insurability of ships, which was its primary business, *Det norske Veritas*, for example, had to keep abreast of the latest production technologies, rate their effectiveness and judge the capacity of particular shipyards to apply them. The collection, assessment and dissemination of this information both created standards for the Norwegian cargo fleet and facilitated performance to the level specified.

Nor was the regulation of the coordination problems of these flexible economies limited to the provision of rules by institutions specialized in the resolution of disputes. Successful coordination of decentralized production units often rested on the provision of a wide range of services to firms which aligned their interests in the sense of disposing them to favor the same, cooperative strategic response to the market. Training of skilled labor was a central example. Small firms acting alone could not bear the cost of providing all-round training to

[37] For collective bargaining as law making by private groups, see Louis Jaffe, "Law Making by Private Groups," *Harvard Law Review*, 51, 2 (December 1937), pp. 201–253.

craft workers; nor could they bear the risk that workers in whose training they had invested would be poached by competitors unwilling to make similar investments. But in the absence of a reliable supply of trained workers able to make flexible use of general-purpose machines the temptation to pursue low-wage, low-quality strategies would have been overwhelming. The solution was collective regulation of the labor market, including communal financing for training, so as to make it impossible to avoid contributing to the pool of skilled labor. Such systems were in place in Solingen cutlery; Lyonese silks; Swiss watchmaking; the Danish wood and metalworking industries; Oyonnax plastics; and Bolognese machinery-building.

A second example were institutions that promoted indigenous technologies, monitored foreign developments relevant to a particular branch of industry, and kept local industry abreast of their findings. Among the many instances of such institutions were the Observatoire astronomique et chronométrique pour le contrôle scientifique de la mésure du temps of Neuchâtel and the many local emulation societies in the Swiss watch industry; the Lyonese Chamber of Commerce; cooperative research institutes in the Danish slaughterhouse and dairy industries, and the Istituto Aldini Valeriani in the Bolognese metalworking industry.

This ensemble of regulatory institutions often for long periods was so embedded in the daily flux of business activity as to appear indistinguishable from it. Indeed, there was not even a clear line between regulation through rules and regulation by means of service-providing institutions. In Lyons, for example, coordination of training and the diffusion of new technology was accomplished principally through a system of rules. Thus apprentices shared in their masters' revenues in a way which encouraged the latter to invest in training and effectively use those in their charge. Intellectual copyrights on new fabric designs allowed manufacturers to capture the gains from innovations in fashion long enough to compensate them for their risks, but not so long as to prevent the district as a whole from profiting from their ingenuity through emulation. Complex regulation of loans for equipment in effect made the manufacturers collectively responsible for maintaining the resale value of highly specialized equipment through constant adjustment and improvement, thereby reducing worries about the security of lending into a market that prospered precisely by constantly changing the organization of production.

But whatever the precise balance between rules and service-providing institutions, when most successfully combined they came to impart a distinctive hue to the local society as a whole. Thus in Denmark the success of the small holder economy during the late nineteenth and early twentieth centuries led to ever more explicit and comprehensive compromises between organized interest associations and political parties regarding, among other things, joint control of vocational training by labor and capital; agrarian reform in favor of the small holders; a shift in the tax burden from landed, particularly agricultural property to intangible capital; and in the crisis of the 1930s a series of elaborate horse-trades and compromises by the various political parties to respect the right of groups represented by the others to maintain their way of life.

Developments in Denmark, because they were so extensive and elaborate, make manifest the essentially political character of these dispute-resolution arrangements. The fundamental purpose of all of these institutions was to offer an existential guarantee to each of the parties assuring a place in the social order by virtue of acknowledging its place in the economy. Consequently, disputes over economic interest became entangled with larger questions about the relative status, social and political, of the parties. The fact that the various groups in this case as in the others acknowledged themselves to be part of a single community did not at all imply complete agreement as to their standing any more than the formation as we argued earlier of an individual within a group implies complete subjugation to it. The constant jockeying by groups (jostled themselves by individuals) for position within a social order whose boundaries were commonly acknowledged thus caused constant disputes, often in the form of strikes, in the borderland between economic affairs and politics *strictu sensu*.

Thus in Solingen during the decades before the First World War, the trade unions' efforts to ensure that piece rates for new types of cutlery were approved by the *Vergleichskammer* led to numerous social conflicts. Similarly, in Philadelphia textiles recurrent disputes over wages, hours and working conditions culminated in an industry-wide general strike in 1903. In Oyonnax, too, the periodic renegotiation of the *tarifs* before and after the First World War was often proceeded by disagreements resulting in strikes. Plainly and neo-classical expectations to the contrary notwithstanding, cooperation in these economies did not preclude conflict or vice versa. Indeed, as a reading of the cases will show, the possibility of the one presumed the possibility of the other precisely because both were rooted in shared understandings of the world.

The most consequential of these disputes concerned the definition of the governance structures themselves and typically embroiled the whole community in a contentious reinterpretation of its traditional identity and institutions which interwove the traditional and the belligerently modern in a way that often confounded their contemporaries, let alone subsequent generations. The reconstruction of the decentralized Lyonese silk fabrique after the French Revolution was as a matter of fact so perplexing to the Parisians of the day that they tried on various occasions to stop it. Before the revolution, the Lyonese fabrique had been regulated by a series of old regime guilds and corporate institutions whose operation had been modified to permit greater innovation in products and production technologies. The revolutionaries' liberal program included the abolition of all such restraints on trade, and they were particularly eager to break the power of the Lyonese corporations which had supported the city's participation in a federalist revolt of the provinces against Parisian domination. The absence of a regulatory regime and the disruption of traditional market connections then paralyzed the silk trade. As they waited for trade conditions to stabilize, the Lyonese undertook a radical reconstruction of their old institutions. Their aim was to offset the power of the mercantile elite which had dominated the old-regime corporation by giving greater weight to representatives of the manufacturers and master weavers. The Lyonese understood this aim as a step in the direction of democracy fully in the spirit of the revolution and consistent with the needs of a revitalized, more innovative commerce. To the prefects and the Parisians, however, this seemed like a cynical manipulation of revolutionary ideals in the interest of perpetuating traditional corporate prerogatives. The Lyonese extended the "industrial suffrage" to encompass the journeymen as well as the master artisans during the great political upheavals of 1831, 1834 and 1848; but despite or perhaps because of the timing of these concessions, Paris remained suspicious of corporate influences well into the century.

Boch tells an analogous story about the Solingen cutlery industry. Napoleon's troops abolished the old-regime corporate system in 1811, and the Prussian government which replaced the French shared their liberal convictions. The resulting regulatory vacuum in turn precipitated a downward spiral of price cutting and quality reduction which destroyed the reputation of Solingen cutlery on the world market for more than a generation. It was only after 1870 that it became possible to revive those elements of the old order which could serve as the armature for unobtrusive regulation of a system of decentralized production which enabled Solingen to reassert its place in international trade.

But communities did not always catch their errors at the eleventh hour. Sometimes the desire to clear the way for a modern economic system by dismantling all vestiges of the old regime regulatory framework destroyed the capacity to make and revise rules for the governance of any kind of self-sustaining economic system. Spitalfields was a catastrophic case in point. Here an alliance of liberal Tories and Ricardian political economists attacked the regulatory regime during the 1820s with the same ferocity as the Napoleonic conquerors of Solingen and their Prussian successors. In the name of *laissez faire* they repealed instead of reforming the eighteenth-century acts of parliament that had established a framework for setting minimum wages in the industry and fixing maximum numbers of apprentices per master. As part of this effort to renew the industry, the liberal forces replaced a prohibition of imports with a tariff, which was then subsequently reduced and finally abolished. The authorities were not deterred from their course by the ensuing rounds of wage cuts, quality reductions and the exodus of skilled workers from the industry. The cumulative effect of these measures was to permit the renascent Lyonese industry to drive its historical rival from its last home bastion.

But even when communities had not been thus stripped of their capacity for self-determination, so the capacity for debate was no assurance that the community would find a workable resolution to its problems. Political stalemate or disorientation was the collective counterpart to strategic confusion. In the Swiss canton of Glarus, for example, the community never found its way out of the crisis of the 1870s. Because of their successful product innovation strategies, the textile printing firms were uninterested in introducing power-driven machinery, which was poorly suited to the production of their specialties. Up to the 1870s their aggressive niche strategy enabled Glarus firms to sell their traditional hand-printed goods all over the world at a premium price. As their production costs rose against those of foreign competitors, manufacturers' efforts to increase productivity by methods which would have put a greater physical strain on workers were then vetoed in a formal public referendum according to the procedures of the canton's direct democracy. The factory owners' inability to find an alternative strategy acceptable to the citizen workforce in turn condemned both to decline.

Ironically, however, it was neither the strategic actors' rigid attachment to tradition nor their blind faith in the future but rather their dispassionate willingness to commingle future and past which most often perturbed the institutional order established by collective deliberation and conflict. As the earlier discussion of hedging and hybrids suggested, the strategic actors were constantly weighing the advantages of incorporating elements of systems that had proven successful elsewhere into their own. Because hybrids were frequently introduced as hedges against subsequent changes in the environment, it was seldom clear, least of all to the actors themselves, when incorporation of some element of the mass production system – to take the most obvious example – was the first step towards a complete break with the system of decentralized organization and when, on the contrary, it would serve to stabilize and reinforce the essentials of that order. The conflicts already referred to in pre-1914 Solingen, for example, resulted in part from fears aroused by managerial experimentation with the new production technology. In the Swiss watch industry, the situation was more complex, almost to the point of comic circularity. An inventor-entrepreneur intent on applying mass-production methods to watchmaking emigrated first to Britain and then to the United States in disgust at the Swiss unwillingness to innovate technically and socially on the lines of his proposal. A generation later, however, the Swiss had rendered his innovations serviceable for their system of production and market strategy and thereby outdistanced the American competition.

Thus the individuals and groups that formed the economies examined in these essays were more supple than anticipated. By constantly reinterpreting their own past and others' interpretations of the future, they created more possibilities for themselves as individuals and communities than our historical accounts and our theoretical traditions have allowed us to perceive. This suppleness allowed them to survive by adaptation changes in environment that would have meant extinction for the single-minded agents of our own more usual model imaginings. Because they did not distinguish between the traditional and the modern or the economic and the political, the historical actors could pursue strategies which in fact elided these distinctions. By plotting economic lives whose drama and pathos was the avoidance of fateful choices, they actually smoothed the course of economic history, as we will see next.

Epochs or orientations?

Economic history, we remarked in our earlier discussion of that discipline's tendency to collapse the actor into the context, is typically timeless in its oscillation between effectively instantaneous events and endlessly long *durées*. History more generally, in contrast, typically narrates a story, and for just the reasons sketched in our treatment of agents capable of strategic choices that foreshadow a future and recollect a past. Usually these stories distinguish between periods of stability and the periods of turmoil which precede and follow them and they draw their narrative drama from the contrast between the one and the other.[38] The case studies in this volume draw on similar motifs in their discussion of the life-cycles of particular regions, sectors and firms as considered in the light of more or less conspicuous choices between competing models of organization. It would seem natural, therefore, to extend these studies into a larger narrative about the changing relations of production systems from the late eighteenth century to the present.

But there is paradox here. On the one hand, the very notion of a grand periodization in historical time requires a distinction between tranquil or stable epochs in which choices are so fixed as to be almost automatic and turbulent transitions in which outcomes are manifestly in dispute. Yet what our analysis of economic actors as knitting together and redefining past and future in their moment-to-moment strategic choices suggests, and what the research reported here reveals, is that even the most apparently tranquil epochs of economic history never have the matter-of-factness that the notion of grand periodization almost inevitably imposes on them. On the other hand it seems odd to think that innumerable overlapping local narratives have no supra-local historically extended analogue. It seems equally evident that the dominant ideas about productive efficiency have changed since the beginning of industrialization and seem more debatable at some times then others. How can we do justice to the impulse to impose a retrospective narrative order on history without belittling the narrative of self-constitution and economic calculation embodied in the notion of persons as strategic actors?

The solution, we think, is to relax the distinction between periods of stability and periods of transition in the same way and for the same reasons that we relaxed the distinction between maximizing actor and constraining context. If the actors assume – correctly – that some of their responses to particular constraints will reshape those constraints and ultimately their own identities, then it becomes impossible to distinguish firmly and finally between periods in which constraints are so constraining as to compel responses that reinforce the

[38] See, for example, the classic account of this type in J. H. Plumb's *The Growth of Political Stability in England, 1675–1725* (London: Macmillan, 1967); and for changing views of the subsequent period as less stable and more contentious than Plumb supposed, see John Brewer, *Party Ideology and Popular Politics at the Accession of George III* (Cambridge: Cambridge University Press, 1976); and *The Sinews of Power: War, Money and the English State, 1688–1783* (London: Unwin-Hyman, 1989).

compulsion and those in which the responses have transformative effects. On balance fundamental propositions will no doubt seem more self-evident in some epochs than in others, but there will be no uninterrupted stretches of self-evidence – in the sense of a world that is automatically and perfectly self-reproducing – within any single period.

From this vantage point it is more useful to distinguish historical epochs according to changing orientations towards the ideas of political and economic regarded as normal or paradigmatic than to divide history into periods where social life was in fact thoroughly organized according to one or another master principle. This notion of changing successive orientations towards paradigmatic or normal ideas is faithful both to the notion that the constraints do change and to our narrative propensity to impose a before and after on historical time. But it also leaves due room for the local narratives enacted in strategic choice; by definition, after all, ideas of normality tend to magnify and thus to increase the importance of dominant conceptions without reflecting or constraining anything like the totality of the behavior they characterize.

One way to develop the connection between these general ideas and the problem of narrating the history of industrialization is to reconsider our earlier discussion of shifts between technological paradigms in the light of the broader debate about scientific paradigms from which we and other students of technology drew their inspiration. One position in that debate, Thomas Kuhn's,[39] reproduces the familiar historian's distinction between periods of continuity and periods of change in a way which precisely captures what we find problematic in its use. This is because, in Kuhn's view, long periods of "normal" science dominated by problem solvers are punctuated by scientific revolutions dominated by "philosophizing" heretics in a successful revolt against orthodoxy. Even though the balance of opinion may shift from time to time, individual participants seldom change their minds. Since the believers always believe the dominant ideas and the doubters always doubt everyone but themselves, belief and doubt – or stability and change – coexist without requiring any account of how they might be connected. That is plainly one of the attractions of this kind of explanation. Its principal defect is to leave unanswered the question of where doubting ideas or change more generally come from and how they can ever result in a revision of the canonical order.

The alternative view of science and, relatedly, change, is associated with the work of Feyerabend, Lakatos, and Popper.[40] Just as we argue that continuity and change are intertwined in every historical epoch, they argue that orthodoxy and heterodoxy, problem-solving and philosophizing, and "normal" and "revol-

[39] See Kuhn, *The Structure of Scientific Revolutions*.

[40] See Paul Feyerabend, "Consolations for the Specialist"; Imre Lakatos, "Falsification and the Methodology of Scientific Research Programmes"; and Karl Popper, "Normal Science and Its Dangers"; all in Imre Lakatos and Alan Musgrave (eds.), *Criticism and the Growth of Knowledge: Proceedings of the International Colloquium in the Philosophy of Science, London, 1965, Volume 4*, (Cambridge: Cambridge University Press, 1970).

utionary" science are much more closely related than Kuhn's sequential depiction suggests. Their claim is that there are always different schools of thought, that their adherents are mindful of one another, and that *all* good scientists, "revolutionary" or not, are picking and solving puzzles with some awareness of the larger "philosophical" issues that separate the schools. Without such awareness, the puzzle solver is a drone, just as the economic agent oblivious to or daunted by the possibility of strategic choice and the recognition of possibilities on which it depends becomes a victim of circumstance.

Good scientists, then, like successful economic agents, depend on learning to depend on ideas while assessing their dependability: Feyerabend, for instance, speaks of a "principle of tenacity" to capture the rules of thumb and heuristics by which a scientist determines to maintain a certain belief *despite* indications of its infirmity.[41] In this view scientific "revolutions," shifts in technological paradigm, or strategic *volte-faces* in business are no mystery, because real science, real engineering, and real economic decision-making is always guided by the anticipation – welcome or alarming – that they might occur; the cognitive wonder is rather that scientists, engineers and business people go on wondering fruitfully about interesting puzzles as though they did not know that.[42]

Against this backdrop we cast the history of industrialization as a sequence of economic orientations rather than making an attempt at periodization in the conventional sense. Because these orientations contain elements of what went before and what is to come, they are by nature coarser-grained and less sharply defined than conventional historical periods. By definition, indeed, a key problem of standard historiography, the dating of the beginning and end of an epoch, becomes meaningless in a world of overlapping and interpenetrating orientations. Thus instead of a counter-narrative of industrialization that retells the familiar story with a more complex cast, a different sequence of acts, and a different *dénouement*, we suggest an alternative dramaturgy in which the action shifts back and forth across the stage, the passage from one act to another is indistinct, and every *dénouement* is undercut by the suggestion of a new beginning. For some this may seem as degraded as the substitution of the

41 Feyerabend, "Consolations for the Specialist," p. 203.

42 In response to these and other criticisms, Kuhn has qualified his original view of the "incommensurability" of successive scientific theories or paradigms. He allows that individual scientists may learn competing scientific languages, although their ability to do so does not imply the possibility of perfectly translating any language into any other. He also believes that changes in scientific language are better analogized to biological speciation than to mutation, suggesting a connection between old and new conceptualizations denied in his earlier writings. See Thomas S. Kuhn, "The Road Since Structure" in *Philosophy of Science Association Proceedings, Volume II, 1990*; and "Dubbing and Redubbing: the Vulnerability of Rigid Designation," in C. Wade Savage (ed.), *Minnesota Studies in the Philosophy of Science, Volume XIV: Scientifc Theories*, pp. 298–318 (Minneapolis: University of Minnesota Press, 1990). But it is unclear what implication, if any, this elaboration of the relations among paradigms has for that aspect of Kuhn's initial position rejected here: the juxtaposition of normal and revolutionary science as cognitively contrary activities.

daytime soaps for Shakespeare;[43] but for what it is worth, we observe that almost all efforts to come to grips with everyday vitality have been initially perceived as an assault on some heroic capacity of humankind, and the discipline of history itself has already cried scandal innumerable times on this account.

The rough account of industrialization we propose proceeds through four overlapping parts, and this partition shapes the grouping of the papers as well. The first part, the modernity of tradition, revolves around efforts to determine which old regime institutions were compatible with mechanization. The answers ranged from the belief that economic modernity required the complete extirpation of mercantilism, as in Great Britain, to successful efforts to adapt such institutions to serve as the regulative framework of flexible economies, as in Lyons or Solingen. This part begins in the eighteenth century with the liberal critique of old regime regulation of markets through chartered corporations. This critique assumes clear form with the Enlightenment attack on the guilds, reaches its apotheosis as a political principle with the French Revolution, and spreads throughout Europe with Napoleon's armies. As we saw, the axial debate concerned a question, as current today as it was then: is the precondition for economic advance the virtual absence of rules regulating competition or more extensive regulation of a particular sort?

The second part might be called the battle of the systems in recognition of the opposing paradigms of production embodied in, respectively, the decentralized DC power-generating system and the centralized AC grid. The economic orientations in this part have affinities with the positions in the preceding one insofar as the dispute about the definition and preconditions of efficiency is still connected to a battle over the meaning of modernity. But it is different in that the contenders, the flexible regions, reflecting back on the earlier struggles, now pit their own alternative idea of modernity against that presented by the mass producers rather than merely asserting the vitality and elasticity of tradition. This part runs roughly from the mid nineteenth century through World War I. Often, as in American textiles, this contrast ended badly for the flexible producers, who are either absorbed into the mass production system as subcontractors or compete with it through sweating. Sometimes, as in the cases of British engineering and the Ansaldo firm, the result was a stalemate. More rarely, as in Swiss watchmaking and most extensively Denmark, the outcome was an open triumph for the flexible system.

The third part, the triumph of mass production modernism, running from the 1920s through the 1970s, figures everywhere and nowhere in our studies. It is everywhere because the triumph of mass production forms the backdrop directly or indirectly for all the studies and for the volume as a whole. It is nowhere because there is no study which concentrates exclusively on this epoch, and those

[43] For an elegant demonstration that such a distinction would have appeared less evident to nineteenth-century audiences than it does today, see Lawrence W. Levine, "William Shakespeare in America," in his *Highbrow/Lowbrow: The Emergence of Cultural Hierarchy in America* (Cambridge MA: Harvard University Press), pp. 11–81.

that traverse it demonstrate repeatedly the incomplete grip of the mass-production model on manufacturing practice. The idea of a triumph of mass production itself is being reassessed to the point of obliteration as study after study comes to similar conclusions. Even Ford's Highland Park plant, the birthplace of the moving assembly line is now seen, not without justice, as a precursor of Japanese systems of continuous manufacturing reorganization more than the rigid American system of mass production.[44]

The fourth part brings the story to the present, by evoking a new battle of the systems. As in the first such battle, one vision of modernity was pitted against another, but this time it appears as though the flexible systems, broadly conceived, will come to be regarded as the norm. Indeed, the more successful the flexible economies, such as those depicted in the case studies of Bolognese packaging machinery, Norwegian shipbuilding and – potentially – Oyonnax plastics, and the greater the appreciation of their antecedents in the previous epochs of mechanization, the more tempting it is to turn the old history of industrialization on its head: the classic centers of mass production such as the Ruhr, the Rust Belt, or the British Midlands would be portrayed as victims of a kind of industrial traditionalism that now blocked the path to the higher and more plastic forms of modernity. Our hope is that the essays in this volume will make that temptation understandable and resistable in equal measure. The outcome of the second battle of the systems, we are convinced, will be even less definitive than the first because the actors themselves, ever more aware of the contingency of their choices, are still less likely than before to regard any result as conclusive.

[44] For a forceful account of Highland Park in these terms, see Karel Williams, Colin Haslam and John Williams, "Ford versus 'Fordism': The Beginning of Mass Production," *Work, Employment & Society*, 6, 4 (1992): 517–555. Williams *et al.* demonstrate that substantial productivity gains were achieved at Highland Park through continuous improvement in workflow via changes in layout. But by ignoring Ford's own vision of mass production as a technological paradigm, they fail to account convincingly for his abandonment of these practices at the River Rouge plant during the 1920s, or for his misplaced optimism about the direct transferability of Detroit methods to the manufacture of complex military products such as river boats and heavy bombers during the two world wars. If contemporary observers and subsequent historians misinterpreted the lessons of Highland Park, so too, on this reading, did Henry Ford himself. For Ford's difficulties in applying mass-production techniques to military goods, see David A. Hounshell, "Ford Eagle Boats and Mass Production during World War I," in Merritt Roe Smith (ed.), *Military Enterprise and Technological Change: Perspectives on the American Experience* (Cambridge MA: MIT Press, 1985), pp. 175–202; and Irving Brinton Holley, Jr., *Buying Aircraft: Matériel Procurement for the Army Air Forces* (Washington DC: Office of the Chief of Military History, Department of the Army, 1964), pp. 326–327, 518–531.

Part I

The modernity of tradition

1 Fashion as flexible production: the strategies of the Lyons silk merchants in the eighteenth century

Carlo Poni
Translated by Patrick Leech

Historians of the industrial revolution and capitalism have emphasized the importance of mass production, a concept almost invariably associated with a series of related notions such as "low prices," "modest or inferior quality" and "extensive market." These notions come together towards the end of the eighteenth century with cotton, which has been recognized as playing a fundamental role in the early phase of the Industrial Revolution in Britain. Competition over prices and the meeting of supply and demand in a free competitive market are the most widely-held conceptual parameters of an interpretative model tied to the categories of classical economics.[1]

To a large extent ignored, or at least underestimated and underused by historians, are the concepts of competition over product quality and differentiation, monopolistic competition, barriers to entry, flexible production, market segmentation and transactions costs. This is due perhaps to the view that these concepts have validity only for contemporary capitalist societies and that they are irrelevant with respect to the economies of the past.[2]

This paper has benefited from the criticisms and suggestions of many colleagues and friends, and in particular Fernand Braudel, Immanuel Wallerstein, Mario Mirri, Vivian R. Gruder, Peter Laslett, and Emma Rothschild. In the final phase, I discussed the text both with Jonathan Zeitlin and colleagues in the Dipartimento di Scienze Economiche of the University of Bologna. Thanks go in particular to Roberto Scazzieri, Patrizio Bianchi and Gianpaolo Rossini for their stimulating suggestions and criticisms. The present version, then, is the fruit of numerous exchanges and of different layers of reading and research in libraries and archives, the latter made more complete with the help of Leonard Wetzler, who allowed me to compare the notes and photocopies I had made in the Archives Nationales de Paris with his own.

1 According to E. L. Jones, economic historians of the industrial revolution, with only a few exceptions, have seen market and demand as a reflection of "the rising tide of mechanical production" ("The Fashion Manipulators: Consumer Tastes and British Industries (1660–1800)," in L. P. Cairn and P. J. Uselding (eds.), *Business and Enterprise* [Kent OH: 1973], pp. 198–226).

2 Cf. P. Sraffa, "The Laws of Return under Competitive Conditions," *Economic Journal*, 36 (1926), 535–550; H. Chamberlain, *A Theory of Monopolistic Competition* (Cambridge: Cambridge University Press, 1933); M. J. Piore and C. Sabel, *The Second Industrial Divide* (New York, 1984); C. Sabel and J. Zeitlin, "Historical Alternatives to Mass Production: Politics, Market and Technology in Nineteenth-Century Industrialization," in *Past and Present*, 108 (August 1985), 133–176; O. E. Williamson, *Markets and Hierarchies* (New York, 1983); and *The Economic Institution of Capitalism* (New York, 1985).

In this chapter I want to show that these concepts can be used to throw light on a variety of phenomena which have hitherto remained obscure and whose importance in terms of the formation of the capitalist market and of entrepreneurial strategies has still to be fully appreciated.

It is within this framework that I will examine silk fabrics and fashion in eighteenth-century Europe. The silk industry is distinctive in the contrasts it offers to the cotton industry. Prices are high rather than low, markets are small rather than extensive. Changes in fashion can best be understood using the concepts of flexible production, monopolistic competition and barriers to entry. Towards the end of the seventeenth century the combination of silk and fashion led to a fundamental shift in the mercantile practices of the textile industry. These new practices were based on the annual or seasonal differentiation of production planned in advance and carried out according to deliberate and systematic strategies.[3]

Fashion has often been associated with luxury. A certain interrelation between the two does exist, but it is nevertheless important to make some fundamental methodological distinctions. The luxury of official court ritual and the ceremonies of the ecclesiastical world were often archaic and old-fashioned, and a piece of clothing considered to be luxurious might also have been old-fashioned.[4] Perhaps it is also true that a fashionable piece of clothing was not necessarily a

[3] Fashion has been the object of historical interest for some decades. See D. E. Robinson, "The Styling and the Transmission of Fashions Historically Considered," *Journal of Economic History*, 20 (1960), 576–587; and "The Importance of Fashion in Taste to Business History. An Introductory Essay," *Business History Review*, 37 (1963), 5–36; E. Robinson, "Eighteenth-Century Commerce and Fashion: Matthew Boulton's Marketing Techniques," *Economic History Review*, 16 (1963), 39–60; D. C. Coleman, "An Innovation and its Diffusion: The New Draperies," *Economic History Review*, 22 (1969), 417–429; N. B. Harte and G. Ponting (eds.), *Textile History and Economic History* (Manchester, 1973); J. Thirsk, *Economic Policy and Projects. The Development of a Consumer Society in Early Modern England* (Oxford: Oxford University Press, 1978); N. McKendrick, J. Brewer, and J. H. Plumb, *The Birth of a Consumer Society. The Commercialisation of Eighteenth-Century England* (Bloomington IN: 1985; U. C. Pallach, *Materielle Kultur und Mentalitäten in 18 Jahrhundert* (Munich, 1987); D. Roche, *La culture de l'apparence. Une histoire du vêtement XVIIe–XVIIIe siècle* (Paris, 1989); T. Kusamitsu, "Novelty, Give Us Novelty: London Agents and Northern Manufacturers," in M. Berg (ed.), *Markets and Manufacture in Early Industrial Europe* (London and New York, 1991), pp. 114–138.

[4] Thus the "mantille de Cour," an old, black ceremonial cloak with white decorations which the French ladies of the court wore on the day they were presented to the king, was almost immediately discarded "le jour passé" and "tout ce qui était noir se change en étoffes de couleurs ou d'or" (M. de Garsault, "Art du tailleur," in *Description des Arts et metiers faites ou approvées par Messieurs de l'Académie Royale des Sciences de Paris*, 2nd edn, vol. IX [Neuchâtel, 1780], pp. 107–108). This means that the role of the court at Versailles in launching new fashions must be clearly distinguished from the rules of ceremony. From the Regency onwards, the "Société de Cour" increasingly became a cultural and social space which included Versailles and Paris, with Paris in a dominant role. It was not by chance that the ladies of the court hurried to Paris for their wardrobe. Cf. N. Elias, *La Società di corte*, Bologna, 1969. The court of the King of Prussia, renowned for an etiquette that required simple, spartan clothing above all for the men, offered little scope for the new fashions from France. Cf. A. Ribeiro, *Dress in Eighteenth Century Europe (1715–1789)* (London, 1984), pp. 66–67.

sign of luxury.[5] Unfortunately the centuries-old polemic against luxury has constituted a major epistemological block to the analysis of the complex phenomena which come together under the name of fashion.

A great deal has been written about fashion. Psychologists, psychoanalysts, sociologists, linguists and historians have asked themselves and continue to ask themselves about the nature of fashion. Is it an expression of the spirit or the taste of a certain historical epoch? Is it the desire for pleasure, a means of communication and symbolic distinction? Or a symptom of the instability of human nature?[6] There is no lack of complex, rich and meaningful definitions, but they rarely explain why fashion (as rapid change) should have triumphed exclusively in Western European countries and not, for example, in the middle or far east. This problem was posed by Fernand Braudel in the early 1960s. According to Braudel, as long ago as 1609 an official of the Shogun at Yedo complained to the Spanish of "la variété de leurs costumes, domaine dans lequel il sont si peu constants que tous le deux ans ils sont vétu de façon différente" while in Japan "il y a plus de mille ans que cette nation n'a pas changé de costume."[7]

[5] Indeed, the compulsive need to enlarge existing markets and conquer new ones led the entrepreneurs to replace high quality fabrics with less costly ones. Thus in 1750, the Sieur Delasalle, one of the major manufacturers and designers of silk fabrics in Lyons, asked the Paris Conseil de Commerce to be allowed to manufacture fabric "en grand et petit dessein avec du trait faux doré et surdoré . . . filé sur soie" instead of with an entirely golden thread. According to Veron de Forbonnais, who supported Delasalle's request, the legislators should be concerned above all with preventing the "faux fil" from being fraudulently sold as golden thread, and should not prohibit its production altogether. It was important to avoid producing "une perfection chimerique." It was enough to protect the consumer with the help of appropriate markings so that "le faux et le fin" could not be confused. Veron de Forbonnais predicted that once the "fabrication en faux" had been perfected, the "fabrication en fin," much dearer, would decline (Veron de Forbonnais, *Lettre à M+++ Negociant de Lyon sur l'usage du trait faux filé sur soie dans les étoffes* [no place of publication, 1759], pp. 18–22, 30). A few years earlier, in 1756, Forbonnais had declared himself against solid gold snuff boxes in favour of those which, although they had not an ounce of gold in them, were perfect imitations. According to Forbonnais, gold snuff boxes were bought by "l'omme de luxe . . . prodigue sans égard," while the "goût générale" was instead by now in favour of "commodité" and "au bon marché" (*Lettre de M. du T. . . . negociant a Paris a M. Rissch, negociant à Francfort sur les ouvrages et les bijoux d'or et argent* [no place of publication, 1756]). The Venetian silk merchant and writer Antonio Zanon thought that fashion was "necessary to the Arts and useful for commerce when based on its own [national] products." He thus proposed that a distinction should always be made between "luxury and fashion and between indifferent and damaging luxury" (where the adjective "damaging" should be understood as "imported") (*Dell'agricoltura, dell'arti e del commercio*, vol. III [Venice, 1764], pp. 153–154).

[6] Cf., for example, A. L. Kroeber, "On the Principle of Order in Civilisation as Exemplified by Changes in Fashion," *American Anthropologist*, 21 (1919), 235–263; P. N. Nystrom, *Economis of Fashion* (New York, 1928); E. B. Hurlock, *Motivation in Fashion* (New York, 1929); P. H. Nystrom, *Fashion Merchandising* (New York, 1932); I. Brennikmeyer, *The Sociology of Fashion* (Winterthus, 1962); K. Gibbins and T. K. Gwinn, "A New Theory of Fashion Change: A Test of Some Predictions," *British Journal of Social and Clinical Psychology*, (1975), 1–9; M. R. Salomon (ed.), *The Psychology of Fashion* (Lexington MA: 1985); R. König, *Menscheit auf dem Laufsteg. Die Mode im Zivilization Progress* (Munich, 1983).

[7] F. Braudel, *Civilisation materielle et capitalism*, vol. I, *Les Structures du quotidien* (Paris, 1967), pp. 240–250.

How can this difference be explained? It may be said, as an initial hypothesis, that fashion as short-term, periodic, programmed change exists in societies characterized by a certain degree of social mobility, in which hierarchical structures are flexible enough to allow for the imitation of the behavior of the higher social groups by those lower on the social scale. *Vivere more nobilium* did not only mean buying land and living off the income from it. It also meant living and dressing as the nobility did. But as soon as the visible distinctions of the social orders are no longer regulated and established by custom or religion, and when sumptuary laws fall into disuse, the conditions which tend to modify the behavior of the elite are created. According to the well-known theory of Georg Simmel, in reply to efforts at imitation on the part of the rising classes, the elite formulates new fashions and new models of consumption, preparing the ground for new efforts at imitation, each of which are destined to be periodically frustrated. A specific process is thus set in motion: a limitless and endless chase in which the social level of a given social group or class can be measured by the number of years it lags behind the fashion.[8] This mechanism, in time, tends to restructure and confuse the traditional distinctions between social groups. Impoverished nobles are unable to follow the changes in fashion in the same way that well-off merchants and tax farmers do. Fashion, in short, is a symptom, albeit an ambiguous one, of the slow passage from a society based on status to another based on wealth. It frees the individual from sumptuary and corporate laws, proclaims its own sovereignty over dress (but only for the wealthy urban classes),

[8] G. Simmel, *Die Mode* (1895), in *Gesammelte Essais mit einem Vorwort von Jürgen Habermas* (Berlin, 1986), pp. 38–63. Even those who have criticized Simmel's approach for its stark dualism, which leaves no space for action on the part of intermediate social groups, and for its inability to explain the dynamics of present day fashion, nevertheless recognize its validity for the study of changes in fashion in Europe in the period between the sixteenth and eighteenth centuries. Cf. G. D. McGracken, "The Trickle-Down Theory Rehabilitated," in *Psychology of Fashion*, pp. 42–43. This author has refined Simmel's theory, taking account of the results of new research on topics such as recent changes in women's fashions, now aimed at expressing social or professional roles, and the influence of the mass media in the launching of new fashions. On anti-sumptuary legislation, cf. G. Bistort, *Il magistrato alle pompe nella Repubblica di Venezia. Studio storico* (Venice, 1912); V. Baur, *Kleiderordnung in Bayern von 14 bis zum 17 Jahrhundert* (Munich, 1975); L. C. Eisenbart, *Kleiderordnung der deutschen Städte zwischen 1350 und 1700* (Göttingen, 1962); H. Freudenberger, "Fashion, Sumptuary Laws and Business," in *Business History Review*, 37 (1963), 37–48; N. B. Harte, "State Control of Dress and Social Change in Pre-Industrial England," in D. C. Coleman and A. H. John (eds.), *Trade, Government and Economy in Pre-Industrial England. Essays Presented to F. J. Fischer*, (London, 1976), pp. 132–165 and bibliography. Changes of clothing, even on a daily basis, already practiced, although only exceptionally, in the sixteenth century, must be distinguished from annual and programmed change in fashion. I refer to that Milanese noble woman who, according to a French jurist, wore new and different dresses every day. And so had 365 dresses. (B. Cassaneus, *Catalogues glorial mundi* [Venice, 1769], pp. 25–26, 324). In Cassaneus' writing, the colours were more or less noble according their closeness to white or to black. On noble coats of arms the noble colours must be placed "in nobilioci loco." The theme of "Mediolani matronam" is taken up by S. Kobierzycki, *De luxu romanorum* (Lovani, 1628), p. 112.

and changes visibly and irreparably the forms of civil society, creating new forms of sociability. It is not surprising, then, to discover that fashion was seen as an insidious agent of social change.[9]

But what is the precise relationship between the "well-off" and those who create fashion? Here we must challenge those interpretations which see fashion as being created by poor, humble artisans of the lower classes.[10] Fashion, seen in the context of silk manufacture, is the product of relatively centralized strategic decision-making, of choices made by powerful economic interests already integrated into the elites of power and taste. These interests, which exercised considerable influence far beyond the borders of a national market, engaged in a dialogue with the taste of consumers, who became increasingly refined and demanding, in turn influencing the strategies of the producers.

The Lyons silk merchants were the first, as far back as the closing years of the seventeenth century, to use annual product differentiation as a strategic weapon to create barriers to entry, to capture important shares of the international market and to outmaneuver firms in competition with them. This change in the rules of production heightened competition and gave a new dynamism to both production and the market. Quoting Adam Smith we may say that at the beginning of every year, the "work to be done" would no longer be the same as the "work done."[11] This in turn led to the consolidation of new power on the market which was destined to upset old equilibria. This cluster of changes constituted a real turning point. The introduction of annual fashions in the production of fabric, planned and programmed in advance, delivered a sharp blow to old market practices which were "embedded in society" and regulated, at least formally, by the sumptuary laws. The introduction of short-life products had another consequence as well. It split production volumes, distributing the new fashions over an increasingly diversified demand. The entrepreneurs were now obliged to work in the short term, to interiorize uncertainty as a structural factor. They were no longer conditioned by long-term expectations. Lyons, in short, was a protagonist of the "great transformation" proposed by Karl Polanyi, even if its

9 Luxury, too, as an ostentation of wealth on the part of high finance and as a bourgeois imitation of the nobility and of the Court of Versailles, could be seen as an attack on the social hierarchy to be re-established with new sumptuary laws. Cf. A. J. Chevalier du Coudray, *Le luxe. Poème* (Paris, 1773), pp. 10–14, 21, 39 *passim*. In fact, in the eighteenth century the imitation of the luxury of the upper classes is nothing other than fashion. Erasmus Jones wrote that "people, where they are not known, are generally honoured according to their clothes." For this reason in a "large and populous city" many people wear clothes "above their ranks" (E. Jones, *Luxury, Pride and Vanity. The Bond of the British Nation* [no place of publication, 1736], p. 36).

10 T. Veblen, *The Theory of the Leisure Class* (1899; reprinted New York, 1975), and *The Instinct of Workmanship and the State of the Industrial Arts*, intro. J. Dorfman (New York, 1964), pp. 180–186, *passim*.

11 A. Smith, *An Inquiry into the Nature and Causes of the Wealth of Nations*, ed. E. Canaan (London, 1961), I, ch. VII. Cf. the comments made by P. Bianchi in his *Produzione e potere di mercato* (Rome, 1991), pp. 29–32.

role in the "modern world system" as promoter of profound social changes has as yet gone unrecognized.

The anonymous author of the report presented to the British parliament in 1713 entitled *The Case of the Silk Weavers Humbly Offered to the Consideration of both Houses of Parliament* wrote that French silk fabrics were preferred to the English ones because "the fashions are or are likely to be taken from France." The result was that the British merchants "cannot make provision for a Spring Trade for fear that a New Fashion should come from France, and render ours despicable." In order not to incur a defeat on the market at the hands of a differentiated product, the English merchants preferred to imitate the French products, with the result that they were left with the residual share of the market (a certain amount of time was needed in order to examine the fabrics which were to be copied, prepare the looms, imitate the dyes, train the weavers and so on). But as soon as the British entrepreneur began to reproduce the new fabric, "another fashion comes from France," and so the game began again. The result was that the French products conquered "the first of the market, and the English the Fag-End" and thus the latter had to be content with selling at much lower prices.[12]

In short, competition over quality and product differentiation rather than price enabled the French merchants to capture the most profitable share of the market, with high prices protected by high barriers to entry, while the English products sold only on the residual share of the market, with low prices. This is a clear case of "time-based" monopolistic competition, each operator with a product in some degree different from his rivals. Innovations had to be not only rapid but also appealing. They had to anticipate and satisfy immediately a demand which was often hesitant but in search of new products. It was by means of this intelligent innovation that Lyons was able to make a victorious exit from the crisis of the seventeenth century.[13]

"French fashions pernicious to England" is the appeal we hear in the pages of

[12] This pamphlet should be seen in the context of the fierce struggle of the English commercial interests against the eighth and ninth articles of the Peace Treaty of Utrecht which reduced duties and removed the prohibition of French commodities. See also *The State of the Silk and Woolen Manufacture considered in Relation to a French Trade* (London, 1713). The anonymous author of this broadsheet recognized that the French "excel in the inventing of patterns, and their skill is so great . . . that by the great esteem they have got in the world will sell 20 or 25 percent above those of any other nation in Europe. For this reason we in England are obliged, for the most part, to copy their pattern." Already in 1679, an anonymous English writer had denounced the import from France of enormous quantities of silk fabrics (velvets and satins) "made in Lyons" (J. B. *An Account of the French Usurpation upon the Trade of England* [London, 1679], pp. 2, 5, 10). One century later A. Goudar wrote ironically about the clothes worn at a ball at the court of London: "En générale les robes étaient superbes et d'un grand goût. Mais ce goût n'étoit pas du crû du pays. Il était copié de la poupée de Paris" (*L'espion françois à Londres* [London, 1779], p. 92).

[13] Supply and demand, rather than being considered in opposition to each other, must be explained within a single discourse, without pretending that they are always in harmony with each other.

Joshua Gee. According to him, the new fashions were cleverly invented by the French merchants in order to prevent the sale of English goods.[14] For Defoe, such was the unpatriotic passion of the English ladies for the French products that the London merchants were forced to pretend that the Spitalfields products had been imported from France.[15]

Something similar happened in the Po Valley. In 1755, a group of Bolognese merchants, having begun to produce fabrics in imitation of French ones (lustring, *taffettas, façonnés* and *péruviennes*) had to suspend production because their products, not having "that shiny and bright quality found in French fabrics, are rejected by almost everyone and end up on the back shelves of the shops." But even if the imitations were successful, there was still the problem of overcoming "the considerable disorder caused by the fact that the French change their designs . . . from one season to another which is, in fact, what attracts the buyer." As there were no "designers capable of these new inventions" in Bologna, the manufacturers had to take "the same French designs" and then set up the looms for the market of the following year. But at that precise moment, a "new invention" with new material and new designs would arrive from France and all the imitation Bolognese products would remain unsold (despite being protected, as were the British ones, by laws which forbade the introduction of foreign fabrics).[16]

In 1764, the silk merchant and writer Antonio Zanon, with somewhat greater analytical rigour, voiced the angry disappointment of the Venetian silk merchants at the fanaticism of the Italians for all things foreign (and especially French) and their "contempt for their own manufactures." The Venetian merchants, defeated in the struggle for leadership in the fashion sector (an area which according to Zanon had been abandoned after "painful experiences"), were now resigned to the inferior role of imitating French fashions. But this very choice, open to them, says Zanon, because the Italian artisans "have a talent for imitation," forced the merchants to dance to a tune they did not know. Like the Bolognese merchants, Zanon notes that "the Italians will struggle to imitate French fashion, but before one suit can be finished another completely different one comes out and . . . all the new Italian fabrics become

[14] J. Gee, *Trade and Navigation of Great Britain* (London, 1729), pp. 30–32.

[15] D. Defoe, *The Compleat English Tradesman*, II, part II (London, 1727), pp. 154–161. Some decades later, L. Muratori wrote that in several Italian cities, "fabrics ingeniously made there only ceased being the object of disdain and were considered to have been made with the necessary dignity for the clothing of noble persons when people were led to believe that they had been brought across the mountains from Lyon" (*Della Pubblica Felicità* [Lucca, 1749], pp. 218–219).

[16] *Memoriale di G. Canevoli e T. Mattioli all'Assunteria d'Arti per ottenere la moderazione del bando . . . rispetto ai lustrini o sieno taffetà, façonné e peruviennes che non possono fabbricarsi in Bologna*, May 5, 1755 (A. S. B., *Arti*, "Miscellanea," vol. XVI, fasc. 68). Around ten years later, Bologna had become "the French warehouse, given that the French adventurers collect there the out-of-fashion fabrics or those which will shortly become unfashionable, and send them from there to Venice. Indeed, some Venetians actually go there to collect stocks" (A. Zanon *Dell'agricoltura, dell'arti e dell commercio*, vol. III [Venice, 1764], pp. 333–334).

old junk in just a few weeks." What was more, immediately before the introduction of a new fashion, the French merchants drastically lowered the price of the remaining stock, forcing the imitator firms to keep pace with an economic situation of ruthless (and perhaps predatory) competition which tended not so much to beat their competitors as to crush them. Zanon concluded that "this very word 'fashion' is an enormous and long-lasting treasure for France, and is considered by the French as a reputation which amounts to a highly-precious capital."[17] On another occasion, he declared, echoing a phrase attributed to Colbert, that the "empire of fashion" was for France what the Potosi mines were for Spain.[18]

According to Zanon, the strategies of the Venice merchants, of little success at home, fared better "in countries too far away to have strong connections with France" such as the countries of the Ottoman empire, where French and Dutch attempts to compete with Venetian damask had failed. Zanon attributed this failure to two factors: "one derives from the unchangeability of Oriental fashions and ways of dressing; the other from the fact that the Turks are unable to free themselves from their preference for Venetian cloth."[19]

Not even the Dutch silk industry was able to stand up to the competition of

[17] Zanon, *Dell'agricoltura*, III, pp. 342, 334–339, 363–364. The Lyonese producer started lowering prices "while he was about to produce a new fashion. Other price decreases followed after the new fashions had begun to circulate" (p. 364). L. A. Muratori observes that the rapid changes in fashion and the fact that fabric and clothes quickly became obsolete led to an increase in the second-hand market. He wrote: "The fairs of a large city are a curious spectacle where the second-hand clothes dealers make a great display of their wealth in various shops with their *andrienne*, their tunics, their robes, their clothes embroidered with Spanish stitching and other sumptuous laces. Clothes and garments that seem to have come straight from the hands of the tailors can also be seen" (*Della pubblica felicità* [Lucca, 1749], p. 277). On this subject, and also on the theft of clothes, see the excellent descriptions in Roche, *La culture de l'apparence*, pp. 313–346; and Ribeiro, *Dress in Eighteenth-Century Europe*, pp. 60–62, *passim*. A. Goudar maintained with worry that every year the warehouses filled up with unsold items from "de veilles modes," and that they were joined each year by "d'autres qui augmentent continuellement les nombres des anciens." These leftovers were eventually sold in northern Europe, and "le plus souvent au dessous du prix de la valeur intrinsique de la première matière" (*Les Intérêts de la France mal entendus* [Amsterdam, 1761], pp. 217–218).

[18] The Venetian merchant was aware that the harshest competition was by now taking place over the English Channel. He wrote: "The English and French genii battle even over the invention of new designs. And armies will perhaps decide the fate of the fashions. But the French will always be superior in their sales, because their inventions are prolific, their work is more skilful and their prices are lower" (*ibid.*, II [Venice, 1763], pp. 123–124).

[19] Zanon, *Dell'agricoltura*, vol. V (Venice, 1765), pp. 124–125. The Turkish preference for Venetian silks is confirmed by numerous sources. The Lyons merchant Bonnafous had little success when he tried to enter the Constantinople market with imitation Venetian damask produced in Lyons. Cf. the letter of May 25, 1755 in *Archives Nationales de Paris* (ANP), F12, 1443. J. B. Labat wrote in 1735: "Les modes changeantes n'ont point gaté l'esprit ni les usages des orientaux" (*Memoires du Chevalier d'Arvieu*, t. III [1735], p. 209). According to Accarias de Serionne, as well, in the Levant – where the new European fashions had not taken root – Venetian fabrics dominated. (*Le Commerce de la Hollande ou Tableau du commerce des Hollandois*, vol. II [Amsterdam, 1768], pp. 146–147.)

the Lyons fashions. The manufacturers of Amsterdam had enjoyed considerable prestige in the seventeenth century. An example can be found in the story commonly told of "velours à ramage fabriqués a Milan, envoyés en Hollande et de là renvoyés et vendu ensuite à Milan pour velours d'Hollande," together with a sure profit for the commissioners. The word "Dutch" was enough to add "une valeur nouvelle" to the Milanese material which could cover the transport costs from and to Milan via Amsterdam.[20] However, around 1720, the great materials of Amsterdam (damask and velvet) began to lose ground on the international market.[21] The Haarlem products fared better as they specialized in imitating

> differents desseins d'étoffes de Lyon, car on n'a jamais possédé dans ces manufactures l'art du dessein qui embellit, qui enrichi tous les ans celles de Tours et surtout et principalement celles de Lyon, par une extrême variété, par les graces du goût et de la nouveauté.[22]

The differentiation of fashion in Lyons, then, gave rise to strong market dynamism, established new hierarchies of power and, around the end of the seventeenth century or at the beginning of the eighteenth, brought about an important process of general restructuring in the new and old silk centers and in the international silk fabric market. In Europe, three different sectors can be distinguished. The first, in the vanguard, was the Lyons area itself which dictated the annual fashions in flowered silks (*façonné* and *grand façonné*). Its productive system, based essentially on putting-out, released on to the market each year a wide range of fabrics manufactured contemporaneously.[23] The second was that of the areas and industries which imitated the Lyons fabrics, above all in colour and design. These often included firms benefiting from extensive help from their governments (following mercantilist policies) to produce import substitutes. This sector, encouraged in England and Holland by the Huguenot emigration, included at different times and with varying success Valencia, Berlin, Venice,

[20] Accarias de Serionne, *Le Commerce de la Hollande ou Tableau du commerce des Hollandois*, I [Amsterdam, 1768], p. 294.

[21] J. G. Van Dillen, "Economic Fluctuations and Trade in the Netherlands (1650–1750)," in P. Earle (ed.), *Essays in European Economic History (1500–1800)*, (Oxford: Oxford University Press, 1974), p. 205. From a letter sent by Esaie Gillot, silk fabric merchant from Amsterdam, to the firm Caret et Cie of Lyons, it would appear that the Haarlem manufacturers were still active towards the middle of the eighteenth century and that they continued to copy the "nouveaux desseins" as they arrived from Lyons (A. D. du Rhone, *Procedure de la Conservation*, Livres de commerce (faillites), Caret et Cie, Liasse VII, Provinces Unies, 1). On this source, see *Papiers d'industrie et de commerçants Lyonnais. Lyon et le grand commerce au XVIII*, under the direction of P. Leon (Lyons, 1976), and the bibliography.

[22] Accarias de Serionne, *Le Commerce de la Hollande*, 1, p. 293. According to Veron de Forbonnais, "L'habilité et l'invention des nos dessinateurs sont tout le mystère de nos succès" (*Lettre à M+++ Negociant de Lyon . . .*, p. 24).

[23] Each entrepreneur presented each year a range of new products, taking account also of the economic possibilities of their particular clientele.

Haarlem, Spitalfields and others.[24] The third sector consisted of areas which had abandoned or avoided the market for fashion fabrics and occupied sectors in the market with products which were often traditional, not subject to rapid change or in any case not "fashionable." For this sector, after the shock of the initial impact, the annual fashions did not constitute direct competition. This was the case of the Venetian damask exported to the Ottoman empire, of Genoese velvet (used in many countries of Europe by the higher public officials, judges and magistrates); of Bolognese crêpe which adapted itself, as decoration, to all fashions (sales of which would rise during the frequent periods of court mourning); and of the damask and velvet of Amsterdam which held on the Paris market "pendant un très grand nombre d'années . . . à cause de leur solidité et surtout parce qu'ils ne changent point de mode." According to Accarias de Serionne,

> telle a été sur cet article la bizzarerie de la mode, qu'on la fait consister dans l'uniformité permanente de l'étoffe, pendant que la mode exige tous les ans imperieusement, des fabriques de Lyon, un goût nouveau et des nouveaux desseins.[25]

Apart from these three major groups, which together cover almost all the forms of the international market, we should also consider those materials (not from Lyons) which suddenly became fashionable because they were able, largely though not exclusively through their low cost, to stimulate the capricious taste of the public. This was the case for a "strong taffeta," also called "nobilata" produced in Rome around 1780 which had notable success on the market. Roland de la Platière bought a piece while travelling in Italy, spending the modest sum of 110 livres "de France" for a suit which was "très jolis, plus que complet . . . en couleur, doublé de blanc."[26] The success of the "nobilata" was probably

[24] On Valencia, cf. S. Rodriguez Garcia, *El arte de las sedas valencianas en el siglo XVIII* (Valencia, 1959); M. Garzon Pereira, *La industria sedera en España* (Granada, 1972), pp. 89–92; A. Domingo Ortiz, *Sociedada y Estado en el siglo XVIII español* (Barcelona, 1976), p. 271; V. M. Santos Isern, *Cara y cruz de la sederia valenciana (siglos XVIII–XIX)* (Valencia, 1981); V. Ribes, *Los valencianos y America. El commercio valenciano con Indias en el siglo XVIII* (Valencia, 1985), pp. 61–92. On Berlin, cf. Acta Borussica. Denkmäler der Preussischen Staatsverwaltung im 18. Jahrhundert, *Die Preussische Seidenindustrie im 18. Jahrhundert und ihre Begründung durch Friedrich der Grossen im 18. Jahrhundert*, 1–3 (Berlin, 1892). Volumes 1 and 2 contain a collection of documents edited by G. Schmoller and O. Hintze. The third volume, written by Hintze, is instead a historical monograph. On the limits of the Prussian silk mercantilism, see the extensive survey of H. Kisch, "Prussian Mercantilism and the Rise of the Krefeld Industry," *Transactions of the American Philosophical Society*, vol. 58, Part 7 (Philadelphia, 1968), pp. 3–15, and the bibliography. For Spitalfields, cf. F. Warner, *The Silk Industry of the United Kingdom. Its Origins and Development* (London, 1921), pp. 53–63 *passim*.

[25] Accarias de Serionne, *Le Commerce de la Hollande* I, pp. 294–295.

[26] Roland de la Platière, *Lettres écrites de Suisse, d'Italie et de Malte*, vol. V (Amsterdam, 1780), p. 100. According to Roland, the *nobilata*, like the other Italian fabrics "a le défaut d'être composée de matières écrus ou mal écroués, d'où le corps, qui paroît en resulter, n'est que la roideur qui empêche les fils de s'approcher assez en rend les plis tranchants, fait qu'elle s'éraille et se coup." In any case, Roland notes, "le goût français est toujours le goût dominant à Rome, comme à Turin, à Parme, à Modene. On n'y a pas, ainsi qu'à Naples, la ridicule manie de singer les anglois" (p. 101). But it was only the gentlemen, not the ladies, who

short-lived, but it is evidence of a new sector which would benefit from further research.[27] Another sector which would have considerable success in the future was that of the *melangés* fabrics (mixtures of silk with wool, cotton or linen), in the right price range for the urban middle classes who tended to imitate the clothes of the élite but with inferior fabrics.

The analysis offered here is a first attempt to classify silk manufacturing and weaving in the era which saw the hegemony of Lyons, an attempt which, however, must be refined and developed. It is necessary in particular to reconstruct the importance of the various imitation industries in relation to Lyons, explain the growing success of *melangé* and account for the greater or lesser resistance of the "traditional" sectors. The latter continued to sell the same fabrics which, although probably not insensitive to changes in fashion and taste, registered these changes on a different time scale.[28]

Up to now, Lyons has been considered as a homogeneous industrial center, but this was not the case. In fact, the imitation industry was of considerable importance here too. Indeed, it had actually originated in Lyons. As far back as 1711, the city authorities ordered all the members and employees of the Communauté des Marchands et Maîtres Ouvriers en Soye not to use designs made by others, nor to sell or lend their own.[29] This order also had the effect of hindering the social and professional mobility of the *maîtres ouvriers*, the large majority of the weavers (hereafter referred to simply as *ouvriers*) who worked on piece rates for the *marchands fabricants* (hereafter *fabricants*), from whom they received the designs and the raw material. If they wanted to begin "à travailler ou faire travailler . . . pour leur compte" they would have had to be enrolled on the register of the *Communauté*, paying 300 livres (reduced to 200 livres for their sons). This rule, introduced, it was said, to revive the financial fortunes of the Communauté, constituted a crucial turning point. It reinforced and formalized a

followed the London fashions. Roland de la Platière, later to be the leader of the Girondin faction, was involved in commerce in Lyons before becoming "inspecteur des manufactures" (Bacquier, *Les inspecteurs des manufactures sous l'Ancien Régime* (Paris, 1927), p. 384). On the interest of Lyons in the scouring ("décreusage") of silk, cf. "Observation sur le sujet suivant proposé par l'Académie des Sciences . . . de Lyon: Trouver une manière nouvelle de décreuser la soie sans alterer ni la qualité ni son lustre," in *Journal de Commerce* (November 1760), pp. 134–135.

27 In 1777 the *failles* of Anvers (silk fabric "a gros grain") became fashionable in France. Orders were so numerous that "les fabricants avaient de la peine à y suffire" (C. Van Cauwenberghs, "L'industrie de la soye à Anvers depuis 1532 jusqu'à nos jours," *Bulletin de la Société royale de geografie d'Anvers*, (1887), 119.

28 For the history of the Lyons silk industry, see E. Pariset, *La Chambre de Commerce de Lyon (1702–1791)* (Lyons, 1886); J. Godart, *L'ouvrier en soie. Monographie du tisseur Lyonnais. Etude historique, économique et sociale* (Lyons-Paris, 1899); E. Pariset, *Histoire de la Fabrique Lyonnaise. Etude sur le Régime sociale et économique de l'industrie de la soie à Lyon depuis le XVIe siècle* (Lyons, 1901; and the more recent analysis by M. Garden, *Lyon et le Lyonnais au XVIII siècle* (Paris, 1970). The work of J. Godart continues to be a fundamental guide to any research on the Lyons silk industry.

29 *Lettres patentes portant homologation de l'Ordonnance Consulaire du 25 Octobre 1711 servant de Règlement pour le Communauté des Marchands et Maîtres Ouvriers en Soie de la ville de Lyon* (Lyons, 1717), pp. 5–9.

Table 1.1 *Organizational rules of the* Communauté des Marchands *and of the* maîtres ouvriers *(1732)*

GRANDE FABRIQUE			
Putting out system		Kauf system (petite fabrique)	
Marchands fabricants (putting-out entrepreneurs)	around 70	*Maîtres marchands*	around 700
Maîtres ouvriers (dependent weavers)	around 8,000		

Source: *Mémoire des Maîtres Gardes*. Access to the market was permitted only for *marchands fabricants and maîtres marchands*, and was prohibited for *maîtres ouvriers*.

process of social differentiation tied to the introduction of annual change in fashion and the growing importance of the *fabricants*. As can be seen from table 1.1, in 1732 the members of the Communauté, also called the *Grande Fabrique*, were grouped into three categories: the *fabricants* (in all less than 100 rich entrepreneurs), real putter-outers; the *ouvriers* (around 8,000 in 1732) who carried out the weaving at home on piece rates for the *fabricants* from whom they received the thread and the designs; and the *maîtres* (around 700 in 1732), small independent weavers who, working on their own, had, like the *fabricants*, a direct relationship with the market, one denied in principle to the *ouvriers*.[30] In the same year 1732, the 1711 proposal was represented by the *Maîtres Gardes* (the heads of the Communauté) with the aim of striking a blow at the independent weavers (the *maîtres*), accused of maintaining their place on the market through fraud, cornering silk and designs stolen by the *ouvriers* from the large merchants who claimed to make up the *Grande Fabrique*.

According to the *Maîtres Gardes* the independent weavers (relegated to the

[30] The term *Grande Fabrique* does not indicate centralized manufacture but rather the network of putting-out firms which gave work to thousands of weavers at home or else the whole range of organizational structures. Each *ouvrier* might have from two to four looms, on which would work apprentices, *compagnons* and members of his family. These social and professional classifications were not constant. The meaning of the words changed according to who used them and the more or less conflictual nature of the context in which they were uttered. The case of Lyons, like that of other old industrial centers, shows the importance of studying urban as well as rural putting-out systems. Cf. C. Poni, "Proto-industrialization: Rural and Urban," *Review*, 9 (1985), 305–314, and "Per la storia del distretto industriale serico di Bologna (secoli XVI–XIX)," in *Quaderni Storici*, 73 (1991), 93–167. It should be remembered that every year thousands of women arrived in Lyons from the surrounding countryside to work as "tireuses de cordes," an operation necessary in order to transfer the design onto the cloth. The introduction of perforated cards (the Jacquard loom) freed the women from this burdensome task. For reasons of space, I will not go into the issue of the gender division of labor here. See Daryl M. Hafter, "Women Who Wove in the Eighteenth-Century Silk Industry of Lyon," in Hafter (ed.), *European Women and Preindustrial Craft* (Bloomington IN, 1995), pp. 42–64.

Petite Fabrique), unable to employ "bons dessinateurs" (paid as much as 8–10 thousand livres a year by the larger merchants) used "la plus part du temps des desseins volés au bonnes fabriques." This double theft (of the design and the silk) had distorting effects. It pushed up enormously the cloth production costs of the large merchants, who employed weavers working at home. On the other hand, it encouraged opportunistic behavior on the part of the independent weavers who were able to offer at low cost their imitation cloth, which was necessarily of inferior quality as it was produced with a mixture of different types of silk found here and there.[31] These accusations were hotly contested by the independent weavers, who claimed that the international success of the *Grande Fabrique*, from which the *Maîtres Gardes* would have liked to exclude them, was due to the high professional quality of their work and that of the *maîtres ouvriers*.[32]

The weavers (*maîtres* and *ouvriers*) often proudly attributed the commercial success of Lyons to their own professional excellence. The *fabricants* tended to downplay these declarations. While claiming not to want to minimize the merits of the weavers, they declared that all the weavers did was set out "sur le dessein de manière qu'ils n'ont qu'à lire et suivre exactement ce que leur est prescrit." The perfection of the *façonnées* was essentially due to the intelligence of the entrepreneur. It was he who chose and decided upon "la qualité des soyes, la beauté de la dorure, la delicatesse du dessein, la sage distribution des nuances, le bon goût." These were capacities, knowledge and sensitivities which could not be acquired "ny en travaillant sur le métier ny restant dans la boutique d'un ouvrier à façon." The professional training of the *fabricant* merchant could not be limited to weaving. He had to take care of correspondence with foreign merchants; obtain orders from them; encourage them to visit the warehouses of Lyons; induce them to send "par anticipation ses commissions pour avoir les loisirs de les effectuer"; entice them with the prospect of large profits so as to persuade them to prefer

[31] "*Mémoire des Maîtres Gardes de la Communauté des maîtres ouvriers et marchands d'étoffes d'or, d'argent et de soie servant de réponse aux requests presentées au Roi . . . par quelqu'un des maîtres ouvriers et marchands de la même Communauté opposans à l'Arrêt du 8 mai 1731* (Lyons, 1732 (AML, Chappe, HH. 132). These memoires include a "Mémoire preliminaire" (pp. 9–26), two "Requêttes" (first and second) presented to the King by the *maîtres marchands* with the "Réponses des Maîtres Gardes" printed opposite (pp. 27–48 and 49–64), and a "Mémoire des négocians de la ville de Lyon" also with the "Réponse des Maîtres Gardes" printed opposite (pp. 65–90). The order of May 8, 1732 forced the *maîtres marchands* (independent weavers) to choose between the profession of *maître ouvrier* and that of *marchand fabricant*. Whoever opted for the latter would have had to enrol on the register of the Communauté, paying 200 livres (and 100 livres for his sons). This repeated attempt to abolish the independent weavers as a separate group, carried out because they were the considered receivers of the silk and designs stolen by the *maîtres ouvriers*, failed, however. The quotations are taken from the "Mémoire préliminaire" (pp. 12–16, 18–19) and from the "Réponse des Maîtres Gardes . . . à la Mémoire des négocians de la ville de Lyon" (p. 75 and *passim*). The majority of the *Maîtres Gardes* were always, or nearly always, on the side of the *marchands fabricants*. The best designers worked for the *fabricants* on a contractual basis, but there were also freelance designers who sold their designs to the highest bidder.

[32] "Première Requêtte . . .," pp. 33–39. Cf. also the "Mémoire des négocians de la ville de Lyon . . .," pp. 83 ff.

"notre manufacture aux étrangères"; take advantage of the moment at which the exchange rate was favorable to foreign correspondents and ask them to pay their debts "adoucissant par un moyen si simple le prix d'une marchandise." It was only this type of knowledge that could ensure "une consommation plus considerable pour l'avenir." But these special skills could only be achieved working alongside the *fabricants* in their warehouses as apprentices.[33]

In another pamphlet, the *Maîtres Gardes* (after defending in similar terms the *fabricants* from the accusations of the *ouvriers* who depicted them as wholesale merchants "inutils à la manufacture") came up with an original analogy, comparing the *fabricants* to the architect, seen as "le chef, l'âme, le créateur perpetual des differens ouvrages qu'il fait construire," and the *ouvriers* to the mason who carries out "son travail purement méchanique" with no idea of the beauty produced by his work.[34]

Returning to the theft or circulation of stolen designs, it should be added that some Paris merchants were also involved in this illicit traffic. Having bought "les premières pièces d'un assortissment d'étoffes," they had the designs copied and then sent them to those independent weavers of Lyons who were willing to supply them. In this way they were able to get hold of the cloth "bon marché" and resell at a price which was perhaps no different from that of the original cloth.[35] Unfaithful traders ("commissioners") did likewise. If a certain type of cloth began to be successful, they would have the design copied and then secretly pass it on to the weavers. Once they had the cloth, the traders would sell these imitations instead of the original, cheating both their clients and the "bonnes fabriques" whose trust they had betrayed.[36]

[33] *Mémoire pour les Maîtres Marchands d'étoffes de soye, d'or et d'argent de la ville de Lyon* (n.d. [post. 1737],) pp. 19–20 (AML, Chappe HH 133). The major *fabricants* were active not only in Lyons. They also operated as entrepreneurs in southern France where they owned reeling plants and silk mills. Out of the many cases of such merchants, I will cite just two examples: Joubier and Montessuy.

[34] *Réponse des Maîtres Gardes et Adjoints Marchands aux observations . . . sur 43 articles du Règlement arreté . . . le 19 juin 1744 pour la fabrique de Lyon* ([Lyons], n.d. [but 1759]), pp. 7–8 (Bibliothèque Municipale de Lyon, *Catalogue Coste*, 10594). M. Paulet pointed out other aspects: "Je toujours vu une difference sensible entre les étoffes fabriquées par *un bel ouvrier*, et celles fabriquées par un ouvrier médiocre. On nomme *bel ouvrier* celui qui prende soin de régler tous les mouvements de son métier et non pas celui qui ne cherche qu'à faire une grande quantité d'ouvrage" (M. Paulet, *L'art du fabriquant de d'étoffes de soie*, VII section, I partie [Paris, 1776], p. 91).

[35] *Mémoire préliminaire* . . ., pp. 14–18. The Conseil d'Etat du Roi only approved the law protecting property in designs on July 14, 1787. See *Arrêt du Conseil d'Etat du Roi portant Règlement pour les nouveaux dessins que les fabricants d'étoffes, de soieries et de dorure du Royaume auront composés où fait composer*. This *arrêt* made explicit reference to requests coming from the *fabricants* of Lyons and Tours.

[36] *Mémoire préliminaire*. J. Savary identified five types of commissioner (*Le parfait négociant*, II, Paris, 1675 p. 143). The commissioners referred to in *Mémoire préliminaire* could be the "négociants qui demeurent aux lieux des manufactures ou dans les villes où il se fait grand commerce, qui achètent de marchandises pour le compte de négocians demeurant d'autres lieux" or else "négocians à qui d'autres anvoyent des marchandises pour vendre pour leur compte, suivant les ordres qu'ils en reçoivent" (Savary, *Le parfait négociant*, p. 144).

This traffic, forbidden by the 1737 Statutes, was not limited to the Paris–Lyons axis. Merchants and traders, encouraged by the profits to be made, also sold designs to foreign competitors, thus encouraging imitations. Even though they were contested, these practices were relatively unharmful as long as foreigners copied the designs and fabrics of the previous year. Indeed, the imitations were predicted and anticipated by the challenge–response mechanism upon which the very success of the *Grand Fabrique* was based. An anonymous *marchand fabricant* wrote in 1761:

La concurrence . . . nous a forcé de chercher la préférence . . . Le goût a dû se perfectionner plus rapidement, se renouveller entièrement et varier toutes les fois que nous avons pu être imité . . . Que l'étranger nous copie encore cent ans, il ne formera pas des dessinateurs. C'est notre goût, c'est notre imagination que lui manquent: l'imitation éteint l'un et ne donne pas l'autre. Eh, comment donneroit-elle ce goût qui regne cette année, et que la mode, une convention générale et le caprice anéantiront l'année prochaine?[37]

After 1730, busy with fierce corporative conflicts and engaged in limiting the illegal commerce in designs and stolen silk, the Lyons entrepreneurs did not recognize the damage that new trading practices, seemingly advantageous, had begun to inflict. The rapidly spreading practice of sale by *échantillons* (by sending out collections of pre-production samples) had enabled the trade network of the *Grande Fabrique* to thicken and spread. This new sales method permitted the large merchants to customize their fabric in accordance with the taste of their buyers with only minimal costs and to start the looms working only after receiving the order. The advantages were clear: a reduction in end of season warehousing, "just in time" production, personalized products, decrease in capital employed compared with the size of sales, and profit increases.

However, this practice could also be seriously damaging. The three months between the sending of the *échantillons* and the arrival of the fabric ordered gave a

[37] "Mémoire sur l'envoie des échantillons de la fabrique de Lyon," *Journal de Commerce* (1761), pp. 108, 128–129. The article was published in serial form. The founder and editor of the journal, printed in Brussels, was Accarias de Serionne. Cf. H. Hasquin, "Jacques Accarias de Serionne, économiste et publiciste français au service des Pays-Bas Autrichiens," in *Etudes sur le XVIII siècle*, I (1974), 159–169. For a long time it was wrongly thought that the editors of the *Journal de Commerce* were the merchant Camus and the Abbé Roubaud. The "Mémoire sur l'envoie des échantillons," the importance of which it would be hard to overestimate despite the fact that it has remained more or less unknown, was reprinted as a pamphlet with the same title in Turin in 1761. This text, as will become clear later, shows how the major silk merchants of Lyons were fully aware of the peculiarities of their practices. From this point on, I will quote from the Turin edition. On the "goût" of Lyons fabrics see also P. Bertholon, *Du commerce et des manufactures distinctives de la ville de Lyon* (Montpellier, 1787), pp. 25–26, 46–49. In a rich and stimulating article, J. M. Jones has dated the supremacy of taste over luxury as late as 1787, although in fact taste had been dictating changes in fashion from the first half of the eighteenth century. Jones' dating can be explained by her sources: the fashion journals published in the last decades of the century. See J. M. Jones, "Repackaging Rousseau: Femininity and Fashion in Old Regime France," in *French Historical Studies*, vol. 18, no. 4, 1994, pp. 939–968.

considerable advantage to the imitation industries. The author of the pamphlet *Sur l'envoie des échantillons* (1761) told how a Viennese merchant, having received the *échantillons* from Lyons, sent them to Venice to have them copied. He was thus able to sell imitation cloth in Vienna even before the original arrived from Lyons. In cases such as these, the original French fabric (and we should remember the talent for imitation on the part of the Venetian weavers) "n'étoit plus . . . qu'un article passé, ou du moins un article soumis à la concurrence et exposé au rabais."[38]

The anonymous author did not limit himself to describing this perverse mechanism. At the end of his pamphlet, he also formulated a twenty-article proposal for a set of rules prohibiting trade by means of *échantillons* (and thus also their expedition). This project was integrated in the nine-article *Règlement* approved a few years later. Not all the spokesmen, however, found themselves in agreement. Adamoli Fils, in favor of the *échantillons*, was of the opinion that the new imitation industries would find difficulty in reaching the size that enabled the Lyons industries to carry out a high division of labor with consequent reductions in costs. He wrote: "Les manufactures compliquées, telles que sont celles des étoffes de goût demandent . . . la réunion d'une grande quantité de métiers" so that all the workers with their different jobs could be employed throughout the year. The speed and perfection of the assembly of the loom was dependent precisely on the specialization of workers and on the division of labor. "Si l'ouvrier qui doit fabriquer la pièce vouloit aussi monter le métier, il y employeroit beaucoup de temps et le monteroit mal." Thus it would not have been easy quickly to achieve the criticial mass and competitive level that the *Grande Fabrique* had acquired "par une longue pratique." No European ruler had so far managed to defend successfully the internal market in favor of local industry and give satisfactory support to the flow of exports through incentives. Finally, in the field of fashion fabrics, with or without *échantillons*, foreign firms would continue to flounder "dans la nuit de l'obscurité."[39]

This analysis, which emphasized what we would now call barriers to entry and the competitive advantages of Lyons, was not enough to counter the arguments against the sending of the *échantillons*, which was prohibited on March 15, 1765

[38] "Mémoire sur l'envoie des échantillons . . .," pp. 82–83. In order to send the *échantillons* the entrepreneurs had to finish the manufacture of at least one piece and then cut it up. If no orders arrived, the loom was dismantled. Often the *fabricants* made mistakes in their predictions. A fabric on which they had "conçus les plus belles espérances" might not meet with "aucun succès." On the other hand, "ses pressentiments contre les desseins d'un goût hazardé et fantastique" were sometimes contradicted with "la plus heureuse réussite" (Adamoli Fils, "Lettre à M. Vial marchand fabricant à Lyon beau-frère de l'auteur," in *Journal de Commerce*, December 1761, p. 70). Production times varied considerably according to the quality of the design and the lengths ordered. The "péruviens" were woven at the length usual for men's clothing, in four days. To produce fifteen *aunes* (the length for women's clothing) of the simplest *broché* ("*nué*"), four of five weeks were necessary. *Broché*, used for women's clothing with complex and rich design, could only be produced "dans l'espace de deux mois et plus" ("Lettre à M. Vial," p. 74. 1 *aune*=1.18 metres).

[39] Adamoli Fils, "Lettre à M. Vial," pp. 76–77, 93.

by the assembly of the Communauté. On this occasion, one of the *Maîtres Gardes*, Jean Marie Duperrel, reported to the assembly that of the thousands of *échantillons* sent every year to foreign merchants, only relatively few orders were obtained, sometimes only from one *échantillon* out of ten sent out, while the other nine were passed on to be copied by the local weavers.

These abuses were widespread. In Hamburg, one so-called merchant had received from a Lyons trader several cases of *échantillons* and had not placed a single order. In Naples, a Lyons merchant who had taken with him a complete assortment of *échantillons*, gave them to the supervisor of the local "Royal Manufacture" to be copied, getting in exchange "quelques commissions légères." In London, another Lyons trader, when he was leaving to return to France, sold a collection of *échantillons* whose value amounted to 1,800 livres for only 13.10 livres. This was probably an exaggeration, but *Maître Garde* Duperrel was stressing, also on behalf of his colleagues, how it was much less dangerous to send designs than to send *échantillons*

> dans les quels on voit non seulement la composition du dessin, mais encore son effet, les nuances, le tissu de l'étoffe, l'armure du métier, la qualité des matières, la quantité des portées, cette art enfin du melange et cette gradation des couleurs qui produisent le belles nuances et qui donnent prix aux étoffes.

However, if the sending of *échantillons* were prohibited, would it be possible to stop foreign firms copying material after buying a single item? J. M. Duperrel replied to this objection that as foreign firms had to buy at the beginning of the season, they would be at a disadvantage of several months with their copied material and would in any case have to buy two *aunes* of fabric. Beaten in terms of timing, they would be forced to carry out their imitations with a delay of one season or one year. "Mais alors les desseins de l'année precedente seroit tombés, et ne soutendrons pas la concurrence avec nos étoffes nouvelles."[40]

However, the trade by *échantillons* (from this point onwards prohibited but never entirely eradicated) gave way to other manipulations favorable to the *Grand Fabrique*. In countries where French cloth was banned (as in Britain and Ireland) some merchants were able to bring in real pieces of cloth under the name

40 *Arrêt de la Cour de Parlement qui homologue la deliberation de la Communauté des Maîtres Marchands et Maîtres Ouvriers fabricants en étoffes d'or, d'argent et de soie de la ville de Lyon du 15 Mars 1765 concernant la prohibition de l'envoie des échantillons des mêmes étoffes hors la dite ville, 14 mai* (1765), pp. 5–10. In this text, the term *maîtres marchands* indicates also the *marchands fabricants. Echantillon* trading had been practiced at least from the seventeenth century. Immediately after the publication of this *Arrêt* prohibiting the sending of échantillons, some Lyons merchants proposed to send short pieces of fabric (from 2 to 3 aunes long), in return for payment, to their foreign correspondents; the latter refused to pay and rejected the offer. Cf. the letter sent by Fabre and Defez from The Hague to M. Carret of October 3, 1764. They wrote: "La deffence du Parlament sur l'exportation des échantillons nous a fait de la peine. Nous nous étions flattés de prendre plusieurs commissions . . . nous ne pouvons pas nous charger des échantillons de trois aunes que vous nous offrez, et qu'il vous est permis d'envoyer" ("Carret et Cie," Liasse VIII, in A. D. du Rhone, *Procedure de la Conservation, Livres de commerce*).

of samples imported for the production of imitations. In 1763, for example, Mr. Cottingham, a large Dublin silk merchant, imported, in his words, "a number of new patterns in pieces in order to serve the kingdom with the newest fashions from his own looms." The Dublin weavers contested this, claiming that he had imported cloth to sell directly on the market. Frightened and embittered at the prospect of losing their jobs, the weavers massed in front of Mr. Cottingham's house and attacked it twice. The imported fabric was cut up, his looms were destroyed and he was burnt in effigy. According to the reporter who narrated the episode, the rebellious weavers "would have assassinated him, if he had fallen into their hands." The crowd dispersed only after having extorted a promise from Mr. Cottingham "that he would never import any more silks" and after he had given over a bond of "1,000 pounds for performing the covenant."[41] Other merchants used the *échantillons* for other purposes. Those of Frankfurt, for example, who did not buy fabric "de la première fraîcheur" as they traded in "vieux desseins," cleverly used the new samples as benchmarks to devalue the year-old fabrics.[42]

The feverish and worried discussion over the theft of and trade in designs and *échantillons* did not stop the entrepreneurs of the Lyons fashions or rather, as one of them put it, of the "modes de Paris faites à Lyon," from celebrating taste in highly poetic language: "Le goût," said the already-mentioned anonymous merchant,

> comme la beauté fragile n'a qu'une saison, commes les graces il n'est soumis à aucune règle, et ne s'acquiert point avec une froide imitation . . . Dans le façonné, le succès de notre fabrique est fondé uniquement sur le goût. C'est par lui qu'elle se soutiene malgré l'envie et la rivalité.[43]

But copying material by designs and *échantillons* was not sufficient if the colors were not imitated as well. This was a difficult task, given the technologies available which did not permit the isolation of the ingredients. It would not even have been enough to use the same materials (mordants and colourings) in the same proportions, as the physical and chemical properties of the water used also made a difference. For their blacks, the Lyons merchants used the water of the Rhône mixed with that of the Saône. For the colors such as crimson, deep red, violet and purple, the water of the Saône alone was used. For all the other colors (white, green, grey, yellow and so on), the water of the Rhône was used.[44] But

[41] *The Gentleman's Magazine*, 33 (1763), 410–411. During the riot, the weavers wounded "several soldiers . . . Their leader, with an astonishing intrepidity, when a loaded musket was at his breast, called to the mob to fire, saying: 'I desire nothing more than to die in this cause.'" In the same year, an anonymous author published a harsh, sarcastic pamphlet against Mr. Cottingham, *The Weaver's Address to the Citizens of Dublin or Dean Swift's Advice to the good People of Ireland together with his Apparition occasioned by the Importation of Wrought Silk from France* (Dublin, 1763). The author sets up an imaginary trial of Mr. Cottingham, although he is never mentioned by name.

[42] Adamoli Fils, "Lettre à M. Vial," pp. 78–79.

[43] "Mémoire sur l'envoi des échantillons . . .," pp. 34–37.

[44] The history of dyeing techniques in France and especially in Lyons, based on a vast information network stretching from India to Europe will be treated in a later article. On the

apart from the inimitable (and perhaps imaginary) dyeing virtues of the rivers of Lyons, how could foreign merchants get hold quickly of the dyeing formulas and the additives used to produce not only the new combinations but also those delicate shades for which the Lyons materials were justly famous? There was one short cut: to buy thread already dyed in Lyons, as the Tours merchants did, although this export was explicitly forbidden by the Communauté.[45]

More pernicious than the illegal trade in designs, *échantillons* and dyed yarn was the emigration of weavers, designers and entrepreneurs, often enticed to move to other states or foreign cities with promises of rich bonuses on employment, or of generous financing and high profits and wages. Sometimes waves of emigration were sparked off by economic crisis and unemployment as in 1752 and 1759. On other occasions, religious persecution and the desire to follow the Protestant religion prompted emigration to freer and more secure countries. In England and Holland, silk manufacturing benefited enormously from the flight of religious minorities after the Revocation of the Edict of Nantes in 1685. In Spain, and in particular in Valencia, the extensive and highly-qualified French emigration was encouraged by the prospect of high wages and profits.

Lyons replied to what it saw as dangerous desertion in several different ways, ranging from the exclusion from apprenticeships of those professing Protestantism to the imprisonment and punishment for criminal activity of those who were discovered making preparations for emigration or already on their way to the frontier. The news that entrepreneurs, designers and technicians were getting ready to open workshops in foreign countries to reproduce the fashions of their home country was looked on with preoccupation by Lyons and gave rise to elaborate reactions. In 1755, the Chamber of Commerce denounced the entrepreneurs Bachelard and Parisien, accused of setting up a silk manufacture in Naples, and retaliated by forcing the "corps de fabricants . . . a envoyer aux negocians qui sont à Naples plusieurs assortissemens de belles étoffes avec ordres de les vendre à bas prix au danger d'y perdre quelque chose." The profit which

importance of the "Art of Dying," an anonymous English author wrote in the "Preface" to the translation of French treatises on coloring: "Few people can estimate the intrinsic value of manufactured woolens, silks or cottons. But men, women and children can judge of their colours, on the beauty of which therefore the first sale of a new manufacturer must depend. And the continuance of that sale will also depend more on the permanency of the colours than on the stuff" (*Art of Dying Wool, Silk and Cotton*, by M. Hellot, M. Macquer and M. le Pileur d'Apligny (London, 1789), p. v).

45 The prohibition "d'envoyer ou de favoriser la sortie hors du Royaume des soyes teintes propres à fabriquer des étoffes" was issued on July 13, 1725 by the *Prevôt des Marchands* on the request of the *Maîtres Gardes* (AML, Chappe, HH, no. 147). According to an anonymous author, the prohibition of the export of dyed silk had forced foreigners to "s'appliquer aux teintures." And they had succeeded so well in this that they had reached the "même degré de perfection, si l'on excepte la couleur de rose, qui est encore inférieur à celle de Lyon, mais qui aura sans doute à son tour le même sort que les autres couleurs" ("Consideration sur le finance et sur le commerce de France," in *Journal de Commerce*, January 1762, pp. 21–23). In fact, smuggling ensured the continuation in the trade in dyed yarn. But the growth of the Lyons chemical sector was hindered by this prohibition. New dye works were set up in the country as a reaction to the mercantilist policy of the city.

was lost would be small in comparison with the benefits "qui en resulteroit pour tous." As well as the dumping of fabric, it was decided to order all the creditors of Bachelard and Parisien (including Soubay de Combes, Daudé, Nicolas, Guerignon and others) to form a consortium and demand together that the two black sheep pay for their misdeeds. This threat, it was thought, would be enough to make them abandon "promptement leur entreprise qui les oblige à une grande dépense." Similar threats and rulings had been issued on previous occasions "lors de l'établissement de la manufacture de Hollande."[46] This shows that the *Grande Fabrique* kept up an effective control over a large area and was able to benefit from a good information and command service capable of functioning even over long distances.

It is by no means certain, however, that the emigration of weavers and entrepreneurs had the results that were feared. Vincent de Gournay, the *Intendent de commerce* who visited Lyons in the 1750s as the official responsible for the French silk industry for the Bureau de Commerce of Paris, smiled "bitterly" when some merchants told him of the efforts that were made to stop the emigration of weavers and merchants. His views in favor of "laisser faire" and hostile to corporate restraints, even after his death in 1759, were caricatured and ridiculed by an anonymous author who claimed he had proposed "d'amener l'étranger sur nos métiers ou les leur envoyer tout montés avec nos ouvriers," offering in return only "son system favori de la concurrences sur le prix." According to this author, competition over prices was not always possible and it should only be resorted to "quand on n'a plus autres resources"[47] (when it was not possible to have recourse to product differentiation). A defender of de Gournay made the criticism that the suppression of corporative restraints would not annul all the competitive advantages. There were certain structures and values that could not be transferred such as "l'ensemble d'une manufacture, le goût, les desseins, l'économie, l'amour du travail."[48] These ideas, similar to

[46] Cf. *Moyen qu'il paroit que l'on pourroit employer pour empescher le succès de l'établissement que se fait à Naples d'une manufacture d'étoffes à l'imitation de celle de Lyon. L'Avis de la Chambre de Commerce* [n. d. (but 1755)], in BML Cotte, no. 10428 or 1002. Cf. also the *Mémoire* of the designer Jean Renard of Lyons to the King of Naples in ASN, "Pandetta comune," fascio 73, fasc. 1380/bis. Renard promised to teach the "composition et la lecture des desseins" and the procedure "pour le blanc, cremoise et ponceau de Lyon" (pp. 1, 13–14). Roland de la Platière met a Lyonnais in Rome who managed an "entreprise" which produced, although with little success, fabric "de soie brochés et en dorure" (*Lettres* . . ., V, pp. 99–100). The imitating manufactures of southern Italy never substantially developed. A. Goudar, who was in Naples in 1767 for the wedding of the Bourbon King Ferdinando IV with Maria Carolina of Hapsburg, was quick to notice the huge expenditure on clothes and jewelry, which amounted to "une grande portion des richesses générales, qui passeroient en France de manière que le mariage se feroit à Naples et les noces à Lyon" (*Naples. Ce qu'il faut faire pur rendre ce Royaume florissant* [Amsterdam, 1771], pp. 111–112).

[47] "Mémoire sur l'envoie des échantillons," in *Journal de Commerce*, pp. 125–136. I do not quote from the Turin edition because it does not include the attack against Germany.

[48] "Observations sur un article du Journal de Commerce . . . concernant feu M. de Gournay Intendant du Commerce," in *Journal de Commerce* (April 1761), pp. 71–85. On the ideas of de Gournay, cf. G. Schelle, *Vincent de Gournay. Laisser faire, laisser passer* (Paris, 1897); G.

those of Adamoli fils, are evidence that, with the lively industrial atmosphere of Lyons, some merchant entrepreneurs began to move beyond the defense of corporative restraints and see the possibility of new protective barriers that were compatible with "laissez faire."

To thwart the setting up of import-substitute industries in foreign countries, the merchant-entrepreneurs of Lyons began to use other methods (which we would now call horizontal product differentiation), adapting designs and colours to the different national tastes. For the German and English markets, light colors were used; for the Spanish some black threads were added. Sometimes some decoration necessary for specific markets was added to the French designs.[49] The correspondence between merchants, preserved in the Archives Departmentales de Lyon, documents the "open" orders, those based on *échantillons* and the variations in colors and designs asked for by foreign traders.

The information system was used in the same way that contemporary entrepreneurs prepare the launch of new products. Absolute secrecy surrounded the phase in which the new annual fashion was conceived and chosen. Once the new fashion had been launched, maximum publicity was given immediately using all possible channels of communication: the *échantillons*, the network of traders, mercantile correspondence, "*pupées*" (dolls) and, from 1781 onwards, the fashion journals. Even the foreign travellers who had themselves made one or more suits in the latest fashion became, once they returned to their country of origin, unconscious bearers of the new Lyonese or Parisian tastes.

According to some, the wholesale buyers were better served if they bought directly from the shops of warehouses in Lyons. Negotiating directly with the *marchand fabricant*, the buyer could ask to "reduire le dessein de trois quarts . . . changer les nuances . . . donner plus de contrast à ce groupe, plus de jeu à cette guirlande . . . substituer ici un fond cannelé leger, passer cette fleur de cette manière." A third party would then enter the discussion when a designer was called upon to make a design out of ideas that were still uncertain and confused.

Sénestat-Escande, *Les idées économiques de Vincent de Gournay* (Bordeaux, 1911); S. Meyssonier, *La balance et l'horloge. La genèse de la pensée liberale en France au XVIIIe siècle* (Paris, 1989), pp. 167–200, 232–239. In 1786 M. Mayet recognized that it was easy "aux Fabriques étrangers de contrefaire nos étoffes et de copier nos dessins. Le bas prix de notre main d'oeuvre est le seul avantage qu'ils ne peuvent nous derober" (M. Mayet, *Mémoire sur les manufactures de Lyon* [London, 1786], pp. 37–38). In a previous statement he considered "Le bas prix des étoffes" as one of the advantages of Lyons. The others were: "la qualité de leurs fonds, le goût qui regne dans leur dessins" (p. 18).

49 *Mémoire sur l'envoie des échantillons . . .*, p. 75. According to Abbé Morellet, "quoique nos modes et notre goût dans les étoffes de soie ayent beaucoup d'empire sur les étrangers, il faut cependant convenir qu'ils ont leur goût auquel nous nous conformons en partie, et auquel nous continuerions de nous conformer, quand même nous ne ferions pas autant des étoffes de soie" (*Réflexions sur les avantages de la libre fabrication et de l'usage des toiles peintes en France* [no place of publication, 1758], p. 50. See also Veron de Forbonnais, *Elemens du Commerce*, Première Partie (Leyde, 1754), p. 252. For criticisms of the obsession with French fashion, harmful to trade and to the morals of the Spanish, see B. G. Feijoo, *Theatro critico universal o discoursos varios en todo genero de materias para desengano des errores comunes* (Madrid, 1728), II, pp. 139–149 *passim*.

The *ouvrier* (the weaver) would then join the group and try to imagine the difficulties of putting the new design into concrete form. "Et s'il ne peut exécuter d'une manière, il en propose une autre."[50] But the opposition between the two types of sale was not total. The merchants who bought by *échantillons* went more or less regularly to Lyons and to Paris to make the most important purchases and to get a feel of the market.

The press too tried to influence the taste of the public, exalting the superiority of the *Grande Fabrique*. "En Espagne," wrote the anonymous merchant, "où la concurrance est si rare et le goût si borné on copie très lentement et très exactement chaque année nos desseins de l'année precedente." In London, "l'on conserve dans l'étoffe le plat, massive et le bizarre goût chinoise. On ne songe guère à nous copier. Si on le fait quelque fois, on voudrait se déguiser à soi même. On nous defigure en nous copiant."[51]

The same author gives confirmation to the practice of lowering by up to 50 percent the price of fabrics that went out of fashion, and informs us that a large part of these fabrics was sent to the provinces, to the West Indies, to the fairs of Leipzig and Frankfurt on the Oder where "les grossiers" opened their emporia to a picturesque crowd of Poles, Danes, Russians, Swedes, Hungarians, Walachians and Turks. This was a frontier world, in direct contact with countries whose customs were more or less immobile, characterized by "différents goûts," little influenced by the fashions and tastes of Paris. In this area, where there was little sensitivity to the "quality" of the product, it was easy to do good business with products which were either in or out of fashion. It would seem almost as if fashion had fixed the Oder (or somewhere not far to the east of it, with the exception of the court of St. Petersburg and Moscow) as the frontier of a certain kind of Europe whose capital was Paris.[52]

But while taste and fashion triumphed, the quality of the material declined. The rapid differentiation of the product and the annual appearance of new fashions led to a decrease in quality. There was, in fact, no reason to produce long-lasting materials. Daniel Defoe wrote in 1728: "French cloth, with all its superficial French gloss upon it is fine, but thin and spongy and will do the weaver

50 *Mémoire sur l'envoie des échantillons . . .*, pp. 78–79. These sales methods, practiced on an international market, indicate a type of flexibility which was personalized and original. We should keep in mind that on local markets goods were often bought on order. I am thinking here of the sale of shoes, for example.

51 *Mémoire sur l'envoie des échantillons . . .*, pp. 57–58. When necessary, the *Grande Fabrique* worked also "sur des desseins d'Angleterre" as happened in 1759 (*Réponses des Maîtres Gardes et Adjoints-Marchands aux Observations particuliers et interessantes des Maîtres Gardes et Adjoint Ouvriers . . . sur 43 articles du Règlement . . . arrêté le 19 Juin 1744* ([Lyon], 1759), pp. 28–29. On fashion reviews, cf. A. Kleinert, *Die frühen Modejournale in Frankreich. Studien zur Literatur der Mode von den Anfangen bis 1848* (Berlin, 1980). Right from the beginning of the eighteenth century dolls dressed in the latest fashions were sent annually from Paris to the principal cities of Europe.

52 *Mémoire sur l'envoie des échantillons . . .*, pp. 69–71. As A. Carraciolo wrote, "Les seigneurs moscovites sont devenus presques parisiens à force de les copier" (*Paris. Le modèle des nations étrangères ou l'Europe française* (Venice (Paris), 1777, p. 44)).

neither credit nor service. While the English cloth wears to the last like a board, firm and strong, and has a kind of beauty even in its rags."[53] Towards the end of the century, an anonymous Bolognese nobleman wrote sarcastically: "Today an *andrienne* lasts, apart from the honour of the fashion, for just three years. Then [at the beginning of the eighteenth century] it lasted forty without dishonour."[54]

A look through the *Règlements* of the Communauté also testifies to the decline in the quality of the fabrics. The 1737 *Règlements*, which the *maîtres* and the *ouvriers* managed to impose on the *fabricants*, established the precise width, density and length of the fabric, and sometimes even the weight each type of fabric should have per *aune*, conforming to the spirit of the *Règlements et Statuts* of 1720. This policy came up against the stiff opposition of the *fabricants*, who, overturning the policy approved in 1737, managed to get a new set of *Règlements* approved in 1744 reducing the width, length and weight of several types of fabric.

The counterattack of the *ouvriers* was spirited. They complained that the density of the *taffetas brochés* had been reduced by 40 percent, with the result that

[53] D. Defoe, *A Plan of the English Commerce* (London, 1728), pp. 178–179. According to Veron de Forbonnais, the production of fabric was "plus parfait en Angleterre qu'à Lyon" (*Lettre a M+++ Negociant de Lyon* . . ., p. 25. M. Paulet, *Maître ouvrier* in Lyons until 1752, also recognized that the British had taken silk weaving "à un très haut degrée de perfection" and attributed this to the high specialization of looms and weavers. In Spitalfields, he wrote, "les métiers une fois consacrés à telle ou telle étoffe ne sont jamais monté pour une autre. Ainsi tel ouvrier qui fait du satin ne fera jamais de taffetas ou de velours . . . Par ce moyen chacun d'eux acquiert dans son genre une précision à laquelle nul autre ne peut attendre, parceque le fabricant ne change jamais la qualité de la soie" (M. Paulet, *L'art du fabricant d'étoffes de soie*, 1st and 2nd edns, [Paris, 1793], p. xv, cited in Alain Cottereau's chapter in this volume). Unlike M. Paulet, M. Mayet criticized foreign manufacturers "où les ouvriers passent tous les jours d'un genre à un autre, par caprice ou par necessité," while in Lyons no one would have asked a weaver of "gros de Tours" to begin work "à la gaze ou au taffeta", because he would not have been able to give to either of these fabrics "cette légèreté, cette delicatesse, ce brilliant qui en constituent la perfection" (*Mémoire sur la manufacture de Lyon*, pp. 19–20). The contrast between these two texts gives rise to intriguing questions.

[54] *Opusculi quattro sopra il lusso* (Bologna, 1787), p. 239. The author regretted "those rich swirling gold and silver fabrics which adorned the shoulders of our ancestors and which from the first days of marriage onwards promised immortality." Today instead, "the flowers to be found on the lightest of silk . . . are so fragile and ephemeral that they seem to be about to lose their leaves" (*ibid.*). Abbé Jacquet's astute judgment was that fashion had removed "à la richesse sa pesanteur naturelle" to give it instead "avec la variété de la nature la semplicité des graces" (*Sur les moyens de faire refleurir les manufactures de Lyon*, December 22, 1784, in Academie des Sciences de Lyon, Ms. 110, p. 52). Zanon pointed out the fragility of the fashion fabrics: "The French fabrics can be worn only for a few hours and in good weather. As soon as they are taken off, they are folded up carefully in white, clean linen and put away in the wardrobe with the utmost prudence. They are only rarely shown and, like illustrious relics, are only touched with great delicacy . . . Quite the opposite is the case with our fabrics. They are worn without care in all weathers and on all occasions, they are hung up on old nails" (Zanon, *Dell'agricoltura*, V, pp. 251–252). Around 1740, an anonymous Venetian attacked the "inventors of fashion," who one year imposed "fabrics covered with flowers, woven with an art that was more apparent than real," and the following year took away "all the merits of the previous fashion" with fabrics "of a different form and design" on which appeared "natural fruits" (*Saggio di una nuova filosofia con cui insegna il modo di far denari* [Venice, 1740], pp. 158–159).

it was impossible to make them "passablement bons," while the density of the "simpletés et doubletés" taffetas had been lowered by around 20 percent. For velvet, the decline in quality was justified and encouraged by the introduction of new (conventional) ways of counting the warp, which resulted in making the fabrics thinner by a third. All fabrics, then, as regards the thickness (and weight) of the warp, were by now "au dessous de tout ce qu'il y a de moindre" and only sold because of "l'apprêt qu'on leur donnent."[55]

Particularly stiff opposition amongst the dependent weavers (the *maîtres ouvriers*) was provoked by an operation called *lustrage* which they had renamed *étirage*. *Lustrage*, carried out by a polishing machine, probably introduced in the second half of the seventeenth century by *fabricants*, was intended to make the warp smoother (before handing it over to the weavers to set on the looms), but it also had an important side effect – that of lengthening the warp by around 8–10 percent. This enabled the entrepreneurs also to draw distinct advantage from the new weight–length ratio, the practice of selling by weight having already been substituted by selling by length.[56] This certain profit for the merchants corresponded to a considerable inverse loss for the weavers. The thread, subject to excessive tension and thus losing much of its natural strength, was easily broken when working with it on the loom. The increase in the number of breakages led to more frequent interruptions (to tie up the broken threads) and thus to an increase in drudgery and a decrease in pay (the weavers being paid on piece rates). This was the reason behind the defense of the quality of the product by the weavers, who accused the merchants of a doubly illicit profit, selling bad quality fabric under the pretense that it was good quality and increasing demand by trading in products which did not last and thus had to be replaced after only short periods.[57]

[55] *Mémoire pour les maîtres ouvriers à façon des étoffes d'or, d'argent et de soye de la ville de Lyon* (1749), pp. 65–69. In this text, the term *maîtres marchands* indicates *marchands fabricants*. There are a number of reports for and against the *Règlements* of 1744. I have made use of this *Mémoire* because it is the most important for the aims of this paper. For the history of the conflicts and negotiations between the various sides in the struggle, cf. J. Godart, *L'ouvrier en soie* . . ., pp. 276–296 *passim*. The decrease in the number of threads in the warp is related to the thinning of the thread thanks to progress made in reeling and throwing techniques in Piedmont. In Bologna (whose reeling and throwing techniques were by the eighteenth century some way behind those of Piedmont), it was calculated that with the new thinner threads it was possible to produce fabrics up to 50 percent longer. Cf. C. Poni, "Misura contro misura. Come il filo di seta divenne sottile e rotondo," *Quaderni Storici*, 47 (1982), 389–390. Nearly all the Piedmontese production of organzine (yarn used for the warp) was exported to Lyons where it was used in the production of high-quality fabric.

[56] In France, sale by length was a characteristic of Lyons fashion fabrics. In the more conservative Tours, sale by weight continued for much longer.

[57] *Mémoire pour les maîtres ouvriers* . . ., pp. 52–58 *passim*. The *maîtres ouvriers* calculated that with bad quality warp, the time necessary to weave a given quantity of fabric would be multiplied five times (*ibid.*, p. 71) with a consequent decrease in wages paid according to piece rates. M. Paulet also opposed the use of lustrage because its high waste ratio, due to the breaking of the thread, was damaging to the *fabricant* (M. Paulet, *L'art du fabricant d'étoffes de soie*, section VII, Part I [Paris, 1776], pp. 416–418). On the other side Bertholon praised this practice invented by Octavio May in 1665 (*Du commerce et des manufactures* . . ., p. 40). According to Veron de Forbonnais, if the Lyons *fabricants* had tried to use the "traits faux" (false golden thread) in the luxury fabrics rather than the "traits fins," they would have been reported by

After 1744, no new *Règlement générale* was approved. A new phase got underway: that of the pressure of the large merchants on the *Bureau de Commerce de Paris* to obtain exemption from the rules and practices that had been approved. An example is that which laid down that "raw" silk (made up of leftover silk, useful only for sewing) should be mixed with "cooked" silk (the best quality silk obtained from the cocoons by means of boiling water). This mixing of different quality silks, although prohibited under the 1744 *Règlements*, was not entirely new (it was practiced, for example, in the production of "petits velours") but was not allowed in the production of golden or silver materials "figuré à la tire ou au bouton," known also under the name of "étoffes façonnées et grandes façonnées." The request to use raw silk in these fabrics – supported by the great designer Jacques Revel – led to a deep division in the ranks of the merchant entrepreneurs. Some feared that excessive (and perhaps fraudulent) lowering of the physical quality of the product might disgust the buyer. The *Maîtres Gardes* thought likewise, and it was only after a long and protracted struggle that, with certain conditions, they authorized the use of these mixed silks in the *façonné*.[58]

According to the merchant Raymond Vial, the use of "raw" silk in *façonnées* gold and silver fabric, "demandé par le consommateur et autorisé par quelques permissions" should have been left to the responsibility of the individual entrepreneur. "Il est si fort," he wrote,

> notre intérêt à bien faire que la plus grande perfection de l'étoffes auroit été la suite naturelle de cette liberté. C'est au genie seul à prescrire les règles de l'art. Mais on a voulu l'assujeter à des lois gênantes aux quelles il a fallu déroger lorsque des productions nouvelles et imprevues en ont fait réconnaître l'inutilité.

This claim for liberty of production was not only theoretical. Vial had on his looms a large number of articles to be sold on the Paris market to which he would not have been able to "donner la légèreté et la main que l'acheteur desire" without using the prohibited silk.[59]

their own *ouvriers*. It is the workers who thus appear as the guarantors of quality (*Lettre a M+++ Negociant de Lyon . . .*, pp. 6–11). Amongst the various contributors to the *Règlement* of 1744, the *maîtres ouvriers*' hostility was particularly directed towards the entrepreneur J. Claude Montessuy and the engineer Jacques Vaucanson. The first was accused of hav-ing produced bad quality fabric (later confiscated), the second for having declared on several occasions that the Lyons fabrics were too good, and that if "ne dureroient que trois mois ce seroit toujours assez" (*ibid.*, pp. 30–31). On the business relationship between Vaucanson and Montessuy, cf. A. Doyon and Liagre, *Jacques Vaucanson, mécanicien de génie* (Paris, 1966), pp. 175–203, 256–261 *passim*.

58 The use of raw silk (*filoselle*) was justified because it gave both softness and *fermeté* to the fabric. In order to manufacture cheap fabrics for export, the entrepreneurs successfully asked for over 145 special permits granting exemption from the *Règlements* of 1744 in the period 1754–62 (J. Godart, *L'ouvrier en soie . . .*, p. 85).

59 Letter from Raymond Vial of 1763, in ANP, *Bureau du Commerce*, F12, 1446. The language of Vial demanding "liberté" from the "lois gênantes" incorporated the discourse of de Gournay, who had declared himself to be in favor of the use of raw silk. R. Vial was the brother-in-law of Adamoli fils. It was only in 1779, after the abolition of the corporations by Turgot in 1776 (later revised to a certain extent) that merchants and weavers were allowed to weave fabrics of whatever quality they wanted, despite the fierce opposition of the *Maîtres Gardes*.

According to another note, signed by a large group of merchants and addressed to the *Intendant du Commerce* De Cotte, the "raw" silk made the golden and silver materials stronger and lighter, while it also made the job of the weavers easier. The perfection of the materials and the ease with which they could be worked would thus coincide with lower production costs and higher profits. The innovation, then, would be to the advantage of everyone, even if it would lead to envy on the part of those who were tied to customs and rules which were both out of date and which gave smaller profits.[60] According to the *Maîtres Gardes*, on the other hand, stern measures should be taken against those merchants who seemed not to have "d'autre but que de vendre une chose pour une autre."[61] They opposed those who argued in favor of the "raw" silk on the grounds that it was a "speciaux pretexte" in order to cheat the consumer. How was it possible otherwise to explain the behavior of those merchants who ordered their weavers "de fabriquer les deux premières aunes de la tête des pièces et les deux dernières de la queue avec de la soie cuite et d'en melanger le corps avec de la soie crue sans aucune lizière distinctive"? This was the origin of the order to give mixed fabrics a special "lizière," a label specifying that it was "étoffe en soie crû" in order to protect the consumers.[62]

It was not wholly true, however, to imply that the decline in the quality of the fabrics was due solely to the immorality of the merchants. Merchants of this type had always existed. At the end of the eighteenth century, the decline in quality was rather the result of the way in which the fashion market had come to be structured. In 1771, Isaac de Pinto, comparing French and English material, wrote:

Les manufactures angloises sont trop étoffées, trop riches . . . durent trop long tems, on les trouve trop chères. Le luxe et la mode veulent du changement. On a trouvé le secret en France de faire des étoffes de soie à grand marché. Il est vrai qu'elles n'ont point de corps, elles s'usent dans peu temps, mais on n'en veut presque que pour une saison. La mode change et pour le même argent on est bien aisé d'avoir deux robes ou deux habits neufs. Ce qui favorise beaucoup plus les manufactures françaises.[63]

Nevertheless, the decrease in quality should not be seen as an absolute decline but rather as a process of vertical differentiation which involved not so much

[60] 1763 letter, in ANP, *Bureau du Commerce*, F12, 1441.

[61] Letter of January 17, 1764, in ANP, *Bureau du Commerce*, F12, 1441.

[62] Letter of December 1764, in ANP, *Bureau du Commerce*, F12, 1441. On August 12, 1766, the *Maîtres Gardes*, in reply to a letter from the Intendant De Cotte, made heavy accusations against the entrepreneur Buisson, "le quel a debuté par la fraude en tramant de galette dans une étoffe precieuse . . . Une anedocte capable de détruire la reputation du fabricant le mieux établi." ("Réponse des Syndics et *Maîtres Gardes* de la Communauté . . . de Lyon," in ANP, *Bureau du Commerce*, F12, 1443). A few years earlier, the merchants had denounced "les manoeuvres que pratique le nommé Buisson, actuellement à la tête des manufactures de Naple, pour avoir les desseins avant qu'ils ayent paru et débaucher les ouvriers de toute espèce pour en rendre l'exécution aussi exacte qu'à Lyon," in *Réplique des Maîtres Marchands aux observations des Maîtres Ouvriers* (AML, Chappe HH 133).

[63] I. de Pinto, *Traité de la circulation et du crédit* (Amsterdam, 1771), p. 26. The English too should have learnt to produce "des étoffes voyantes et moins chères" (*ibid.*).

abandoning high quality production as a greater attention to market segmentation. An observant commentator on the relationship between quality and fashion, Veron de Forbonnais, wrote: "Dans chaque pays il est differentes classes de peuple et de facultés. Pour les seduire et les contenter, il convient de leur offrir dans chaque espèce particulière des assortimens proportionnés aux differentes facultés." This strategy, aimed at producing "assortiments de divers prix dans une même étoffe," would be appreciated by the shopkeepers as it would lead to increased profits. Fabrics of different qualities, ranging from the best to the worst, would be perfect if they were all worth the price asked.[64]

The relative decline in physical quality was in some ways compensated for by the uniformity of the fabric. The Lyons *fabricants* brought the organzine for the warp from Piedmont, the best in Europe, following a commercial technique, probably invented by the silk merchants of Turin, which guaranteed the quality of the product by means of a quality indicator called the *titolo* (count). The *titolo* indicated the weight of a given length of thread (400 *aunes*) in "deniers" (1 denier=1/20 of a gramme). This sales technique signaled a decisive reduction in transaction costs. Merchants identified the thickness of the thread by its number. The count 24 (that is 24 "deniers") indicated the thinnest organzine. Thicker fabric was indicated with numbers from 24 to 60. However, these numbers were only approximate. The number 27, in fact, when referring to a "ballot" (about 50–60 livres), was used to indicate thread between 24 and 30 "deniers." Once the variation rate was indentified (with the help of specialized consultants paid in Lyons by the livre or *ballot*) the threads (in skeins) of different thicknesses were separated and used for different types of fabric. Alternatively, they were separated in such a way that on the looms the thinner threads were always next to the thicker ones. This meant that the streaks which lowered the quality and the price of the fabric could be avoided.[65]

The Lyons silk industry, although Lyonese, was also Parisian. Many of the Lyons merchants had a shop or workshop in Paris.[66] The capital was also the most important outlet market for the Lyons material, and it was in Paris that the sending of goods to foreign markets often took place.[67] The expression "modes de Paris faites à Lyon," quoted above almost in passing, would seem to indicate

[64] Veron de Forbonnais, *Elemens de commerce, Première Partie* (Leyde, 1754), pp. 254–255, 260–261. According to Forbonnais, women from the middle and lower ranks of society gave little thought to quality as long as they were dressed "d'une manière aussi éclatante qu'une femme de plus haut étage ou plus riche."

[65] M. Paulet, *L'art du fabricant d'étoffes de soie*, Section VII, Part I, pp. 413–416. According to M. Mayet this technique – "mettage en main" – was practiced only in Lyons (*Mémoire sur les manufactures de Lyon*, 1786, pp. 20–21).

[66] Claude Louis Couchard, for example, *marchand fabricant* in Lyons, was at the same time "maître des six corps" in Paris (L. Gueneau, "Paris. Les industries et le commerce de la soie," in *Revue d'Historie Moderne*, 4 (1926), p. 293. Philippe de Lasalle also lived sometimes in Lyons and sometimes in Paris.

[67] According to L. Gueneau's estimate ("Paris. Les industries et le commerce de la soie," pp. 288–289), of the 9,000 pieces of silk which came into Paris in January 1779, 7,500 came from Lyons, 500 from Tours and 300 from Avignon.

an even stronger tie, even though it would imply relegating Lyons to a purely executive role. The social practices underlying the Paris–Lyons relationship, though, need some explaining. One of these practices was the mediating role of the merchants and traders who operated on two markets at once. Even more decisive was the role of the over 100 designers, often merchants or partners of merchants. This double professional and social role is proof of the high reputation the designers enjoyed both in Lyons and in France as a whole. This reputation, along with the high number of designers, is something which cannot be found in any other European silk city. In Italy, the fabric designers, when they existed, do not seem to have occupied a position of prestige with respect to the merchants.[68] In England the importance of having first-class designers was never fully appreciated by the silk entrepreneurs engaged in a continual effort to imitate the French patterns.[69]

In Lyons, the studies to be completed before becoming a designer were long and costly. After learning the rudiments of design, the prospective designer was sent to a workshop of a flower painter. Training usually finished in Paris, sometimes in the workshop of one of the Gobelins painters. After returning to Lyons and finding employment in one of the merchant firms, the young designer would continue to spend long periods in Paris every year, where he would make visits, in the company of older and more experienced colleagues, to various print and art collections, the Gobelins and porcelain factories, the Versailles gardens, the Jardin des Plantes, the Cabinet du Roi, the most famous churches, the royal palaces, theatrical performances and so on. This immersion in the atmosphere of the capital of elegance, of luxury and of refinement, where he would converse with fabric sellers, *couturiers*, *faiseuses de modes*, tailors, goldsmiths and embroiderers, refined his taste and civilized his art. But the inspiration and crucial information necessary for the design of the fabrics for the following year would be gained visiting the most famous silk warehouses, those of the famous silk traders Bourjol, Barbier, Saurozat, Nau, Despignes, de Courcy, David, Le Roux, Doré, Mercier, Buffault, Martin and so on, all to be found in the same area of the city around Châtelet and Palais Royal. It was here that the most important customers of the *Grande Fabrique*, together with the designers, developed "chaque jours ce que la Fabrique de Lyon produit de merveilleux toutes les années."[70] These visits

[68] This subject has so far gone unexamined. In 1764, F. Griselini started working on a manual for the training of fabric designers, although it seems to have been left unfinished. Cf. F. Griselini, "I principi dell'arte del disegno per ogni genere di stoffa di seta," in *Giornale d'Italia attinente le scienze, lettere e arti*, (December 1756), pp. 178ff. The Venetian Pietro D'Avanzo, silk fabric designer, had written some years earlier (sometime around 1753) his *Regole per la meccanica del telaio da seta*, which remained unpublished. It has now been published with a brief introduction by I. Chiappini di Sario.

[69] P. Thornton, *Baroque and Rococo Silk* (London, 1965), p. 57. This work, which I came across rather late and only after the first manuscript version of this paper had already been circulated, is to be highly recommended for anyone researching into silk and silk manufacture.

[70] Cf. Joubert de l'Hiberderie, *Le dessinateur pour les fabriques d'étoffes d'or, d'argent et de soie* (Paris, 1765), pp. 86–113. See also E. Leroudier, *Les dessinateurs de la fabrique Lyonnaise au*

gave rise to substantial interaction between the producer firms and the needs and expectations of the market, something which would not have been possible if the firms had been represented by salesmen or by commissioners, not always able to relay the right information to the designers at the critical moment at which the designs and colors were invented.

In a recent article entitled "Implementing the Marketing Concept through a Program of Customer Visits," two American scholars have emphasized that the firm as a unit is always on the brink of losing contact with its customers, and that similarly the marketing functions are always on the brink of losing contact with other fundamental sectors of the firm (the Research and Development functions and the organization of the actual production). The solution to this problem, the authors Edward F. McQuarrie and Shelby H. McIntyre argue, is that the firm should make a deliberate decision to send at regular intervals a team of "decision makers" (including designers) to visit the customers. According to the authors, "face to face communication has a unique capacity to transfer complex and ambiguous information among parties."[71] This information procedure is presented as an original phenomenon, which is in a certain sense true. However, even in the eighteenth century it was an international practice among the entrepreneurs of the *Grande Fabrique* who correctly entrusted the designers with the task of drawing up the new fashions in collaboration with the most important customers.

Each year hundreds of designers returned from Paris to Lyons with thousands of designs. Although this was costly and in a sense redundant, it was nonetheless functional in terms of the sophisticated and pressing requirements of the market. The final choice of design and color for the year – the bouquets of flowers arranged in elegant garlands, the harmonious combinations of different brightly colored stripes, the exotic chinoiserie – was made in Lyons in collaboration with the merchants.[72]

XVIII siècle (Lyons, 1908), and *Le livre de Raison de Jacques Charles Dutillieu publié . . . par F. Breghot du Lut* (Lyons, 1886). Since the age of Louis XIV France had dictated styles of dress to the other courts and cities of Europe. It could be said that the fashion in fabrics, with new designs and colours every year, developed alongside an adjacent sector that had already consolidated a leading position.

[71] E. F. McQuarrie, Sh. H. McIntyre, *Implementing the Marketing Concept through a Program of Customer Visits* (Cambridge MA: 1990). The two authors make reference to a "technology company."

[72] According to L. E. Miller, the overall number of designers in Lyons oscillated between 180 and 270 in 1758–59. They were divided into three categories: freelance, salaried and partners of the large merchants. Cf. "Design, Technique and the Division of Labour in the Lyons Silk Industry," paper presented to the "Workshop on Design, Commerce and the Luxury Trade in the Eighteenth Century" (1988). On April 27, 1749, journeying from Paris to Lyons, a Lyons fabric designer remarked to his companion, Abbé Nollet, that "le fabricants de Lyon se cachent les uns aux autres les desseins qu'il font faire, et qu'ils ne font jamais voir leurs nouvelles étoffes avant d'avoir fait leur principaux envoies à Paris et ailleurs. Et la plus part des dessinateurs passent deux ou trois mois à Paris pour avoir connaissance des desseins des autres et pour savoir ceux qui ont les mieux réussi" (J. A. Nollet, *Journal du voyage de Piedmont et d'Italie en 1749*, Bibliothèque Municipale de Soissons, Ms. 150).

Sometimes, when the new fashion had already been launched, the designers returned to Paris to check how the market was reacting. It should be remembered that each individual entrepreneur was in competition with all the others and that the fabric designs, even if they came from the same cultural–industrial matrix, were of varying qualities. There were, for example, differences in color, decoration and taste, degrees of perfection in the design and of precision in the weaving, not to mention the names of the firm or designer.

These innovative practices made a clear separation between the designing of the fabrics and their actual production. But as soon as the decision to set up a new design on a loom was made, the conditions for creative cooperation were established. The designer copied the design onto squared paper; working sometimes with the help of a second designer, he began to prepare and organize warp and weft and to arrange *maillons*, *lacs* and *aiguilles*. In this he got help from a trusted technician, from a *liseuse* with the specific task of transposing the design from paper to loom, and from a weaver. The problem was how to transpose the artistic innovation onto the loom; how to reconcile the design, and its imitation of "la belle nature . . . avec les règles austères de la mécanique"; how to overcome the obstacles "que la mécanique oppose au genie de l'artiste."[73] In fact, to overcome these obstacles the designer should have acquired as part of his training at least "une teinture du métier" which could have been intensified later with innovative experiments. According to Joubert de l'Hiberderie, writing in 1765, the designers were the first to create order out of "le chaos dans le quel l'art de fabriques des étoffes étoit encore plongé au commencement du siècle."[74]

Without doubt the designers also contributed to the intellectual–industrial atmosphere which influenced the great Lyons technological innovators who transformed the machinery of the loom. Here I will mention only a few: Baptiste Falcon who, using an idea of Basile Bouchon, introduced perforated cards into the loom, by means of which designs could be memorized and mechanical impulses could be sent, transferring the designs onto the material; Jacques Vaucanson to whom the invention of the automatic loom is attributed; and Joseph-Marie Jacquard, who, at the beginning of the nineteenth century, concluded a series of long and laborious innovations by inventing the loom which bears his name and which, along with the invention of the Falcon loom, was the precursor of the machine with the numerical controls and the robots which characterize flexible production today.[75]

[73] Joubert de l'Hiberderie, *Le dessinateur*, pp. xvii, xxii, 15, 21, 28.

[74] De l'Hiberderie, *Le dessinateur*, p. x. According to M. Paulet, the transposition of a design should be left to a "monteur de métier. Cet artiste doit . . . être le conducteur du dessinateur, parceque sur cent dessinateurs, à peine en est-il un qui conoisse le métier; et c'est cependant ce que tous devroient connoître. C'est la base du métier." Paulet, self-taught weaver and designer, emphasized his own ability to combine "les effets des dessins avec la possibilité de les rendre en travailland l'étoffe" (M. Paulet, *L'art du fabricant d'étoffes du soie*, Section VII, Part, II [1789] p. 884).

[75] C. Ballot, "L'évolution du métier Lyonnais au XVIIIe siècle et invention de la mécanique Jacquard," in Ballot, *L'introduction du machinisme dans l'industrie française* (Paris-Lille, 1923).

But designers also played important, direct roles in innovation, above all making possible product differentiation through the support of the necessary technological flexibility. In 1770 Philippe Delasalle, well-known designer and entrepreneur, constructed a mechanism, certainly based on perforated cards, by which one design could be substituted with another *without resetting the loom* so that orders from clients could be met as quickly as possible. A costly, difficult and time-consuming operation (the previous design had to be destroyed in order to set up another) became, if we can believe Delasalle, something which could be done in an instant. He wrote:

Souvent le terme pour le quel on veut l'étoffe étant très court, on est obligé de rénoncer la commission parce qu'une partie de ce temps nécessaire pour la fabrication seroit employé a remonter le dessin, ce qui n'arrivera point avec le nouveau méchanisme puisqu'au moment même qu'on reçoit la commission on peut le faire travailler.[76]

This problem was solved satisfactorily only by Jacquard. The words of Philippe Delasalle, however, suggest that the innovations in Lyons in the eighteenth century should be seen also as a reply to the challenge of annual product differentiation and the simultaneous offer of a large number of designs.

During the eighteenth century, these technological innovations probably did not have a wide diffusion, but it is certain that the Lyons merchants (and the designers) were the very first to exploit fully and systematically the flexibility offered by the artisan or putting-out organization of production. In the past, this

These new looms were also intended to substitute the "tireuses de cordes" with a mechanical device and to discipline the work of the weavers. Cf. H. Clouzot, *Le métier de la soie en France* (Paris, 1914) and Bianchi, *Produzione e potere di mercato*, pp. 83–84. Insufficient attention has been paid to the extent to which the general development of methods of production is dependent on inventions and innovations originally aimed at improving the production of luxury goods. Cf. Robinson, *The Importance of Fashion*, pp. 9–11. Following the practice which developed under Colbert, the inventors received bonuses and compensation for their innovations. The funds for this compensation were financed by the duty on foreign silk (collected in Lyons). Abbé Bertholon wrote: "Il est just que celui qui enrichit la société par d'utiles découvertes, jouisse du fruit de ses labeurs, recouvre ses avances et partage un profit que d'autres ne manqueront pas de faire . . . Le Souverain doit donner des recompenses: des titres d'honneur à ceux qui ont de la fortune, et des richesses à ceux qui en ont besoin. Tel est l'usage des revenus de l'Etat" (*Du commerce et des manufactures . . .*, p. 94). See also Ballot, "L'évolution du métier," pp. 340–359 *passim*.

[76] Cf. the bundle of letters in the ANP, *Bureau du Commerce*, F12, 1444A (1772–75). Claude Millet, a turner and mechanic, accused Lasalle of having stolen from him the invention. Cf. *Mémoire pour Claude Millet maître tourneur et machiniste de la ville de Lyon, pour justifier qu'il est l'inventeur du métier pour la fabrique en soie contre le Sieur Lassalle*, [1769], in ANP, *Bureau du Commerce*, F12, 1444A, 1769–70. The new frame took the name *reversible*. Cf. also Belle M. Borlan, *Philippe de Lasalle* (Chicago, n. d.). The Jacquard loom was also "reversible," it being possible to change the pattern in a short time. Cf. N. Rothstein, "The Introduction of the Jacquard Loom to Great Britain," in V. Gervers (ed.) *Studies in Textile History* (Toronto, 1977), p. 282. The introduction of perforated cards encouraged large firms to set up archives to stock the patterns created, so that they could be recalled when needed. In this way the construction and production of new patterns could have rapid access to past experience.

(potential) flexibility had been exploited only partially. It is true that each textile center produced contemporaneously tens of different types of fabric. For example, in 1544, the "sayetteurs" and the "bourgeteurs" of Lille were authorized to produce more than ten different fabrics (without counting sub-types) as well as fabrics which were "inventés presentement ou à inventer ci-après."[77]

Even in the most important of the old silk cities there was considerable variety of product. But this multiplicity often hid high specialization of production in the few sectors or types of fabric which had good access to and a well-established reputation on the international market. This was the case of Genoese velvets, Venetian damasks, Bolognese crêpes, Florentine satins, Milanese gold and silver fabrics and so on. These names, however, rarely corresponded to a single, unchanged and unchanging pattern. The merchant entrepreneurs had known for centuries that supply could be differentiated in order to defend their market share from a dangerous competitor or when entering a mature market. Neither, of course, was demand stable. At the end of the sixteenth century the Bolognese crêpe merchants realized with alarm that German buyers were beginning to require heavier crêpe than that which was being offered. In the eighteenth century, the Italian merchants who sold silk fabrics in Poland tried to satisfy the oriental tastes which predominated amongst the Polish nobility, offering fabric "ben lustri," the largest possible gold and silver flowers and Persian patterns or Arabic designs.[78] It would not be difficult to cite other examples. But it was only in Lyons, or Lyons and Paris, that the strategy of annual programmed fashions in design and color were discovered, that the opportunities of product and process differentiation and flexibility were explored to the full and used consciously as an annual market strategy.

In 1764, Antonio Zanon, certainly not someone who was sympathetic towards the much envied French competitors, praised highly their superior skills. He wrote: "To do justice to the French [for 'French' read 'Lyonese'] it should be admitted that they have brought more perfection and charm to silk fabrics in fifty years than the Greeks and the Venetians have in seven hundred."[79] Nevertheless, these extraordinary results and their enduring superiority was never sufficient to

[77] Cf. N. Vanhaeck, *Histoire de la sayetterie à Lille*, vol. II (Lille, 1910), pp. 58, 102–103, 106–107.

[78] A. Manikowski, "Il commercio di tessuti di seta in Polonia nella seconda metà del XVII secolo," in *Fasciculi Historici*, 10 (1982), 160–161. The history of the new draperies can also be considered in the light of differentiation processes. See M. Malowist, "Les changements dans la structure de la production et du commerce du drap au cours du XIVe et du XVe siècle," in *Croissance et régression en Europe, XIVe–XVIIe siècles*, Paris, 1972. See also E. Coornaert, *Un centre industriel d'autrefois. La draperie-sayetterie d'Hoondschoote (XIVe–XVIIIe siècles)* (Paris, 1930); J. Craeybeckx, "L'industrie de la laine dans les anciens Pays-Bas méridionaux de la fin du XVIe au début du XVIIIe siècle," in *Produzione, commercio e consumo dei panni di lana (nei secoli XII–XVIII), Atti della "Seconda settimana di studio (1970) dell'Istituto internazionale di storia economica "F. Datini" di Prato* (Florence, 1976); and D. C. Coleman, "An Innovation and its Diffusion: The 'New Draperies'," in *Economic History Review*, 2 (1969).

[79] Zanon, *Lettere dell'agricoltura, dell'arti . . .*, III, p. 160.

protect the second city of France from the ups and downs of economic conjuncture. Wars which shut off access to external markets, court mourning, the instability of the production and price of raw materials, the mercantilist policies of governments aimed at bolstering the national imitation industries all struck cruel blows to *Grande Fabrique* employment, wages and profits. Count d'Albon wrote in 1789:

Avec vos fabriques de luxe vous ne pouvez pas répondre aux ouvriers de deux jours de vie. Aujourd'hui vous les occupez, demain une recolte manque, le goût des consommateurs change, un deuil de Cour de quelque durée interdit l'usage de vos fabrication, une autre manufacture vous enlève la preference, une guerre s'élève, vos métiers s'arrêtent, vos ouvriers n'ont plus du pain.[80]

It might be asked when the production of the "modes de Paris faites à Lyon" began. This question cannot be answered with any precision. Economic, social and technological transformations, particularly the complex ones which involve mentalities, deep-rooted practices and modes of production have no precise beginnings, no immaculate birth, and do not spring from intellectual creative impulses deprived of concrete roots. We might be tempted to indicate the year 1713. But *The Case of the Silk Weavers*, printed in that year, describes mercantile practices which had already become established by this date. For how long? Since the second half of the seventeenth century? Since the late seventeenth or early eighteenth century? I have no definitive reply to this question, but research into this point should probably start from a letter sent on January 6, 1685 from the *Prévôt des marchandes de Lyon* (the mayor) to Colbert. This letter included a brief report on the progress made by Lyons weaving, which had by this time almost completely replaced the imported Italian fabrics previously sold on the Paris market. But these new fabrics did not circulate under their own name, as Lyonnais fabrics. The preference of the French consumers for rich Italian fabrics (from Genoa, Milan, Florence, Venice) was still so strong that, in the words of the *Prévôt*, "les marchands de Paris . . . obligent nos ouvriers a mettre sur leur étoffes les planches et les marques des pays étrangers." In a later letter of November 3, 1665, the Prévôt was even more explicit, stating that "tout ce qui s'envoie de cette ville à Paris s'y vend sous le nom des fabricants d'Italie."[81]

Italy and the Italian textile centers, then, were still the reference point, the crucial yardstick. Just ten years later, however, a significant novelty appears. In the work of Jacques Savary there is an explicit mention of "étoffes façonnées et de

80 Comte d'Albon, *Lettre sur le commerce, les fabrications et les consummations des objects de luxe* (Lyons, 1789).

81 These two letters are quoted in E. Pariset, *Histoire de la fabrique Lyonnaise* . . . , pp. 82–84. In the letter of January 6, 1665, the *Prevôt* declared with pride: "Il n'y a point d'ouvrage de soie, d'or et d'argent de quelque endroit d'Italie qu'il vienne que nos ouvriers n'égalent (à la reserve des velours noirs de Gênes . . .) et que même ils ne surpassent." For an introductory comparison between the Italian and French silk industries, cf. S. Ciriacono, "Silk Manufacturing in France and Italy in the Seventeenth Century," *Journal of European Economic History* (Spring 1981), 167–199.

couleurs dont les modes changent d'une année à l'autre." Unfortunately, this reference is nowhere elaborated upon, apart from a comment that with this fabric "l'on y peut gagner considerablement parce que il y a quelques fois à perdre la moitié de ce qui reste." Nor is there any reference either to Lyons or to the silk sector. Despite these limits, the importance of Savary's note should not be underestimated, and may be taken as a profitable starting point for future research.[82]

That the Lyons merchants had a considerable degree of awareness of their own commercial practices emerges in any case from the way in which they managed and manipulated the rules of the Communauté at the beginning of the eighteenth century (as regards the protection of design copyright, for example). It was only towards the middle of the eighteenth century, though, that this awareness matured sufficiently to be aired publicly. This took place during the often highly conflictual debates over strategic choices, debates which were often backed up with arguments that were highly original and stimulating including at times views that foreshadowed theories. Yet in their own day these ideas did not generate an analytical school, also because from the mid eighteenth century onwards, the attention of the public and of the principal economic thinkers was captured by the physiocratic model that was hostile to luxury and fashion, seen as taking work away from agriculture and depressing population growth.

It might be asked whether the production of traditional textiles and fabrics came to an end in Lyons after the invention of annual fashions. The answer would be no. The production of imitation fabrics continued in Lyons throughout the entire eighteenth century: "false" Venetian brocade, Bolognese crêpes, Genoese velvets. The same was true for plain fabrics (whether traditional or not),

[82] Jacques Savary in *Le parfait negociant*, vol. 1 (Paris, 1675), made a sharp distinction between the entrepreneurs who imitated and those who invented. After listing the many difficulties encountered by those who tried to imitate foreign or local fabrics, he weighed up the advantages and disadvantages of inventing new fabrics. The buyers would have hesitated before buying because their quality would only become clear after they had been worn for some time. This disadvantage was balanced out by the fact that little money would be lost "sur un ouvrage nouvellement inventé parceque les Français, qui aiment naturellement le changement, ne manquent jamais à l'acheter, quelque méchant qu'il soit, pour paroître à la mode." Besides this, he advised the merchants who had "trouvé l'invention de quelque marchandise" to produce small quantities "jusqu'à ce qu'il ayent reconnu qu'elle puisse avoir le debit" in order to avoid the danger of heavy losses. Savary was not writing out of hearsay. He himself had previously launched three new invented fabrics, and these had met with various degrees of success. Without going into too much detail, the first fabric was in fashion "seulement pour cinque ou six ans"; the second was "un droquet façonné" (imitated also in Rouen) in fashion for three or four years; and the third was a gold and silver *droquet*, sold in Paris for only a few months because it had imperfections despite being very attractive. It seems clear that none of these inventions involve annual change or innovation in either design or color. These were isolated attempts, different from the fashion system of Lyons (*Le parfait negociant*, pp. 15–23). Nevertheless, before Savary no one (as far as I know) had described and understood better the rules to follow for inventing new products and new processes. According to A. Geijer, in 1667 Colbert ordered the manufacturers "to produce new patterns every year" (*A History of Textile Art* [Stockholm, 1979], p. 154). Unfortunately, the author does not cite a source for this information, nor have I been able to find one.

whose production increased from 1770 onwards in relation to the *façonnés*. These sectors (the traditional and plain fabrics) probably supported employment in the months between the spring and winter fashions, or when, because of court mournings, there was little or no demand for annual fashions. Perhaps this was another reason why the production of fabric with a short life-cycle, although in expansion, was not able to take over the entire silk sector, even in Lyons.[83]

Up to this point, I have assumed that fashion fabrics had a short life, as is maintained by the French, English and Italian sources which reveal the commercial strategies of the *Grande Fabrique*. But the important research of Peter Thornton, Donald King and Natalie Rothstein have shown that annual fashions can also be considered as part of a longer cycle, following the dominance of particular styles or tastes. According to Rothstein, between 1700 and 1825, some ten distinct periods can be identified, including the "bizarre" style of the years 1700–1712; the "luxuriant phase" of 1713–1719; the "lace patterns" of 1720–1732; the quest for three dimensional form of the years 1733–1742; the rococo of 1743–53; the retreat from naturalism of the period 1754–1770; the advent of neo-Classicism of 1770–1780; stripes and spots in 1780–1790; the dark stripes of 1790–1795; the eclipse of woven design (1796–1810); the revival of woven design (1810–1825).[84] These authors, and Natalie Rothstein in particular, present these medium-term phases, the result of careful and meticulous research, with some caution. But these periods do lengthen the life of the fabrics. As long as one style or taste had not completed its own life-cycle, it was no doubt easy to adapt materials, but above all clothes, to the new annual fashions, using trimming, embroidery, frills and flounces.[85]

The success of the strategies elaborated by the Lyons merchants in the eighteenth century was dependent on the presence of two fundamental conditions. The first was the preference for *façonné* (flowered silks), the second the preference for silk in fabrics and in fashionable clothing. The latter was based on the mercantilist exclusion (although this exclusion was challenged and never absolute) of calicoes and *indiennes*, cotton with bright colours and designs imported from Asia. The conflict for control of the home market was particularly harsh in England and France, where the vested interests of wool and silk feared the competition of these exotic, attractive, resistant and washable fabrics, very

[83] *Ibid.* In 1739, out of 8,289 looms, 4,874 (around 58 percent) were being used to weave *façonné*. In 1769, 5,500 of the 11,000 looms produced *façonné*. Cf. H. Clouzot, *Le métier de la soie en France (1466–1815)* (Paris, 1914), p. 76. Cf. also the statistics of Abbé Jacquet, *Sur les moyens de faire refleurir les manufactures de Lyon* (1784), in *Académie des Sciences, Belles Lettres et Arts de Lyon*, Ms. 110. The number of looms does not appear to be compatible with the number of weavers, around 8,000 in 1731, each of whom had the right to use up to four looms. All these statistics require further analysis.

[84] N. Rothstein, *Silk Designs of the Eighteenth Century* (London, 1990), pp. 32–62.

[85] Cf. Ribeiro, *Dress in Eighteenth-Century Europe*, pp. 60 *passim*. With embroidering, it became possible to transform *plein* into *façonné* fabric without having to use complicated looms. This technique was first developed in Lyons. As it was easy to imitate, it spread to other countries in Europe who used it in competition with the *Grande Fabrique*. See also M. de Saint-Aubin, "Art du brodeur," in *Descriptions des arts et métiers*, XIV, pp. 175–176.

popular because of their low price and used both for clothing and for decorating furniture and walls. In both countries, the battle against calicoes was won (at least in legal terms). In 1721, the English Parliament prohibited their import and consumption, although not their production, as this was aimed at foreign markets. France went one step further, prohibiting the production as well as the consumption of *indiennes*, at least up to 1759 when the restrictive legislation was removed.[86]

These were ambiguous victories, though. The same vested interests that managed to exclude calicoes and *indiennes*, nevertheless used them, re-elaborating their designs and exotic colors, especially in silk fabrics, in a successful campaign against the conservative taste which characterized Italian silks.[87] It may have been this unexpected blow, repeated later by means of recourse to other "folkloric" tastes, which revealed the weakness of the Italian products in a crucial phase for the formation of the new fashions.

The preference for silk and *façonné* began to weaken towards the end of the 1770s when a sharp change of taste towards the simple, the natural and the spontaneous began to favor plain silks and cotton. The work of Jean-Jacques Rousseau is both the highest expression of this new sensibility and a crucial element in its growth and diffusion. The greatest refinement became the exhibition of simplicity. In the 1780s, cotton conquered Versailles. Marie-Antoinette began to dress in cotton, and with her the ladies of the court.

In 1789 the brilliant pamphleteer L. A. Caraccioli wrote that by now all the Parisians were wearing white *indiennes*:

Une assemblée de françoises etoit autrefois un magnifique parterre ou l'on voyoit l'émail de toutes les couleurs. Ce n'est maintenant que le tableau le plus monoton, et qui n'offre aux yeux que du blancs et du blanc.[88]

[86] Cf. for France, E. Depitre, *La toile peinte en France au XVIIe et au XVIIIe siècle* (Paris, 1912); H. Clouzot, *Histoire de la manufacture de Jouy et de la toile imprimé en France*, vols. 1–2 (Paris-Brussels, 1928); S. Chassagne and S. D. Chapman, *European Textile Printers in the Eighteenth Century. A Study of Peel and Oberkampf* (London, 1981); S. Chassagne, *Le coton et ses patrons (France 1760–1840)* (Paris, 1991). For an analysis of the literature on this subject, see J. K. J. Thomson, "State Intervention in the Catalan Calico-Printing Industry in the Eighteenth Century," in *Markets and Manufacture in Early Industrial Europe*, ed. M. Berg (London and New York, 1991), pp. 57–59. Along with the calicoes, the import of Indian silk fabrics was also prohibited.

[87] R. Grönwoldt has shown that there was a certain innovative activity also in Italy ("Notes on Italian and Spanish Textiles of the Seventeenth Century," *Studies in Textile History*, ed. V. Gervers (Toronto, 1977), pp. 126–132). On the influence of the far eastern products on English taste in the seventeenth and eighteenth centuries, see Jones, *The Fashion Manipulators*, pp. 204–207.

[88] L. A. Caraccioli, *Lettres d'un indien à Paris*, vol. 1 (Amsterdam, 1789), pp. 235–236 *passim*. The preference for white *indiennes* (not produced in France) gave rise to a crisis not only for silk fabrics but also "les toiles peintes et imprimées," which had fallen to "un point qui est à peine concevable: point de demandes, point de consommations" (*Réflexions d'un négociant electeur de 1790*, no date or place of publication, pp. 2–4). Cf. also E. Leroudier, "La décadence de la fabrique Lyonnaise à la fin du XVIIIe siècle," in *Revue d'histoire de Lyon*, 10 (1911), 343–361, 415–444. Abbé Jacquet wrote that even for *façonnés* competition had

In the same year, a number of French newspapers reproached French ladies for being "tributarie de l'industrie angloise, par l'usage immoderé des étoffes de coton blanc." Cotton had become the fashion.[89] French newspapers were making the same criticism of their ladies, of being unpatriotic, that from Addison and Defoe onwards had been made of the British ladies, overfond of French fashions.[90]

The defeat of silk and *façonné* and the victory of plain fabrics and cotton led to a major crisis in Lyons where, in the late 1780s, also due to the collapse of the production of raw silk, unemployment and poverty reigned, only partly mitigated by the activity of several generous (but insufficient) charities. This deep crisis, much longer and more destructive than those preceding it, reveals the extent of the presumption of the great Lyons silk merchants. They had believed that they dominated the fashion fabrics market, but this dominion was tied to a certain taste, a certain style. As soon as there was an abrupt and unexpected change in taste, when Paris (and Versailles) "betrayed" Lyons, the great city on the Saône found to its dismay that it had its own Achilles heel. Already unable to maintain its place on the market, the coming revolution was to strike another severe blow at a fabric now seen as aristocratic and antipatriotic.

The beginning of this paper proposed a sharp opposition between silk and cotton. The conclusion suggests a different relationship. When the annual fashions moved into cotton, as Abbé Morellet had predicted, cotton fabrics were subjected to similar product and process strategies already applied with success to silk fabrics.[91] At the same time, through cotton, fashions became popular with larger and lower social groups. The victory of cotton over silk, at least in the late eighteenth century, was also a victory of fashion over luxury.

It was only after the revolution that silk fabrics came back into fashion. But they were unable, in a market that was by now in enormous expansion, to regain the dominance they had enjoyed in the early modern period. Silk fabrics had to share the field with wool and cotton, which had conquered what was now a bourgeois (English-dominated) market for men, while silk, expelled from jackets and trousers, was forced to limit itself to women's clothing.

become fiercer. "A peine une belle étoffe sort-elle de notres manufactures, qu'elle est contrafaite par les fabriques rivales, et il ne leur coûte presque rien pour s'approprier les plus beaux fruits de notre génie." This led to the reduction of outlets "qui empêchent à l'inventeur de se recompenser de ses travaux et qui l'obligent de partager avec le plagiare le moisson qu'il devroit receuillir seul." This in turn resulted in a major reduction of gains for the *fabricants*, fabric designers and workers of the *Grande Fabrique* (*Sur les moyens de faire refleurir . . .* , f. 54).

89 *Affiches de Dauphiné*, June 20, 1788, p. 38. According to Abbé Jacquet, the worst blow to the Lyons manufactures came from fashion itself, "qu'elle avoit si bien servi, la mode qui leur devoit son plus belle éclat, ses plus brillantes parures, oui la mode, le tyran capricieux leur a porté le coup le plus mortel" (*Sur les moyens de faire refleurir . . .* , f. 57).

90 At the same time the market for silk fabrics for wall coverings and interior decoration diminished due to competition from wallpaper. Cf. H. Clouzot, Ch. Follot, *Histoire du Papier peint en France* (Paris, 1935).

91 Abbé Morellet, *Réflexions sue les avantages de la libre fabrication*, p. 111.

I hope that by now I have succeeded in giving a satisfactory and detailed explanation of the reasons behind the extraordinary and long-lasting success of Lyons, its unique combination of complex elements that made it so difficult to imitate. In more general terms, this combination can be said to have brought together a rich and intricate superstructure; a marked dominance of software over hardware; seemingly endless creative capacities; rapid decision-making; systematic management of information (at times withheld, at others diffused); and a high degree of culture, taste and imagination. It was made up, too, of technological innovations of the significance of the Jacquard loom which memorized designs and transferred them to the cloth in ways which were already in some sense "computerized." As Jonathan Zeitlin and Charles Sabel have already suggested, Lyons can be seen as a unique instance of an industrial district linking a "pre-industrial" past with a "post-industrial" future.[92]

[92] See Sabel and Zeitlin, *Historical Alternatives*, pp. 133–176.

2 The fate of collective manufactures in the industrial world: the silk industries of Lyons and London, 1800–1850

Alain Cottereau
Translated by Cyprian Blamires

In 1822 a *Lyonnais* who was worried about how his own silk industry would cope with the threat of the British industrial revolution composed an imaginary debate on its future.[1] In this work the case for the inevitability of concentration in factories is put forward by an American visitor, while the essential soundness of the Lyons homeworking tradition is argued by a merchant from the city. The two protagonists visit a famous factory, *La Sauvagère*, the first real large-scale silk-weaving establishment to be established in Lyons: does some inexorable law of progress mean that this is a foretaste of the fate in store for the whole silk manufacture?

The American: I have been visiting England, Germany, Prussia, Holland, Russia, and Italy with the intention of making a special study of silk manufacturing in all those countries: I was particularly keen to find out more about the situation in Lyons [. . .] I had heard of the *La Sauvagère* factory: it has the same highly impressive method of administration and distribution of work that you find in big factories in Germany and England. [. . .] But I was astonished to find that this factory is the only one of its kind in Lyons, and that the work of the Lyons manufacture is spread over a whole quantity of small workshops.
The Lyons merchant: It is a very old tradition in our town for manufacturing work to be carried out by workmen who are independent of the heads of the manufacturing businesses, who negotiate the prices of the pieces with them. This method was introduced by the founders of our manufacture and it has been the norm for three centuries.[2]

Then the merchant put forward his main argument for dispersed manufacturing:

[1] This text follows a whole series of debates with C. Sabel and J. Zeitlin. We first met in London in 1981 at a History Workshop on their "historical alternatives to mass production," and since then further discussions have helped me to deepen and broaden the analyses involved. So I wish to emphasize here my debt to C. Sabel and J. Zeitlin as well as to each of the members of the working group from which the present volume arose. In addition to this crucial contribution I am also particularly indebted to J. Zeitlin and M. Sonenscher for their advice, analyses and documentary information about the situation in Britain over a period of four years.

[2] Comte F. M. Fortis, *Voyage pittoresque et historique à Lyon, aux environs et sur les rives de la Saône et du Rhône* (Paris: Bossange Frères, 1822), II, pp. 444–445.

decentralization in small urban workshops is the right system for mobile specialization – a "commission-based industry" was the term used in those days – in the competitive market of constantly fluctuating fashions. Centralization of capital would be a disaster for the entrepreneurs:

Unlike other branches of manufacturing, silk manufacturing is of its very nature subject to relentless variation, and it will sometimes grow, contract, stagnate and revive all in the course of a single year. The manufacturer has no way of escaping these continual fluctuations and he can only make fabrics he can be certain of selling. When the economy is booming he has hordes of workers available and can double or even quadruple his output, which means he can meet even very substantial orders within a month or two; but then he will slow down or stop, depending on the circumstances. If he has been imprudent enough to produce too much merchandise for current consumption, the chronic instability of taste will mean huge losses for him.[3]

This is followed by reflections about social relations and on the fact that workmen can continually nourish hopes of promotion through outperforming their fellows in terms of the quality of their work, something that would be lost in a move to a large-scale factory system on English lines – as it already has been in the *La Sauvagère* factory. The American comes back with a long counter-argument clearly reflecting Fortis' own opinion to the effect that England's success has shown the way forward. He points out that in the case of the silk industry market variations are much more fatal to the small companies than to the big ones, causing frequent failures, unemployment, nomadism and revolts. Factory concentration on the other hand stabilizes production rhythms and makes possible the rational management of stocks to make the best use of capital and facilitate the adoption of mechanization and innovations. He concludes with a warning that if the manufacturers refuse to go down the trail blazed by English industry there can be no other outcome but the destruction of the Lyons industry.

A success story that flies in the face of the "large-scale industry" vision of history

This was but one among a whole series of dire prophecies warning that if manufacturing in Lyons did not adapt to the English factory model it would die: the irony is that Lyons did refuse to accept that model but it was the British industry that died first. Throughout the nineteenth century *dispersed* manufacturing in Lyons was a success story while silk manufacturing in London and then throughout Britain went into steep decline. A comparison of these two experiences subverts determinist and evolutionist visions of industrialization. It raises fundamental questions about methods of organizing production, about basic economic concepts, about the coordination of action and about the differentiation of the economic system from the public sphere. I shall tackle these

[3] *Ibid.*, pp. 445–446.

questions by means of a detailed analysis of the Lyons and London experiences in the first half of the nineteenth century. But first some essential preliminary remarks about the context, with an overview of developments in the silk industry in the nineteenth century, which I want to contrast sharply with *retrospective* visions of its destiny.

When Fortis published his dialogue in 1822 the Lyons and Spitalfields industries were still comparable in size, organization and type of product. In 1815 London had about 12,000 looms and Lyons about 14,500 (see appendix 2.2, *comparative table*). The two industries had both grown substantially in the eighteenth century, with Lyons consolidating its undisputed global leadership on the level of the volume of its production and the perfection of its products. Then the French Revolution had seemed to offer a chance for revenge to its traditional rivals – London, Florence and Bologna. The *sans-culottes* dictatorship had prohibited silk, an aristocratic symbol, destroyed the industry's organization and closed off external markets. London had seized the opportunity to muscle in on the international markets and taken over from Lyons as front runner in the fashion arena.

The London and Lyons industries shared the same basic organization. Manufacturers bought the silk and marketed the fabrics and in between they organized the making-up of the silks, which they subcontracted to specialized workshops or firms. In the eighteenth century manufacturers and workshop heads had soon developed what was quite unambiguously a wage relationship, as those involved and the tribunals clearly recognized. The historian J. H. Clapham noted this in his polemic against fanciful views about Spitalfields based on the ambiguity of the term *artisan*: "The Spitalfields industry of the Eighteenth Century was mainly capitalistic. The normal weaver was, and always has been, a wage-earner."[4]

Even before the Revolution the dispersal of work had already been challenged and condemned as archaic: the argument was that it needed to make way for an economic "revolution" considered historically inevitable. For example in 1787 the *intendant* Terray had advised the government to take *authoritarian* measures to suppress the intermediate class of master-workmen in order to *laisser faire* a *natural* historical trend towards concentrations of waged employees in big factories; he claimed that the obvious benefits of centralized ordering and economies in intermediate costs would soon spell the end of the system of dispersed work.[5] By the early 1820s the inexorability of industrial progress had become a cliché in British political economy. It fused the three spectacular components of the cotton revolution – mechanization, concentration in facto-

[4] J. H. Clapham, "The *Spitalfields Acts*, 1773–1824," *The Economic Journal* (December 1916), 459–471.

[5] The texts of memoirs and correspondence relating to the 1787 project may be found in two complementary files in the Archives Nationales, F 12 766 and F 12 1441. Terray uses the expression "necessary revolution." See also a commentary in Maurice Garden, *Lyon et les lyonnais au XVIIIe siècle* (Paris: S. E. Les Belles-Lettres, 1970), pp. 581–582.

ries, and mass production – into the classic model of industrialization. Typical of innumerable such examples are these brutal remarks by a liberal Tory in 1826:

> Is not the London silk manufacture on the decline, and must not those engaged in it seek employment elsewhere? . . . There is no prospect of the hand-looms ever being able to compete again with the power-looms. This must throw an immense population out of employment, and be the cause of appalling distress, till the individuals interested shall have been dispersed and engaged in other pursuits.[6]

The years 1824–26 were a key period for London manufacturing on account of the implementation of deregulation measures – the abolition of import restrictions and the suppression of legal protection of wage rates. Impressed by the arguments of the Ricardian economists the government intended such measures to pressurize the Spitalfields workers into modernizing in the face of domestic and overseas competition. It was hoping to launch the silk industry onto the same triumphant trajectory as the mighty cotton industries, but in reality the deregulation experience proved a disaster for Spitalfields, as we shall see in more detail below, and it was nothing more than a deceptive half-success for the other silk manufacturing areas. A huge gap was to open up between the Lyons and Spitalfields industries: in Lyons the number of looms doubled in the fifteen years from 1816 to 1831 (to 30,000) and then doubled again over the following fifteen years (to 60,000 [with 30,000 of them in the rural areas around Lyons] in 1852), while the London industry went into a continuous decline after 1826, with numbers falling below 5,000 by 1853.

With hindsight the most astonishing thing is that in Lyons in the 1830s and 1840s there was such a widespread anxiety that London and the other United Kingdom silk-producing areas might eventually crush local silk manufacturing by extending their industrial revolution to the industry. A highly alarmist piece by Arlès-Dufour that created quite a furore in 1834 will serve as a good illustration of such anxieties. This young silk broker, a disciple of Saint-Simon, but also close to the radical English Benthamites, was an ardent propagandist for the free market and the "laws" of political economy. In this piece he noted the progress made in England between 1826 and 1834 as indicated by the substantial rise in the consumption of raw silk:

> . . . this is certainly a worrying development, especially when one thinks of the huge capital resources available to the English industrialists, their willingness to work together, their entrepreneurial spirit, and the superiority they have hitherto indisputably enjoyed in the mechanical arts. [. . .] In my opinion our [silk] industry must sooner or later go the way of the cotton industry, and that is why I am asking my fellow-citizens to pay particular attention to what is happening in Manchester and Lancashire in particular . . .[7]

6 M. Berg, *The Machinery Question and the Making of Political Economy, 1815–1848* (Cambridge: Cambridge University Press, 1980), p. 26.

7 Arlès-Dufour, *Un mot sur les manufactures étrangères* (Lyons: Mme S. Duval, 1834), pp. 56–57, 132–133.

The type of argument put forward by Fortis and Arlès-Dufour was frequently to be repeated by local notables and publicists and became a subject of constant controversy in the Lyons press between 1831 and 1853.[8] The classic model of industrialization had already been constructed by the 1820s on the basis of the first revolutions in the English cotton industry. For the silk industry the problem was that this powerful model was not simply an explanatory schema that was verifiable or refutable: such was its fascination that nobody managed to disentangle from the general picture those features that had a purely limited reference to other merely local situations. Its deterministic element was not limited to a particular set of historical causal connections; *for it was applied without any rational distinctions as if it were a universal metrical coordinate*: concrete situations were assessed according to the degree to which they departed from this standard model. The one unique model of industrialization came to be seen as an essential and obvious benchmark for evaluating all industrialization situations: they could only be understood when their degree of similarity to the standard model became clear. Thus for example nineteenth-century global descriptions of the Lyons situation treated silk-manufacturing there as a "*survival*," or an archaism that lay under sentence of death. If we take the term *survival* seriously it implies that ageing or death are biologically imminent, with the implication that the field will be left open for younger entities. But in reality contemporaries did not have any overwhelming sense of there being *one single possible option – that of industrialization*: instead there was a polemic in which *other possible options were rejected*. It is the difference between prophecies warning of coming disaster and the kind of retrospective judgments based on beliefs about historical fatality that have been made since by supporters of the classic model of industrialization.

Polemic was very much to the fore in discussions about deregulation between 1824 and 1826 and statistics were frequently falsified by statisticians working hand in glove with the economists to spread the gospel of *free trade*. This was especially true of Porter's official publications: his 1847 classic *The Progress of the Nation* presented a much too optimistic account of the deregulation measures

[8] Among contemporary books and pamphlets, see especially: Camille Beauvais, *Essais sur quelques branches de l'industrie française* (Paris: Delaunay, 1825); J. A. F. Oznam, "Mémoire présenté à MM. les fabricans d'étoffes de soie, sur l'état actuel des manufactures de soieries en France, et les moyens d'en prévenir la décadence . . .," *Annales de l'industrie manufacturière, agricole et commerciale; de la santé publique et des beaux arts*, 1, nos. 1, 2, 3 (1827), 74–92, 191–206, 298–308; E. Beaune, *Essai sur les moyens de faire cesser la détresse de la fabrique* (Lyons: Baron, 1832); Arlès-Dufour, "Angleterre," and other anonymous articles in *L'Echo de la fabrique*, nos. 25–42 (April 15–August 12, 1832); and *Un mot*; and *Importance de l'industrie des soies et des soieries* (Lyons: impr. de L. Boitel, 1842); C. Beaulieu, *Histoire du commerce, de l'industrie et des fabriques de Lyon* (Lyons: Baron, 1838); Alex Devillier, *Nouveau manuel complet de soieries*, 2 vols. (Paris: Roret, 1839); W. Kauffmann, *Des causes locales qui nuisent à la fabrique de Lyon* (Lyons: Midan, Giraudier, 1845); Ph. Thierriat, *Du malaise de la classe ouvrière et de l'institution des prud'hommes appliquée à l'organisation du travail dans la fabrique lyonnaise* (Lyons: Typographie J. Nigon, 1848). The reports of the discussions in the Chamber of Commerce provide evidence of similar anxieties. For the Lyons press, see appendix 3.1 on material available for research.

taken in 1824–26, an account used later as a source by statistical collections and numerous secondary works.[9] Porter took as his chief indicator the substantial growth in the import of silks: but in his extrapolations he failed to take account not only of a small but growing element of re-exports but also of a revolution in the quality of the raw materials, which meant a fall in added value that brought about a fall in the values of products weight for weight.[10]

In actual fact, far from tripling or quadrupling between the 1820s and 1850s, British production only doubled, as can be seen from table A3 in appendix 2.2.[11] Nonetheless, the growing gap between the respective national production levels became significant from the 1840s on: British production was lower than French by a quarter in the 1840s, by a half in the 1850s, by two thirds in the 1860s, etc. If the table were to be corrected in the light of the substantial rise in the price of silk threads at a time when their quality remained the same – a rise that took place between the 1850s and the 1870s – a reduction in British silk production would appear after the 1850s, a reduction consistent with the trend in the size of the workforce as shown by censuses.

On the level of foreign trade the two economies developed in increasingly divergent directions. In the late eighteenth century British silk exports were minimal and subsequently declined even further; they were mainly to the colonies and occasionally also (in sporadic bursts) to the United States. The

[9] G. R. Porter, *A Treatise on the Origin, Progressive Improvement and Present State of the Silk Manufacture of England* (London: Longman & Co., 1831); new edn of 1847, 2nd edn in Porter and F. W. Hirst, *The Progress of the Nation* (London: Methuen & Co. Ltd, 1912); quoted in part in Clapham, "The *Spitalfields Acts*," and Donald Cuthbert Coleman, *Courtaulds: An Economic and Social History, 1, The Nineteenth Century, Silk and Crape* (Oxford: Clarendon Press, 1969), and cited more critically by Maurice Lévy-Leboyer, *Les banques européennes et l'idustrialisation internationale dans la première moitié du XIXe siècle* (Paris: PUF, 1964). Quite fantastic figures were added by Hirst in the 1912 version of Porter, speaking for example of 24,000 looms in London in 1824 (p. 350), when the estimates of parliamentary inquiries in 1831–32 put forward the figure of 14,000 and the figure of 16,000 put forward by some had been considered excessive. Likewise, his retrospective figure of 150,000 silk looms in England (p. 350), also proposed by Arlès-Dufour, should be divided by two according to parliamentary inquiries and the merchant's own professional press (cf. the observations made by the Commission for customs assessments, 1868, 1873, 1884, which have the best documentary evidence; see also Natalis Rondot, *L'industrie de la soie* (Lyons: Pitrat aîné, 1875).

[10] Porter and Hirst, *Progress of the Nation*, p. 347. There is a similar presentation in Arlès-Dufour, *L'importance de l'industrie*, pp. 5–6. In reality, weavers and exporters were condemning the manipulation of export figures from the time of the parliamentary inquiries of 1831–32 and 1834. Cf. a synthesis in PP. Hand-Loom, 1834, pp. 4–5, by Ballance.

[11] I have taken as reliable those estimates of the value of national production for France and Great Britain, that took account of their extrapolations of falls in added value. Arlès-Dufour, following Porter, overestimated the number of British looms and the size of its domestic production, extrapolating in turn the official import figures without taking note of the fall in added values, and without making the numerous necessary corrections to customs declarations. Cf. for example his reference to the 1851 Exhibition, p. 8. His colleague in the Chamber of Commerce, N. Rondot, was often led to rectify his data publicly (cf. the printed reports of the Commission of Customs Values, listed in the bibliography). He also rectified them in his historical retrospectives of 1875 and 1885 (*Exposition universelle internationale de 1878 à Paris, Groupe IV, Classe 34, Rapport sur les soies* [Paris: Impr. Internat., 1885].

expansion of the years 1820–40 was chiefly down to growth in domestic consumption, which was stimulated by the reduction of tariffs on raw materials. But after the lowering of customs tariffs in 1845 and especially after the Franco-British free trade treaty of 1860 there was an increasing disproportion between British domestic production and consumption, so that in the 1860s imports outweighed domestic production and by 1900 more than 90 percent of domestic consumption was imported. The situation in France in general and Lyons in particular was exactly the opposite at the beginning of the nineteenth century and the gulf between the two countries grew steadily. In the eighteenth century, silk had become France's leading export and it remained so throughout the nineteenth century, reaching record levels in 1913 (cf. appendix 2.2, tables and observations 4–5). France consumed only a quarter of its production, and 60 percent to 90 percent of Lyons output went to exports. In 1900 Lyons and Saint-Etienne production for export to Great Britain was greater on its own than the latter's entire output.[12]

The economic space of the silk industries

If nineteenth-century large-scale industry categories are too inadequate and too distorted by polemic to cope with such counter-examples, some kind of positive reassessment is clearly required. I shall attempt to make just such a reassessment by focussing comparatively on the cases of Lyons and London in the first half of the nineteenth century – in other words just at the period when the gulf was opening up between their respective performances. There will be two angles to my approach: on the one hand, I shall reconstitute the local ways of thinking of practitioners, manufacturers, merchants and workers, contrasting the different presuppositions underlying such ways of thinking in London and Lyons, while on the other hand, I shall where necessary re-conceptualize seductively "self-evident" elementary notions such as that of "a company," "an economic space," "home working" or "factory working," etc., drawing on recent new thinking in the area of theories of action and theories of institutional economics. I will also try to show that in reality the two angles are reducible to just one – the comparative restitution of old local presuppositions, something that will hopefully provoke a fresh debate about the assumptions we have inherited from the nineteenth

[12] Armand Cosson's unpublished 1982 thesis, "Fabriques et ouvriers du textile à Nîmes. Crise – renaissance – déclin (1790–1850)," University of Paul Valéry, Montpellier, analyzed the parallel decline of another silk industry, that of Nîmes, which began in 1826. Such provincial declines, together with the growth in French consumption between 1820 and 1840 and the expansion of smuggled exports, are a perfectly adequate overall explanation of the statistical contrasts between the marked expansion of the Lyons industry and overall lack of global expansion in French exports between 1815 and 1845. (Pierre Cayez, "L'industrialisation lyonnaise au XIXe siècle. Du grand commerce à la grande industrie," unpublished thesis, University of Lyons II, Lille, 1977, in the face of the statistical contrasts, had felt obliged to reject the French customs statistics for the export of silks.)

century. It is my contention that many such misleading assumptions embedded in our thinking need to be reassessed.

A key element in my discussion will be the rediscovery of the notion of "*a collective manufacture*," (*fabrique collective*), an old expression that I disinterred fifteen years ago. It was current in eighteenth-century France, it appeared in the judicial enactments of the tribunals, and it was theorized by Le Play, who made it into an analytic concept of economic space.[13] In its broadest sense *a collective manufacture* is quite simply a localized ensemble of production systems that are linked by the use to which the merchandise they produce is put. The truth is that collective manufacturing served as a framework for capitalist industrialization more frequently than the allegedly "classic" path whose "universality" is purely mythical. Collective manufactures differed according to the composition of the production systems they integrated, according to the inter-company connections they involved and according to the extent of the common domain that they regulated.

The collective manufactures of London and Lyons were archetypal because they combined three closely connected mutually-reinforcing dynamisms:

> They provided a *framework for the shifting organization of production* under the authority of the manufacturers.
> They manufactured *varieties of individual products.*
> Their markets operated *to order* and gave rise to *auction prices*: the operators were "*price makers*" and not "*price takers.*"

The collective manufacture: a space for the shifting organization of production

According to a venerable terminology, the collective manufactures of London and Lyons were classifiable as examples of the *domestic system*, or *home working*, on the pattern of a mode of organization common to Western Europe in general. This system has often been described and commentators have noted that it involved large-scale production operations and big markets without any self-consumption (unlike varieties of "*household economies*"), in a wage relation and not in industrial subcontracting relations.[14] However, these traditional descriptive criteria have the disadvantage of focussing on short-lived production and

[13] On Le Play, see Alain Cottereau, "Vie quotidienne et résistance ouvrière à Paris en 1870," in *Le Sublime* (1980), pp. 63–81, "The Distinctiveness of Working-Class Cultures in France, 1848–1900," in Ira Katznelson and Aristede Zolberg (eds.), *Working-Class Formation: Nineteenth-Century Patterns in Western Europe and the United States* (Princeton: Princeton University Press, 1986), pp. 111–155; cf. especially "Problèmes de conceptualisation comparative de l'industrialisation. L'exemple des ouvriers de la chaussure en France et en Grande Bretagne (1848–1914)," in Susanna Magri and Christian Topalov (eds.), *Villes ouvrières, 1900–1950* (Paris, L'Harmatton, 1990). This last text includes clarifications and rectifications of the essay "The Distinctiveness," published in 1986 but written in 1981.

[14] Duncan Bythell, *The Sweated Trades* (New York: Cambridge University Press, St. Martin's Press, 1978), pp. 12–23.

ordering flow charts, whereas a detailed comparison of manufacturing economies in Lyons and London with their internal rhythms leads to an emphasis on the dimensions of organizational mobility, polyvalency, and specialization. These were the areas that proved decisive for the fate of the two manufactures, and they cannot be understood unless we study connections and permutations in time with their sequential patterns and potentialities.

Let us first of all follow through a manufacturing cycle for "*novelties*" in Lyons and London in the early 1820s.[15] The manufacturer begins by making samples three months before each of the two main annual seasons. He tries to anticipate what will be the next fashion and consults designers. In London the latter are a small number of independent artists who sell exclusive rights to their patterns. But before the Lyons manufacturer makes his choice he calls for rival tenders from the designers he had recruited for several seasons and from the Lyons "*design workshops*" (small subcontracting firms that employ designers and specialists in the translation of pattern into manufacture). Thus the Lyons manufacturer produces a new seasonal catalogue of 200 or 300 samples while the London manufacturer adds a few designs to his catalogue of very specialized products in one single type of fabric, a catalogue that is carried over for several years. Next comes the preparation of fabric samples as a test of the quality of the pattern in real materials and its industrial feasibility. In Lyons, the seasonal making up of thousands of samples is a task for the weaving workshops: in London the samples are very much less numerous and are made up by a specialist sample-maker who is most often a permanent employee of the firm.

The next stage is the commercial phase of orders on national and international markets. Between a fifth and a tenth of the samples will be selected for ordering – between October and January for the spring and between April and July for the autumn. These orders determine the fashion for the season to come. The manufacturer buys his silks in the light of his orders and has them prepared by subcontracting throwing firms. At this stage he chooses the qualities, provenances and throwing specifications as a function of each of the fabrics he wants, of their cost price, their qualities and consistencies and colors. The threads have still to pass through several more hands, in particular for the essential operation of dyeing before weaving, an operation that is subcontracted to small dyeing firms. The threads are subject to further dressing operations at the manufacturer's premises and are warped there or in "*mistress-warpers*" workshops (female piece workers, workshop heads) in the case of Lyons, or, in the case of London, on the manufacturer's premises.

The warped threads are delivered to the weavers' workshops for the actual weaving. There is negotiation and then the orders are shared out among the different workshops with which agreements have been reached. In both manufactures workshop heads use employees working at home or elsewhere assisted

[15] The main descriptions in this essay come from the two collections of materials referred to in appendix 2.1 unless otherwise stated.

by members of their own family: winding is carried out by wives, children or employees (in Lyons mistress winders are in charge of their own workshops with female winders); there are casual staff who specialized in setting up the looms or making up fabrics and in various operations connected with weaving. The warps are delivered to the manufacturer as they are completed and he puts the final touches to them in his workshops before delivering to the customers who have ordered them. In the course of a year two complete cycles of seasonal fabrics can thus be completed and several other cycles can be intercalated in between them or combined with them for the production of non-seasonal fabrics, plains, figured stuffs and brocades – going through the same phases but without such tight deadlines. Thus at any given moment a chart of the hierarchy of production would show different operators working on orders under the authority of the manufacturer:

internal employees for administration, handling, dressing of threads and fabrics, and designers.

subcontracting firms, medium-sized firms (throwing, spooling of warp threads) and small firms (design workshops, warping and dyeing workshops).

casual freelancers: (designers) and a few odd specialists like chemist-dyers, treated and remunerated in Lyons as consultants, or mechanics.

those paid by the piece: weavers who are workshop heads, mistress-warpers.

employees of employers and their employed families, in the workshop or outside it: journeymen and apprentice weavers, spoolers of weft, tiers, folders, etc.

But a chart like this would be misleading. It would confuse this organization with a more common model of the stable firm surrounded by subcontracting and piece-working satellites. The key feature is organizational mobility. For each of his production operations, for each fabric line, the manufacturer does not merely have the option of rearranging his raw materials. He can also rearrange all his operatives. The testimonies of the research collections all show that he makes considerable use of this option. In Lyons at the *Conseil de Prud'hommes* (industrial tribunal) the cases and the discussions always presuppose that each weaver's workshop makes a legitimate habit of dealing with several manufacturers at once and indeed often changes manufacturers. In London changes are just as legitimate but seem to be less frequent; weavers testifying before parliamentary inquiries almost always dedicate all their looms to one single employer, but they change employers according to the season and the economic situation. In both cases the normal reasons for making a change are the same: on the employer's initiative the most suitable person must be found for the new line of fabrics; on the initiative of the workshop head the most suitable time and type must be found to fit in with his availability and with the commitments of the moment. In the case of a bad economic situation the employer chooses the workshops he wants to keep for himself, but without any mutual commitment in law beyond the particular order in question. Likewise the choice of subcontractor seems to be reassessed every time in terms of the particular features of each fabric.

This type of *ad hoc* organization driven by the compatibilities of men with techniques suggests a surprising comparison; it has many analogies with the type of organization known as *adhocracy* – a neologism that was poorly conceived but well marketed, invented by A. Toffler in 1970 and taken up by Mintzberg.[16] According to these authors the highly valued *adhocratic* model represents the maximal flexibility of mobile organizations, brought together to solve a problem, in a context of perpetual non-stabilized innovations. Typical examples are the mobile Silicon Valley brains trusts, cinema teams, software manufacturers. The model has been taken up again today in prophetic utopias imagining the cooperation that will be made possible by generalized informatics connections.[17] But this twenty-first century utopia was realized in its essential outlines by the collective silk manufactures of the nineteenth century regarded by the prophets of the day as traditional.

There is however a difference between those manufactures and the anarchic-type utopia often involved in the model of adhocracy: in the silk industry of Lyons, London and all of Europe, there was a very powerful hierarchy, incarnated by the manufacturer and his salesmen. But still this was not the authoritarian hierarchy of the nineteenth century that we imagine in retrospect. It was a *mobile hierarchy negotiated* at the time of each deal with subcontractors and workmen. From the most formal point of view, this principle of shifting organization may be conceptualized as the dynamic of the network, an alternative to the more classic model of the permanent hierarchical network. It presupposes a *virtual common space* in which further connections are always possible in addition to the temporary ones in operation. The difference may be rendered schematically in the following manner. Suppose that A1, A2, A3 and A4 are four silk manufacturers. Following the convention let us choose three subordinate specialities in the order that the work is done: designers (B1 . . . B4), dyers (C1 . . . C4) and workmen who are workshop heads (D1 . . . D4). The classic chart would permanently attach at least one representative from each

16 Alvin Toffler, *Future Shock* (London: The Bodley Head, 1970), ch. 7, pp. 112–135; Mintzberg, *Structure in Fives: Designing Effective Organisations* (Englewood Cliffs: Prentice-Hall, 1983). The expression "adhocracy" involves an illogicality since it is not the situation that governs here but a preconstituted hierarchy; having authority over temporary subordinates would give even less power than authority over permanent subordinates unless the intercompany space was specifically regulated. This is the whole problem. The social possibility of such a space has not been sufficiently analyzed. So the model may turn into a dream of unbridled power possessed by the pilots of such *ad hoc* formations flying above relational realities. We find here one of the ambiguities of the notion of *flexibility*. It may express a will to power of the kind that management discourse wants to operate with, filtering out any undesirable limitation which is seen only as a "rigidity." In this case, *flexibility* is only a new descendent of *laisser passer*. It dazzles with the dream of unconstrained action in an institutional void.

17 Th. Malone and F. Rockart, "Computers, Networks and the Corporation," *Scientific American*, 265, 3 (September 1991), 92–99.

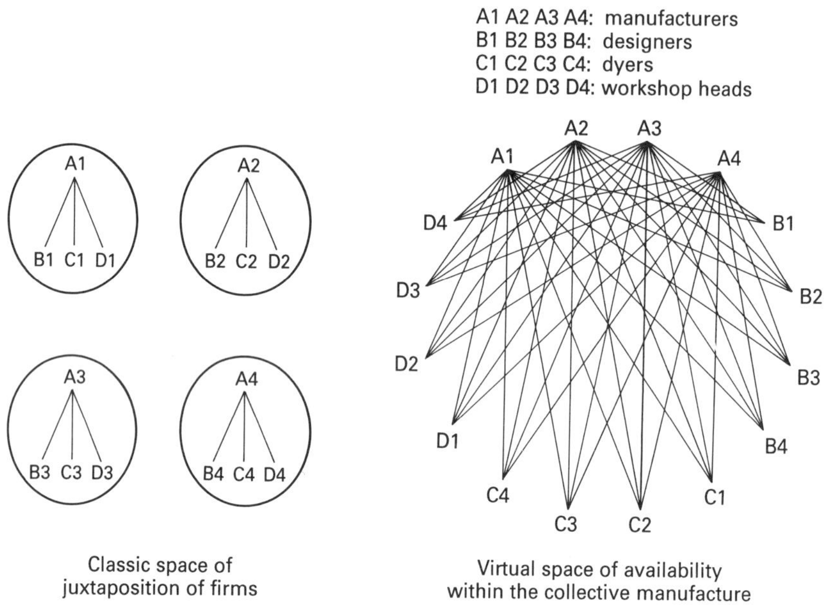

Figure 2.1 Firm space and the space of the collective manufacture.

speciality to the sphere of authority of each manufacturer. But in collective manufacturing each speciality remains virtually available for each manufacturer. When any industrial operation is set up any manufacturer may choose from B1 . . . B4, C1 . . . C4, D1 . . . D4. After a while all the specialities may ultimately have been connected with each manufacturer. Figure 2.1 sets out the difference in the two spaces in schematic form.

These vital negotiated subordinations condition all the situations occurring in our two manufactures: bargaining, settlement of prices, effective powers. There is a need for more precise analytic concepts to study these phenomena, for the available concepts are too vague: "external economies" leave the exterior of firms too undifferentiated and the advantages of "industrial districts" or "industrial fabrics" often remain too empirical, too heterogeneous, confined to the geographical dimension. Voluntary inter-company strategy models are useful but still do not get to the heart of the matter: what we really lack is the idea of a specific level of inter-company economic space external to the alliances and combinations of firms, but itself inserted into wider economic ensembles, markets and political entities. A space like this, characterizing certain *collective manufactures*, may be formally defined as *a condition of the possibility of shifting hierarchical connections*. It involves a specific level of regulation and administration of relations between the economic units of the manufacture.

In order to disentangle these interwoven dimensions I shall first of all study requirements for quality of craftsmanship and fabrics so that I can then explain the particular modalities of negotiation, of the market, of the organization of work and of regulation.

The quality of the silks

Woven fabrics hold a particular place in the history of human products, as has been emphasized by the anthropology of techniques: down the centuries all possible types of textures have been tried out, and the industrial revolution, which made woven fabrics one of its specialities, did not introduce any major modification in this area – quite the contrary in fact:[18] as far as cotton was concerned the method of the industrial revolution was to impoverish textures in accordance with the logic of mass production. Its only significant innovation with respect to earlier textures had to do with uniformity, but whether the latter may be considered a positive quality is far from self-evident. It meant that manufacturing techniques were focussed on the acceleration of work as a means of productivity. However the upheavals in productivity and costs had a huge impact on systems for categorizing fabrics; the assessment of silks diverged increasingly from the assessment of cottons, while woolen cloths were sometimes compared with the former and sometimes with the latter.

At the beginning of the nineteenth century silks already had behind them a two-millennia-long intercontinental East–West mercantile tradition and an old culture of perfectionism was still in place. This emerges clearly from a study of the accounts of industrial exhibitions in France and the accounts of inquiries on silks and design in Great Britain. The example of shawls is particularly striking.[19]

This figured and brocaded fabric was introduced in 1795–89, first in fine woolens, then diversified in mixtures of wool and silk and in pure silk, then imitated in printed fabrics including cotton. From 1806 it became a real fashion hit. In 1838 a jury report from an exhibition observed that shawls had become a common feminine garment, worn by all classes of society, and even incorporated into typical rural village clothing.[20] In Lyons, 4,000 looms were kept busy on shawl fabric. In a context of the constant renewal of the types and fashions of

[18] Leroi-Gouran, *L'Homme et la matière* (Paris: Albin Michel, 1971), pp. 278–296.

[19] On shawls, the detail of qualities, fashions, pieces and quality–price differentiation may be followed through all the reports of exhibition juries from 1798 on (see the listing of Expositions at the end of this chapter). On the beginnings of shawls and the creation of a fashion, see Magnien and Deu, *Dictionnaire des productions, de la nature et de l'art* (Paris: Bailleul, 1809), article "châles." For Great Britain, see especially, among the parliamentary inquiries in the collections of materials, the whole range of inquiries on design, and the account by the Norwich manufacturers, PP. Hand-Loom, 1840, II RAC, J. Mitchell, pp. 301–302; an earlier controversy with France was developed in Exposition [1851] . . . *XVe jury . . . Châles* . . . 1854, IV, p. 7. On the technical level, see Monique Levi-Strauss, *Cachemires* (Paris: Biro A., 1987); Jean Bezon, *Almanach historique et industriel des manufactures de Lyon* (Lyons: Ballay and Conchon, 1857), v. 4+5; Amédée Couder, *Analyse du dessin des cachemires et moyens de rendre les schalls français supérieurs à ceux des Indes* (Paris, chez auteur, 1834). Progress in productivity is discussed below.

[20] Exposition [1839] *Rapport* . . ., p. 132.

shawls, public choice between qualities and prices was always made with reference to traditional Indian shawls. The main line of technological innovation conformed to this: in India prestige shawls were made by techniques related to carpet-making; in the West techniques looked for an approximation to the same effects in figured and brocaded fabrics, using old techniques for brocading on a semple loom and then the techniques of the Jacquard loom; for the most sophisticated shawls there were also specific methods of brocading.

At the top of the hierarchy imported Indian shawls could sell for 4,000 or 5,000 francs. People were much impressed by the variety and brightness of the colours, considered inimitable, and by the finesse of the patterns, which represented the work of thirty to fifty persons for 6 to 7 months.[21] In France the top of the range, recognized as undisputedly the finest in the West, was an imitation of cashmere, it normally sold for 300 to 500 francs in 1838. It incorporated the equivalent of two and a half months to three months of a weaver's work, against 300 months for the original Indian cashmere. The French experts admitted that the best shawls made in their own country were inferior to the Indian shawls in respect of solidity, relief and the effect of coloring, where European technical procedures could not really compete. On the other hand, according to Bezon, a professor of weaving, the most demanding of the authors consulted, "our shawls have an elegance and harmony of design – eminent qualities from the point of view of taste – that far outdo the Indian shawls."[22] Bezon also defines the discernment required by the international silk buyer, as the final public consumer, to discriminate amongst very wide variations of quality and price: "It is quite indispensable to have a perfect knowledge of the value of the materials and the difficulties that may obstruct the setting-up of the loom or the weaving; you also need to be able to assess and judge the value of a design right away by examining its composition, the finesse of the fabric and the type to which it belongs."[23]

Such competence was fairly commonplace and it defined a range of quality/price judgments in all the silks markets. Assessments claimed to make objective judgments that were common to East and West. There was such a close connection between assessments and the difficulties to be overcome that quality was expressed in terms of production costs and reflected in the range of prices. To a retrospective inquiry this convergence initially seems somewhat paradoxical when compared with fashion phenomena. The contemporary press had long been describing vogues and fads as races between originality, imitation and banalisation that reflected old social hierarchies in a new way. It described as a perversion what have more recently been called self-fulfilling prophecies.[24]

[21] These times are given in a report by syndics of shawl makers from the Kashmir Province, a report translated and sent by a French consul, and printed in Exposition [1839], pp. 133–136.

[22] Jean Bezon, *Dictionnaire général des tissus anciens et modernes*, 2nd edn, 8 vols. and atlas (Lyons: Imp. de T. Lépagnez, 1859–63), IV, p. 360. [23] *Ibid.*, V, p. 23.

[24] Robert M. Merton, "The Unanticipated Consequences of Purposive Social Action," *American Sociological Review*, 65 (March 1936); Bernard Walliser, *Anticipations, équilibres et rationalité économique* (Paris: Calmann-Lévy, 1985).

Instead of importance being given to the intrinsic qualities of objects, what counted was the likelihood of adoption by others.

Responses to this apparent paradox are supplied by the debates and analyses that took place in the course of British parliamentary inquiries in the 1830s. With each debate on international trade and silk fashions, participants less familiar with the profession imagined that France's leadership was to be explained by a snobbish preference for things French and that this excused British failures. But this simplistic view was consistently subject to correction in the debates. Importers protested indignantly and showed that the issue was not about prejudice but about judgment of intrinsic qualities in open competition. Witness this deposition by the most notable of them, the importer James Morrisson, MP, representative of the great City traders, who followed others in correcting the opinions of the Commission president: "I very much doubt whether that preference [for French products] is given for foreign articles merely because they are foreign; I think the public are always ready to purchase our own goods when they are really equal to foreign."[25]

Very detailed debates elucidate the criteria by which competing qualities were established. The notion of fashion, to which is connected the stereotyped expression "*caprices of fashion*," leaves room for a fine criteriology of the elements of the excellence of fabrics, which analysis of the materials can group around five terms utilized by professionals: *good taste; novelty; style; invention; finish* or *execution*. Only one of the five elements – novelty – relates analytically to the phenomenon of imitation or self-fulfilment. This alone is a component of value directly and literally connected with the ephemeral originality of the moment. All testimonies stress this fleeting character of fashion articles that sell well only for the first few weeks and fall in price as the season progresses, then at the end of the season can at best be sold off at rock-bottom prices ("*jobbing*"), while at worst they are unsalable. The "novelty" effect on prices is even quantified exactly; a seasonal article may lose up to 25 percent of its value in ten days.[26] A novelties market of this kind thus acquires the characteristic features of a perishable goods market. But – and this is the constantly emphasized crucial point – a fashion article is never an article with just the single "novelty" dimension. It is an article that combines the five qualities listed. While assessments of novelty arise from imitation or the paradoxical "caprice" of self-fulfilling prophecies, the four other qualities (good taste, style, invention, quality of execution) all relate to criteria of excellence shared by the global market.[27] Hence the claim to objectivity in evaluations, an essential piece of data to integrate when we approach markets and bargaining.

[25] PP. Arts and Manufactures, 1835, SC, no. 2339, p. 16.

[26] PP. Silk, 1831–2, SC, R. Clay, no. 6803, p. 387.

[27] This point was noted almost unanimously not only in French analyses but also in British, although to admit such a thing exposed a British inferiority in competing qualities. Among others at the 1831 inquiry, depositions by Bagallay, Ballance, Bottrell, Bouillon, Bowring, Brunskill, Cheeper, Clay, Cox, Dillion, Doxat, Jacombs, Merry, Prout, Schwabe, Sedgwick, Smith, Stephens, Stone, Wallis.

Perfection of workmanship and the artisanal ethic

There are such depressing accounts of the *canut's* job that I initially assumed it to be archetypically monotonous, with its repetitious back-and-forth movement of the shuttle between 8,000 and 30,000 times a day.[28] But as I familiarized myself with the technical manuals and with the Prud'hommes' discussions I came to see the content of the job very differently. For those involved in it, the completion of a piece of fabric invariably seemed like a victory in a struggle with the hazards of the environment. I was encouraged in my new view of matters by the very few manuals giving pedagogical advice about silk weaving. Manual weaving of silks was a test of skill and attention which required lengthy training and mobilized numerous techniques of mind and body for dealing with the material: touch, eye coordination, skill at handling the shuttle, coordination of shuttle and pedal movements, and the anticipation to integrate the independent variables in the series of sequences.[29]

One set of independent variables related to the unpredictability of the raw material. Silk was not like cotton and wool, in lengths it was much trickier to handle. Its variations in consistency, elasticity, solidity, shine and touch, as a function of the raw material used, as a function of previous preparation, dyeing and dressing before weaving, as a function of climatic changes during the work, etc., made it essential for the weaver to be able to improvise during the weaving and to adapt his hand movements to any one of 100 unforeseen occurrences. He was for ever having to anticipate, to interpret what his fingers, his ears and his eyes were telling him, then correct the "irregularities" (of tension, torsion, solidity, thickness, etc.) so as to recover the "regularity" of the fabric that was indispensable for the attainment of the required results. All these requirements were emphasized by the users and builders of looms.

British parliamentary inquiries did not descend to this level of analysis, but they did touch on its strategic consequences for negotiating. In the 1823 Inquiry the Lords began by putting technically stupid questions before homing in on the crucial situations which made it possible for them to profit from the pedagogical talents of the workers' leaders. Witness this explanation as to why the length of

28 Statistics based on the actual quantity of daily production, measured by length, taking into account the number of threads per unit of length. According to the silk-making manuals shuttle rhythms could be anywhere between 30 and 120 a minute, depending on the period, the fabric and the batten techniques adapted to it. If we multiply by the minutes of days of between 12 and 14 hours, we get a theoretical number (if speed is continuous throughout the day) of 20,000 to 86,000 moves. The gap between the continuous theoretical rhythm and the actual rhythms of shuttle use is of the order of 60 percent to 80 percent. It represents the proportion of slow periods and necessary interruptions to the manufacturing process.

29 The majority of eighteenth- and nineteenth-century manuals simply define the fabrics to be made and the methods required for their manufacture. The best ergonomic and pedagogic descriptions that I have found are in Bezon, *Dictionnaire*. Precious information may also be found in Devillier, *Nouveau manuel complet*, and Paulet, *L'art du fabricant d'étoffes de soie*, 7 vols. (Impr. L.-F. Delatour, 1773–89). For Great Britain, I have used Clinton G. Gilroy, *The Art of Weaving* (London: Wiley and Putnam, 1845).

time that completion of a given fabric would require could not be foreseen at the moment the process was set in motion, as given by the weaver James Baker:

Question of a lord: You know when you take them [silks] into your hand whether they will work well?
Baker: I do not pretend to be a judge of silk in the raw; I am not a manufacturer but an operative weaver; I know when I get a cane in the room whether it is good or bad; it is delivered to me always in a dyed state.
Q: Before you put it into the loom, do you know whether it will work easily or with difficulty?
Baker: No, not till we get it into the loom and begin to try it; it is a question with me, whether the manufacturers themselves are not often deceived in that; whether they do not buy silk as very good, which turns out to be very different afterwards, when it comes into work.[30]

A completed piece was subjected to two tests: the first on delivery by the weaver to the manufacturer, the second on delivery to the customer who had given the original order. Each time there was an *evaluation of the success of the piece* and of its faithfulness to the sample. Here we find one of the five quality criteria mentioned above, that of the *degree of perfection* or *the finish*. To expert eyes, the quality of the craftsmanship is reflected in the regularity of the textures, the regularity of the tensions, the clarity of the colors and the firmness of the motifs; for silk or wool it is to be seen more specifically in certain sensual qualities of *feel*, suppleness and softness, all those metonymic qualities that are reminiscent of the human caress; finally for silk alone, for plains and figured stuffs, quality is reflected in the different effects peculiar to each texture, effects of light, shimmer, grain and relief.[31] These criteria of excellence, sometimes formulated positively (in technical manuals and parliamentary inquiries), were more often expressed negatively, especially on the occasion of disputes at Prud'hommes' debates on the quality of the pieces delivered. There was an abundant local taxonomy of defects, connecting aspects of the fabric with technical causes of defects; for example, a *piedfailli* denoted an irregularity produced by a mistake in the order of the pedal movements.

Silk fabrics thus had a position at the opposite pole from cotton fabrics among great developments in manufacturing culture. Industrial processes aimed to achieve the old artisanal perfections by quicker and less costly means, but at the same time, especially where the notorious "calicoes" – flagships for the cheapest kind of industrial mass-production – were concerned, industrial processes modified design and subordinated aesthetics to serial mechanical production. One of the consequences of this was a difference in the development of the weavers' contribution to aesthetics. In the case of calicoes, the industrial revolution quickly eliminated any relation between the weavers' work and the final appearance of the product. The greater uniformity of the threads, largely

30 PP. Lords' Committee, 1823, CLVI, 57, p. 71.

31 E. Chevreul, *Théorie sur les effets optiques que présentent les étoffes de soie* (Paris: F. Didot frères, 1846); Henri Algoud, *Grammaire des arts de la soie* (Paris: Schemit, 1912).

imputable to mechanical processes, of itself facilitated automation and the acceleration of weaving rhythms with mechanical energy. The definitive disjunction between weaving and aesthetics took place with the printing of calicoes; it was all the more radical in that the printing, initially done by hand, was later mechanized and ceased to make any attempt at matching texture to printed pattern, exploiting only the passive quality of the new calicoes – their receptiveness to rapid printing. In the middle of the century it was just such sacrifices of quality that made it possible to have calico prices that were one-tenth of the price of the cheapest plain silks.

A consideration of the aesthetic requirements of modern silk manufacturing leads to a recognition of the fact that it constitutes an aesthetic domain that is properly artisanal in the heart of the world of contemporary industrial objects. Here the word "artisanal" does not refer directly to the way production is organized but to a particular relation between aesthetics and manufacture, between the work and the weaver's ability to manipulate the loom. The requirement for "*perfection of execution*" is clearly not a matter of aesthetic representation, which is thought of today in terms of the *gratuitousness* and autonomy of the work of art. This is an excellence that results from a paradoxical ascesis: skill with the loom and mastery of the material are manifested in a quality of *purification*, by virtue of which the maker's manipulative skill is all the more spectacular in that it reaches the stage of *total forgetting* in the finished work. In handloom weaving, the ascesis was constituted by the imperative of "not leaving any traces." But this was not a matter of mere obliteration, for such "perfection" celebrated the adequation of manual dexterity to material, giving it concrete form, and made it available to inspection by any and everyone. As Wellmer recently noted, a perfectly finished work derives from an aesthetic life, a "craft culture" midway between a pure utilitarian event and a cult of autonomous art. To acknowledge this phenomenon of *craft culture* is to go beyond the quarrel between functionalists and believers in decorative arts, for *craft culture* transcends industrial–functional reductionism as well as the decoration of raw products by the additions of *taste* or *fashion*. It offers a habitable world where day-to-day living does not involve any conflict between instrumental expertise and aesthetic enjoyment. In correlation with the *act of dwelling*, craft culture objects invite us both to use them instrumentally and to treat them poetically, and this without any incongruity.[32]

In the case of silks, this remarkable phenomenon of integration was particularly evident in the evolution of style. While the industrial printing of calicoes caricaturally reinforced productivist instrumental logic (uniform raw fabrics) with pattern differentiation left to the last minute according to the taste of the moment (mechanical printing), silk-making mobilized technical and human resources to make each individual phase of production work towards the final perfection, something that was very impressive to purchasers. Such differences

[32] Albrecht Wellmer, *The Persistance of Modernity. Essays on Aesthetics, Ethics and Post-Modernism* (Cambridge MA, London: MIT Press, 1991), ch. on "Art and Industrial Production."

in development had repercussions on all aspects of industrialization. In the case of cotton, the spinning/weaving relationship was one between production of "*raw material*" and first embodiment in a *form*; the crude fabric was then itself raw *material* receiving its *form* from the printing. Here the productivist race led to the stage-by-stage eradication of individual features, of *first forms*, to fit in with the relentlessly repetitive templates of the machines. But in the case of silk the long chain of manufacture involved each phase of development being focussed on the final appearance. Thus for each type of fabric the threads had to be prepared differently and had to be made from different types of raw silks from different places. Likewise the nature of the desired result in terms of the desired effects of color and light dictated certain aspects of each of the preceding processes. This explains one of the advantages of the organization of manufacture in shifting negotiated hierarchies: in a market where new types of fabric are constantly being introduced and where the success of each fabric depends on *specific* contributions to the desired result, it is important to choose the persons, materials and processes best able to revamp the coordination and arrangement of *first forms*. In this symbolic case the industrial style did not just mean some pattern added in at the end. Far from establishing a (passive) matter/(active) form relationship, *it implied a first form/second form relationship* that persisted throughout the process. In the more classic terms used in his metaphysics by the later Aristotle, we could speak of a *hylomorphic* style, i.e. a style which enables each creative act of form to highlight the potential of the material to which it is applied.[33]

The weavers' skill and their investment of self-esteem

This fundamental difference between silk and cotton fabrics also enables us to understand the difference in the quality expected from mass production weavers and the quality expected from silk weavers. In the first case there is a tendency to repetitiveness, with experience and the ability to improvise sacrificed to pure speed and cost reduction. By contrast the kind of ability to produce a quality finish expected from the *canut*, the *silk weaver*, could come only from a wealth and diversity of experience. In Lyons this was the fundamental principle of the economy of apprenticeship, reaffirmed many times, and I shall be looking more

[33] This term is appropriate only to the extent that *hyle* does not have the purely passive modern sense and that its receptivity has a crucial role in the genesis of forms. In a postmedieval sense, to avoid contradictions, it would be better to refer to a *first form–second form* relation, a *morphomophic* style. This criterion could also be applied to the great classifications of the industrial sectors to distinguish between industries of *n* transformations and industries of final consumption in a different way: on the one hand, sectors of the production of raw materials in the strict sense, able to be used for mass production to the extent that they exploit the eradication of specific forms in order the better to lend themselves to later transformations or to a final consumption that is indifferent to forms since it destroys as it consumes, on the other hand all the industries where specific forms themselves are actually exploited – whether *first forms, intermediate forms, or final forms*. But to go down this path leads to heresy for it raises doubts about long-established distinctions between production, exchange and consumption.

closely at it below. Though similar principles were advocated in London, the fact was that they were less and less in tune with the reality. Many of the weavers testifying before inquiries explained that it had taken them years to acquire their skills. In the two manufactures a hierarchy of levels was established according to a weaver's ability to make such-and-such a type of fabric, and to make it in first or second choice of quality. Each assessment of a piece on delivery was a test of skill for the whole professional milieu, employers and workmen alike. For each pattern the degree of success of each workman was compared with that of the others. Manufacturers were always ready to make comparisons and workmen had frequent opportunities to do so for themselves whenever deliveries of pieces to the warehouses took place.

The system of collective manufacture was able to bring out the best in the weavers in this first half of the nineteenth century: the discussions that took place on the occasion of the acceptance of orders and the assessments of pieces at the time of delivery resulted in intense investments of personal pride in levels of quality achieved and speed of production alike, to the extent that skill and speed came to be equated. In the research materials I have used this equation is always implied and sometimes overtly stated, for example in this testimony by the manufacturer William Hall to the House of Lords:

> There are many journeymen that, from dint of application and from being expert and ready at their work, can do a greater quantity of work, and also turn it out better, make it more saleable; and the consequence is, that the hands that can make their work the quickest, generally speaking, make it the best; the color is not tinged, it is not injured; and we find that those hands who are very regular and very constant, not only make more, but generally speaking, if they are employed, turn their work at best, consequently we give them the best works, those that will bear the best price.[34]

When asked at parliamentary inquiries between 1818 and 1835 what was the normal production of fabric in a given time, London weavers would often launch into tales of record performances and individual tales of prowess: such-and-such a length of "cane" (a warp of fabric) completed in so many 16/17-hour days. This equation of skill with speed was taken for granted in the Prud'hommes' debates and in the columns of the workshop chiefs' newspaper: to prove that you were a good workman it was simply necessary to state that you could make such-and-such a quality of fabric in a given length of time. Bosses and workers frequently alluded to the advantages of this game of self-esteem, though they were also aware of the dangers of the competitive spiral it could suck them into. In the *Echo de la fabrique* (nos. 32–36) in 1833 workers and journalists reacted to a press campaign for factory concentration by showing that the decentralized bargaining system involved a productivist race which was actually advantageous to the manufacturers, a race that would be eliminated by factory concentration.

This investment of self-esteem had a powerful impact at the Prud'hommes.

[34] PP. Lords' Committee, 1823, CLVI, 57, p. 9.

There was a classic type of angry scene provoked by merchants' criticisms of defective craftsmanship. Confrontations occurred when such criticisms cast doubt on a workman's reputation for skill. Sometimes the protagonists would then meet subsequently at tribunal conciliation or judgment sessions. Typical of workshop legends of the time was the celebrated heroic tale related by Bezon:

A man called Thomasset had been a velvet weaver for thirty years and his skill had earned him the sobriquet "king of the velveteers." For many years he had specialized in very demanding work but his advancing years had deprived him of the necessary dexterity and he could no longer be trusted with orders requiring great delicacy of touch such as those for white velvet. The manufacturers were happy for him to keep on working on less demanding fabrics so he was asked to prepare a piece of black velvet. Thomasset was so sensitive to what felt like a demotion and he felt so upset at having to come down from the pedestal on which his reputation as a skillful weaver had placed him, that soon after starting work on his piece of black velvet and with just half a metre of fabric woven, he rushed off in a fit of fury and cut his own throat with a razor [. . .] Manufacturers must really get to know the characters of the workers doing jobs for them and they must soothe those susceptibilities of the workers that have their origin in self-esteem; for well-placed self-esteem does a great deal to help make good workers.[35]

The bargaining process in which the weaver was involved, the rhythm of his working pattern and his identification with his work thus all in some sense led in the same direction, and this was true, remarkably enough, of both manufactures, though not of all collective manufactures in general. In silk manufacturing rhythms of commercial pressure always dictated work rhythms and the process of negotiation always cranked up the tempo of competition. As a result there was a considerable ambivalence reflected in the assessments of the workers' leaders, something worth stressing as a significant phenomenon although we do not need either to praise or to condemn it. Sometimes the passion for work was presented positively as an exalting aspect of artisan existence: sometimes it was presented as an exaggerated over-competitiveness that degraded weavers to the level of "beasts of burden" or "mechanics."

The polyvalence of the Lyons weavers: an inversion of the classic terms of the division of labor

While the two manufactures of Lyons and Spitalfields ran on the same fundamental principles of shifting organization, design of products and respect for quality, they nonetheless differed in numerous respects in all these areas. Over the first half of the nineteenth century Lyons manufacturing consolidated the advantages it had as a collective manufacture, while London manufacturing weakened its position to the extent that it caved in to pressures towards mass

[35] Bezon, *Dictionnaire*, II, p. 81.

production, from which it in fact derived no advantage whatever. This contrast will become especially evident when I look at differences in styles of regulation, but it can also be seen in the modalities of the division of labor which I am now going to study, as well as in manufacturers' managements strategies.

My comparative inquiry will show that two opposite approaches to the division of labor operated in the London and Lyons workshops for the whole of the period under investigation: in London the principle of strong specialization and in Lyons the principle of polyvalence. The contrast had been implicitly stated in 1773 by the Frenchman Paulet from Nîmes in the course of an admiring description of the organization of Spitalfields silk manufacturing:

Only the English seem to have brought their manufacturing to a really high degree of perfection. London alone contains about eight thousand looms, and the reason for this great number is that as the workers who specialize there in one type of fabric never make any other, once a loom is dedicated to a given fabric it is never converted for different fabrics; thus a particular worker who makes satin will never make taffeta or velvet; in this way each of them acquires a unique precision in his field. Manufacturers never change the quality of the silk, so that there is always the same dressing and the same denier for the thrown silk and the same size of tram, and some would even refuse orders for satin or taffeta if they were required to change the quantity or quality of the silk they normally use. An example like this gives a better idea of the genius of this estimable nation than would a long dissertation.[36]

Specialization by fabric applied to manufacturers as well as weavers. In the course of the parliamentary inquiries between 1818 and 1840, practitioners often referred to the workforce of the manufacturers and weavers specializing in such-and-such a type of fabric, pointing out that a crisis in a given type brought on bankruptcies for manufacturers and unemployment for weavers. We are not talking here just about the great subdivisions linked to the various weaving loom techniques (plains, novelties, figured stuffs, velvet, ribbons, tulle), but about types of fabric within these categories: for example, in plains, "bombazines, satins, poplins." The same was the case in other English silk manufactures and in the old cotton plains. It seems that the Spitalfields manufacture was subdivided into relatively compartmentalized sub-manufactures for weaving operations. It was a system that remained unchanged in its general lines up to the industry's demise.[37]

By contrast the Lyons manufacture had always had a much greater degree of polyvalence, which applied both to merchant-manufacturers and to workmen. With the merchants the great houses always covered huge ranges. A comparison of the publicity material for each industrial exhibition, nominally devoted to the award-winning companies, shows a movement of perpetual renewal of the particular types in which the companies were recognized as excellent. At any given date the directories of manufacturers and workshops make it possible to

[36] Paulet, *L'art du fabricant*, I, p. xv.

[37] Charles Booth, *Life and Labour of the People in London*, 2nd edn, 17 vols. (London, New York: Macmillan, 1902–3), 1st series, IV, pp. 241–251.

identify a few principles that governed combinations of specialization with polyvalence. A diversification that aimed at the best rate of continuity of production and integrated the complementary rhythms of fashions would go with particular specializations in particular levels of excellence of quality. Thus in the case of black silks, there were to be found in Lyons, not as in London specialists in one kind of black fabric, companies specializing in a very highly reputed black which they applied to various fabrics in plains and novelties, guaranteeing continuity at any reason. Likewise the companies might excel in a particular mixture of such-and-such a silk with such-and-such a wool, producing it in plains, novelties, and figured stuffs.[38] In general traders and workshop heads had specialisms that combined not only several types of fabric but even several great families of types (plains, novelties, velvets, figured stuffs).

The number of Lyons workshops specializing in plains was very limited when compared with London, where the great majority specialized exclusively in them. By contrast the number of London workshops specializing in novelties, already smaller than in Lyons at the start of the nineteenth century, continued to fall even further, while in Lyons after 1819 the majority of the workshops included novelties as an element in their diversification strategy. We find further multivalency on the level of individual Lyons workshop heads and journeymen, where it was regarded as an essential qualification. All the skills mentioned above were cumulative in that skill acquired for such-and-such a type of fabric became valuable for a completely different type. Ability to cope with the constant unpredictabilities of weaving was largely a product of variety of experience.

Here we encounter an aspect of skilling that directly contradicted Adam Smith's principle of the division of labor (shared by Paulet in his admiration for Spitalfields). In London nobody doubted the classic "law" by virtue of which *specialization implies greater skill.* But in Lyons a completely opposite logic was taken for granted, of a type unknown to the classics of political economy, to the "philosophy of manufactures" or to industrial evolutionism, as expounded by the likes of Ricardo, Ure, Babbage, J. S. Mill or Alfred Marshall. In the

[38] A statistical analysis of the most detailed directory, the Indicateur annuaire de la fabrique d'étoffe de 1843, confirms these combinatory principles. Only 18 percent of companies, all fairly small, advertise a single speciality. One single tiny speciality out of a possible 154 is practiced without being mainly associated with others, "umbrella taffeta" (3 percent of traders make it exclusively, against 2 percent who make it in association with other fabrics). A few fabric qualities are made as sole products by a number of companies: plain velvets and several shawl specialities. By contrast, the fabrics that enter into the most varied combinations are very common novelty fabrics, i.e. dresses (17 percent of traders) and waistcoats (22 percent). Their connections make it possible to identify a kind of very simple grammar based on two rules internal to novelties and one external rule of complementary association. Among novelties, dresses and waistcoats are with very few exceptions mandatorily associated with one complement at least – either a fashion article for the complementary sex (masculine with feminine and vice versa for example: dress–waistcoat, dress–tie, shawl–waistcoat); or else with a complementary seasonal fashion article (summer/winter, seasonal/continuous). The same rules apply in workship polyvalence but less strictly and to smaller ranges.

generalizing scientistic language of such writers this Lyons principle could be formulated thus: *specialization reduces skill, variety of experience improves efficiency*. The relationship between multivalency and efficiency is manifest in the superiority in individual physical productivity of the Lyons workers over the Londoners. In Lyons this conception had normative evidential status in apprenticeship relations,[39] for workshop heads who made their apprentices specialize in one single fabric betrayed their promise to "*show the craft*." Those who indulged in such short-term speculation would be brought before the Prud'hommes the relatives of the apprentice and either enjoined to rectify their conduct or punished. If we are to grasp the difference between levels of efficiency in the two manufactures, we need to compare them in respect of the polyvalence labor and the mobility of their organization on the manufacturing level. In Lyons in 1830, a new type of fashion fabric could start as many as 5,000 or 6,000 looms out of 30,000 working within a few days, while in London the figure would be less than 500 out of 12,000. This unheard-of capacity for quasi-instantaneous mobilization was emphasized by all the authors (including de Fortis, cited at the start of the present essay). One of the most eminent French silk-making notables, Camille Beauvais, who himself introduced numerous fashion fabrics in the years 1818–24, remarked retrospectively how in 1810, at a time of changes in the market, hundreds of weavers previously working in light fabrics managed to convert (themselves and their looms) to rich velvet for court clothing.[40] At the time of the 1831 inquiry, commercial agents for Morrisson's (importers of French articles) noted that the Lyons workshops had a polyvalence that seemed quite extraordinary to them and quite inconceivable in Great Britain, with workers in figured stuffs alternating seasonally between very wide rich brocades and figured ribbons of a complexity not found anywhere else. It was observed that these Lyons weavers worked on Jacquard looms that were modified so as to be convertible in the space of a few hours from grand brocades to figured ribbon, and that 1,500 to 1,800 looms worked in this way.[41]

Many Lyons manufacturers were fully aware of the superiority of this form of shifting subcontracting hierarchy and knew that workshop heads had to be free to change employers quickly when they so desired. But in London analyses were already being governed by the factory model. The Smithian conception of the division of labor called for the concentration of ordering and the hierarchical coordination of odd specialities on a fragmentation–central coordination–fragmentation pattern. At the time of the 1831–40 Inquiries, many London manufacturers had the cotton industry in mind when they described the difficulties of their sector in terms of *a lack of the attributes of factory efficiency*. Their ideal was mechanization and although its application to the silk industry seemed to grow ever remoter, there was a definite tendency to increase the

[39] Norbert Truquin, *Mémoires et aventures d'un proletaire à travers la Révolution* (Paris: Maspéro, 1977), p. 133. [40] Beauvais, *Essai*, p. 43. [41] PP. Silk, 1831–32, SC, p. 445.

production of the lowest quality plains. Dependency on fashion was felt to be an irrational obstacle to the expansion of stocked series.

The economy of availability in collective manufacturing, as against the model of factory concentration

The growing influence of the mass production model in thinking at the local level reflected a movement of concentration and integration among factory throwsters and manufacturing weavers, in imitation of the cotton sector: fixed capital was sunk into homogeneous production that yielded increasing returns. Weaving was integrated to guarantee outlets but this integration ran counter to the advantages of collective manufacturing: instead of the best resources of the whole manufacture in raw materials and labor being available to each unit, each had to use his own raw materials, with very restricted and specialized uses, of a quality often much inferior to the thrown threads imported from Italy. Here again the imitation of the cotton model led to a mistaken view of silks as merely a raw "*material*" rather than a first "*form*" requiring appropriate handling skills. But in Lyons the warehousing business was reconstituted with the explicit aim of extending the range of possible choices from among all the world qualities for each variation in fabric. In the years between 1810 and 1850 skills produced in France improved in quality all the time and eventually caught up with Italy to share the first rank internationally. But the fact that Britain was supplied by her colonies and the movement to integrate weaving and throwing both became handicaps for the British silk industry, although access to colonial supplies was no hindrance to the cotton industry.

One of the main British throwsters, J. Hall, himself admitted the mistake he had made in attempting to integrate weaving and throwing at this Coggeshall base, when he replied to a question on the quality of the plain fabrics he made in Coggeshall:

The president: Is that done in the best kind of work?
Hall: I could not boast; I think Spitalfields would make it much better; if I had to begin a broad silk manufactory I would begin in Spitalfields in preference to any other part of England."[42]

Like his colleague T. Sayer,[43] J. Hall demonstrated the incompatibility of management strategies for throwing, orientated towards stocks, uniformity and economies of scale, with those for silk-weaving, orientated towards varieties of rhythms, forms and volume.

The factory model also fascinated certain people in Lyons, including a businessman who was a great innovator and adventurer and whose failures were almost as impressive as his successes, Charles Depouilly. He emerged as a top-rank creator in Lyons and Paris in the field of fashion and mixed fabrics. In

[42] *Ibid.*, p. 380. [43] *Ibid.*, pp. 124–127.

1824, anticipating the official opening of the English frontier to French silks, he tried to establish a house "à l'anglaise" in England itself – i.e. a business involving the construction of a big weaving shed on the supposed cotton model in addition to the use of domestic workships. The best looms were to be bought and put under the direct management of the investor – and these happened to be Jacquards, which were just then being introduced into the country. His new novelty fabrics that were at the height of their global commercial success he would produce for stock, in order to establish an unassailable market position through the cheap prices made possible by large-scale manufacturing. He kept more than 300 looms busy in the Manchester area and only a few less in Spitalfields. Among them were 60 Jacquard looms. In two years his business collapsed disastrously, and the causes of the catastrophe were outlined in a Commission of Inquiry by one of his former agents, Emile Menetrier: his strategy of mass factory production for stocks had led to the creation of a mountain of unsold stock because he was not able to anticipate fashion trends in England or diversify his products in time. His failure resulted from excessive investment tied to a poor strategy.[44]

The baleful influence on British silk-making of the cotton mass production model was also evident in the areas of the spooling of warp threads and warping. After 1806, Spitalfields manufacturing integrated these operations within the establishment of merchant manufacturers, while in Lyons the spooling and some of the warping continued to be subcontracted out to workshops. British integration looked like a rationalization, saving on transactions and time, and Arlès-Dufour saw it as a model for Lyons. But this supposed saving on transaction costs was not without its disadvantages, for it resulted in a loss of skill on the part of the spoolers and a deficiency in the area of the matching of threads to the peculiarities of the further stages in the processing.

By contrast Lyons' efficiency, based on shifting linkages pointed up the inadequacy of Marshall's (1949) model of economies of scale, which in fact only rationalized the British fascination with the classic paradigm of mechanization–concentration–mass production as established in the 1820s. This paradigm was quite inapplicable to collective manufacturing where there was no continuous linear relation between size of firms and profits. In reality increases in the size of firms and the growth of a hierarchical centralism was actually liable to bring about a diminution in the scale of industrial operations and to cause a catastrophic fall in productivity. The contrast between big and small businesses had no meaning in this type of industry because in it industrial assembly could be effected through temporary cooperation between production units of all sizes, and on a bigger scale than would be the case with the integrated hierarchical firm.

[44] *Ibid.*, pp. 469–475; AML, F2 Fabr. de soies, aff. div., ACCo, Reg. des lettres.

Manufacturing styles, reputations and collective responsibilities

In the semantics of the five dimensions of qualities referred to above, the collective nature of reputations was embodied in expressions of the form "the Lyons style," "the London style," "the Nîmes style," "the Manchester style"; in other words, "the style of X (this or that collective manufacture)." Formally, X plays the part of an agent and functions as an identifier of the author. One does not speak of the styles of such-and-such a manufacturer, still less of the styles of individual weavers. This very particular modality of identification common to manufacturers of every type of fabric and regularly encountered in the documents under study, is highlighted for example in a dialogue between a commission president and a weaver of plain cotton from Glasgow. The weaver refers to fabrics made by British collective manufactures (in this case Glasgow and Bolton); he calls the fabrics "the same," first because they are from the same manufacturers though on several different sites, and then because they are of the same description. The president corrects him:

Do you consider that the sales at the Leipsic fairs are sales made generally on account of the manufacturers at Glasgow, or are they rather made on account of the Glasgow merchants or shippers?
The manufacturers in Glasgow . . . are only half shippers at best [*by purchasing Bolton's goods*] [. . .]
Then you conceive there is a competition in the Leipsic market of Bolton goods and Glasgow goods; . . .
I do not mean to say Bolton goods . . . I mean the Bolton manufacturer, whether he made the goods at Bolton or purchased them at Glasgow [. . .], from Perth or Aberdeen . . .
Are you aware that there is a distinction recognised in the Leipsic market, among the buyers of British manufactured goods, between Bolton and Glasgow manufacturers? [the emphasis is mine]
No, I never heard that there was any distinction; I mean to say that they are the same description of goods.
You mean when the goods are exactly the same quality?
Yes.[45]

The attribution of styles to the collective agents that were the manufactures of a place was of course not based on artistic criteria. But as in artistic production, style designated a relation of quality between product and author which left a *trace peculiar to* the agent. In the case of the silk industry in France and the British Isles, the treatment of fabrics by persons acting almost as individual agents was based on what amounted to local collectivization with a lack of differentiation between the *traces* of operations in the various figured products. Once the manufacturers addressed themselves in a very mobile way to the various dyers of the locality, to the various local designers, pattern-makers and weavers, etc., the

[45] PP. Hand-Loom Weavers, 1834, SC, p. 48.

particular trace of the operators in the product became a style common to the whole manufacture of the place, different from that of other manufactures, but indistinguishable within the manufacture as between one firm or weaver and another – though this indistinguishability was of course relative to the functioning of the markets.

The result was a reputation peculiar to the manufacture and a responsibility that local authorities were obliged to take on themselves. In fact, the better the reputation for quality of a manufacture, the greater the pressure to support that reputation. Reputations take a long time to acquire but are easy to lose. Manufacturing milieux were awash with edifying anecdotes about collective catastrophes engendered by the irresponsible behavior of one particular individual. It was sufficient for a manufacturer, profiting from the good reputation of such-and-such a product of the manufacture, to sell a product inferior to what was expected *just once* for the whole of the manufacture to lose a market for good. A reputation was clearly then a common good of a collective manufacture, not imputable individually to the economic units that composed it. In the language of institutional economies, it was a *specific asset*. In the theory of action it supposed a normative coordination by common orientation as opposed to the strategic mediation of the market. Its effects were visible on the level of the competition between collective manufactures on international markets.

Collective manufactures could be considered as quasi-agents, deciding to invest in reputation as long as marginal benefits exceeded marginal costs. But we shall then have to argue on the basis of quasi-accounts whereas there actually were no real institutionalized accounts. For example collective investment would lead to a common increase in the costs of training within the manufacture with the benefits being perceptible in the coefficients of the probability of orders for the samples presented to buyers. But the parallel must stop there, for the economic sanction has nothing to do with the sanction of real accounts.

On the contrary, the fragility of a reputation for excellence is manifested by the disproportion in strength between the real accounts of firms and the quasi-accounts of manufactures. A firm can make enormous profits if it goes for a "coup" – e.g. the lion's share of the orders for a type of fabric in demand, obtained by undercutting on prices and concealing a poorer quality which will only become apparent to consumers with use. The loss of a reputation common to a whole manufacture – a major factor in a common quasi-account – would make a minimal impact on the accounts of the firm in question in the light of the huge profit resulting from its coup. That is the danger of tactics that are parasitic on a reputation for quality, and no amount of guarantees or economic sanctions on individual firms can prevent them. Under the old regime in Lyons, respect for this discipline was one of the *raisons d'être* for corporative institutions, and very explicitly so. Immediately after the French Revolution, as in the British regime of the same period, only a certain unofficial discipline sanctioned by a coherent manufacturing milieu could avoid the danger. Such manufacturing reputations cannot be adequately analyzed in terms of the image of the *monopoly*, as has sometimes been done for a firm's investment in its reputation. Leaderships remain exposed to competition according to judgments of quality, quality that is constantly tested with each presentation of samples to buyers. The consolidation of Lyons, market leadership in fashions was not a mere monopoly of reputation; it had to do rather with that subtle economy of product variety expressed in sales on samples, something that we now need to look at more closely.

The economy of variety and its regulation: dynamism in Lyons, collapse in Britain

In 1835 the head of a great London luxury goods company on Regent Street, John Howell, importer of silk from Lyons, who liked to whip up competition between visitors from Lyons, Paris and London, explained the difference between the French and the British in respect of their systems for ordering silks:

> . . . it is usual for the Lyons manufacturers to come twice a year to England, that is, in the spring for the autumn, and the autumn for the spring, and they produce perhaps 200 or 300 patterns, not paper patterns, but silk patterns or gauze patterns, or whatever it may be, and from these patterns we make our selection; and it sometimes happens that we have so good an opinion of certain patterns, that we say: "Now you must withdraw that, it must be made for us only," and for 20 or 30 pieces they will do that. Now the English manufacturers never give us that advantage, they think it very expensive to put to work a pattern to show us the effect of it, for it looks so different on paper to what it is in reality, that we cannot decide whether we shall have it or not, and we often urge them to bring us a little piece ready, to see the effect of it; sometimes we want colour, sometimes we want a little change in the disposition; but there has always been an objection to the expense incurred, and therefore we are obliged to bear the expense if we are content to order from a paper pattern . . .[46]

This divergence between the systems of production and sale on orders in London and Lyons told the tale of a recent collapse in London manufacturing in the face of competition from Lyons. Before the deregulation of 1826, the two manufactures were based on the same marketing principles. They could both be described as industries *of commission* or *of orders* as opposed to industries *of speculation*. According to this distinction that was current at the time, industries of commission did not produce stocks but waited till they had the orders before commencing manufacture. Industries of speculation on the other hand manufactured in anticipation and then had to dispose of their stocks.

Markets based on samples were a solution particularly appropriate to the constraints of fashion and the commission system. From the point of view of the markets the political economy of fashion presents at first sight as the very type of the domain of *uncertainty*, in the precise sense in which uncertainty is contrasted with probabilist anticipation (according to Knight's definition, 1921); it is impossible to know whether the caprices of fashion will or will not adopt a given

[46] PP, Arts, 1835, SC, Howell, pp. 30–31; also Gibson, pp. 27–30, S. Smith, p. 23.

fabric pattern. The success or otherwise of a particular sample pattern was something of a lottery, as emerged from the British inquiry on design in connection with silk manufacturing.[47] However, the principle on which the great companies of merchant-manufacturers in silks operated consisted in going beyond this uncertainty and acquiring a power of investment sufficient to make up a large quantity of samples and to write them off against the small number of articles that would actually be noticed and ordered. Above a certain minimum level – here, 200 or 300 units – the collections of samples would anticipate a volume of orders sufficient to guarantee a minimum continuity of business. The rates of success of the varieties in a collection went from 1/5th to 1/10th. In London and in the other British silk manufactures the same system of production to order, without stock, had also predominated up to 1826. But it required of each firm a smaller variety by reason of the specialization by types of fabric; each pattern in each type was produced in greater quantities.

Orders from samples were not made simply on the basis of novelty value, but with regard to the five dimensions of quality referred to above (*taste, novelty, style, invention, perfection of execution*). *Pace* current theories of economies of varieties, models of competition and arbitrage between elements of quality and price,[48] these qualities were perfectly congruent and operated together to make products special; it was not the case that one of them could only be acquired to the detriment of another. Product differentiation was not the result of something arbitrary added to the pattern or of something arbitrary about its reputation but was based on an equation between the level of difficulty involved in the figuring and a judgment as to the perfection of the finished piece.

Such ideas were reflected in the particular forms of the markets. First of all the settling of fabric prices was the outcome of haggling, known in economic theory as *price making*. Then complex organizational provisions were developed, involving the establishment of benchmark *quotations* and *rates*. While the Lyons fairs were destroyed in the Revolution, other great international fairs, especially the two main seasonal Leipzig fairs, retained a leading role in the international comparison of offers on the basis of samples, and in the arbitrage between the *rates* of the different localities. Apart from the great fairs, comparisons of supplies with demands took place on networks of intermediate buyers or of manufacturers and wholesalers. The first of these networks had been in existence for a long time and lasted for the duration of the nineteenth century in Lyons. The second had begun to supersede them, as the geographical concentration of collective manufactures made possible unilateral visits by buyers to local manufactures. Such encounters between buyers and manufacturers are reminiscent of the way town center squares served rural markets: the public square was visited by buyers and was the location for assessments and quotations at auction prices; in a collective manufacture, visits and comparisons of supplies from different sources

47 PP, Arts, 1835, SC, p. 14.

48 Kevin Lancaster, *Variety, Equity and Efficiency: Product Variety in an Industrial Society* (Oxford: Oxford University Press, 1979).

by prospective clients took the place of advertisements of supplies and prices of transactions. At the same time in the years 1810–50 there also developed a system of visits by producers to wholesalers.

Worries about the final quality of products led to a second comparison of supply and demand on the markets after the execution of orders, when wholesalers or buyers checked the conformity of fabric rolls with the original sample. It is an important sign of the subjection of markets to quality that the rejection rates at delivery were high, and this created a climate of anxiety about the quality of the end product right up to the moment the price was accepted by the client. Rejected pieces went with remaindered fabrics to a secondary market held by specialized warehousemen, or more frequently were offered for sale at reduced prices. Here again the surprise of the laymen in the British commissions of inquiry led to clarification of this decisive point in dialogues like the following in 1831:

The president: Did I not understand you to say, that the trade of Lyons was an order trade?
J. Bowring: It is, to a very great extent.
Pr.: Then, how can that apply; if it be an order trade of course the order must be given to a certain number of pieces?
J.B.: It constantly happens, though the order is given, that the buyer refuses to take the goods, sometimes on account of delay in production, sometimes on account of change of fashion, sometimes for other causes which the dependence of the manufacturer on the buyer prevent his appealing against; there is a vast quantity of uncertainty about the matter . . .[49]

In the years 1792 to 1804 the commercial decline of France benefited English fashion and Spitalfields manufacturing. This was a period when the West saw the emergence of new conventions in male attire. Lyons silk manufacturing soon revived rapidly and recovered its global leadership in numerous sectors of fashion. Up to 1826, Spitalfields, protected by the prohibition of imports of French silks, did manage to retain its dominance of the English market. In the meantime the novelties sector expanded until it made up one half of the entire production of both manufactures.

The policy of deregulating British silk manufacturing in 1824–26 had as one of its goals the creation of an export trade like that of the cotton industry under the stimulus of international competition. But the inquiries of 1831, 1834, 1835, and 1840 revealed a picture of disaster: an influx of French products on the English internal market combined with the loss of the few outlets for English exports. After 1834 practitioners had to recognize that there was no tangible sign of the curative powers supposedly associated with deregulation. Instead, it was all too clear that the economy of variety had collapsed, and that there was no

[49] PP, Silk, 1831–32, pp. 560–561. Another witness, R. Clay, a wholesale importer, explained at the same inquiry that his rates of rejection for bad workmanship were of the order of 25 percent in ribbon and a little less in broad silks (p. 387).

possibility of investing in innovations in the areas of style, design and fashion. Only two areas were still in the hands of the British: firstly the least variable types of fabric such as certain furnishing fabrics linked with other London sectors; and secondly and more importantly very poor quality types such as printed bandanas that were caught up in the spiral of destructive competition, as the weaver J. Prout so rightly called it.[50] Where the English provinces were concerned Lyons now took precedence over London and the inquiries demonstrated how the Macclesfield and Lancashire weavers were now content to produce imitations of Lyons, in printed fabrics, as they had previously imitated London innovations (though in smaller quantities).

Two diagnostic elements are particularly revealing on this collapse of the London economy of varieties:

the absence of effective legal protection of patterns and novelties, a disincentive to investment in creativity and style, which was completely the reverse of the situation in Lyons.

the creation of a situation of industrial incapacity to produce samples and establish an economic design sector that could resist Lyons innovations.

A few elliptical comments from a London manufacturer sum up the gulf between legal design regimes in Lyons and London and its consequences for Franco-British competition:

President: At what time have you gone to France to purchase for your summer trade?
The month of February or March, occasionally in June, August and September.
President: If you could have a [French] pattern copied [in England] in three weeks, why do you not have it copied earlier in the season?
If I had sent a pattern to Coventry to be copied, it would have been all over the town before I could have introduced my own from France.[51]

In other words, the non-protection of patterns in England meant that it was in the interest of traders in figured stuffs and novelties to give their orders to

[50] In the years between 1800 and 1820 bandanas were either imported from the Indies or woven in superior qualities with dyed threads, especially in Macclesfield. But then competition with Lancashire drew them into a race for lower costs by the lowering of qualities: printed imitations, less and less fine warps, etc.: according to testimony from the years 1818–31, that led of necessity to a loss of the initial prestige and to the abandonment of the type by consumers. This is, in fact, what happened after 1845. (PP, Silk, 1818, II, Critchley, p. 81, Foster, pp. 87–88, Keefe, p. 117; PP, Silk, 1831, Brunskill, p. 33, Clay, pp. 385–389, Bowring, p. 532; Prout, p. 809, a Macclesfield weaver, formulated the dynamics of the process of destructive competition and at the same time gave an example of the fall in added values: "a great number of looms were employed in weaving a light and valuable article, which consumed from 16 to 20 ounces weekly, for which the workmen received from 20s. to 30s. per week; that branch of our trade is nearly annihilated, and instead of this description of goods, we are now employed in weaving bandanas and other petty articles, which consume from 40 to 50 ounces weekly, for which workmen receive as wages, from 7s. to 9s. weekly . . ."

[51] PP, Silk, 1831–32, p. 41.

manufacturers in countries where patterns were protected. Inquiries on design and copyright set out the reasons for the inadequacy of British legal provisions and expressed amazement at Lyons' efficiency without showing much sign of understanding it. All these are elements that throw light on a comparative analysis. For underlying all these differences was a growing divergence between the various collective disciplines of the manufacturers within the British manufactures, which itself reflected a new cynical individualism on the part of British firms, an *every man for himself* attitude that was not shared by Lyons manufacturers at this time. In fact, the protection of designs in Lyons, though based on specific legislation, did not function effectively except to the extent that there existed an area where manufacturers trusted each other and accepted a shared responsibility for the overall quality of the manufacture's output.

The protection of figured designs dated from old regime regulations while that of novelties had been elaborated by assemblies of merchants between 1795 and 1803, then adopted in a legislative measure in 1806 (law of March 18, section III). It presupposed new legal concepts that were in keeping with practice, for the current British-inspired French legislative principles were not suitable to deal with the problems of fashion novelties. Apart from matters arising from transactions, what was invariably at issue was the partition between a *public* sphere open to all and sundry and *private* areas where individual appropriation could be legitimate. In respect of patents the principle was to make the invention public but at the same time to allow the inventor to exploit it for a limited period – guaranteeing him a fair renumeration for his investment and talents – before *public utility* brought to bear its requirement of availability to all. Where authorial rights were concerned artists and writers naturally sought publicity, so the legislation simply guaranteed them the moral right of their status as authors together with the rewards that might go along with that. But where fashions and novelties were involved was it morally well-founded and economically practicable legally to protect the objects of the arbitrary choices made each season by the public? In particular, how to make compatible the effects of seasonal surprise peculiar to novelties, which were based on the mastery of secrecy and of the exploitation of the element of surprise, with a legal protection for innovations involving public registration and sanctioning the precedence of innovations over each other?

At the juridical level the response from the silk manufacturers was to suggest that the law should adapt to practice. This led to the innovation of legal protection for the seasonal novelty – with novelty status authenticated by experts in the profession – through secrecy: samples of novelties would be deposited under seal with the Prud'hommes for the duration of the period of their protection, and would be examined only in cases of dispute. In practice, this system functioned from the beginning of the *Conseil de Prud'hommes* system in 1806, but with correctives: instead of submitting disputes to the Commercial Tribunal (*Tribunal de Commerce*) after their examination by the Prud'hommes in conformity with the law, the manufacturers agreed with the two tribunals to establish a much speedier and less costly procedure: the Prud'hommes members became judges as well as experts, though submission to their judgments was "voluntary."

Protection of novelties was based on a much broader policy of fabric differentiation. Taxonomy hierarchized fabrics by great families (plains, figureds, novelties, velvets, shawls . . .), then by *types* (for example within plains, *gros de Naples, satins, taffetas*, etc.). Now the *type* was not just a practical descriptive notion utilized by the markets and for bargaining. It signified a normative value essential for the manufacture: the *type* was a kind of fabric that belonged in common to the manufacture, it could not become anyone's private possession either by patenting or by the depositing of samples. The originality of a product, exploitable by *one* maker in *the* manufacture, began only within a *type*. This necessitated a differentiation of products which had to do both with nomenclature and with actual transactions. Though the prohibition on the individual appropriation of types was not stated in the form of a law, it governed access to the status of objects of merchandise. It was consecrated in an unofficial manner by transactions but infractions of it were sanctioned juridically by the Prud'hommes and the Commercial Tribunal: these two authorities shared a determination to prevent individual appropriation of *types*, including what the milieu regarded as the creation of a *type*.

Faced with Lyons' efficiency, British fabric manufacturers in general began to regret the extension of *laisser-faire* to the area of design, for it was not just the British silk industry that capitulated to French novelties – even the proud Manchester cotton *free traders* had the worst of it on foreign markets at the hands of Alsatian makers of printed calicoes as a result of the latters' superiority in design. However the new legislative measures officially inspired by the French experience were quite impotent in the face of the every man for himself rule of the factory industrialists, a rule that was no longer questioned by anybody. This impotence represented a great weakness in the silk industry and its effect was reinforced by that of another – the inability to organize seasonal production of original fabric designs and samples economically.

The political economy of samples

After the deregulation measures of 1826, the traditional London economy of the production of novelties samples could not cope with French competition. In Great Britain this period saw a decline in the industrial energy of the *collective manufacture*. After the crisis of 1823–31, the making of samples became a very rare speciality skill requiring specialist looms that were equally few in number and that were progressively brought under the management of the rival firms. The result was that from now on most of the manufacturers had either no sampler or just one, while each Lyons manufacturer could effectively call on several thousand.[52] Moreover, the Lyons industry profited from the protective regulation of fashion patterns and so could sell several purchasers the same model to order, while the few original British novelties were carefully camouflaged by orders reserved exclusively for one single buyer, so as to avoid the piracy tolerated under the regime of every man for himself. Thus in the British Isles the logic of hierarchical concentration dragged down the level of sample production to such an extent that it was all but destroyed.

By contrast, the skill of thousands of Lyons weavers at sample-making aroused the admiration of British practitioners and researchers, and among them the lyrical eulogies of J. Bowring in particular have remained famous. This economist, a tireless worker in the free trade campaigns, a radical, and a friend of Bentham, established himself in London as an expert in the silk industry and questions of industrial design in the 1830s. Among his frequent and numerous depositions we may note the following:

> . . . the fact that struck me most in France was the way in which taste was formed, and I was exceedingly surprised at finding among the weavers themselves and among their children, and amongst everybody connected with the production of patterns, an attention devoted to everything which was in any way connected with beauty, either in arrangement or in colour. I have again and again seen the weavers walking about gathering flowers, arranging them in their most attractive shapes. I found them constantly suggesting to their masters improvements in their designs and I learnt that in almost every case, where the manufacturer had great success, there was some individual in the fabric who was the creator of beautiful things [. . .] The French manufacturer considers that his pattern is the principal element on which he is to depend for his success [. . .] I visited [. . .] the house of a weaver [. . .]; he turned over some hundreds of patterns, and said I will give you my opinion of these; I was much struck with the correctness of his taste, and I found they were able to prophecy pretty correctly what the success of a pattern would be. In this way is the mind of the silk weavers of France devoted to that to which the mind of those of England is not devoted [. . .] The number of them [their artists] is very great; the manufacturer does not depend on one or two or ten [. . .] the observations of the operatives are invaluable to the masters; I have heard manufacturers again and again say that the masters who stood upon their own sagacity, and had not availed themselves of the hints of their dependants had failed.[53]

The picture might seem too idyllic to be credible, but a study of the workers' press and the stories told at the Prud'hommes enables us to confirm J. Bowring's declaration: in the disputes that arose every week over indemnities for setting up samples, workers would put forward in the presence of experts suggestions they had made at their homes to the manufacturer (in person or else to his designer) about ways to improve the aesthetics of the projected fabric, or to correct costs: in

[52] From 1835 to the 1840s, every parliamentary inquiry into industrial art and design grappled with this impotence of the British in the face of the French when they tackled the silk industry. The best syntheses were formulated by Morrisson, S. Smith, Guillotte, Arts and Manufactures, 1835, PP, SC; J. Bowring, PP, Arts and Manufactures, 1836, SC; T. B. Holdway, Copyright of designs, 1840, PP, SC.

[53] PP, Silk, 1831–32, SC, pp. 532–535.

1835, according to Claude Guillotte, one of the two British builders of Jacquard looms and an expatriate from Lyons, labor costs in France were four to six times lower than in Britain.[54] But above all British compartmentalization between the artist, who drew patterns only on paper, the designer, lower in status and despised by the artist, and the weaver – whom the artist never saw – eliminated the essential quality required of the pattern-designer: anticipation of the effects of weave, touch, light, grain, colors and nuances. All British inquiries on design noted this defect. The diminution and then virtual disappearance of London innovators in design gives us another striking example of the impact of British pressure for the concentration of firms under a single direction: the result was a kind of movement towards the *enclosure of rare skills*, just as in the peasant economy resources under common responsibility in a village space could disappear. Within the Lyons industry by contrast, in the areas of the originality of the fabrics, the style of the manufacture, the quality of the *types*, or the availability of work skills, institutional innovation and unofficial rules of responsibility worked together to define and renew a boundary between a domain exploitable individually by companies, and a domain of normative competence common to the manufacture. Viewed in this context, the creation of schools of design in Lyons just after the Revolution on the initiative of the local notables looks like just one contributory factor among others and must not be overestimated. British inquiries on design saw in the Lyons schools a panacea that could bring an end to the inferiority of the United Kingdom. They led to the establishment of schools on this supposed model, but the results were disappointing, for the new individualism that dominated British firms had blinded the British researchers to the collective normative context of the Lyons schools. Failing to take due account of this, the individualist filter exaggerated the importance of the preliminary training of the young, not just in the silk industry but in the whole of Western "modernization."[55]

"Men of capital" against "men of credit": the social significance of accounts

This kind of methodological individualism could be expressed in the Weberian theory of firms. It reflected a centuries-old tendency towards autonomization of

[54] PP, Arts and Manufactures, 1835, SC, p. 56.

[55] On the importance of industrial design in general and the way it was handled in Great Britain in the years between 1830 and 1850, see Adrian Rifkin, "Les écoles anglaises de dessin: un succès contradictoire," in Collectif "*Révoltes logiques*," *Esthetiques du peuple* (Paris: La Découverte, 1985), pp. 113–129. In the Lyons municipal census of 1810, there were 66 design workshops and 42 "mistress-readers," employing 25 girls; in the 1843 directory there are 117 addresses of "offices" and workshops for pattern-drawing and design, with a workforce that was probably around the several hundred mark: several hundred more were directly employed by the manufacturers. On the competence and polyvalence of the Lyons designers in the eighteenth century, on their research trips in quest of the latest fashions, and on their frequent and difficult contacts with workmen, see the article on the Lyons designer Deschazelles (Louis Trenard, "Un notable lyonnais pendant la crise révolutionnaire: P. T. Dechazelle," *Revue d'histoire moderne et contemporaine* [1958], 201–225), and Claude Cinier, *Considérations sur le droit de propriété des dessins* (Lyons: Impr. C. Rey jeune, 1842).

the economic domain with the systemization of the markets/company accounts combination. From the point of view of current theories of action and coordination the *accountancy* aspect, less often stressed, was just as important as the *markets* aspect. I do not want to dwell over-long on the issues here but I shall briefly summarize a major shift internal to the social significance of accounts. It is another factor that helps to explain the weakening of the British silk industry under the impact of the mass production model. At the beginning of our period, in the years between 1770 and 1810, the Lyons and Spitalfields manufactures were still operating on very similar management and credit systems, but within two decades a divergence had developed; for while these systems continued to hold sway in Lyons till the 1870s, in London they began to lose ground over the years between 1810 and 1830.[56] In this area there was a close bond between the composition of capital, its mobilization and credit.

In both manufactures the fixed capital of the merchant-manufacturers was virtually zero – it amounted purely to warehousing and trading installations – but their circulating capital was substantial – it was essentially made up of the precious raw material undergoing transformation. An efficient practical method of management had been established in the eighteenth century. It followed a fairly classic logic common to several sectors and excellently analyzed in two superb monographs on printed textile manufactures of this period.[57] The source of this efficiency could be summed up in a formula: the velocity of rotation of capital. In fluctuating and unpredictable markets, at negotiated prices, strategic "ratios" were quickly reduced to the speed of transformation of materials and the extreme minimization of stocks. These formulae were similar to those of big business management in general, but with one significant difference: the London and Lyons manufacturers did not seem to give any priority to speculation on raw materials, which for them was subordinate to the rotation of their circulating capital, and shared with their creditor partners, the silk merchants.

In the contexts of the period, it followed that the most practicable formulae

[56] For Lyons this account is principally based on analyses of credit in the press and in local archives (AD, series M partially classified, including series 4 M). Scattered empirical evidence may be found in Cayez, *L'industrialisation lyonnaise*. For London the parliamentary inquiries indicated were utilized. For the political context of London business, we have used G. Ingham, *Capitalism Divided? The City and Industry in Britain* (London: Macmillan, 1984). For the commercial and banking history context, W. M. Clarke, *The City in the World Economy* (Penguin: Harmondsworth, 1965), and G. A. Fletcher, *History of the London Discount Markets* (London: Macmillan, 1976). An instructive comparison with the French banking context may be made with respect to the eighteenth century thanks to the fine monograph by Carrière *et al.*, *Banque et capitalisme commercial. La lettre de change au XVIIIe siècle* (Marseilles: Institut historique de Provence, 1976). On the French and European context in the first half of the nineteenth century, see the classic by Levy-Leboyer, *Les banques européennes*. Another classic, J. Bouvier, *Le crédit lyonnais de 1863 a 1882. Les vingt années de formation d'une banque de dépôt*, 2 vols. (Paris: Sevpen, 1961), bears on a later period, but gives precious insights on credit and banking in Lyons at the beginning of the Second Empire.

[57] Serge Chassagne, *La manufacture de toiles imprimées de Tournemine-lès-Angers (1752–1820)* Paris: Klinsksieck, 1971); Pierre Caspard, *La Fabrique-Neuve de Cortaillod, 1752–1854* (Paris: Publications de la Sorbonne, 1979).

for the mobilization of capital encouraged them to have only a minimum of capital of their own and to maximize recourse to the capital of others through credit systems. This made it possible to adapt the cash flow to the level of orders or, in other words, to have stocks of money as flexible and adaptable to demand as the stocks of products being worked on.

On both sides of the Channel trading credit was guaranteed by a combination of the long credits of silk merchants to manufacturers and the long credits of manufacturers to purchasers of fabrics to order. The Lyons fairs, destroyed in the Revolution, were only partially replaced by warehouse trade. The fairs moved to Germany, Switzerland and Italy and banking activity to Geneva and Paris. However the coherence of the milieu was gradually rebuilt and a system of credits developed that were relatively long on notes and letters of exchange at sixty then ninety days; the credits were often renewed directly (without the involvement of banks) between businessmen (silk merchants, silk manufacturers, buyers in silks). This mode of direct banking activity without bankers has to be understood if we are to understand the management efficiency of the Lyons merchant-manufacturers. A good part of the transactions in silks continued to create a high volume of credit-money negotiable without recourse to discount professionals. In this way the flexible monetary management of silk companies was linked to a direct mastery of the volume of what in those days the French financial milieu called "book money," a notion that could be regarded as characteristic of merchants' direct banking activity, by contrast with "bank money" created by networks with the mediation of professional bankers. (In England at the same period there was a contrast between "book credit" on the one hand and "acceptances" on the other, objects of "discount" organized by professional intermediaries.) The "merchant-manufacturers," heads of the big Lyons silk houses, could more accurately be called "bankers-traders-manufacturers." In 1845 in the course of putting forward proposals for reform a pamphlet defended the profession of silk buyer (124 houses had been set up in Lyons alone), by the synergy of its multiple functions: the grouping of orders, the provision of a system of credit for clients, face-to-face bargaining over transactions on the production floor.[58] Financial histories, obsessed with the unilinear evolutionism of banking "progress" through the specialization of tasks and influenced by the Saint-Simonians, did not see this link between the polyvalence of manufacturers, the manufacturers' efficiency of management and *ad hoc* bargaining on qualities. So the non-division of banking work was interpreted once again as an "archaism," or "backwardness" with respect to the British model.[59]

In London there were the same links between the management of silk manufactures and the mobilization of capital and they persisted up the decade

[58] Kauffmann, *Des cause locales*, ch. "On Long Credits," pp. 64–67.

[59] Especially after Alphonese Courtois fils, *Historie de la Banque de France et des principales institutions françaises de crédit* (Paris: Guillaumin, 1875).

after 1810, then rapidly regressed.[60] The decline was connected with the British monetary crisis of 1810–20. Its initial impact was a disaster for manufacturing, but then some of the practitioners managed to convert it into an opportunity for the future, again integrated into a vision of a historical necessity: the concentration of the control of fixed capital went in the direction of progress, it was connected to an accumulation of funds facilitated by recourse to the discount market and banknotes. It made possible large-scale economies on the costs of transactions. One of its champions, Th. Stone, head of the biggest house constituted in this way, explained this vision in detail, taking as his example the management strategy of his own firm: he simultaneously abandoned production on demand, the commercial circuit of prestige companies (known as "Houses of the West [End]"), the system of credit sales and purchases and the manufacture of fashion fabrics or figured stuffs. He launched into the production of low quality plains kept in stock, paid low wages and sold his production for cash to wholesalers ("warehousemen"). In the teeth of violent attacks from the victims of this system he proudly claimed himself to be one of those men of capital whose own industrial capital gave guarantees in case of bankruptcy, facilitated recourse to the short-term money market, and laid the basis for a new alliance with City Finance.[61] Echoing this "modern" manufacturer, a director of Morrisson's, international silk dealer John Dillon, also defended a new basis of trust in industrial matters against the old esteem for the manufacturer enjoying wide credit; the latter was lowered to the rank of a parasite who used money belonging to others.[62]

The substitution of one form of esteem for another is a shorthand way of describing the transformation that occurred in Great Britain under the impact of the success of the great cotton industry and its factory system. The change was not limited to the sanction of economic efficiency. It substituted one interpretation of social life for another and in public esteem and in debates on national economic policy it downgraded what had been considered a mark of excellence a few decades earlier. This change was observable in debates on British silk-making in the 1830s and was pushed forward in the cotton milieu in the years from 1810 to 1820; it received a theoretical justification and a generalizing evolutionist presentation with Babbage in 1835. Instead of the two systems being regarded as rival alternatives a law of progress was formulated by the process of filtering out the disadvantages of the one and emphasizing the advantages of the other. In a chapter "On the causes and consequences of large factories" Babbage established a convergence between masses of fixed capital, financing by one's own capital, the virtues of large-scale hierarchies, and the increase in the

[60] According to the inquiry of 1831–32, the credits of the silk dealers to the silk manufacturers were still for 14 months around 1807, they were then reduced to 10 months, then to 8 months around 1821, and to five months in 1826. Credits of silk manufacturers to their purchasers were reduced at the same time from 14 months to 5 and 2 months. The main depositions on the changes in the credit system are to be found in PP, Silk, 1831–32, SC, Banbury, Bridges, Dillion, Heath, Martin, Royle, Stephens, Stone.

[61] PP, Silk, 1831–32, SC, pp. 346–349. [62] *Ibid.*, p. 437.

"amount of well-grounded confidence" (p. 219). His remarks implied a rejection of the old system of book money, the system of direct trading credit, which was implicitly transformed into ill-grounded confidence. By contrast he praised the new hierarchical individualism based on the individual's own capital, a new cliché of trust-talk which also reflected a kind of enclosure movement – there was no longer any place for the old spaces of common trust in manufactures and direct credits within business milieux. So the old mutual control became insignificant and invisible – especially the old discipline in the area of qualities made possible by the extensive reciprocal dependence created by direct trading credits. In Babbage's view this meant that areas of confidence could only increase under the strengthening grip of the new captains of industry.

With this implicit commonplace the conditions for the pertinence of theories of "transaction cost"[63] were also created: the *raison d'être* of hierarchical stabilization in firms was to economize on bargaining. For this logic to operate two elements had implicitly to be reduced to non-values:

the *advantages specific to real wage agreements* that are lost by the suppression of bargaining: the irreplaceable functions of joint assessments, the advantages of discussion itself is an appeal to the peculiarly human qualities of autonomy and judgment.

the normative orientations constitutive of a common domain of the manufacture, implemented in the course of transactions and consolidated as necessary by jurisprudential life.

In the case of London in the 1830s, the positive value of these background elements tends to be denied: the new economic theory sees bargaining as a waste of time. Babbage contributed to the transformation. Transactions were treated by him from a purely negative point of view as costs to be economized, in his chapter on verifications, or, in expressions like the following, on the moral advantage of having great industrialists to manage great factories: ". . . high character supplies the place of an additional portion of capital; and the merchant, in dealing with the great manufacturer, is saved from the expense of verification . . ." (p. 219). This purely negative view of transactions, together with a purely positive view of hierarchical concentration, justified not only classic industrial concentration but also what might be called a methodological individualism of accounts. The absence of clearly defined spaces of trust outside the company, or more generally the elimination of any local axiology of cooperation, helped to make normative provisions proper to collective manufactures invisible and illegitimate. All that remained as a place of account-taking, in the most literal sense, was the firm's individual viewpoint, constituted as a semantic filter by the accounts.

[63] R. H. Coase, "The Nature of the Firm," *Economica* (November 1937), 386–405.

Bargaining

Both in Lyons and in London the question of fixing piece prices was a source of deep conflict within the silk industry. The great events embedded in the collective memory were linked to violent revolts over these prices; the Spitalfields riots of 1768–71 led to the Spitalfields Acts of 1773, when regulation of the rates sanctioned by the magistrates was authorized.[64] Later, there was a furious controversy over the abolition of the Spitalfields Acts between 1821 and 1824. In Lyons, official *tarifs* or price lists were decreed in 1779 and again in 1786, but when the central authorities refused to ratify these local agreements in the name of wage freedom riots ensued. These were much less significant however than the "revolt of the *canuts*" in 1831 – once again provoked by controversy over the weavers' rates. Negotiations between workers' and bosses' delegations, supported by the Prud'hommes and the local authorities, had led to the establishment of agreed *tarifs*. But a group of manufacturers complained and the central authorities played down the significance of the agreement, reducing it to an optional commitment, which provoked the fury of the workers and a series of demonstrations; then, as a result of the authorities' inability to contain the peaceful demonstrations, there was a semi-insurrectional situation of local seizure of power until the authorities came back in force (though without a pitched battle).[65] The second and more violent revolt of April 1834, though still more directly political and republican and in certain respects close to the spirit of English Chartism, was a logical outcome of the events of 1831: since a mandatory official *tarif* had proved impossible to impose, there was nothing for it but to band together and impose an unofficial one. However freedom of association had been abolished by the July Monarchy and so the need to guarantee such a price list led to the call for a democratic republican regime.[66]

The polemical heritage of these events has been to leave the significance of *tarifs* poorly understood even today. The antagonistic views of liberals and workers often combined to turn the *tarifs* either into outmoded survivals of antiquated conceptions, or into a heroic foretaste of the conquests of the labor movement. But these symbolic oversimplifications blinded them to the subtle links between the functioning of manufactures, individual bargaining, workshop economy and local public space. A comparison of interindividual bargaining practices, workshop economies and relations with magistrates clarifies this background to the standardization of prices. On this level the old *laisser-faire*/interventionism, free individual discussion/collective constraint, freedom of contract/protective labor legislative dichotomies simply dissolve away.

The capital element, an element too often misunderstood, was the "bargain-

[64] C. R. Dobson, *Masters and Journeymen, A Prehistory of Industrial Relations 1717–1800* (London: Croom Helm, 1980).

[65] Fernand Rude, *L'insurrection lyonnaise de novembre 1831, le mouvement ouvrier à Lyon de 1827–1832* (Paris: Anthropos, 1969).

[66] Robert J. Bezucha, *The Lyon Uprising of 1834. Social and Political Conflict in the Early July Monarchy* (Cambridge MA: Harvard University Press, 1974).

ing" over pieces. *Bargaining* meant highly effective negotiation and not the legal fiction that the "labor contract" came to embody later on. It set over against each other on the one side the "workshop head" or "master workman," and on the other side the "master merchant" or his "clerk." In London bargaining was similarly practiced between the "weaver" ("journeyman weaver" or "operative weaver") on the one side, and his "master," or the substitute for the latter, "the bookkeeper" or "foreman" on the other.[67] The partners negotiated to weave a warp and weft provided by the manufacturer in their home. They arranged the conditions regarding price and quality and delivery dates. There were two moments for negotiation and sealing of transactions: the manufacturer's delivery of the warp to be woven to the workshop head, then the "handing over" of the woven warp by the workshop head to the manufacturer. One opened the period of the completion of the commitment and the other brought it to an end. In the meantime numerous encounters might take place, either at the weaver's home, or at the manufacturer's establishment, in connection with the progress of the work; supervisory visits to the weaver's workshop to check on the way the weaving was going, with discussions about technical solutions of problems; and approaches by the weaver to the manufacturer when difficulties arose concerning the quality of the threads, the delivery of the weft, the stability of the colors; discussion of possible adjustments to deadlines and repeat orders, etc.

That such negotiations were real and meaningful was illustrated in Great Britain for example at the time of the campaign for the abolition of legal price lists. A few industrialists were hostile to the *Spitalfields Acts* official rates on the grounds that their rigidity prevented remuneration from being tied to skill. But their demands for clarification resulted in a clear demonstration of the way that bargaining and reciprocal agreements still took place on wage levels and working conditions even with fixed rates. The following dialogue between the 1823 Lords Commission and a London weaver, I. Grugean, on fabrics under a regime of standard rates, sums up the way different testimonies converged and adds implicit instructive elements that we shall study further:

A member (Mb) of the Commission – Do you consider yourself of the class of the most skilful?
Grugean: I believe I am so considered, I have been employed by one house for more than twenty years.
Mb: Is that house in the best branch of the trade?
G.: Yes, and I believe as good employment.
Mb: Probably during the time other parts of your family have been employed by that same house?
G.: Yes [. . .]

[67] In France the word "marchandage" became widely used between 1800 and 1850 to designate in particular interindividual haggling at the point of acceptance of orders. It was only after 1848 that the expression to some extent took on in the world of social law a pejorative connotation of exploitation which it particularly shared with the English notion of *sweating*.

Mb: . . . is it in the power of the masters to reward and encourage a man who is more skilful than his fellow workmen?
G.: Yes.
Mb: In what way?
G.: By giving him the works that will enable him to provide in the best way for his family. An unprincipled master, was it not for the regulations, might impose upon the public, and pocket a part of the money, by employing an inferior hand upon a superior work, [*allusion to behavior that was deceitful about quality*], but it is not now his interest to do so; and we are so situated, that none of those ever get work without a recommendation from a fellow journeyman; and I would never recommend a man for employment on a rich article, unless he had been employed in it: if he was incapable of it, he would injure me in my employ; and there is no interest on the part of the manufacturer to employ such.
Mb: In point of fact, you know that the most skilful workmen do get the most profitable employment?
G.: That is invariably the rule.
Mb: Have the masters the means of rewarding the men by giving the best employment to their families?
G.: Yes.
Mb: Do you find that the families of the most skilful men are better employed than those of other persons?
G.: No doubt; for the master will take care to give the best employment to the families of such men. There are many journeymen in our business have worked thirty or forty years, and never arrived at that pitch of skilfulness.[68]

Bosses and workmen to whom similar questions were put on "the power of the masters to reward and encourage a more skilful man" all replied like Grugean, that a regime of standard rates was the best for negotiating the best return, the most favorable deadlines, and above all the distribution of orders to the rest of the family or to the weaver's apprentices.[69] The manufacturer Hale summed up a piece of negotiation that was perfectly legitimate in his eyes:

A man who is very skilful, and a very good workman, generally has one or two apprentice boys, and if we find a very good workman, we give him not only the ordinary work we would give to an apprentice boy, but we give him a superior work frequently that is paid a penny, two pence, or three-pence a yard more. They will tell us "If I go to-another, he will give me two looms for my apprentice boys far superior to what you do."

The two testimonies implicitly involve typical elements of bargaining over standard and floating rates. A background factor in the discussions is the

[68] PP, Lords' Committee, 1823, CLVI, 57, pp. 73–75.
[69] *Ibid.*, p. 11. The question of the possibility of remunerating individually according to skill was put systematically to all those called to testify, employers and employees alike, who had experience of it, at the inquiries of 1818 and 1823. See especially: PP, Silk, 1818, Baker, Critchley, Elks, Gibson, Moore, Parker, Wilson; PP, Lords' Committee, 1823, CLVI, 57, Baker, Bennett, Gibson, Grugean, Hale, Jones, Keefe, Moore, Racine, Wilson. The theme was also set out in detail in PP, Silk, 1831–2, Brunskill; PP, Hand-Loom 1834, Longson; PP, Hand-Loom, 1835; PP, Hand-Loom, 1840, II, RAC, Mitchell.

unpredictability and great variability of each weaving operation: the employer, unable to predict the time that work on the warps will take, must accept the weavers' own estimate. The logic of these negotiations only becomes intelligible if we remember the coincidence between skill and speed referred to above. We can then understand how, even under a regime of fixed rates, prescribing such-and-such a remuneration for such-and-such a quantity of such-and-such a kind of fabric, the negotiations bear on tremendously variable levels of remuneration. If the rate for the work is invariable then variable factors become the strategic elements: the skill of the different members of the workshop, variations in the quality of warps and wefts, and especially, in the case of London, the arrangement of tasks that are more or less remunerative for the totality of the craftsmen in the workshop. If there is no fixed rate, the protagonists take going rates from earlier negotiations on similar fabrics, instead of taking them from "the tariff" or "the list." On the micro-economic level of bargaining, the difference is minimal. The mistake in the economists' anti-*tarif* rhetoric was to confuse the establishment of a going rate with the individual establishment of a price–performance link. For practitioners the issue was much clearer. "Free discussion" or "free labour" on the scale of individual negotiations consisted not in fixing the *tarif* for the work each time, but in assessing on each occasion how to fit the individual situation in with a standard rate. The difference was not between the existence or non-existence of a rate but between an official rate and an unofficial one.

The manufacturers were concerned about the quality of execution of the weaving, for they knew that the degree of "perfection" had direct repercussions on the final sale price, which was renegotiable at the moment of delivery, whatever the price agreed at the time of the ordering. This reinforced the weavers' power to bargain in periods when trade was good, if they were skilled at their job. One result was that there was a similar system for the recommendation of workers at times of changes of employer in London and in Lyons. Grugean's observations quoted above are revealing. The Lords were themselves astonished that when a worker changed employers recommendations were not made from one manufacturer to another; but Grugean and other witnesses confirmed that the employers had more confidence in the skill and expertise of qualified workers who were more likely to give an objective account of the skills of their fellows than a manufacturer abandoned by a weaver; for in the latter case, what would be the use of an assessment of a weaver by an employer when the two were at odds after a failed negotiation? Grugean explained on the contrary just how important it was for him to give rigorous assessments without any indulgence, *in order to maintain his own credibility with all his potential employers*. In a very similar fashion, another London weaver explained in 1834, long after the abolition of the Spitalfields Acts, that the same system of recruitment using workers' recommendations to employers, with objective and pitiless assessments, was still operative.[70] Such

[70] PP, Hand-Loom, 1834, John Duce, p. 590.

rigorous assessment in the hands of subordinates was a masterpiece of mobile negotiated hierarchical organization.

If in the blackmail reported by Hale we focus on the form "they will tell us: 'if we go to another . . .'," abstracting from its content, we will be isolating a very general element that is common to negotiation scenarios: the reciprocal presentation of other options, the contradictory assessment of the partner's alternatives to the transaction envisaged, up to resolution on the basis of a compromise – or not, as the case may be. This type of scene of interaction has features in common with price bargaining in public markets. In both cases there is a kind of theatrical minidrama over the conclusion or otherwise of a transaction. The rehearsal of these arguments about agreements in the Prud'hommes enables us to study them in detail from an "interactionist" viewpoint – some of the features of which will be summarized here. There was a whole art of negotiation in the way those involved played their cards over supplies and demands, put themselves in positions of strength in the argument, and dressed up their own proposals while presenting others' proposals in an unflattering light, etc. A system of visits at short intervals for the purpose of making comparisons made the spaces of the collective manufactures look like market squares. In both cases the discussion and settlement of prices was carried out under pressure from the public – in market places directly and in sight of everybody, while in the quasi-market places constituted by urban collective manufactures the system of visits for comparison at short intervals represented a kind of publicity, with a free quasi-general availability of information on the transactions of each.

In fact each individual meeting between weaver and manufacturer was prepared for and followed by the collection of information on the various current rates for fabrics that were either identical or very nearly so. The interest of each in this information was an incitement to the generalized exchange of it. Manufacturers and weavers had to be ready with each deal to say "if I go to such other . . .," "if I take such other . . ." Thus on each such occasion we find references to "fair prices" paid by "the most respectable masters," accepted by "the most honourable silk weavers." In other words we have benchmark prices and agents recognized by the whole milieu. When there was no official rate these typical prices acted as going rates. When official rates were in force, normal conditions applied, fair conditions for the sharing-out of orders and the distribution of income according to the real return on the taxed articles. A contrario, isolated country setups both in the Lyons area and in the English countryside raised problems that were often explicitly discussed about fixing prices "without knowing the rates."

Regulations in favor of freedom of bargaining in Lyons

Bargaining in which jobs were at stake, legitimate in principle both in Lyons and London, did not involve the same options in each place for changes of employer.

It operated in combination with distinct workshop economies and with different legal norms and it differed in the subtler conventions of negotiation. Here again a certain level of detail is necessary to account for the actual historical divergence of the two manufactures. In London it was the invariable norm in bargaining situations for there to be two or three craftsmen in a family workshop working for one boss. In the view of all witnesses – manufacturers and workmen alike – a change of employer was possible and legitimate with each new order. But it was all the looms from the one workshop that took on the new orders from the new employer. Simultaneous work for several employers was never represented as commonplace. The legitimacy of such changes negotiated by whole families was all the more remarkable in that the obvious interest of the manufacturers, frequently noted in commissions of inquiry, was to hang on to the most highly skilled – for reasons of regularity of finish, to ensure that a given article was always woven by the same hand. This system of mobility with its global employment of workshops managed to function coherently with the strong specializations in types of fabric that was noted above.

In Lyons however workshop polyvalence combined with a complex mechanism for the diversification of employers. The option of changing one's employer was not always a legitimate freedom, and in the period of the most severe regulation in the eighteenth century (1744–70) strong restrictions were applied on the basis of the "receipt tickets" (*billets d'acquis*) regime. Their strategic character was linked to seasonal cycles and workshop tactics. A workshop head almost always had several looms and the regulation in force between 1744 and 1789 limited their number under normal circumstances to four. In the mid eighteenth century the most typical bargaining situations as envisaged in discussions about how work was to be shared out presupposed that master merchant and master workman agreed wherever possible to keep all the looms active at the same time for the same order. But a substantial and irreversible development occurred, and discussions in the 1770s show the master workmen preferring to diversify types of fabric and employers as much as possible. Logic dictated a certain fidelity where benevolent masters had every interest in encouraging and rewarding the reliability inherent in a stable relationship, and such relationships continued to predominate in Spitalfields and the other British silk manufactures between 1800 and 1850. But in Lyons – for a century and a half from the 1770s to the demise of the manufacture – there was an autonomy based on combinations of work. Just as peasant family economies diversified by combining crop cultures, speculative alimentary cultures and speculative industrial cultures, so workshop economies diversified the risks of seasonal success and the rhythms of the regulation of accounts; they combined high-risk fabrics, very well-paid but extremely tricky to exploit (heavy start-up costs, uncertainty about covering costs), and risk-free fabrics of poorer quality, less variable, not very profitable, and guaranteeing a minimum of activity.

With the accentuation of the seasonal character of fashions, rhythms and rates, the option of having several employers became a conflictual issue at the heart of the bargaining process. At the start of the season, when the orders deriving from the preceding fair had been completed and buyers were prudently ordering new patterns for the next fair, the commercial destiny of a fabric was unknown, and with it the future rate; at the time of the international fair the whole of the manufacture was on edge to see how many orders the merchants would get. Very quickly everyone knew which patterns "had been appreciated" as news spread of the negotiations between manufacturers and workshop heads, "press" signs on this or that fabric, promises of repeat orders, and promises of premiums on the most urgent orders. At this point workshop heads who had gambled on the wrong patterns had an interest in changing their employer as quickly as possible, if necessary downgrading the priority of orders already being worked on. Those who had gambled on the right patterns and who had already delivered a length of warp were still not necessarily in an advantageous position. If a price had already been fixed before the fabric became a hit, they were bound by *custom* to keep to this price for the whole season, even though sales and labor costs had risen hugely. Hence again the considerable strategic importance of the possibility of changing one's employer. If a master workman could say to his first employer: "I prefer to work on the same type of fabric for another merchant since the rates he pays are one-third higher," the employer was in a weak position and could be obliged to accept a new agreed rate for the work in line with current levels; he was all the more liable to give way in that in silk manufacturing a weave was more "perfect" if it had been woven throughout by the same hand.

There were several ways that master merchants could limit master workmen's changes of employer and reduce their bargaining-power. From 1744 to 1770 they utilized regulatory weapons in combination with the weapon of credit. The regulation of debts, known as the "receipt tickets" regime was a source of perpetual controversy for two centuries. Credit for workshop heads organized within the manufacture had two functions; the encouragement of workshop economies, which was its avowed aim, and the restriction of bargaining options. To understand the changes in its regime between 1744 and 1790 – liberalization followed by the establishment of a regulatory system after 1807 – is to appreciate one of the strengths of Lyons manufacturing, something the British free traders could not or would not analyze when they were making their inquiries into the mysteries of the superiority of Lyons.

Receipt tickets embodied a regime of direct credit between merchants and manufacturing workshops and was analogous to *book money* between merchants. Similar to business credit, this regime organized the circulation of advances and loans with astonishing efficiency through a simple mechanism of signatures. Like business credit it created the possibility of mobility insofar as although the debt remained attached to the person of the debtor, there was a partial depersonalization of debts on the creditors' side, but the parallel stops there since in the workshops the credit system took account of the inherent poverty of the workers and operated without changes and without speculation or discount rates. Unlike with business credit, neither the weavers' own goods nor their moral credit could serve as pledges for advances. In the common regime of the popular milieu, the only possible loans were to cover expenditure on food and rent between paydays. But the economy of the Lyons silk workshops required more than this; it required the possibilities of advances on the price of equipment, the purchase of looms and multiple purchases of complementary equipment for each variety of fabric; in addition there were advances for food and lodging not just for the weaver's family but also for the one or two employees normally taken on by the workshop head. The originality of the system consisted in the way it pledged the reimbursement of the debt on the future work of the workshop head while allowing changes of employer. Unlike book money, circulation was not a mere transfer of the debt of one beneficiary of the right to reimbursement to another; the receipt ticket system meant that the right of reimbursement circulated among manufacturers at the same time as the possibility of reimbursement represented by the promise of work made by the workshop head who was the beneficiary of the advances.

Concretely the provision materialized in the obligatory circulation of small pieces of paper (either separate or in "booklets") called "receipt tickets" at the moment of the change of employers. When a workshop head wanted to have one or several of his looms operate for another employer, he had to present the new employer with a "ticket" signed by the most senior of the employers for whom his looms were working at the time and countersigned by the others; the ticket certified the state of his indebtedness. Either he owed nothing or if he was in debt the amount of the debt was noted and the new employer made a promise to deduct a portion of the future wages of the weaver to reimburse the earlier employers. The old regulations – and then a law of 1806 that remained in force right down to the demise of the manufacture – fixed the rate of the deduction at one eighth of the sum paid for the work. Here is an example of the format:

> I the undersigned Monsieur A, [the most senior among those supplying the orders for the pieces being made up in the workshop] agree that Monsieur C [future employer] give the work to Master O [the master workman, workshop head] on the two looms such and such were operating for me before, declaring that he owes me the sum of so much, on condition that Monsieur C pays me the 8th part of the money for the pieces and it is his responsibility to have this present countersigned by Monsieur B [another employer younger than Monsieur A who has looms working on his orders in the same workshop].

In the eyes of the protagonists the heart of the regulation lay in the workshop head's option to leave an employer freely even when he was in debt to that employer. A way of thinking long hallowed by common law held that an employer who made advances of equipment to a subcontractor or employee could legitimately demand in return the promise to reserve the use of that equipment exclusively for his own orders until the time of reimbursement. According to testimonies given at the inquiries in the years between 1818 and

1840 this was how the equipment advances made in Spitalfields and the rest of England operated. When business was booming employers offered advances for purchases to weavers, on condition that they could have exclusive rights to the weaver's work throughout the period of reimbursement, on low rates of interest. However a great many weavers subsequently changed their minds after having accepted this kind of deal and preferred to borrow more dearly where possible so that they could reimburse the manufacturers immediately and recover their freedom to negotiate; for they wanted to be able to play different employers off against each other on their new equipment. But the majority were not able to redeem themselves and indebtedness for workshop equipment became an endless servitude that put the workshop at the mercy of the lender.[71]

By contrast with these situations Lyons regulations diverged from the common law and evolved towards enfranchisement from debt slavery and a *positive freedom to bargain*. The numerous episodes in this development in Lyons can be presented schematically in four phases.[72]

(a) From 1744 to 1770: extremely severe restrictions on the right to leave place workshop heads in the hands of their creditors and deprive them of freedom to negotiate: in common law master workmen are bound to obtain prior authorization before changing employer, unless they reimburse the totality of their debt (see table 2.1 appended, juxtaposing the different key texts from 1744 to 1806 on the right to leave). Thus manufacturers profit individually from their advances to workshops: they impose a series of orders for articles on the best hands they have selected and keep these to themselves for years on end. From the point of view of the workmen, this asymmetry of the right to leave is usually described as "debt slavery."

(b) From 1770 to 1792: successive regulations and legal enactments favor a symmetrical power of negotiation, emancipating indebted workshop heads from obligations towards their initial creditor: according to the text of a compromise adopted by collective negotiation on April 17, 1770 a master workman, even if indebted, is "free to leave if he is not happy with the orders given to him."

(c) From 1792 to 1803: the industry initially suffered a collapse of markets and credit in the context of the civil war in Lyons (1792–95). Receipt ticket regulation disappears along with other local statutes, but the new theoretical liberty is also a total inability to make the industry work for lack of any basis

71 PP, Silk, 1831–32, pp. 416–417, 846–850; PP, Hand-Loom, 1834, p. 590; PP, Hand-Loom, 1835, p. 174, p. 188.

72 According to the evidence of local courts, *maître-gardes* and *Conseils de Prud'hommes*. As a complement, national and local archives, which for the Old Regime are mainly: AN, F 12 657, AN, F 12 764 A and B, AN, F 12 765, AN, F 12 766, AN F 12 1444. Jurisprudence encourages us to revise the great classic histories of the Lyons manufacture in the eighteenth and nineteenth centuries, Justin Godart, *La jurisdiction consulaire à Lyon* (Lyons: A. Rey, 1905), and Ernest Pariset, *Historie de la fabrique lyonnaise, étude sur le régime économique interieur de l'industrie de la soie à Lyon depuis le XVIe siècle* (Lyons: A. Rey, 1901). In the same revisionist strain, see the work of Carlo Poni and his contribution in the present work.

Table 2.1 *Regulations governing the right to leave where there is indebtedness*

1744 Regulation

Section X, Article 3. **If a master workman who is a piece-worker voluntarily stops working for a master merchant, he will be obliged to pay in cash** to the said master merchant all that he owes him; but if a master merchant stops giving work to a master workman who is in debt to him, after the said workman has paid for the materials, the said master merchant can deduct no more than one-eighth of the price of the labor on the pieces given to him, independently of advances he may have made on account for the said pieces. With respect to what is still owed in cash, the new master merchant who gives work to the same workman will be obliged to pay to his predecessor one-eighth of the agreed price on behalf of the said worker . . .

1770 Compromise

Section X, Article 3. **A master workman who has received,** after the publication of the present regulation, **advances** from a master merchant, **will be free to leave him, if he is not happy with the work being given him to make** by the said master merchant [. . .]
With regard to the sums that the said master workmen owe in cash, if the said master workmen, [. . .] want to leave, [. . .] all the sums which the said master workmen still owe, and all those that have been or will be advanced to them in cash subsequent to the aforesaid publication, can be deducted from the price agreed for work on the pieces that will be made for the master merchants by the said master workmen only to the amount of one-eighth; and if these master workmen should have a prior indebtedness to other former master merchants only one-eighth of the price charged for their pieces may be deducted, to be paid by preference to their most long-standing creditor.

1787 Compromise

Section IX, Article 3. [beginning is identical to Section X, Article 3 of 1744, "If a master workman . . ., agreed" then it adds:
[. . .] Nevertheless **if a master workman has legitimate complaints** to make against the master merchant for whom he has been working and to whom he is indebted and if he could not manage to do the work either because of poor-quality materials **or because of the lowness of the rate,** he must give a warning and must apply to the *jurés-gardes* for a ruling on the legitimacy of his complaints and he must be justified if appropriate by the damages he has to [. . . illegible] and by being made to settle up with his eighth . . .

1801 Compromise

Section XII, Article 4. **If a workshop head voluntarily stops working for a manufacturer, he is obliged to pay the manufacturer in cash all that he owes him**; if it is the merchant-manufacturer who stops giving work to a workshop head who is indebted to him, after the said workshop head has settled the account for the materials provided to him, the said merchant-manufacturer can require for his reimbursement from the merchant-manufacturer who is then giving work to the said indebted workshop head only one-eighth of the price of the work that the latter is doing, and must mention in the account that he will deliver to the new merchant-manufacturer that he is owed money by this workshop head.

Chamber of Commerce project of June 2, 1803

When a workshop head owes money to the merchant-manufacturer for whom he has stopped working, the person who wants to give him work **will promise to withhold one-eighth of the price of the work** on the said piece, in favor of the merchant whose debt is the oldest in the said account, and thus successively, **provided however that the workshop head has stopped**

working for the said merchant with the latter's agreement or for a legitimate reason. In the contrary case, the merchant-manufacturer who gives work to the workshop head will be obliged to pay the one who is a creditor in materials or cash, notwithstanding all this previous debt.

Costaz project of March 25, 1805

Article 7. **If a workshop head voluntarily stops working for a merchant-manufacturer** and does not pay him in cash all that he owes him, **no-one is to give work** to this workshop head, **without paying off the merchant-manufacturer** who is creditor in materials and money, notwithstanding all previous debts, and without prejudice to their settlement.

Project of April 18, 1805 and Law of March 18, 1806

Section IV, Article 31. **When the workshop head is in debt** to a merchant-manufacturer for whom he has stopped working, a person who wishes to give him work **will promise to withhold one-eighth of the price of the work** on the said piece, in favor of the merchant whose debt is the oldest on the said register, and thus successively **in the case where the workshop head has stopped working for the said merchant, with the consent of the latter or for a legitimate reason**; in the opposite case, the merchant-manufacturer who wants to give work to the workshop head will be obliged to pay off the person who is owed materials or money, notwithstanding any previous debt.

for credit between manufacturers and between workshops and manufacturers. Later, in the framework of the attempts at reconstruction, negotiations take place and partial agreement is reached (1795–1803), but they come to grief over the question of the tickets: workshop heads are opposed to projects for local agreements because projects for restoring the workshops' credit seem to them to revive the old arbitrary tutelage of the creditors of the years 1744–70.

(d) After fresh consultations between the Chamber of Commerce and workshop heads, a joint regulation governing the financing of workshops and freedom of negotiation is accepted, put to the Emperor and inserted in the law of 1806 establishing the *Conseil de Prud'hommes*. The decisions of the Prud'hommes consecrate an interpretation of the texts adopted that is the most permissive possible for the workmen: a workshop head may always take the initiative and change his employer, even if he is indebted, as long as there is no question of him concealing his obligation to reimburse. Consequently at the level of the industry's manufacturers, responsibility for the financing of workshop equipment takes on a collective dimension: the manufacturers who are giving the orders are *collectively beneficiaries* of the equipment and *collectively responsible* for reimbursement, without restriction by imputation to any one individual manufacturer's account. By contrast the workshop heads are individually responsible and free in their production and investment decisions. On this interindividual level of employer–weaver bargaining, the corrective action of the regulation takes into account the asymmetry of economic powers. As a result the regulations and judicial enactments restore to the workshop head (except in recessions) the option of rejecting one job in favor of another, at a time when the way his equipment is financed might tend to deprive him of that option.

Regulation and deregulation

At the time of the campaign for the abolition of the Spitalfields Acts rates, a myth became current in the official commissions on *free trade* and *free labour* in Lyons: it was said the Lyons labor prices were quite freely negotiable and that this helped to explain the lower export prices for Lyons fabrics. British officials imagined that the abolition of a *tarif* in 1786 had been irreversible.[73] The same tale was repeated in public discussion in France and Britain after the abolition of the 1831 Lyons *tarif*, and it was then taken up by the historians. But research into Prud'hommes archives and the professional press demonstrates the absurdity of this belief. On the one hand, up to 1831, the Prud'hommes supported by notables among the workmen and manufacturers, had always considered official minimum levels of remuneration for the weavers as necessary and legitimate, at least for plain fabrics, which were the most poorly remunerated. On the other hand, after the government disowned the 1831 *tarif*, the Prud'hommes replaced it by a "market price," a sort of official rate for the best companies, and often decreed individual wage increases in response to complaints from workers, when prices did not seem to have any equitable basis, and in particular when they seemed to arise from an "abuse" by the employer of the workman's necessitous situation.[74]

The regulation of prices as a provision against disaster: positive experience of regulation in Lyons, disastrous experience of deregulation in London

Comparison of the way these *tarifs* were justified in Lyons and London shows that the multiple economic functions of going rates were a known quantity before the lines were all blurred by classical economics. Subsequently the only function historians remembered was that of "protecting" wages: every historical account of social history for a century and a half has described how the price lists arose from a tradition of equity and solidarity defended by the workers in the face of the threat from the new "laws" of political economy. But this is only a small part of

[73] PP, Lords' Committee, 1823, pp. 191–194.

[74] Thanks to the work of Paul Truchon, "La vie intérieure de la fabrique lyonnaise sous la Restauration," *Revue d'histoire de Lyon*, 9 (1910), 409–434, and Rude, *L'insurrection lyonnaise*, based on the Lyons municipal archives, certain silk industry *tarifs* are well known: the *tarif* for plains of December 1, 1802, the *tarif* for plains and figured stuffs of June 18, 1811, renewed on December 29, 1817. My research on the *Conseil de Prud'hommes* has brought to light new ones that were apparently not known to the above authors: an employers' initiative for discussion of a *tarif* in 1795 (though we do not know whether or not it led to any definite result), re-edition of the 1802 *tarif* in February 1807 by the *Conseil de Prud'hommes*, adoption of a *tarif* by the *Conseil de Prud'hommes*.

the story, for there were other equally essential aspects to the picture that were specific to collective manufactures or to the silk industry. In Lyons the manufacturers themselves took the initiative several times of asking for mandatory official *tarifs*, justifying their action in terms of their own interests as well as those of the workers. For example in 1817 (letter of 11 December) at a period of grave economic crisis, a letter from the "principal manufacturers" to the President of the Conseil de Prud'hommes called for a *tarif* on the grounds of a threat to the manufacture arising from falling prices for labor, which were inequitable: for in a recession the less "honorable" manufacturers were tempted to fraudulent behavior in respect of quality, and some of them, "taking advantage of the wretched position of the majority of our workshop heads," imposed scandalous prices that would eventually ruin the workshops and the whole manufacture.

This hypothesis assumed that there were connected processes that led to disaster, a hypothesis that governed the regulation policy of the Lyons manufacture throughout its entire existence. One of these processes has already been described: "destructive competition," unleashed by behavior that was deceitful about quality. Impassioned debates between manufacturers had occurred in Lyons between 1795 and 1818 over the means to prevent this. A majority and then a substantial minority wanted to restore definitions of minimum norms of quality, in addition to minimum price norms for pieces. Finally mutual surveillance by direct credit, the commercial network of buyers and the judicial enactments of the *Conseil de Prud'hommes* and the Commercial Tribunal on imitations eventually established this policing of quality minima without the help of regulations. The contribution made by price lists was not understood by those unfamiliar with the manufacture either in Lyons or in London: the precise grids of minimum rates for pieces by types of fabric made it much less profitable to indulge in fraud over fabric qualities, for attempts at deceit over the habitual definition of articles could not evade the legal obligation to work with the benchmark minimum rates per piece of the authentic quality. By making it illegal to charge lower rates for lower qualities, *tarifs* discouraged practices that were deceitful about quality.

Disincentives to another *process leading to disaster* that was closely connected with the first arose from the nature of the silks market as a perishable goods market: a sales slump with a rash of price cutting could lead to the accumulation of stocks of unsold items and sales at a loss, with catastrophic effects for market equilibrium. This fragility was particularly noticeable when comparisons were made with the cotton market. Cotton fabrics, with a sale price ten to fifty times lower, could easily be disposed of at times of temporary recession by small price reductions, for demand was more durable and very sensitive to price reductions. By contrast in a recession the rate for silk fabrics could fall to a level quite incommensurable with production costs, given the urgent need to sell before the fashion changed. In this situation manufacturers might well be fearful of disaster for the whole sector, for Adam Smith's mechanism for the re-establishment of markets did not operate. It was necessary to prevent the accumulation of stocks that would be sold off at disastrous prices in cases of slump. Just as with the markets in perishable food supplies that are regulated today by the European Community, it was better to limit production and not sell below the *tarifs* than to have the markets collapse because of sales at a loss.

A third catastrophic process leading to disaster, treated as a major danger by the collectivity of Lyons manufacturers, was the bankruptcy of workshop economies. As it happened, the abstraction of a labor market composed of individual supplies and demands had no sense. The Lyons notables and silk institutions took responsibility for the equilibrium of the workshop economies of the manufacture, but not for any mechanism of weavers' supply and demand. It was necessary to avoid the catastrophe of workshop chiefs having to sell their equipment and means of work, purchased with advances paid by manufacturers. The receipt tickets regime described above was one response to this major danger: by giving manufacturers collectively a solidarity with the workshop equipment, and by maintaining the autonomy of workshop heads, Lyons regulation set up an original regime involving the decentralization of tests of economic equilibrium.[75] More precisely, it established a collective responsibility for the conditions of decentralized operation. The responsibility was exercised simultaneously by the Commercial Tribunal, by the Chamber of Commerce, by the municipality and by the voluntary employers' associations, especially the learned societies. The establishment of minimum *tarifs* fitted in with the general logic of the prevention of the collapse of workshop economies.

In London disaster came home to roost on a massive scale: the abolition of the Spitalfields Acts and the opening of frontiers brought about the collapse of the London manufacture and set the other manufactures on the road to decline. Yet the danger of such a catastrophe was proclaimed loud and clear by many voices at the inquiries of 1818 and 1823. In describing the functions of piece price lists under the Spitalfields Acts regime, manufacturers, workmen and traders gave (without knowing it) the same justifications as those put forward by the Lyons milieu. Price lists prevented the development of self-amplifying and destructive disequilibrium processes and collapses of workshop economies, they safeguarded a market for perishable goods, they prevented "destructive competition" over qualities.[76] The accusations of the adversaries of the Spitalfields Acts about their rigidity ("fixity") together with charges of irrationality or economic heresy deflected attention from the three-fold preventive force of the price lists.

[75] This notion of the decentralization of tests of economic equilibrium echoes the interesting article by Michel de Broey, "La possibilité d'une économie decentralisée. Esquisse d'un alternative à la théorie de l'équilibre général," *Revue économique*, 38, 4 (1987), 773–805.

[76] The depositions of manufacturer J. Ballance provided the clearest formulations of the economy of perishable goods and of the need to forestall the build-up of stocks that would be sold off at a loss (for example, in 1823, summary, p. 110). The same deposition explains the difference between silks and cotton markets. The same arguments from manufacturers W. Hale, pp. 10, 15, 75; R. Bennett, p. 85; complementary arguments on the prevention of the collapse of qualities by W. Hale, pp. 31–32 and by weavers Grugean, p. 75 and Keefe, p. 58.

The prevention of a catastrophic fall in rates was transformed by them into a rejection of the mechanisms of market adjustment. The function of policing minimum levels of quality was treated as "absurd."[77] In the language of today's institutional economic theory, the defenders of the Spitalfields Acts were describing the sources of an economy of variety that was incompatible with mass industrialization. But conversions to Ricardian orthodoxy and the conjunction of free trade with industrialism, generalizing the British cotton experience along with others, drove this into an irrational obscurity and hastened its fall.

The significance of the deregulation experience is still obscured even today by legends put about by Ricardian liberals which had been resoundingly disproved by the parliamentary inquiries of the period. The most caricatural example relates to the functioning of the price lists under the Spitalfields Acts regime. The Ricardian camp – with Ricardo himself, who took an active part in the controversies with the minister Huskisson – never tired of repeating that in Spitalfields work was overpaid because of its "artificial" prices (for example, cf. Ricardo's intervention as recorded in Hansard, May 21, 1823). But practitioners were in unanimous agreement that the regulated prices of the fabrics produced, far from being too high, were on the contrary either far too low, where *plains* were concerned, or else perfectly fair in the case of figured stuffs. It was a prominent adversary of the Spitalfields Acts, the manufacturer Thomas Gibson, who in 1823 involuntarily obliged the two employers' leaders in the anti-Spitalfields Acts campaign, Wilson and Moore, to accept this assessment of current prices and so echo the statements of all the other practitioners. Gibson wanted to show that the Spitalfields Acts were unfavorable to the workers because they artificially prevented prices from rising when demand was strong, yet never prevented them

[77] The manufacturers of the anti-Spitalfields Acts campaign subsequently broke spectacularly with the Ricardians and accused them of intellectual trickery when they saw only disaster in place of the promised prosperity. They then confirmed the rightness of the regulations as provision against disaster. Witness the declarations of Moore himself (organizer of the anti-Spitalfields Acts campaign) in 1831, when he contradicted his arguments of 1818–24:

Q.: Has the skill of our artisans improved since [1826]?
Moore: No; I think their workmanship has fallen off: such, at least, is the case in Spitalfields.
Q.: What do you attribute that to?
Moore: I attribute it to the low prices that are paid in Spitalfields, which have induced great slovenliness of workmanship. The Committee need not, I think, travel further than their own minds to know that that must inevitably be the case; whenever you pay a low price, you generally get an inferior article . . .
Q.: . . . is the capital of manufacturers and throwsters greater or less now than it was?
Moore: There is decidedly less capital employed in both branches of the silk trade than there used to be before the alteration of the system; very far less. Most of the throwsters are ruined, and a great many manufacturers also, and some of them are men of capital; but a great many of them are men of no capital.
Q.: Then it follows that the three points, on which Mr. Huskisson is supposed mainly to have relied, namely, the superiority of our machinery, the skill of our artisans, and the extent of our capital, have entirely failed?
Moore: *I consider it is an entire failure*, as far as those three principles are concerned . . .

(PP, Silk, 1831–32, p. 752; emphasis added)

from falling in a slump. On the terrain of workers' interests he failed to convince – in the face of the complete unanimity of the weavers, who were resolutely behind the Spitalfields Acts. By contrast, with the help of his own experience and his own accounts, he showed how a huge free sector had developed in figureds, made up of all the novelties introduced since 1806. This free sector which had never been regulated had much *higher* prices and was much more profitable for the weavers.[78] In the course of the debates and counter-interrogations, Wilson and Moore were obliged to admit that this was perfectly correct.[79] But this element alone was enough to wreck the Ricardian argument.[80]

Another aspect of the Ricardian legend made it seem that under the Spitalfields Acts regime prices were fixed arbitrarily by magistrates. But the 1818 and 1823 inquiries elicited numerous testimonies as to the real functioning of fixed price lists, which in actual fact happened to conform with the relevant theory: prices were fixed by collective negotiations between workers' and employers' delegations, set up according to *ad hoc* public procedures and only then made mandatory by the justices of the peace.[81] While such a procedure was widely rooted in tradition, we might also see in it, with respect to talk of "survivals," principles identical to those fundamental to twentieth-century French law on collective agreements, also found in many other continental European legal systems. The only notable difference relates to the way the power of extension is shared out: in modern France the administrative authorities hold a substantial part of the power of assessment formerly in the hands of the English justices of the peace.

The violence of the controversy may be explained in terms of an issue that people were very aware of at the time though it was never really aired in a rational manner: were the therapeutic virtues of *laissez-faire* infallible? With the most doctrinaire Ricardians experience could do nothing against principles, whatever their results. A caricatural expression was given to this by Arlès-Dufour, a fervent Ricardian convert: in 1834, commenting on the fall in the prices of pieces in

[78] PP, Lords' Committee, 1823, CLVI, 57, esp. pp. 126–128, 134–136.

[79] *Ibid.*, Moore, pp. 147, 148, 151, 156; Wilson, pp. 178, 188, 189, 190.

[80] Nonetheless, the legend of prices that were too high because of some "archaic rigidity" has persisted right up till today, perpetuated by reports from Royal Commissions, and then hallowed by an article of Clapham's in 1916 ("The *Spitalfields Acts*") which is still considered authoritative.

[81] Legally, in cases of disagreement the magistrates could fix the rates themselves. The anti-Spitalfields campaign endlessly repeated that the magistrates were using and abusing this power, while all the evidence of the 1818 and 1823 inquiries showed the contrary. This was one of the areas in which the manufacturers Wilson and Moore, leaders of the repeal campaign, were most often caught red-handed giving false testimony (1818 inquiry, pp. 153–154, and p. 164; 1823 inquiry, pp. 152–154; pp. 170, 171). In a more general fashion, the end of the 1823 inquiry showed that since 1820 the anti-Spitalfields Acts campaign had systematically boycotted collective negotiations in order to prove more effectively that they did not work. A counter-example of successful negotiation on an article at the end of 1822, with rationalization and flexibilization of wage grids, was given by three accounts of the same agreement (1823 inquiry, manufacturer Bott, pp. 91–93, weavers Holt, pp. 94–97, Webb, pp. 97–101).

Spitalfields since 1827, Arlès-Dufour explained that this collapse of labor rates proved that before the Repeal the London weavers were overpaid.[82] The circularity of the reasoning was perfect. Equilibrium arrived at through *laissez-faire* validates prices, prices arrived at by *laissez-faire* validate equilibrium. There is no further possible scope for a test of one by the other, nor for a comparison between action *ex ante* and sanction *ex post*, since a price set under a supposed *laissez-faire* regime necessarily reflects an equilibrium of "natural" prices, and since the distortion observed empirically with the earlier adjustments necessarily reflects the disequilibrium of earlier levels of the market.

This economic liberalism, not falsifiable in the Popperian sense, having affinities with a certain political authoritarianism, was not typical of all liberal economists. Another approach, an empirical one having affinities with a certain kind of political liberalism, illustrated by J. Bowring, did not deny the industrial disaster, nor did it abandon its fundamental principles, but immediately set off to do comparative research on Lyons and Spitalfields; this quest, which sought to analyze the failures experienced, supply an acceptable diagnostic and correct the policies being followed, was never completed. But these inquiries and others like them shed a great amount of light on workshop economies, and a hitherto unsuspected difference was at least partially brought to light – an enormous gap in productivity between the Lyons and Spitalfields workshops.

Workshop economies and their regulation: productivity, the family and the wage-earner

Already in 1818 some Lyons manufacturers had suggested that there was a big gap between the respective efficiency levels of the labor forces in Lyons and Spitalfields, but the Londoners themselves believed their workshops to be just as efficient as their Lyons counterparts until the day when the invasion of cheaper French products began to sow the seeds of doubt. However no one seemed to have any clear idea about precise levels of productivity. What became clear as a result of Bowring's inquiries in Lyons after 1831 was the difference in the manufacturers' labor costs: on figureds it was 50 percent at least and could be much higher, even up to 100 percent, beyond which comparison is no longer possible since the low labor cost and the high quality of Lyons figureds eliminated the possibility of any comparable fabrics being produced in Great Britain. The gap with respect to figureds could be explained in part by differences in techniques, especially by the absence of Jacquard looms in Britain until 1823, and subsequently by the persistent reluctance to improve them. We shall return to this in connection with comparative innovation policies. But in plains a considerable gap was also revealed which could not be explained purely by the automation of equipment. The example most often quoted for comparison was that of heavy-grained plains called *gros de Naples*, the most frequently made in

[82] Arlès-Dufour, *Un mot*, pp. 39–60.

London as in Lyons in the 1830s; their import into Great Britain made possible a simultaneous close comparison of price and quality.[83]

The inquiries of 1831 to 1835 confirmed that there were big gaps between costs in general and labor costs in particular. According to the most reliable inquiries, manufacturing costs of plain *gros de Naples* were 38% to 40% higher in London at the same quality and measured on the basis of current rates of exchange. Raw materials, which accounted for 67% to 75% of the final cost, were 21% to 25% dearer in London. The different processes (including dyeing, warping, etc.) which accounted for 25% to 33% of the final amount cost 49% to 90% more in London. Among them the weaving properly so-called, with the spooling of the warps, which was included in the Lyons weaving price, cost 82% to 100% more in London.[84] A comparison with plain satins gives the same sort of figures with a still wider gap: the cost of bottom-of-the range satin weaves including spooling of warps and "dressing" was 109% higher in London. The costs of bottom-of-the-range figured *gros de Naples* increased the gaps in the final cost in comparison with plains because of the greater input of labor. This part went from 33% to 44% and the price of weaving and spooling of the weft was 95% higher in London, which took the final cost to 47% higher than in Lyons.

The interpretation of this data gave rise to confusion. It seemed paradoxical that the London weavers could be paid so much more than their Lyons

[83] In this area, as with the Spitalfields Acts, a wealth of primary information exists, but it has remained in the shadow of superficial polemic or old misinterpretations recycled at second hand. So I have begun by noting from the data hundreds of occurrences that involved a few partial precise relations between specifications of quality, standard *tarifs* for labor, modes of working, returns, and physical quantities produced in a given time. In addition I have used grids of the London and Lyons *tarifs*, supported by technical specifications given in the research material and in the manufacturing manuals. Very few fabrics were directly comparable once I limited myself to qualities that were sufficiently precise. Among these "gros de Naples" were the object of more than twenty comparative depositions to British inquiries in 1831–32, 1834 and 1835. The qualities taken into consideration were determined by importers and traders, who thus essentially treated as equivalent two benchmark "descriptions," while agreeing that Lyons had the superiority in "execution": the "1000 to 1100 three double" in London, (3,000 to 3,300 threads, 3 threads in the reed, 2 threads in the tram) and in Lyons the "7/16 of 35 to 40 ranges" (7/16 ells wide, 2,800 to 3,200 threads) double or triple with two or three ends (two or three threads per comb tooth for a chain, tram made up of two or three threads).

[84] PP, Silk, 1831–32. The components of cost in Lyons were given chiefly by the depositions of Bowring, in addition by commercial agents of importers or manufacturers. On this basis London manufacturers drew up comparative tables with their own costs. For the figures set out above, recalculated on the basis of those tables after checks on their reliability by reference to meticulous witnesses, I have used Ballance, *gros de Naples unis*, p. 510, Wadden, *gros de Naples unis et façonnés, et satins*, pp. 510, 658, 660. By contrast I have eliminated the fantastic data given by Doxat, who flooded the commission relentlessly with tables but whose witnesses showed a lack of reliability. The gaps in costs shown up by Bowring, Ballance and Wadden, were observations made in autumn 1831, at the moment when Lyons prices were at their lowest. I recalculated the gaps by taking the prices in May 1832, when the London and Lyons *tarifs* were at the closest point they reached between 1800 and 1850. But this situation did not last: the Lyons rates started to fall faster than the London rates with the gains in productivity set out in the following paragraphs.

counterparts when London rates had undergone draconian falls since 1826 – around 50 percent in five years, much greater than in Lyons. London manufacturers and weavers unanimously believed that the level of prices for weaving work had become catastrophic for workshop economies and the 1826 levels subsequently seemed to all the practitioners to have been minima for equilibrium. But in Lyons, once the 1831 slump had passed, the workshops enjoyed much higher revenues, while signs of much better living standards abounded.[85] Of course everyone knew that the cost of food was lower in France, but comparisons of prices show that the extent of the differences in food prices – from 0 percent to 30 percent depending on the economic situation – do not significantly weaken the paradox of the differences. Likewise the general downward trend of food prices in the period 1816 to 1847 only brought the "real" fall in London revenue to one half of the "nominal" fall.

The key to the mystery of the discrepancies between the two industries lay within the grasp of the inquirers but it was not discovered. No doubt this was because at the time neither the Ricardians nor the practitioners managed to develop clear analytical concepts of labor productivity. Indications of productivity are to be found scattered through the research material I have used, but only incidentally, without giving rise to systematic comparisons and analyses. In fact such comparisons produce enlightening surprises. In the case of *gros de Naples*, the simplest comparison that can be made relates to individual physical productivity by loom. For London, in the case of the *three double*, the norm in 1826 was 3 yards of length per day in an ordinary weaver's average week. In 1831, controversies and scattered statistics put it at between 3 and $3\frac{1}{2}$ yards.[86] But at the

[85] The classic on the topic is Émile Villermé, *Tableau de l'état physique et moral des ouvriers employés dans les manufactures de coton, de laine et de soie* (Paris: Renouard), 1840, pp. 342–399, observations made in Lyons in February–April 1835 and May–June 1836. Villermé, very surprised himself by the much better standard of living in Lyons in comparison with all other textiles workers, refutes the pessimistic accounts of their lot in the national press after the 1831 and 1834 revolts. The numerous budgets in my collection of research materials confirm Villermé's impressions, once allowance has been made for variations according to hierarchies of skills and the vagaries of the economic situation.

[86] Speeds were most often given per week. It has been necessary to distinguish between records for a day, a "normal" day's work, an ordinary or continuous week, production on one warp, on several warps of the same pattern, or for long periods including or not intervals for unemployment, preparation, or setting-up. The history of London tariffs, never lowered between their beginnings and 1826, together with evidence of earnings, implies that there was no very perceptible progress in productivity on the level of workshops or individuals between 1769 and 1826. The norm of 3 yards for the *three double* is indicated by Ballance, p. 496. (Norms of speed in 1831–32 are given by Doxat, pp. 213 and 216–217). Speeds over three months can be calculated from Gibson, 1831–32, pp. 301–302; Stone, the biggest producer of *gros de Naples* in Spitalfields, p. 909, gives his tariffs and average payments for the inferior quality, which allow us to calculate the average, after a recalculation of gross gains on the basis of the net gains (or 3.2 yds per day for the *three single* working 6 days out of 7). In general, speeds must be calculated on the basis of weekly tariffs and revenue, taking account of deductions for expenses set out in the budgets. I have used 71 references to weekly revenue in the British inquiries, matched up with the price lists, leading to the speed margins indicated, excluding periods of unemployment but including time devoted to preparation.

same period the Lyons *tarifs* commission took as a basis for the double and triple with two ends (and the same number of threads per warp) an "ordinary day" of $3\frac{3}{4}$ ells to 4 ells (or 4.9 to 5.25 yds, speed 63 percent to 67 percent higher than in London), and the evidence of actual production shows that the everyday performance of "a good worker" was 5 ells (6.56 yds) or a speed 85 percent to 120 percent higher than the $3–3\frac{1}{2}$ yds of the Londoners. Gaps in productivity levels with respect to other qualities of *gros de Naples* and other plains were just as impressive.

It is no use trying to "explain" the collapse of the London workshop economies in the years from 1827 to 1834 with words like "decay" or "wretchedness." If we compare working methods there with the way things were done in Lyons, we can begin to understand how the London industry got caught up in a destructive self-amplifying disequilibrium process instead of a harmonious restoration of equilibrium as imagined by the Ricardians. The reflections of the weaver Wallis in 1834 give a concise summary of the enormous changes that were happening in London:

Wallis: . . . they [the weavers of Spitalfields] have been going down in their circumstances for the last six or seven years, and every year places them in a worse condition; they are less able to bear any depression of trade every year than they were in the preceding one; for when in work, the wages do not allow them to make good the necessary wear and tear in their household goods and apparel and so on, all their wages being expended in the common necessaries of life. [. . .]
Q.: What sort of a person has a weaver to get for an assistant?
Wallis: A girl generally, to prick and quill [. . .] but they do it mostly for themselves, in consequence of not being able to pay the wages to a girl or boy; [. . .] they do it now by one of the branches of their own families, their wife or some one in their own family, in preference to having another loom's work for them to be upon, for they find it more advantageous [. . .] He [the weaver] must work upon the average 16 hours a day every day in the week [. . .] but they would hardly do 6 yards; they would do 30 yards a week; that is about the rate [. . . p. 337]
Q.: What effect have the reductions of wages had upon the supply of goods?
Wallis: To increase [. . .] inasmuch as the necessities of the men, where they could avail themselves of it, cause them to exert themselves to a greater degree, and to employ sometimes one or two branches of their family.[87]

Those familiar with English historiography will recognize here the notorious features of the sweating system as it was described a little later on in connection with the manufacture of clothing and footwear: a family system of labor based on the home, outside the purview of all collective controls, forcing wages into a downward spiral through competition, and making all the members of families work harder and harder.[88] Wallis gives a good account of the symptoms of the

[87] PP, Hand-Loom, 1834, p. 335.

[88] After 1826 conditions in the Spitalfields workshops fit with the different criteria for "sweating" as put forward by James A. Schmiechen, *Sweated Industries and Sweated Labour. 1914* (London: Croom Helm, 1984), introduction. If we wanted to take comparisons further we could look at typical systems similar to the sweating system but regulated and collectively

economic shipwreck of the workshops: the progressively growing incapacity year by year to cope with economic downturns; the failure to modernize plant; and in particular a phenomenon more specific to the organization of the British silk manufactures and which merits a closer analysis: the redistribution of work to family members, linked to the virtual disappearance of employees paid by the workshops. Wallis points out that at times of recession, low prices, endemic unemployment and ferocious competition between workshops, weavers preferred to reduce the numbers of their looms, for if two members of the same family worked one loom it was possible to survive by taking lower rates for orders. Productivity per loom – all that was taken into account by individual manufacturers – clearly progressed because of the lengthening of the working day (16 hours) and week (7 days out of 7 instead of 6), which was made possible on a single loom by help from family members. This physical productivity by loom was of course a different matter from the individual physical productivity of the weavers. Thus the 30 yards per week mentioned above by Wallis were an improvement on the old production level of 21 to 28 yards, but they were produced on a single loom by 2 persons (i.e. 15 yards per person) whereas previously the two persons would have produced on two looms in six days a total of 42 to 56 yards, and for this the only help necessary would have been the part-time assistance of a child. By contrast in this year 1834 in Lyons, workshops with three or four looms, freed from related tasks by the use of outside employees who were much faster workers, could produce weekly, in days of 12 hours and weeks of 6 days, 40 to 50 yards of *petit gros de Naples* per person per loom.[89]

Thus calculations of labor productivity have to be read not just in the context of the external observer's own analytic framework but also and more fundamentally in the context of the *modes of "accountability" institutionalized among economic units*. In Lyons the communal responsibility of manufacturers for equipment and for the funding of looms on the one hand and commercial successes on the other encouraged the industry to incentivize not just returns by loom but also the individual efficiency of each member of the manufacture's workshops. If some of the looms or some of the workshop members were underemployed it would threaten the equilibrium of the whole, an equilibrium that was a matter for the whole professional collectivity and sanctioned by the rates at which the work was paid. Result: while in Lyons the productivist race and the improvement of techniques was the order of the day, the increased return on looms in London under the pressure of deregulation led to a reduction in weavers' individual productivity and in global labor productivity on the level of the whole manufacture.

controlled by local professional milieux. On this see my comparative essay on the footwear industries in Paris and London, "Problèmes de conceptualisation comparative . . .," 1990.

89 For more global analyses cf. the testimony of Spitalfields weaver Duce in the same inquiry (pp. 594–595); the manufacturer Ballance at the 1832 inquiry (p. 707). Evidence from accounts provided by Thomas Gibson (pp. 301–302) reflect the same trend towards family-based work involving several persons working the same loom.

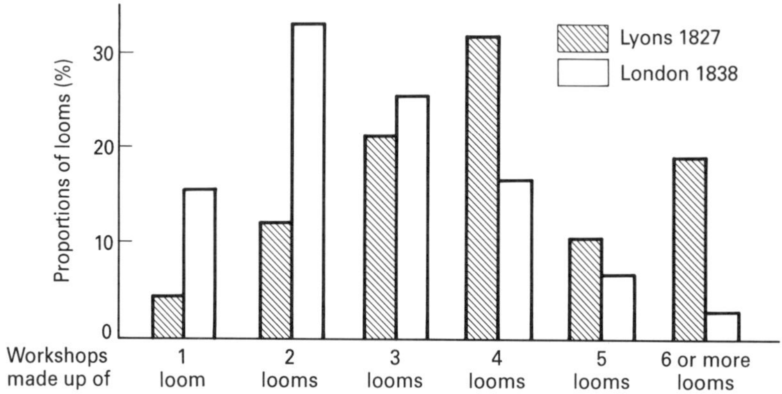

Figure 2.2 Lyons and London: distribution of looms by percentage, according to the sizes of workshops where they were located.

The contrast between the two workshop economies was reflected statistically in the distribution of the looms: the 1838 Spitalfields census revealed a predominance of workshops with just one or two looms, while in Lyons the looms were more often grouped in threes or fours for each workshop (cf. figure 2.2).[90] Such statistics reflect a divergence between a system embodying *an economy of extinction* or *sweating* without any scope for revival on the one hand, and a race for productivity and quality on the other. In Lyons workshop size was governed by the *productivity–skill–polyvalence* combination. The internal arrangement of tasks aimed at continuity of work and diversification of types of fabric, which meant making the best use of the time available and diversifying to cope with the risk of repeat orders not materializing, on the pattern of the style of management practiced by the manufacturers. Unlike in the British sweatshops[91] each of the workshop members' skills were boosted and honed. Workshop management strove to enhance everyone's strengths, taking account again of the speed–skill connection and of the financial and moral worthwhileness of perfection of execution. Wage-earning journeymen were very mobile and were liable to

[90] A table drawn up in accordance with the results of the census of Spitalfields looms carried out in July 1838, PP, Hand-Loom, 1840, II, RAC, J. Mitchell, pp. 219–228. For Lyons the figures come from one of the surveys I carried out in the municipal archives, a fiscal survey of households in 1827 covering 764 looms and 301 workshops. The figures of a survey for *la Croix-Rousse* for the year 1825 (1679 looms in 507 workshops), involving larger workshops, show an even sharper contrast with Spitalfields.

[91] An example of a journeyman's career is to be found in the autobiographical account by Truquin, published in 1977 – he shows pride in his continuously improving qualifications, from velvet plains to figured samples. In London, in addition to the numerous commercial testimonies and controversies on quality already mentioned, a direct testimony to the abandonment of encouragement of quality at the workshop level because of the leveling-down of *tarifs* was given by Duce, PP, 1831–32, p. 595.

change workshops with each order; outside periods of recession they were always in demand and encouraged to reach ever new heights of perfection, to the point that some could be recruited to teach particular types of fabric to the rest of the workshop. Statistically 60 percent to 65 percent of Lyons workshops included at least one waged employee.[92]

Workshops were not built up by progressive acquisitions of one, two or three looms as portrayed later in edifying liberal and republican accounts. The commonest scenario was for a young journeyman to set up his workshop with two, three or four looms. This is what was stated by the workshop chiefs' newspaper and it is confirmed by my statistical surveys: young workshop chiefs were proportionately more numerous at the head of workshops with two or three looms while it was much more often the older heads, elderly weavers or widows, who headed up workshops with just one loom. Far from resembling the sweating of the London silk industry, the small workshops reflected a reduction of activity on the part of elderly weavers who did not have a son or son-in-law to whom they could hand on the workshop and who would give them work.[93]

Maximum workshop size was a function of two rules. The first related to the organization of tasks: the returns on the workshop were optimal as long as the workshop head could devote the greater part of his own time to the actual weaving. This management rule had already been formulated by workshop heads in 1786, when they had worked out that the most effective workshop size was three looms. It was reasserted by the Weichmann inquiry (1848, AN F 12 2203), which took these maxima to four or five looms, when the workshop head could devote only a quarter of his time to the management and direction of the work of the others. Beyond that the workshop became less profitable since the head himself no longer did any weaving. Statistics indicate that workshops of more than five looms were either for velvet or for tulle in conditions that were very particular, less seasonal, less linked to diversifications and multivalencies – or else (in a small number of cases) for plains, employing as boarders ten to forty girls who were apprentices and minor journeymen.

This last important case, for ever being denounced violently and indignantly by workshop heads as the height of immoral exploitation, especially if the boarding hostels were run by the church, also involved the transgression of a

92 In 1810, a period when there was a great shortage of waged workers, in the suburb of Brotteaux 50 percent of the workshops contained at least one employee; in 1825, in Lyons-town, despite the extreme over-crowding, 59 percent of the workshops nonetheless contained at least one employee, while in the suburb of la Croix-Rousse, with big new blocks for workshops, the figure was 65 percent (according to my statistical surveys of the Lyons Municipal Archives and the Archives of the Rhône).

93 The employment of workshop heads by their sons or sons-in-law is noted in fiscal archives. On professional vocations according to kinship, see Yves Lequin, *Les ouvriers de la région lyonnaise* (1848–1914) (Lyons: PUL, 1977); what he calls professional heredity and endogamy are rather more frequent with workers in silk-weaving establishments than in other sectors.

second general rule, foundational for the economy of the manufacture: *The Prohibition of Exploitation of Journeymen and Apprentices by Workshop Heads*. This was perhaps the most remarkable rule in the history of Lyons regulation: it was maintained as taken for granted without ever being formulated and without any basis in a positive legal prescription. It was embodied in a strict custom: money paid for orders was shared out according to a fixed proportion, uniform throughout the manufacture, with no opportunity for workshop heads to interfere arbitrarily in the allocation. After the French Revolution, the rule could at the most be described from the juridical viewpoint as a *professional custom* sanctioned by the Prud'hommes. But like many other tribunal rules it contradicted the juridical doctrines of the judicial hierarchy and could function only by a consensus of the professional milieu. The consensus was made manifest by the fact that the members of the manufacture did not appeal against decisions based on these customs, even when their chances of winning looked very good. This prohibition helps to explain why, unlike many transformations of collective manufactures, Lyons workshops did not develop either towards a direct concentration of journeymen, or towards sweating: outside clandestine apprentice workshops, workshop-prisons and workshop-convents, the professional milieu successfully prevented the transgression of the rule; consequently a possible development in the direction of the concentration of journeymen or sweating lost any of its potential economic attraction.

For journeymen a 50 percent allocation was generally the rule – or sometimes in certain definite conditions a third. Similar customs had been in force in British manufactures but they had disappeared in the years 1810–20, while in Lyons they prevailed right up to the demise of the manufacture. For apprentices a more complex system of rules was imposed after 1806. Contrary to the post-Revolution deregulation decreed in France, the silk industry's apprenticeship rules were minutely detailed and their implementation overseen by the Prud'hommes. The manufacture replaced old regime constraints that imposed long fixed periods of apprenticeship (five to seven years) without remuneration by a mandatory period of four years. It organized the financing of training by credit in kind allowed by workshop heads to apprentices during the first year; reimbursement of this was effected over the following three years under the form of a fixed allocation of proportions of the money paid for orders between workshop heads and apprentices. Once the apprentices were considered ready for productive work they were put on what was called the "task" regime. They had a minimal "task" of two thirds of an "ordinary day" fixed and the workshop head took the money for all the work they did as repayment for his training and accommodation expenses for them. Beyond that, orders were shared out fifty/fifty between workshop heads and apprentices so as to commit the apprentices to advancing their own skills right from the end of their first year. The Prud'hommes peacefully settled thousands of conflicts between apprentices and workshop heads every year. In the context of a constant shortage of labor, this regulation sought to maintain the attractiveness of the craft by protecting apprentices' interests as much as those of workshop heads, and it succeeded in constructing a balanced political economy of professional training.

Once again, comparison with British practice and discussions is instructive: here a general controversy over apprenticeships ended with the famous abolition of statutes in 1814; following that, apprenticeship as regulated by the Spitalfields Acts, with its seven-year duration and its rule imposing a limit of two apprentices per workshop, was presented as an archaism from another age, unreformable and unjustifiable.[94] But the success of the Lyons system showed the possibility of an institutional reform of apprenticeships as advocated by practitioners and economists opposed to Adam Smith. It managed in practice to prevent the dangers of *sweating* and deskilling; dangers that Spitalfields manufactures and workmen pointed out in vain at the 1818 and 1823 inquiries, only to see their worst fears realized after the 1824 repeal.

The progress of handloom weaving in Lyons: a technological pathway where individual skill was at a premium, and one that was completely overshadowed in London by classic mechanization

The success of the Jacquard looms at the start of the Industrial Revolution has not easily fitted in with retrospective evolutionist classifications: where to situate this innovation within the handloom pathway, when the sense of economic progress was directing technology towards the application of mechanical energy, towards the division of labor on specialized machines and towards the hierarchical mediation of the control of machines and their mode of use? Among all the typologies of mechanization there needs to be a separate category for the Jacquards and the technical logic they represent. If it does not already exist, such a category may be created by combining two very simple notions: that of a "tool," i.e. an instrument that extends the intelligent operations of the body on things while allowing a return from the things to the senses; and that of an "automaton," i.e. an apparatus carrying out programmed guiding of operations. The combination of the two notions may seem paradoxical, since the intervention of the automaton between the agent and the object of the operations seems to create a rupture, a rupture directly opposed to the idea of continuity between the body and the sensations arising from objects as implied by the notion of a tool. But the apparent paradox merely reflects a tension that is in fact reduced by tool-automata, while machines embody a bias towards rupture.[95]

[94] Iowerth Prothero, *Artisans and Politics in Early Nineteenth-Century London. John Gast and His Time* (London: Dawson, 1979); John Rule, "The Property of Skill in the Period of Manufacture," in P. Joyce (ed.), *The Historical Meaning of Work* (Cambridge: Cambridge University Press, 1987), pp. 99–119.

[95] The category of *tool-automaton* today represents a prospective ideal for certain micro- and mini-computers, to the extent that their interfaces aim to restore directly to intelligent sensation (sight, hearing, touch) the coming and going between operating orders and signs of effectuation.

The invention of the Jacquards was of the continuity of the progress of the handloom. The "automaton" aspect had already been present for several centuries with the pedal mechanism for raising the threads. Depending on the level of observation adopted by the observer, two, ten or twenty inventions may be found to have preceded and prepared for the automation of the raising of the threads for figured designs according to the Jacquard process. Serious histories of technology all cite at least two inventions: (a) the perforated paper system, using needles to separate instantaneously the lines to be raised from the others with each passage of the weft; it was invented in Lyons by Bouchon in 1725 and perfected and made operational by Falcon in 1742; (b) around 1750, the Vaucanson loom for figured stuffs (never operational) installing above the loom a mechanism for the automatic raising of the threads: a pierced cylinder mounted on a truck and moving backwards and forwards and gradually rotating, struck the rows of needles so as to select the lines to be raised with each passage of the shuttle. The innovation of looms *à la Jacquard* lay in combining the Vaucanson mechanism and the loom *à la Falcon*: the selection of the threads to be raised was carried out by the automatic application of perforated cards using needles with the Vaucanson cylinder transformed into a prism.[96]

The automation of the choice of threads to be raised did not fundamentally affect the mobilization of *savoir faire* among the weavers. These tools still called for the use of basic skills in weaving plain silks and for the capacity to anticipate their multiple parameters. Just as it did before the spread of the Jacquards, the weaving of figured stuffs continued to involve a requirement both for accumulated experience and for expertise with increasingly complex weaves. The perforated cards did not alter the relations of power between manufacturers, designers and weavers. Designs had always been and remained a shared prerogative of manufacturers and designers to which the weavers brought their knowledge on feasibility and their corrections. As with its predecessors, work on the Jacquard required a minimum one- or two-year training followed by the possibility of constant improvement throughout the weaver's career. Hierarchies of *tarifs* reflected this hierarchy of skill by allowing for a progression of corresponding net incomes.

The automatism of the perforated cards was used not to reduce the burden of the old mental calculations, but to increase the power of this same individual mental mobilization and produce much finer qualities of fabric, much bigger designs, more varied nuances of color, and more complex effects of light while

[96] We possess abundant documentation on the innovations of the years 1740–98 in the Archives Nationales, F 12 642, F 12 1441, F 12 1444 to 1446, F 12 2182, F 12 2199, F 12 2201. All these dossiers have been analyzed and were the basis for the best survey still available today, that of Charles Ballot, "L'évolution du métier lyonnaise au XVIIIe siècle, et la genèse de la mécanique Jacquard," *Revue d'histoire de Lyon*, 12 (1913), 1–52. This is also the best reliable synthesis on the invention of the Jacquard, while numerous manuals of the history of techniques in French and in English go on repeating old legends which have no historical foundation except as phenomena of the kind of mythological heroization favored by the history of inventions.

reducing labor costs. The enormous gains in productivity brought about by the Jacquard were not therefore linked to some kind of deskilling, nor to some gain in the power of the manufacturer, unlike what may have happened in certain sectors of the cotton industry. They are of course attributable in the first place – as all the histories of techniques have emphasized – to the elimination of the draw girls. But they are attributable even more to the increase they brought about as individual tools in the power of the work that could be done. Furthermore the technical progress of the Jacquard looms was continuous and the phenomenon must not be portrayed as the timely eruption of a mechanism into the history of machines. A shawl manufacturer explained that eight weavers would have been required on eight 1818 model Jacquards to do the work of one weaver on one 1850 Jacquard, and these 1818 models had already multiplied several times over the capacity for work on the first truly operational models dating from the year 1807.[97]

Continuous innovations affected all aspects of the looms: preparatory design and setting-up operations, hooking systems, automatic shuttle change mechanisms on the battens, and the ergonomics of weaving – diminution of the mental and physical efforts required of the weaver and provision of facilities for the prevention of breakdowns or for their immediate repair by the weaver himself. In the face of these incessant innovations London and Great Britain never managed to follow the lead given by Lyons manufacture. The secret introduction of the first Jacquard into Great Britain by G. Wilson was not known about until 1821 while it did not begin to be used widely till 1823 and then only gradually – sixteen years after it had become commonplace in Lyons. Subsequently this handicap was never overcome – witness the unanimous judgments of universal exhibition reports, confirmed by analyses of commercial failures.

This difference in innovation rhythms does not apply solely to the Jacquards. It applied to all handloom techniques and helps to explain differences in productivity in plains, "armures" and "small fancy" as well as figured stuffs. The 1818 and 1823 inquiries established that there were no notable innovations in London. Wilson even claimed that the flying shuttle gained Spitalfields nothing, while when it was adapted for the Lyons silk industry in 1778 by Lassale as a "shifting box," it went on being perfected, especially by automation and modular adaptation to all types of battens. The main innovation observed in London, the "wooden draw boy" or "duck loom," introduced around 1804 and applied to "fancy weaving" – very much criticized by the Spitalfields weavers – seems only to have been a bad copy of one of the numerous automatic looms for "fancy weaving" developed in Lyons since the 1750s, in continuous use and subjected to continuous improvement up to their replacement by the Jacquards. Wilson even accused the Spitalfields weavers of having failed to adopt certain innovations that were in use in provincial manufactures. But from a comparative point of view what is striking is the backwardness of all the British handlooms, both in

[97] Exposition [1851] . . . XVth jury . . . Gaussen, p. 5; and Bezon, *Dictionnaire*.

the provinces and in London, compared to the Lyons looms. For example, a major innovation adapted by all the Lyons looms in the years between 1810 and 1820 was the "regulator" designed to make the yarn beams move forward automatically and gently, taking account of the growing thickness of the roll. In London it does not figure in lists of innovations, and Gilroy's technical treatise locates its introduction from France between 1835 and 1838, with its application being limited to power looms.[98]

The explanation of this innovation gap brings us back to the differences between the two collective manufactures and their regulations. Once again the London manufacture was subject detrimentally to the growing force of the classic model of cotton industrialization with the technical logic it involved. The idea of mechanization that was starting to become the norm in Great Britain, its model of factory exploitation, and the entrepreneurial appropriation of mechanical inventions left less and less room for collective innovation by the manufacture.[99] On this terrain the difference in the treatment of Jacquard looms in the two countries symbolizes the growing divergence of the economic spaces. The introduction of the Jacquard into Great Britain gave rise to a frantic race. W. Hale pointed out the importance of these looms to S. Wilson in 1816 (the two great manufacturers were still friends); he publicly announced his intention to import the loom and make it available in his own country. Both took measures to set up imports, but Wilson beat Hale to it and set up a temporary monopoly on the use of them, called a *patent*, in March 1821, in accordance with the law of the period.[100] The new *every man for himself* of the companies allowed Wilson to keep the secret of the machine and exploit it for himself out of sight of the rest of the world in his Streatham factory. He refused to pass it on and sell it to his fellow industrialists as long as he could keep it under wraps, which put a block on its diffusion for several more years, while William Hale had already been planning to make it widely available back in 1820.

In Lyons on the other hand local regulations put major innovations directly into the public domain of the manufacture. In a manner strictly comparable with the regulation of *types* of fabrics and its treatment of the border between what was free and *common* to all and what was sellable and *original*, the Lyons manufacturing chiefs followed a policy of immediate sharing of major inventions and mechanical processes, limiting their private appropriation to variations on each piece of machinery. This sharing was stimulated by the shiftirs organization of production. The activity of constructing the looms, perfecting them, selling them

98 Gilroy, *The Art of Weaving*.

99 This did not prevent the collective innovation of the manufacture from having kept an important place there: this may be deduced from numerous case studies, and from the global presentation by Raphael Samuel, "The Workshop of the World: Steam Power and Hand Technology in Mid-Victorian Britain," *History Workshops*, 3 (1977), 6–72, or from suggestions by Charles Sabel and Jonathan Zeitlin, "Historical Alternatives to Mass Production: Politics, Markets and Technology in Nineteenth-Century Industrialization," *Past and Present*, 108 (August 1985), 133–176. But in British textiles king cotton crushed silk.

100 *Journal*.

and maintaining them, constituted on its own a small collective manufacture that was structured on the pattern of the bigger one. Dozens of specialized workshops for mechanics, locksmiths, comb manufacturers, batten manufacturers, shuttle manufacturers, etc., operated in symbiosis with the polyvalence of the workshop heads. The logic of *tools-automata* in the service of productive skill was an encouragement to develop mechanisms such that the weavers could always be at one with their looms; that they could keep a constant watch on each of the operations by sight, hearing and touch; that they could anticipate problems or make them good immediately; and that the looms could be adapted to suit each person and each other.

In this kind of operational context specialized competition between the tool manufacturers turned on ability to control and adjust the assemblies that the workshop chiefs put together. As a complement to the specialized mechanical workshops hundreds of workshops also grew up for multivalent weaver-mechanics, making fabrics and marketing this or that technical process invented in the workshop.[101] The regulation of equipment credits produced a collective solidarity over loans and created a collective responsibility for the quality of the machinery while preventing anyone from trying to get exclusive rights to use. Unlike in the technological sectors of cotton, there was no necessity to put a watertight hierarchical constraint into the programming of automatisms. In the practice of innovation in Lyons there was no provision designed to bypass the weavers' *savoir-faire*.

Yet in order to fool the government that is just what the leading manufacturer Wilson claimed the Jacquard could do when he gave a deceptive description of it to the 1823 inquiry. He attributed to it all the official virtues associated with the cotton sector, and announced that the new machine would bring about a revolution comparable to that of Arkwright. It would be labor saving if there was a readiness to suppress price lists for the weaving. It would give rise to mass consumption, not only by virtue of the prodigious reduction in the times taken by individual operations, but also in that the skilled Spitalfields workers could be dispensed with and factory production launched throughout the Kingdom: "I think it [the French Jacquard loom] enables every plain weaver in the Kingdom to become a figured weaver, to make worts which we have the greatest difficulty in making in Spitalfields."[102]

In Lyons on the institutional level the management of the frontier between the communal and the privately exploitable was protected by complementary local

[101] In the directories there were 150 to 200 addresses for specialisms in the constructions of looms and there are indications of the multivalency of the weaver-mechanics, given as annotations of fiscal returns, as well as in the small ads in the professional press.

[102] (PP, Lords, 1823, p. 174 cited here). Other points about the Jacquard from the same witness pp. 173–175, 180–182, 184, 188–189, 192. To the peers involved in the inquiry, Wilson's bad faith on deskilling was all the more manifest in that in support of another argument he gave examples to show that his most skillful and best paid workers on figured stuffs outside regulation were those working on the Jacquard machines (pp. 188–189). But the Ricardians and then the historian Clapham believed Wilson's statements which were in full conformity with their philosophy of the history of mechanization.

authorities. At the center was the municipality. It alone could use a traditional law, brought back by legislation on patents, to buy innovations from individuals and put them into the public domain straightaway, instead of letting them be exploited privately by individuals – in the Lyons manufacture, great technical innovations were thus treated as true communal goods, with the town council accepted as having a right to manage the common patrimony. This was the manner of things in the management of the *Grand Fabrique* under the old regime and it was so again after 1795, until the liberal individualist wave of the Second Empire put an end to it. The agreement between the town council and Jacquard himself was based on these principles: Jacquard gave up the right to patents on his past and future innovations and left the fruits of his art to the community. In return the community provided him with a pension and with various advantages (AML, F2 Silks, Inventors).

At the same time there were active institutions for research policy, public expertise and the sanction of rules and customs. Research policy was dictated by voluntary or official learned societies.[103] These did not restrict themselves to the diffusion of knowledge and innovations; they provided a place for debates between the leading inventors and even attempted a policy for the orientation of work, launching research programs in the shape of public competitions with prizes given to the first to solve particular problems. Public expertise was assured by these societies, by the Chamber of Commerce and by the Prud'hommes. Finally the Prud'hommes constituted the main authority for arbitration and the sanction of rules and customs. It also followed a policy of efficient technological modularity. By imposing calibers of all kinds it made compatible or interchangeable a great number of elements of the looms in the manufacture. Thus it guaranteed technological conditions favorable to the polyvalence of the machinery and to its continuous incremental renewal at the lowest cost.[104]

The ability of all these bodies to contribute to a coherent common policy was facilitated by the rotation of the silk notables through all of them. In the case of Jacquard, for example, it was the learned societies who took care to ensure the public availability of his process. They encouraged competition among loom-builders within the trade and brought public recognition for the respective

[103] In particular the Société des Amis du Commerce et des Arts, a voluntary society, and the Conseil du Conservatoire des Arts, a municipal institution for teaching and a living technological museum created on the model of the Conservatoire National des Arts et Métiers.

[104] The Council adopted calibers and standardization for the loom components, for the velvet irons, for the silk floats etc. The most spectacular result was without doubt the standardization of perforated cards, noted by a former Prud'hommes councillor, the manufacturer Paul Eymard, *Historique du métier Jacquard* (1863): "The most convincing proof that these successive inventions were borrowed from one another is that a Jacquard card in use today may be applied both to Vaucanson's planchette with needles and to Falcon's, and the match is so good that Falcon's initial matrix must have fixed the dimensions . . ." The effects of this mode of card-piercing, supplemented by the standardization of *lisages* and supports, were comparable to what could easily have been the case today if computer operating systems had been standardized from the start and made cumulatively compatible as they progressed.

contributions of each. For its part the Prud'hommes protected workshop economies, regulated their equipment credits and kept a check on the fairness of contracts between mechanics and workshop heads. In exercising these functions it was led to condemn Jacquard many times, for one very simple reason that hagiographers of the inventor have never faced up to: the looms built by Jacquard did not work.[105] He never managed to make his invention properly operational. It was down to other loom-builders, freely incited to emulation by those same notables, to be the first to succeed in making Jacquard-type looms that actually worked. They sold hundreds of them between 1807 and 1811, while Jacquard built only fifty-seven, many of which he had to take back, paying damages and interests to the workshop heads: the Prud'hommes did not allow deception of workshop heads on the quality of machinery. By eliminating poorly-designed Jacquards and facilitating credits for the most efficient looms, the Prud'hommes, in the name of equity, gave a decisive impetus to innovation and to the retooling of the Lyons workshops under the First Empire.

Such are the main features of the collective silk manufactures in Lyons and London in the first half of the nineteenth century. They had common forms of coordination. But the classic path of industrialization in Great Britain based on the cotton model weakened the London manufacture to the point of causing its collapse. The real dynamics of the manufactures have been lost from view behind a smokescreen of old polemics and it was necessary to rediscover their originality and the conditions of their success. In the process many other intellectual issues have emerged. Profiting from a renewal of the economics of institutions, commentators have indirectly suggested how the links between so-called classic political economy and the triumph of the cotton model could have provided the illusion of the economic, the illusion of a *purity* of its laws in an institutional void. The recovery of the experience of the silk industry is an invitation to relativize this purity and relate its more general concepts to individual institutional and material conditions proper to the cotton industry in the British context.

On a more implicit level the details of economic coordination lend themselves to a major debate in the theory of action. According to Habermas' *Theory of Communicative Action* (1986), modern forms of coordination of action are split into two: firstly, systemic coordination, which economizes coordination by the

[105] A plethora of subsequent mythological accounts from the 1830s on fantasized about an imaginary controversy over the invention of the Jacquard looms, which were supposed to have aroused a hostility on the part of the workmen which in fact never existed and which has no other foundation than the heroic literary genre on the great inventors as cultivated among engineers. The only real basis for the myth in the history of the Lyons workshops was the episode of the condemnations of Jacquard for abuse of trust. But the hostility was directed against the poor quality of certain looms, which is quite a different matter from hostility to their diffusion. The best account of these questions is still today the article by Ballot, "L'évolution du métier lyonnais." However Ballot was not able to study the local archives and knew nothing about the *Conseil de Prud'hommes*. The present study has relied on previously unknown archival materials that can now be added to the dossier on this old controversy, to be interpreted in the framework of the Prud'hommes' jurisprudence; the details will be set out in a work currently in preparation.

mediation of money and law, thus limiting itself to strategic information, and leaving orientation to the free choice of personal preference. In my comparative study, transfers of competence to the market and to civil liberty, in a climate of emancipation with regard to traditional authorities, was related to this dynamic. The second form of coordination, *communicative action*, rooted in the "lived world," far from economizing intercomprehension, aims at a common understanding and orientation in action, this regulatory aim including the acceptance of conflicts. This was the case with normative life in the Lyons manufacture, constitutive of a real "local public space."[106] But historical inquiry encourages us to take questions back to the model: the old British failure of the liberalization of the silk industry, attached to the success of the British cotton industry like its shadow, echoes contemporary doubts on the systemic self-sufficiency of the market. Moreover the equilibrium invented by the Lyons industry between decentralized coordination and a common normative orientation, is also not unconnected with current questioning about the system/lived world disjunction. While the Spitalfields manufacture was dismembered under the impact of authoritarian technocratic liberalism, the Lyons manufacture successfully carried out an original experiment with the domestication of law and the market, which the revolts of the *canuts* actually consolidated in the teeth of the doctrinaire economists.

APPENDIX 2.1 RESEARCH MATERIALS

The following appendix covers research materials on Lyons and London that have been subjected to database analysis. Thousands of files have been assembled on every aspect of the life of the manufacture, incorporating extremely precise details on interactions, labor and techniques. The general approach was not to code meanings initially in order to assemble them in frescoes but rather to explore meanings in context without making any prejudgments at the moment of entering the data. Informatics makes it possible instantaneously to rearrange and compare thousands of details and lends itself to an exploration of local modes of thinking and the meanings of local interactions in context. One part of the "sources" that could involve tens or hundreds of references for each segment of analysis could not be included in the notes and is noted globally here. This way of referring to sources, which is a consequence of my working methodology, is admittedly not very satisfying; better methods will no doubt be available to us in a few years, when alongside printed editions there will no doubt be electronic editions of collections of materials as a "hypertext," so that the reader will have the option of reconstituting, modifying and testing the researcher's mode of proceeding with the data.

The Lyons collection, apart from the manuscript archive sources referred to in the notes, is made up chiefly of fourteen workshop heads' local newspapers published between 1831 and 1853, and containing accounts of Prud'hommes' sessions, retrospective articles, supporting evidence for the claims of workmen or merchants, debates with the national press, technical articles and professional small ads:

[106] Alain Cottereau, "'Esprit publique' et capacité de juger – la stabilisation d'un espace publique en France au lendemain de la révolution," in A. Cottereau and P. Ladriere (eds.), *Pouvoir et légitimité – Figures de l'espace publique* (Paris: Ed. de l'EHESS, 1992).

L'Echo de la fabrique,	1831–1835
L'Echo des travailleurs,	1833–1834
L'Indicateur, journal industriel de Lyon,	1834–1835
La Tribune prolétaire,	1834–1835
Le nouvel Echo de la fabrique, journal industriel de Lyon,	1835
L'Union des travailleurs,	1835
Le Vigilant lyonnais,	1835–1836
Le Lyonnais,	1837–1843
L'Echo des ouvriers, journal des intérêts de la fabrique et de chefs d'atelier,	1840–1841
L'Echo de la fabrique de 1841,	1841–1845
L'Echo de l'industrie,	1845
La Tribune lyonnaise,	1845–1851
L'Avenir,	1846
Le Moniteur de la fabrique, journal spécial des chefs d'atelier, écho des prud'hommes	1852–1853

The Spitalfields collection consists mainly of the following parliamentary inquiries (abbreviations on the left):

PP. Silk, 1818. I. Great Britain Parliament. (*First*) *Report from the Committee on Silk Ribbon Weavers Petitions*. June 3, 1818.

PP. Silk, 1818. II. *Second Report of Minutes of Evidence heard before the Committee Appointed to Consider of the Several Ribbon Weavers Petitions Relating*. April 30–May 8, 1818.

PP. Lords' Committee, 1823. CLVI. 57. *Minutes of Evidence before the Lords' Committee on the Bill to Repeal Four Acts Relating to the Wages of Persons Employed in the Manufacture of Silk*, Lords' sessional papers. CLVI. [Hansard 1823] (H.L.36). CLVII.

PP. Silk, 1831–2. *Report from Select Committee on the Silk Trade with the Minutes of Evidence* 1831–1832.

PP. Hand-Loom, 1834. *Report from Select Committee on Hand-Loom Weavers' Petitions: With the Minutes of Evidence* . . . 1834.

PP. Hand-Loom, 1835. *Report from Select Committee on Hand-Loom Weavers' Petitions: With the Minutes of Evidence* . . . 1835.

PP. Arts, 1835. SC. *Report from Select Committee on Arts and Manufactures, House of Commons*, 1835. V.

PP. Arts, 1836. SC. *Report from Select Committee on Arts and Manufactures, House of Commons*, 1836. IX.

PP. Hand-Loom, 1839. I. RAC. *Report from Assistant Hand-Loom Weavers' Commissioners*. Part I.

PP. Hand-Loom, 1840. II. RAC. J. Mitchell *Report from Assistant Hand-Loom Weavers' Commissioners*, Part II. Report by J. Mitchell . . . on the East of England.

PP. Hand-Loom, 1840. II PP. RAC. A. Austin *Report from Assistant Hand-Loom Weavers' Commissioners*, Part II. Report by A. Austin . . . on the South of England.

PP. Hand-Loom, 1840. III. PP. RAC. S. Keyser *Report from Assistant Hand-Loom Weavers' Commissioners*, Part II. Report by S. Keyser . . . on the West Riding of Yorkshire, Macclesfield, etc.: and on Germany.

PP. Hand-Loom, 1840. III. PP. RAC. *Report from Assistant Hand-Loom Weavers' Commissioners*, Part III.

PP. Hand-Loom, 1840. IV. PP. RAC. *Report from Assistant Hand-Loom Weavers' Commissioners*, Part IV.
Hand-Loom Weavers, 1840. IV. PP. RAC. J. Fletcher *Report from Assistant Hand-Loom Weavers' Commissioners*, Part IV, by Joseph Fletcher.
PP. Hand-Loom, 1840. V. PP. CR *Report from Assistant Hand-Loom Weavers' Commissioners*, Part V.
PP. Copyright, 1840. PP. SC. *Report from Select Committee on Copyright of Designs*, 1840, vol. VI.

APPENDIX 2.2 COMPARATIVE STATISTICAL DOSSIER

Table A1 *Number of looms, estimates for the urban areas of London and Lyons*

Year	*London*	*Lyons*
1802	10,000	11,200
1807	11,000	12,000
1815	12,000	14,500
1824	14,000	24,100
1833	13,000	31,000
1838	11,750	29,000
1856	5,000	33,000
1861	4,500	33,000
1881	2,200	26,000
1885	2,000	25,000
1891	1,800	19,000
1901	900	11,000

Sources: On Lyons, cf. table A2. On London, 1802 to 1838: the parliamentary inquiries cited, in particular the 1838 census of looms and weavers, the only really reliable data, since the other figures are manufacturers' estimates based on the total number of looms they were providing work for; it was an approximative procedure which encouraged overestimates because the same looms got counted several times over. The 1838 census was a surprise: the old method had resulted in estimates of 14,000 to 16,000 looms, whereas only 11,500 to 12,000 were found in the urban area (PP, Hand-loom, 1840, II, RAC, J. Mitchell, pp. 219–228). For 1901, Charles Booth, Life and Labour of the People of London (London, New York: Macmillan, 1902–3), 1st series, vol. 4, p. 255. For 1861, 1881, 1891, extrapolations based on the weavers involved in the census. For 1856, PP, 1856, Masters and operatives, XIII, no. 11710 (referred to by Duncan Bythyell, *The Sweated Trades* (New York: Cambridge University Press, 1978). Cf. also Prothero's estimates in *Artisans and Politics* in *Early Nineteenth Century London. John Gast ad his Time* (London: Dawson, 1979).

Table A2 *Lyons, number of looms*

Year	*Looms in Lyons town*	*Looms in in suburbs*	*Urban area*	*Looms in the country*	*Total no. of looms in "The Manufacturer"*
1739	8,381				
1752	9,404				
1768	11,007				
1788	14,777				
1802	10,434				
1807	10,960				
1810	13,294	1,400	14,694		
1815			14,500		
1817	17,024				
1820	18,631		19,200		
1822	19,292	3,411	22,703		
1825	20,101				
1829	16,932				
1830	18,000	11,278	29,278	8,265	37,543
1832			30,000	12,000	42,000
1833	16,857	12,918	29,775		
1836			30,000	16,000	46,000
1840			26,750	30,750	57,500
1853				60,000	
1861				116,000	
1866	12,646	20,497	33,143	87,000	120,143
1873				120,000	
1880				119,000	
1900			10,980	86,600	
1905				93,000	
1914				58,903	

Sources: Figures that are not rounded come from municipal censuses of looms carried out directly, the remains of which are to be found in AML, F2, Silks, censuses; in ADR, 6M, unclassified industrial statistics; in AN F 12 1583, AN F 12 2390; the round numbers between 1810 and 1861 all come from Chamber of Commerce estimates, either directly or cited in secondary sources; for the year 1840 the figures come from *Statistique de la France, 1840*, tome 5, Industrie; for the year 1802 they come from the official census carried out by Deglize (AML, F2, and Registers of the Lyons Chamber of Commerce); figures for 1866 come from the census and are cited by Natalis Rondot, *Exposition universelle internationale de 1878 à Paris, Groupe IV. Classe 34. Rapport sur les soies* (Paris: Impr. Internat., 1885), p. 96. Figures for 1873 to 1914 are taken from Yves Lequin, *Les ouvriers de la région Lyonnaise (1848–1914)* (Lyons: PUL, 1977), p. 84. Note that the reduction in the number of looms after 1880 does not reflect a fall in production but the introduction of mechanical looms, here added to the handlooms. Many fantastic figures were put into circulation in other secondary publications and have often been picked up by academic works; such figures were based on inadequately adjusted extrapolations from the statistics for the consumption of silks.

Table A3 *National production of silks, Great Britain and France, estimates*

Year	*Gt. Britain £ millions*	*France £ millions*	*France FF millions*
1770–1790	3.40		
1788		5.20	130
1812		4.32	108
1800–1820	4.50		
1832		6.12	153
1829–1933		8.48	212
1834	9.50		
1844		12.00	300
1847		14.00	350
1841–1850	10.80	14.00	350
1851–1860	11.50	20.00	500
1861–1870	9.60	29.80	745
1871–1880	7.10	28.80	720
1881–1887	6.43	29.40	735
1900	3.20	26.00	650

Sources: The figures for decades and those for 1881–87 come from Michael G. Mulhall, *The Dictionary of Statistics* (London: Routledge, 1909) as far as possible I have checked the plausibility of his extrapolations on the basis of national consumption, by comparing his extrapolation coefficients with partial ratios "prices of silks/added values." His approximations are very risky but seem to be the least unreliable and fit in more or less with the most trustworthy international data as calculated from 1861 in France by the Commission Permanente des Valeurs des Douanes, textile section (as printed in 1868, 1873, 1883, 1884, 1897, 1892, 1902). Figures for 1900 are taken from him and suggest that Mulhall overestimated the French figure for 1881–87; he probably overestimated the continuous fall in rates of added value. French production entered a phase of expansion after 1887 and in 1912–13 broke all previous records. The other old figures for France are taken from Tolozan for 1788, Chaptal for 1812 (cited by Rondot,. *Exposition universelle internationale*), Arles-Dufour for 1829–33 (*Un mot sur les manufactures etrangères* (Lyons: Mme S. Duval, 1834, pp. 133–136), John Bowring, *Second Report on the Commercial Relations between France and Great Britain. Silk and Wines* (London: HMSO, 1835), for 1834, Exposition . . . 1844, *Rapport du jury*, for 1844; Pariset, *Histoire de la fabrique Lyonnaise*, p. 309, for 1847. The figure for British production in 1834 comes from Ballance, PP. Handlooms, 1835, p. 663.

Table A4 *British imports and exports of silk fabrics*

Year	*GB Imports in £ millions*	*GB Imports in £ millions*	*GB Exports*
1820		372	
1825		297	
1826	448		
1827	556	174	
1828	677	179	
1829	596	221	
1830	535	521	
1831	550	500	
1835		972	
1840		793	
1845		766	
1849		998	
1850	1,924	1,041	
1854	2,280	1,440	
1860	2,799	1.587	
1865	7,326		
1870		1,450	
1875	12,265		
1880	13,325	2,031	
1885	10,269		
1890	11,319	2,229	
1895	15,237		
1900	15,377	2,064	
1910	13,521	2,279	

Sources: Figures found in PP. Silk, 1831–32, Hume's deposition, p. 14; and in Porter, *Silk Manufacture in England*, p. 347 and p. 351.

Table A5 *French imports and exports of silk fabric*

Silk Fabrics Thousands of francs, yearly averages		
Year	*Imports*	*Exports*
1827–36	2,800	121,400
1837–46	5,500	134,700
1847–56	7,500	274,700
1857–66	7,500	414,000
1867–76		430,000
1877–86		251,000
1887–96		243,000
1897–1906		280,000
1913		435,000

Sources: found in Direction générale des Douanes, *Tableau décennal du commerce de la France, 1827*, . . . etc.

Table A6 *Number of persons active in the British silk trade and silk manufacture, estimates given by Armstrong and based on censuses*

Silk, silk goods, ribbons
Occupied, thousands.

Year	*M*	*W*	*T*
1841	[28.2]	[33.0]	[61.2]
1851	52.6	76.7	129.3
1861	39.9	72.1	112.0
1871	25.9	53.1	79.0
1881	18.5	40.9	59.4
1891	16.1	31.8	47.9

Sources: Extract from W. A. Armstrong, "The use of information about occupations," in E. A. Wrigley (ed.), *Nineteenth-Century Sources. Essays on the Use of Quantitative Methods for the Study of Social Data* (Cambridge: Cambridge University Press, 1992), p. 264, rectified results of censuses. Note that no distinction has been made between two very different sectors, the spinning and throwing of silk on the one hand and the weaving of silk on the other. The figures for 1841 are not directly comparable to the others since we are not told which of the 99,000 persons active in the textile industries was involved in cotton, which in wool, which in silks, etc.

3 The rise and decline of flexible production: the cutlery industry of Solingen since the eighteenth century

Rudolf Boch

Introduction

The production of cutlery of all sorts in the industrial area around the city of Solingen in the lower Rhineland of Germany can be seen as a "classical" example of a historical industrial system based on "flexible specialization."

In many aspects its inner structure as well as its development over a long period of time seems to have been typical of several other export-orientated trades, located in neighboring districts: files and edge-tools in Remscheid, ribbon-weaving in Barmen and Ronsdorf, or the diversified silk industries of Elberfield, Cronenberg and Krefeld.[1] All of them grew out of fairly small artisanal trades, producing for regional and supra-regional markets in the seventeenth century. But they were shaped by the economic and social conditions and experiences of the eighteenth century, becoming large export industries, designed to serve the then existing "world market." By the late eighteenth century they were seen by many contemporaries as the most important industries of the western parts of Germany.

During the first decades of the nineteenth century the collapse of the old institutional framework, as well as fast changing market-conditions influenced by rising national tariff barriers and – in some cases – superior British competition, brought severe distress to the trades. Only in the second half of the nineteenth century, however, did most of these "flexible specialized" industries reach the peak of their development, now integrating new mechanical devices into their system and logic of production. By contrast, the twentieth century became a time of continuous crisis. By the 1960s – at the latest – these industries had changed substantially; they had lost their peculiar character and much of their former economic importance.

[1] For a short survey of the economic and working-class history of these export trades – most of them sited in the Bergisches Land, the former Duchy of Berg – see R. Boch and M. Krause, *Historisches Lesebuch zur Geschichte der Arbeiterschaft im Bergischen Land* (Cologne, 1983), pp. 13–90 and my recently published article: "Tarifvereinbarungen und Ehrenräte zwischen Unternehmern und Arbeitern im Regierungsbezirk Düsseldorf während der Revolutionsjahre 1848/49," in B. Dietz (ed.), *Bergisches Jahrbuch* (Wuppertal, 1990), pp. 4–51. The most important – mainly older – literature about these trades is cited there.

The following case study of the Solingen cutlery industry tries to set out the reasons and conditions for the rise of this regional system of "flexible production" up to the time of its greatest economic success in the last third of the nineteenth century as well as the causes of its decline after the First World War.

In this chapter I argue that rather than an "immanent logic" of technological change, external forces led to a substantially altered industry orientating itself towards a model of mass production: two world wars and a world economic crisis as well as a shift in orientation among "social actors" – on the one hand the entrepreneurs, following a certain vision of future industry, and, on the other, a working population reacting to the deprivation of its former privileged status in an unforeseen way. The article describes how the continuous political and economical disasters of the first half of the twentieth century fostered the emergence of "national industries" – based on semi-skilled labor and "single purpose" machinery – as well as trade protectionism as a necessary precondition. It shows how "flexible production" as a form of industry not designed for such a world but heavily dependent on international trade was most severely hit. I argue that it was not the greater efficiency of standardized production on "single purpose" machines that led to a breakthrough in the direction of mass production of cutlery; instead, the destruction of the traditional apprenticeship system left no other alternative. Furthermore, I point out that the entrepreneurial debate about the future of the industry after the turn of the century had a decisive impact on the destruction of the traditional training system for skilled labor – the basic precondition for a system of "flexible specialization." Proclaiming the death of the "old system" more than a generation before it really happened, this debate became a self-fulfilling prophecy.

Although the present chapter – aiming at a compressed description of 200 years of history of a "flexible specialized" industry and the related public debates in the crucial phases of its development – is necessarily a sketch, the arguments proposed are based on many years of original research[2] concerning the period between the late eighteenth century and the 1920s and on a recently published detailed study of the period up to the 1960s.[3]

In what follows I shall describe "flexible specialization" not only as a peculiar system of production, but also as a specific "social system", i.e. a systemic relation between capital and labor. Only with this "social dimension" in mind

[2] See my book, *Handwerker-Sozialisten gegen Fabrikgesellschaft. Lokale Fachvereine, Massengewerkschaft und industrielle Rationalisierung in Solingen 1870–1914* (Göttingen, 1985) and my various articles on the Solingen cutlery industry, e.g.: "Was macht aus Arbeit industrielle Lohn-Arbeit? Arbeitsbedingungen und-fertigkeiten in Prozeß der Kapitalisierung: Die Solinger Schneidwarenfabrikation 1850–1920," in *Sozialwissenschaftliche Informationen für Unterricht und Studium*, 9 (1980), 61–68; "Changing Patterns of Labour Conflict and Labour Organization in the German Cutlery Industry, Solingen 1905–1926," in L. Haimson and G. Sapelli (eds.), *Strikes, Social Conflict and the First World War* (Milan, 1992).

[3] J. Putsch, *Vom Ende qualifizierter Heimarbeit. Entwicklung und Strukturwandel der Solinger Schneidwarenindustrie 1914–1960* (Cologne, 1989); and by the same author: *Vom Handwerk zur Fabrik. Ein Lese- und Arbeitsbuch zur Solinger Industriegesellschichte* (Solingen, 1986).

can the history and the decline of "flexible specialization" as a distinct model of industrialization be fully understood.

The "golden age" of flexible specialization in cutlery production, 1870–1914

The historical structure of the Solingen industry[4]

Shortly before the First World War the system of "flexible specialization" in the Solingen cutlery industry had reached its greatest importance. At the beginning of the twentieth century the industrial region around the city of Solingen was the centre of German cutlery production. More than 80 percent of the annual cutlery production of the German Empire (i.e. knives of all sorts, scissors, razors, scalpels, swords etc.) and more than 90 percent of all cutlery exports came from this small district south of the Ruhr area.

In 1907, a workforce of nearly 18,000 people was engaged in this regionally concentrated industry. Most of them were men; only some 900 female workers were employed on a few specialized work processes. But, out of the whole workforce of 18,000, only 7,000 were typical factory workers, while nearly 11,000 were either semi-independent outworkers or artisan workers owning their own workshop or sharing it with a dozen (or even more) colleagues.

The manufacture of cutlery in the district of Solingen can be traced back well into the middle ages. One of the main reasons for the rise of a specialized metal industry in this hilly area seems to have been the abundance of water power. The waters of the river Wupper as well as of the many small rivulets drove the numerous wheels, which gave power to the forging hammers and grindstones required in the process of manufacture.

During the eighteenth century the growth of a world market fundamentally altered the social structure of the production system as it had historically developed. The medieval cutlery professions changed into sizable, capitalistic export-oriented trades. The new production system centered around the so-called *Verleger-Kaufleute* (merchant-capitalists), who not only controlled commerce but organized manufacture in a "putting-out" system. By the mid eighteenth century the formerly independent smiths, grinders and hafters had become mere artisan workers, getting their raw material or semi-finished products from the merchant-capitalists in advance and being paid by the piece. Well into the second half of the nineteenth century this production system remained unchallenged. Only the forging process from the 1860s onwards was increasingly mechanized. By the 1890s the traditional trade of the hand smith had nearly died out. The forging of the blades for scissors, knives and razors was

[4] For further details of the economic structure of the Solingen cutlery industry as well as the work situation and the social status of the diverse groups of workers (i.e. forgers, grinders, hafters etc.) within the industry, see Boch, *Handwerker-Sozialisten*, pp. 28–43, 94–122, 181–188.

now done under huge steam hammers by a largely semi-skilled workforce. Some forty forging factories supplied the whole cutlery district with this semi-finished product. But in the subsequent production processes of grinding and hafting (i.e. the integration of the finished blade with the handle), the "putting-out" system was still the dominant feature.

For a proper understanding of the underlying industrial structure one has to realize that the cutlery grinders of Solingen were not domestic outworkers in the narrow sense of the word. In contrast to the cutlery hafters, who worked in domestic isolation scattered over the whole district, the grinders used to work in groups of between two and fifty grinders in the grinding mills, which were placed along the rivers and rivulets of the district. This work situation continued, after the 1850s, when steam-driven grinding halls started to supplement the existing water-mills. Up to 100 grinders now worked under one roof, but not as wage laborers for a single entrepreneur. Each grinder still owned his tools and worked on his own account for different merchant capitalists, only "paying for steam" power in the halls. Even in grinding halls, which were, as it happened, frequently built next to mechanical forges in order to use the surplus steam power, this system of "paying for steam" remained intact. While in the last decades before World War I a considerable number of forging factories developed into important cutlery firms, often with higher turnover than the old established merchant-capitalists, the grinders on the very premises of those firms remained independent. The rise of steam power – in sharp contrast to its effects on the forging sector – neither integrated grinding within centralized factory production nor led to a revolution in grinding techniques. Only the main source of energy had changed.

Solingen's performance on the world market in the early years of the twentieth century

The cutlery industry of Solingen reached its peak economic expansion on the world market in the last decade before World War I. In 1907 Solingen obtained a 52 percent share in the world trade in cutlery, driving it up to nearly 60 percent by 1913.[5]

Solingen produced a wide range of products to suit the needs of highly differentiated regional markets abroad. In 1903 the industrial district exported cutlery of all sorts to forty-seven countries all over the world.[6] Taking account of smaller export orders under 10,000 Reichsmark which are not covered by the above figures of the local Chamber of Commerce, the number of countries served by the industry was even substantially higher.

During the 1880s Solingen exporters managed to break up the dominant position of the English cutlery industry located in and around the city of

[5] See H. Möhle, *Der Exporthandel der Solinger Schneidwarenindustrie* (Frankfurt, 1931), p. 39.
[6] *Annual Report of the Solingen Chamber of Commerce* (1906), pp. 3–4.

Sheffield. By the late 1890s Solingen matched the annual volume of exports of its English competitor on the world market. Not long afterwards Sheffield had fallen to a poor second in the league table of cutlery exporters, seldom reaching more than half of Solingen's exports in volume and value up to 1914.[7]

One explanation for this dramatic victory is to be seen in Solingen's greater readiness and ability to produce cutlery in small and even very small series economically. Solingen's distinguishing feature during this epoch was its "babylonic variety" of types and models of cutlery. Many Solingen firms not only offered hundreds, but thousands of different models of cutlery.

A second explanation, closely linked to the first, may be found in the fact that the mechanization of forging in Sheffield started comparatively late and assumed a different shape, a different social framework. In contrast to Solingen no independent mechanical forges developed, producing only a single product – the semi-finished, stamped blade – for a plurality of cutlery firms. In Sheffield nearly every cutlery firm of some size maintained its own department for mechanical forging, whose production capacity in most cases far outdistanced the possibilities of subsequent treatment, such as grinding and hafting. Standing idle for some days a week, these mechanical devices produced high overhead charges for the average Sheffield firm.[8] It was for this reason, that small firms and long-established large firms in the high-quality sector still engaged traditional hand-smiths. In 1908 their numbers stood well above 600,[9] while in Solingen this profession had nearly died out. The continued existence of a large group of hand-smiths, though "economical" for many individual firms, proved to be as disastrous for the competitive power of Sheffield's cutlery exports as the application of modern forging machinery in scattered firms, attempting to become "fully fledged" factories.

Solingen entrepreneurs, meanwhile, enjoyed the advantage of relying on only a few dozen of mechanical forges, which – by competition – were forced to keep a thousand or more dies for common and uncommon models of cutlery in their stores. These comparatively few modern forges served, though privately owned, as a kind of common institution for the whole cutlery district, creating a huge rationalization effect without losing the ability for "flexible specialization."

Long before the advent of the computer, these forges managed to supervise a complex and diversified inventory of patterns and semi-finished products

[7] For the competition between Sheffield and Solingen before World War I, see O. Beyer, *Die handelspolitischen Hemmungen der Solinger Industrie* (Gummersbach, 1929), and the brief summary in chapter 9.2 of Boch, *Handwerker-Sozialisten*, p. 189.

[8] This is confirmed by interviews with prominent Sheffield entrepreneurs in 1910. See Royal Commission on the Poor Laws and Relief of Distress, Appendix Vol. VIII, London 1910, pp. 475.

[9] S. Pollard, *A History of Labour in Sheffield* (Liverpool, 1959), p. 333. Despite some blind spots and a partially outdated perspective, this book still offers a good account of the Sheffield cutlery industry, its outworkers and its trade-union movement. See also G. Tweedale, *Sheffield Steel and America: A Century of Commercial and Technological Interdependence* (Cambridge, 1987).

enabling the whole cutlery industry to base their diversified final production on specialized intermediate goods. These some dozens of forges became the most expansive and innovative part of the whole region's history from the 1890s onwards, using the latent overcapacity of their mechanical forging devices for the production of other goods, such as hardware, household equipment or bicycle parts. On the eve of the First World War some of these Soligen firms had become important subcontractors of the German bicycle and automobile industries already. In contrast, no development and diversification of this kind took place in Sheffield.

Furthermore, the Solingen firms could – to an ever greater extent than their English competitors[10] – rely on a dynamic, indigenous artisanal "brains-trust" in the next stage of the production process, the grinding of the forged blades. Thousands of semi-independent, highly-skilled grinders, working on "universal" or "general-purpose" machines – as one can call the water- or steam-driven grindstones – could realize any buyers' wish. The grinders' skill and their traditional tools provided the "flexibility" the dominant market structure needed, being able to produce small series nearly as economically as bigger ones.

But the economic prosperity of the Solingen cutlery industry in the last thirty years of the nineteenth century was not only due to its ability to produce for small markets with high differentiated tastes. The economic prosperity of the Solingen district was closely linked to the rise of the US market: the huge, constant demand for household wares it provided decade after decade.[11] In the 1860s and 70s the United States held the biggest share of the Solingen cutlery exports. This constant demand helped the Solingen industry through the following period of the so-called "Great Depression," when growing tariff-barriers in many countries endangered the future development of the district. Despite severe tariff increases by the American administration during the 1890s aiming at fostering

[10] The stagnation of Sheffield cutlery industry in the last third of the nineteenth century was caused not only by wrong entrepreneurial decisions, e.g. the adoption of the new forging techniques, but seems to have been closely linked to an ongoing destruction of the traditional apprenticeship following the decline of the local trade union movement, initiated by the spectacular report of the "Royal Commission on Trade Outrages" in 1867 (see S. Pollard, "The Ethics of the Sheffield Outrages," *Transaction of the Hunter Archeological Society* 7 (1957), 118–139, and S. Pollard (ed.), *The Sheffield Outrages. The Minutes of Evidence and Report of the Royal Commission on Trade Outrages 1867.* Reprint with Introduction, Sheffield 1971). From the 1870s onwards, the old cutlery unions lay in agony. They lost their grip on the trades and permitted the emergence of a system of sub-entrepreneurs, the so-called "little mesters." These "little mesters" – in contrast to the old restricted apprenticeship system – relied on an unlimited number of (half-trained) apprentices and soon began to experiment with the introduction of piece-work (the further subdivision of labor) in the grinding of cutlery. This development in the long run not only ruined the wages, but also the traditional skill, the artisanal "brains–trust" as a basis of "flexible specialization." In the early years of the twentieth century only a few large, well known firms producing for a high-quality market, maintained the old apprenticeship system and paid wages well above the average. (This destruction of crucial traditional patterns of industry is confirmed not only by English contemporary sources, but also by highly interesting reports of Solingen trade unionists, who visited Sheffield during the last decade before World War I.)

[11] For further description see chapter 9.2 in Boch, *Handwerker-Sozialisten*, p. 192.

the development of a national cutlery industry, the Solingen firms managed to export large amounts of cutlery, mainly high-quality products. But due to the growth of American cutlery production Solingen's exports to the US stagnated at a high level until the outbreak of World War I. Instead, exports to Great Britain and its Empire – the last redoubt of "Free Trade" – grew in importance during the last decade before the war, overtaking the USA as the biggest single market for German cutlery by 1913.[12]

The "social system" of "flexible specialization"

The economic success of the Solingen cutlery production before World War I was due in part to a "social contract" between the entrepreneurs and certain groups of workers, mainly the grinders, which had developed since the early 1870s.[13] This "social contract" effected a limitation of competition within the district and secured high standards of quality. It contained crucial elements from the old institutionalized *Fabrikverfassung* (industrial code) of the Solingen industry during the *ancien régime* which had been nullified under Napoleonic rule in 1811.

In the beginning, the contract was not a voluntary agreement, rather the workers' movement of Solingen had to force it through against the will of a majority of the entrepreneurs. The highly skilled cutlery grinders in particular had been able to organize themselves in powerful local unions, once the prohibitive laws against freedom of association were abolished in 1869. During the favorable boom period of the early 1870s they succeeded in forcing the entrepreneurs to accept precisely defined wage rates for the different types of cutlery, substantially higher than the average wages paid previously.

This system of minimum wages was intended to prevent – as had been the case in previous decades – the competition among the hundreds of Solingen cutlery firms from working to the workers' disadvantage, i.e. resulting in ever lower wages. After a while the grinders' unions' project obtained the support of a number of larger, established firms. Given that under the prevailing market-conditions and general political circumstances they had to pay higher wages anyway, these firms were interested in fixed minimum wages across the whole industrial district. Thus the grinders' unions and the established firms which at this point founded "trade associations" (*Fabrikantenvereine*) came to have identical interests, i.e. action against the so-called *Preisdrücker-Firmen* ("price cutters") in the cutlery industry: usually small firms, only recently established by

[12] From 1903 to 1913 the exports of Solingen cutlery to the UK and its colonies increased from 2.5 to 7.45 million Reichmarks. In 1913 the UK and its colonies took 19.4 percent of all Solingen exports, while the US market's share was a mere 18.5 percent. See Beyer, *Hemmungen*, p. 9 and C. Heitland, "Der Handel mit Solinger Erzeugnissen," unpublished manuscript, Cologne, 1923, p. 128; see also Tweedale, *Sheffield Steel.*

[13] For the renewed "social contract" between capital and labor as a necessary condition for a fully-functioning system of "flexible-specialized" production, see chs. 3, 4 and 5 of Boch, *Handwerker-Sozialisten*, pp. 44–93.

upwardly mobile artisans and "putting-out" merchants. These firms notoriously lowered wages – especially during slumps – in order to undercut the asking prices of the older firms maintaining long established commercial contacts. The number of newcomer firms had risen rapidly since the official introduction of "Freedom of Trade" (*Gewerbefreiheit*) in 1811 whereas the number of established firms had likewise decreased. The established firms were therefore under pressure and the organizational power of the independent workers movement was indispensable for the assertion of the joint interests of grinders and established firms against the newcomers. A significant feature of this situation was that only the unions were able to enforce the fixed minimum wages on those many entrepreneurs not organized in the *Fabrikantenvereine*.

Minimum piece rates – in so complicated a production system as the Solingen cutlery industry – could only be determined through very detailed negotiations between trade union officials and members of the employers' association: For every traditional type of cutlery, for every model, for every different work process within the production of the diverse models a special rate, i.e. a minimum piece rate, was agreed on and written down in published booklets called *Preisverzeichnisse* ("piece price lists").

For the supervision of these agreements and future negotiations a kind of arbitration court called the *Vergleichskammer* was formed, consisting of employers and workers in equal number (cf. *Conseils de Prud'hommes* and English "joint boards").

Not only were the entrepreneurs forbidden by sanction of the trade unions from paying less than the minimum price-rate but in order to prevent the subversion of the complicated wage-system no Solingen firm was allowed to introduce a new sort of cutlery or to reduce the quality of an existing sort, without the agreement of the *Vergleichskammer*. By means of these agreements – an important feature of the social conflicts during the last decade before World War I – the trade unions got a grip on the quality of the cutlery, the methods of manufacture, and the mode of production.

By no means did the fixed and comparatively high wages and the rights of co-determination granted to the unions lead to "the ruin of the Solingen cutlery industry," as had been feared by the local Chamber of Commerce in the early 1870s. On the contrary, already in 1877 – a slump year – the Chamber of Commerce wrote in its official report on the continuing favorable development of cutlery production:

> This favorable result is definitely due to the constancy of our high cost of grinding, the main labor factor, since 1875. Thus it was possible to secure the quality of the end product to a greater degree then had been the case under the previous conditions, much to the disadvantage of its reputation . . . here the blessed results of a close cooperation of workers and employers can be clearly and undoubtedly seen.[14]

[14] *Annual Report of the Solingen Chamber of Commerce* (1877), p. 4.

The majority of the entrepreneurs – as a leading Solingen merchant expressed it in 1883 – "had to be forced into their luck."[15] Price-lists and arbitration courts were nothing new for the Solingen industry: they had been in existence two generations earlier at the end of the *ancien régime*. However, their positive effect on the system of production had been forgotten by the dominant new type of entrepreneur of the 1870s. This type of entrepreneur believed, totally and completely, in the dogma of *freies Speil der Kräfte des Marktes* ("free play of market forces") and indeed, was himself a historical product of the "unbound competition" after the destruction of the old "industrial code" (*Fabrikverfassung*) of the trade.

Fixed minimum piece rates and arbitration courts had developed with the complex system of flexibly specialized cutlery production during the eighteenth century. It had been their function then to protect this system from the specific dangers of its tangled, diversified and complex way of manufacture and dependency on export as well as to regulate its adaptation to a quality-determined market.

The system of production as well as the structure of the markets served by the Solingen cutlery industry had not changed substantially since the abolition of the old "industrial code": Solingen cutlery firms still frequently competed with each other in export markets and constantly ran the danger of underselling each other. Since all the firms engaged the "self-employed" outworking grinders, and since all firms profited from rising productivity – such as effected by the introduction of mechanical forges – competition among the firms often resulted in lowering wages.

Furthermore, the competition between the main European cutlery centers – Solingen, Sheffield and to a certain extent Thiers (France)[16] – was still based on a high level of quality. During the whole nineteenth century – as in the centuries before – cutlery and even razors or spring-knives were not regarded as cheap consumer goods for everyday life. Even in better-off strata of the European working-classes or the *petit-bourgeoisie* they were bought "once in a life-time," on the understanding that they could be passed down from generation to generation in the family.

The working techniques, local industrial structure and social relations between capital and labor in Sheffield, Thiers and Solingen had much in common – all three still standing in the tradition of a supra-national "old European" organizational pattern of mercantile export trade. Under these conditions, and, given a more or less unaltered production system, a wage level which satisfied the traditional needs of the workers still had a positive impact on the quality standard of the products and could lead to a competitive advantage.

15 Carl August Bick, Chairman of the arbitration court (Vergleichskammer) in October 1883, see Boch, *Handwerker-Sozialisten*, pp. 57, 312.

16 The vast literature on the – in comparison to Solingen and Sheffield – fairly small cutlery industry in and around Thiers takes as its principal source the encyclopedic study by C. Pagé, *La coutellerie depuis les origines jusqu'à nos jours*, 6 vols. (Chatellerault: H. Rivière, 1896–1904).

This lesson, learned already in the 1770s by Solingen entrepreneurs, a time when the minimum-wage system had broken down for several years, with the outcome that the cutlery workers gradually lowered the quality standards of the products in order to earn their living or rather to survive in a downward spiral of wage-pressure, had been subsequently forgotten in the first half of the nineteenth century.[17] Only the renewal of that old experience, that fairly high and equal wage rates could lead to a competitive advantage – now enforced by organized labor – brought prosperity back to Solingen in the last third of the century.

Flashback: a first crisis of the system of "flexible specialization" after 1815

In the early years of the nineteenth century the dramatic breakdown in quality standards during the 1770s, leading to a severe export crisis, was still well remembered within the circle of old-established "privileged" merchants and the official representatives of the cutlery guilds.[18] But the influence of this local elite on the policies of the administration of the Duchy of Berg, to which Solingen as well as other important neighboring export-trade cities belonged, was diminishing rapidly. Since the 1790s a new type of young civil servant was beginning to set the fashion within the administration in nearby Düsseldorf on the Rhine.[19] Followers of the recently translated works of Adam Smith, these new administrators believed that only the total abolition of all old regulations, privileges and guilds could secure the future of the export trades of the Duchy.

In the summer of 1801 the government urged the local administration and magistrates to examine existing privileges, guilds and other regulations and to make proposals how to abolish or at least curtail their functions "for the public good." This internal administrative instruction was made public in December 1801 by the *Westfälische Anzeiger*, the first regional bourgeois paper of wide

[17] See my article "Tarifvereinbarungen und 'Ehrenräte'" in Dietz (ed.), *Bergisches Jahrbuch*.

[18] The last Obervogtsverwalter of the Solingen cutlery trades Adam Edler von Daniels gave a precise analysis of the crisis of the system of trade regulations in the 1770s and the disastrous effects in his widely read pamphlet: "*Vollständige Abschilderung der Schwert- und Messerfabriken fort sonstigen Stahlmanufakturen in Solingen*" (Düsseldorf, 1802; rptd. Remscheid, 1981), pp. 91–96.

[19] Theodor Ark, born in 1755 and from 1798 in the highest ranks of the regional administration, becoming Privy Councillor the same year, can be seen as a "prototype" of these new, "Smithian" civil servants. In 1807, time now fully on his side, in a survey on the Solingen cutlery industry he launched a furious attack against the "privileged despotism" of the traditional "industrial code" citing the *Wealth of Nations* page after page. In 1809 he wrote – in the same spirit of "unconditional freedom of trade" – an official report on the industry of the whole Grand Duchy of Berg, which served as standard information for the new Napoleonic reform legislation as well as for the Prussian administration in the years after 1814. See B. Dietz and F. Hoffmann, "'Das Fabriken' und Manufakturwesen des Großherzogtums Berg zu Beginn des Jahres 1809. Eine ökonomische Zustandsbeschreibung des Staatsrats Theodor Ark," in *Zeitschrift des Bergischen Geschichtsvereins*, 92 (1986/87), 173–195.

circulation.[20] The publication provoked contradiction and – unusual for Germany of that time – a public debate.[21] Not only the officials of the still-existing guilds but also a substantial section of the regional merchant-capitalists did not want an overall change. They were in favor of preserving at least those old institutions and rights which had regulated internal competition and controlled the standards of quality.

The conflict was not fought out, but resolved by external forces when French troops invaded the Duchy in 1806. In 1809 the Duchy of Berg had become an enlarged Grand Duchy under strong French influence and a Napoleonic dynasty. All privileges, guild-orders and trade-legislations, whenever and for whatever reason established, were abolished at once. The official deputies of the Solingen business profession (*Kaufmannschaft*), backed by some senior civil servants in the administration, petitioned against this decree. However, they only succeeded in postponing its commencement – for the cutlery industry – until 1811.[22]

[20] *Der Westfälische Anzeiger*, Dortmund no. 85 (December 1801), pp. 1354. The article, after having cited the internal instruction in full length, expressed the hope that only the small town guilds (e.g. butchers, bakers, tailors) would be abolished, but not the "industrial codes" of the large export-trades: "One can not think of the abolition of these important privileges, existing for centuries." Though the author admitted that abuses had become notorious even in these flourishing trades over time, he pleaded for abolishing merely these abuses while preserving "industrial order" (*Ordnung*) as a whole.

[21] For the public debate which followed, see December 1801 and the whole year 1802 of the *Westfälische Anzeiger* (1801) p. 1,533; (1802), pp. 22, 192, 256, 545, 1,041, 1,057, 1,073. This important regional bourgeois newspaper has not been scientifically evaluated up to now. The debate centered around the further existence of the privileged "Garnnahrung" of the Wupper Valley textile-trades, involving first of all merchants and factory-owners of the Duchy. The general one – despite two or three pleas for "freedom of trade" – was similar to the arguments proposed in the first article in no. 85 (December 1801): reforms "yes," total abolition of the inner, historically developed regulations of the trades "no." It seems as if only some big and influential merchants and bankers in the cities of Elberfeld and Remscheid as well as the small "unprivileged" merchants on the fringe of the small-iron industry fully supported a system of "freedom of trade." The future of the "industrial code" of the Solingen industry stood in 1802 not in the foreground of public interest, though the above-mentioned (n. 18) booklet by A. E. v. Daniels can be seen as a contribution to the debate. At the same time the same debate about the "reconstruction d'un ordre industriel" took place in the flexibly specialized silk-industry of Lyons. While the Lyons merchants were in favor of such a reconstruction, Paris officials and bankers tried to defend the "principes de liberté" established in the first years after the revolution of 1789. See the contribution by Alain Cottereau in this volume.

[22] The effort to annul the decree of November 1809 is described in the old, but well-informed and sensible work of A. Thun, *Die Industrie am Niederrhein* (Leipzig, 1879), II, pp. 62–63. While A. E. v. Daniels and the well-known factory-inspector (General-Fabrikeninspektor) K. Eversmann wrote supporting "expert reports" for the case of the Solingen merchants, the young Solingen mayor K. L. Goebel enclosed a letter to the petition in which he tried to neutralize it from behind, arguing that only deregulation and the introduction of machinery (!) could meet the challenge of English competition in cutlery production. Goebel's letter is printed in F. Hendrichs, *Geschichte der Solinger Industrie* (Solingen, 1933), p. 174–178. A. Thun in 1879 summarized the outcome of the abolition of the Solingen "privileges" as follows: "the most delicate of all industrial codes was replaced by nothing; nothing else but the naked despotism of the individuals." See Thun, *Industrie*, II, p. 63.

The Prussian administration, taking over the Grand Duchy of Berg in 1814, followed the resolute French line in economic policy. The Prussian reform-orientated civil servants, many of them convinced Smithians, regarded themselves as custodians of the "freedom of trade," always endangered – as they saw it – by "backward-looking local merchants and stubborn artisans" who desired to tread the beaten path and to cling to the old *Schlendrian* ("jog-trot").

The final abolition of the old Solingen industrial code in 1811 was not at once followed by disastrous effects. The Napoleonic Wars, by creating a constant demand for swords and bayonets, led to a boom in the manufacture of weaponry which for several years attracted many artisan workers from other sectors of the cutlery production. For a while it seemed that the then apparent winners of the deregulation of industry, the former unprivileged merchants, and the well-to-do artisans who could now become merchants without legal restrictions determined public opinion. They were backed by the experience of many out-working grinders, hafters and smiths whose wage level was on average higher than during the previous period of regulated minimum wages – though due only to the economic boom.

This situation, however, changed dramatically when business returned to normal after 1815. As the result of an unrestricted influx of labor from neighboring agrarian districts into the trade and the fierce competition between old established merchant-capitalists and "newcomers" described above, the wage level fell rapidly and sharply. In 1826 – a year seen by contemporary merchants as a "good business year" – a mayor of an industrial village within the cutlery districts reports to Düsseldorf on behalf of the wage earners "that despite the cheapness of victuals the workers are unable to earn enough to satisfy their most necessary needs."[23] The so-called truck system – i.e. payment in overvalued kind, instead of cash money – also became notorious in the Solingen industry of this period.[24]

Between 1815 and 1836, however, the work force doubled and the volume of regional cutlery production tripled.[25] But this fast economic growth was paid for

23 Report of the mayor of Höhscheid K. Höfer, February 1826 quoted in M. Henkel and R. Taubert, *Maschinenstürmer* (Frankfurt, 1979), p. 188. Several other official sources confirm the dramatic fall of the average wage level, e.g. the reports and publications of the district magistrate G. v. Hauer in 1822, 1830, 1836. See Boch, *Handwerker-Sozialisten*, p. 316 and Putsch, *Handwerk*, pp. 75–77, 84. After a short recovery in the second half on the 1830s wages fell again – as contemporaries estimated – to well below the level reached in Napoleonic times.

24 These bizarre abuses even led to a debate in the provincial diet in 1843. A nearly unanimous address of the deputies to the king and the government in Berlin pleading for strict legal action was nevertheless turned down: the means of payment were declared a "private matter." See G. Croon, *Der Rheinische Provinzialtag bis zum Jahre 1874* (Düsseldorf, 1917, rptd. Cologne, 1974), p. 204. A further description of enforced wage-cutting by the means of payment in kind in a deregulated system of "flexible specialization" is given in Henkel and Taubert, *Maschinenstürmer*, p. 166; Boch, "Tarifvereinbarungen und 'Ehrenräte'"; Boch, *Handwerker-Sozialisten*, pp. 83 and 320.

25 Estimate by the mayor Höfer in 1936 in a report to the district magistrate, quoted from Hendrichs, *Geschichte der Solinger Industrie*, p. 190.

dearly: wages, profits and quality standards all fell dramatically. It was during this period of rapid growth that "Solingen" became a synonym for "cheap and shoddy wares." Furthermore, the diminishing quality of the bulk of Solingen cutlery together with the enforced "sell-off" of the district's industrial heritage and the continuous production "not for, but against the direct producers,"[26] turned temporary growth into structural crisis, cumulating during the 1840s. The systematic impoverishment of the artisan workers had made them unable to reinvest in tools and workshop facilities. The quality of technical equipment – from grindstones to waterwheels – not only stagnated, but deteriorated between 1815 and the Revolution of 1848. The merchant-class was unable or unwilling to invest in the sphere of production either. The high-quality sector of the international cutlery market was totally lost to Sheffield, which, due to the English infrastructural advantages after 1815, such as cheap transport, cheap coal, and cast steel was able to compete successfully with Solingen's low wages even in the low-quality sectors.

The flexibly specialized cutlery industry of Solingen had not been founded in and was not designed for a market situation shaped by consolidating or rising national states – many of them trying to build up a fully fledged "national industry" complete from staples to specialized trades. Solingen suffered not only from the prohibitive or high custom-duties of the neighboring European countries after 1815[27] but also from the strong political attempts of the German *Zollverein* to which Solingen belonged, to build up its own effective iron and steel industry. These political decisions for a certain path of industrialization deepened the structural crisis of the deregulated system of "flexible production" described above.

In the late 1830s proposals by merchants from the Solingen and Remscheid industrial districts to lower the existing tariff on steel and iron from Great Britain and Sweden were rejected by Prussia and the *Zollverein*. The flexibly specialized industries of these cities depended heavily on these imports since the charcoal-based production of iron in the neighboring Siegerland was technically outdated and the German cast steel production hardly existed. Contrary to the direct interests of the Solingen and Remscheid industries the tariff was instead increased substantially in 1844 in order to foster a future national heavy

[26] The conservative merchant Johannes Schuchard, member of the Chamber of Commerce in the neighboring Wupper Valley and deputy of the provincial diet used this expression in a programmatic article in the *Westfälische Anzeiger* (no. 34 [1833], p. 168) to describe the economic changes since the abolition of the regional "industrial codes."

[27] Take for example France, one of the most important markets for the cutlery and small-iron industry of the Duchy of Berg in the eighteenth century: While in the second half of the eighteenth century and over most of the Napoleonic period the import-duty on these products was only 10 percent by value – a figure not fitting very well into the image of a "prohibitive" mercantilist eighteenth century – the tariff rose to nearly 100 percent in the average in the years after 1816! See *Bericht des 13. Ausschusses an die Plenarversammlung der Rheinischen Provinzialstände über die Verhältnisse des Ackerbaues, der Fabriken und des Handels in den Rheinprovinzen* (Elberfeld, 1827), p. 41.

industry:[28] the result of constant political pressure from Rhenish bankers and wealthy industrialists, who had just begun to invest in the Ruhr area and to experiment with a – still comparably costly – coal-based production of steel. Thus, the flexibly specialized industries had to pay the price for a new paradigm of "national independent" industrialization consciously pushed by certain groups of entrepreneurs from Cologne, Aachen and Elberfeld aiming to imitate the "English model" of industrial production at all costs.[29]

Ever since the first severe effects of deregulation and the deteriorating market situation began to be felt in Solingen groups of merchants and artisan workers had attempted repeatedly to put the reintroduction of some measures of the old regulation system on the agenda: between 1820 and 1824 former privileged merchants and artisan workers submitted several petitions to the administration in Düsseldorf and to the Prussian court in Berlin. In 1826 the cutlery grinders even went on strike – forbidden by Prussian law – and demanded the reintroduction of a system of minimum wages.[30] Though the grinders were forced back to work and the strike ended without results it had long-lasting effects: the whole working-class movement of the Solingen district from then on centered around this very issue: the reintroduction of minimum wages.[31]

In 1830 the artisan workers, led by their former guildmasters and backed by established firms, again petitioned the Prussian administration. They formulated their demands – now ten years old – in four points: 1. The reintroduction of an arbitration court (*Fabrikengericht*), to be charged with the additional duty to supervise the quality of the products. 2. Reintroduction of the "master-piece" (*Meisterstück*) in order to secure a high standard of artisanal qualification. 3. Reintroduction of a minimum wage system with flexible adjustment to the rise and fall of raw material prices. 4. The legal abolition of the truck system.

These demands were seriously discussed at a conference in Düsseldorf in November 1830, to which merchants, master artisans and mayors of the cutlery district were invited. But only a few weeks later three of the four points, including the central demand for a system of minimum wages, were rejected by the higher administration as "running counter to the existing laws of the *Rheinprovinz*."

[28] See Croon, *Provinziallandtag*, pp. 198–199, and the highly interesting debates in the original protocols of the provincial diet in 1839, 1841 and 1843 in: Archiv des Landschaftsverbands Rheinland, Provinzialarchiv files no. 280 following, Brauweiler. The "Eisenzoll-Debatte" of 1843 was published by the Kölnische Zeitung on July 22 and July 23, 1843 without citing the names of the speakers. Many merchants and industrialists were members of the diet, the protagonists of the "new, national path" to industrialization as well as those of the traditional "flexible specialized" industries.

[29] R. Boch, *Grenzenloses Wachstum? Das rheinischen Wirtschaftsbürgertum und seine Industrialisierungsdebatte 1814–1857* (Göttingen, 1991).

[30] A detailed description of the strike is given in Henkel and Taubert, *Maschinenstürmer*, p. 145. For the story of the petitions between 1820 and 1848, see Henkel and Taubert, *Maschinenstürmer*, p. 185; Boch and Krause, *Historisches Lesebuch*, p. 62; W. Bruckmann, "Die Einrichtung des Fabrikgerichts zu Solingen und seine Tätigkeit in den Jahren 1841 und 1842," unpublished manuscript, Münster, 1976.

[31] This is a central argument of my book *Handwerker-Sozialisten*.

Only the abolition of the notorious truck system was taken into "closer consideration."[32] This "closer consideration" took another eighteen (!) years. Only in 1848 did the social upheavals in the industrial districts of the Rhineland force the government to serious legal action against the fraudulent paying in kind instead of cash money.

In this very year of 1848 the social pressure of the working population for a while overwhelmed the resistance of the government and its allies, the thrifty entrepreneurial "newcomers" and the politically ambitious big capitalists. Some of the old regulations were re-established, not only in the Solingen district, but in all neighboring export trades with a similar structure of production.[33] In spring 1848 the Solingen cutlery workers formed – with the help of some corporately-orientated entrepreneurs[34] – guild-like "*Bruderschaften*" which instantly tried to negotiate a minimum wage and "closed-shop" system with firms of their respective production sectors. In a few sectors commissions representing artisan workers and entrepreneurs in equal numbers were formed in the traditional manner and equal piece rates, i.e. minimum wages, for every product and every working process stipulated. Such was for example the case in the sectors of the forging and grinding of weapons still dominated by the old established firms. But in other sectors, many firms at first consciously delayed negotiation and later on – in the summer of 1848 – refused to take part in a process of wage-fixing.

While in neighboring districts – the textile trades of the Wupper Valley and the Krefeld area – the artisan workers successfully carried through minimum wage systems and arbitration courts, which lasted well into the reactionary period of the 1850s, the efforts of the Solingen cutlery workers were abortive. In Solingen the social changes within the commercial class seem to have been more severe and the structural crisis deeper since the industrial deregulation of the Napoleonic period than in other neighboring export regions. In the winter of 1848/49, the direct revolutionary pressure for political and social concessions had vanished. Meanwhile even established firms in the weapon industry refused to pay the wages fixed the previous spring, while established firms in other sectors of production, having paid higher wages voluntarily, now reduced them substantially. There were rumors about industrial action, discussed in secret meetings, by the artisan workers. Strikes as a last resort in solving a wage conflict were not absolutely unfamiliar to them, already having been used by the guilds in the eighteenth century. Strike regulations, though described in flowery language, formed an integral part of the statutes of their 1848 *Bruderschaften*, too.[35]

32 Boch and Krause, *Historisches Lesebuch*, pp. 63–64.

33 Boch, "Tarifverträge und 'Ehrenräte.'"

34 In Solingen and the Wupper Valley the merchant-capitalists who favored a corporate solution, already a minority within the commercial world, were themselves split into two extreme political wings: a royalist conservative group of older merchants and a left-wing, democratic group. Despite all differences both groups had central economic aims in common: minimum wages, rudimentary forms of "trade unions" and a system of "trade-based" arbitration. For further information, see my article "Tarifveträge und 'Ehrenräte.'"

35 See Boch, *Handwerker-Sozialisten*, ch. 5, p. 89.

But the hope for a traditional compromise between capital and labor, supported by the poor prospects of a strike under conditions of economic crisis, was stronger still. The artisan workers still believed in those few patriarchal entrepreneurs who had become their leading figures over the years, and hoped for the good will of the Prussian administration. Only gradually did they begin to realize that the administration – despite all reform proposals of the years 1848/49 – did not take their specific interests into account. Further their traditional leaders had become more powerless than the workers themselves: they had even lost the influence on the "public opinion" of the district they had possessed as businessmen of integrity before and during the first months of the revolution.

But the experiment of the *Bruderschaften* and the shortlived system of fixed wages in the weapons industry was not in vain. Many of those artisan workers who took part in it were still alive in the early 1870s, when the local trade unions were founded. The statutes and regulations of the "new" trade unions were very similar to those of the *Bruderschaften* and their central demands the same as in petitions of the 1820s and 1830s or in the "wicked" year of 1848. In the early 1870s, the Solingen cutlery workers looked back on a sixty-year tradition of struggle for the reintroduction of certain features of the old "industrial code" of the eighteenth century. But, for the reinstitution of these elements they now trusted in their own strength alone.[36]

Their success in this project led to the new unforeseen upswing of the district's cutlery production. It was supported by the international political atmosphere favoring "Free Trade" (at least until the late 1870s), and by the ever-growing American market whose constant orders strengthened the bargaining position of the unions as well as their power to reinforce the quality standards. "Made in Solingen," like "Made in Germany," now became a label for products of excellent quality at a reasonable price.

The last decade before World War I: the spread of a "sense of crisis"[37]

At the beginning of the twentieth century, the Solingen cutlery industry held a strong position on the world market. It had reduced its Sheffield competitors to second place and was praised by foreign observers for its technological capacity.[38]

[36] See my recently published articles: R. Boch "Zunfttradition und frühe Gewerkschaftsbewegung," in U. Wengenroth (ed.), *Prekäre Selbständigkeit. Zur Standortbestimmung von Handwerk und Kleinbetrieb im Industrialisierungsprozeß* (Stuttgart, 1989) (Veröffentlichungen des Instituts für Europäische Geschichte) and R. Boch, "Die Entstehungsbedingungen der deutschen Arbeiterbewegung: das Bergische Land und der Lassall'sche 'Allgemeine Deutsche Arbeiterverein,'" in A. Herzig (ed.), *125 Jahre deutsche Arbeiterbewegung* (Hamburg, 1989).

[37] The following is based on chapter 12 of Boch, *Handwerker-Sozialisten*, pp. 230–250.

[38] For example by the contemporary cutlery expert Prof. G. I. H. Lloyd in his book *The Cutlery Trades* (Toronto, 1913; rptd. London, 1968), p. 381 or in the English trade union weekly *The Metal Worker*, nos. 17, 18, 19 (1908).

Despite these facts, a sense of crisis spread among the entrepreneurs. In September 1908 the most important local newspaper published an article with the programmatic title: "On the crisis of the Solingen industries."[39] The article was written by the chairman of the Solingen Employers' Association (Arbeitgeberverband), an organization consisting not only of cutlery-producing firms but also of firms from the metal industries, which very often had developed out of the innovative mechanical forging plants. The organization consciously distanced itself from the corporate orientation of the old established "trade associations," the so-called *Fabrikantenvereine*. The article demanded that fixed wages for individual models and workprocesses should be given up altogether. Otherwise, the position of the Solingen cutlery industry was endangered. It was not foreign competitors such as the Sheffield and US cutlery industries that the author of the article feared, but rather that Solingen's leading position among the cutlery industries of the German-speaking area was in real danger. Within this area, a number of traditional producers in Württemberg, Saxony and Bohemia had recently changed to specialized production of, for example, pocket-knives or table-ware only. In contrast to the Solingen mode of production – i.e. formally independent artisan workers and the "putting-out" system – these specialized goods were produced by semi-skilled workers engaged on piece-work, and their production took place in factories. Furthermore, many firms in the German Empire bought forged semi-products from Solingen firms and ground them on a piece-work basis in their factories. Both types of competitors, the author argued, were able to sell their products much cheaper than the Solingen industry was, and could therefore satisfy and thrive on the "growing demand for cheap cutlery." In this "growing demand" he thoroughly believed.

As a matter of fact the specialized production mentioned in this article was not unknown in Solingen either: a number of firms had produced cutlery in this way since the 1880s. In Solingen, however, these firms had so far been socially isolated. They were not part of the established corporate organization of capital and labor and their growth had been prevented by joint political action by the old entrepreneurs' organization, the *Fabrikantenvereine* and the unions.[40] The production of cheap cutlery had been assigned a marginal place in the Solingen industry, as the bulk of Solingen products were to be of high quality.

In his article the chairman of the new entrepreneurs' organization, the Arbeitgeberverband, challenged this very policy. He threatened that unless the unions agreed to a general lowering of the wage standards, the members of his organization were determined to force a breakthrough for the piece-work, factory-based production of cheap cutlery.

Before 1908, such polemics against the "excessively high wages" and long-term attempts of individual entrepreneurs to introduce piece-work in purpose-built factories had not been heeded by the majority of Solingen entrepreneurs. A

[39] The article was published in the regional *Kreis-Intelligenz-Blatt* on September 12, 1908.

[40] For the emergence of these workshops for cheap cutlery – nicknamed *Bärenställe* (bear-pits) – on the fringe of the production system and the story of their social proscription by the Solingen public, see Boch, *Handwerker-Sozialisten*, pp. 65, 73, 148, 168, 173, 226, 235.

change of this dominant attitude occurred as a reaction to a serious export crisis. From 1907 to 1908, the export of Solingen cutlery dropped from 26.3 million Reichsmark in 1907 to 22 million Reichsmark in 1908/09 and returned to the 1907 figures only in 1911.[41] This crisis was generally – by "old" and "new" entrepreneurs alike – regarded as a structural crisis. The growth-rates of the previous decades were regarded as unrepeatable. The times in which it had been possible to take over the market share of an important competitor on the world market, as had been the case with Sheffield, were gone. In recent years, a number of countries, e.g. Japan, Spain and Sweden, had started to increase tariffs and to develop national cutlery industries. They, too, like their competitors on the domestic German market, produced large series of cheap, low-quality products.

In this situation the stagnation of sales on the US market became a focal point of attention for the owners of large and small firms alike. Though the Solingen cutlery industry was much more efficient than the American and Japanese industries, and though the American cutlery industry was able to hold its ground in the domestic market only on the basis of very high protective tariffs,[42] these very tariffs enabled it to compete successfully with Solingen for the rapidly growing lower-priced segments of the American domestic market, and indeed frequently to undersell Solingen products.

In the US domestic market the Solingen export firms could observe a dramatic change in market structures. The enormous increase of the production of cheap cutlery was only possible on account of certain basic preconditions still largely unknown in Europe: a completely new system of sales and distribution, i.e. the chains of department stores, fixed-price stores (Woolworth, etc.) and mail-order sales had developed since the 1880s.

Only a few larger Solingen firms such as Henckels had already gone into the marketing of their own products by establishing retail outlets in large European and American cities and producing catalogues. The majority of the small firms and putting-out merchants still manufactured to the varying orders of the wholesale traders and held minimal inventories. The bulk of the producers thus relied on either a huge number of small but established markets for "traditional" types of knives and cutlery, which had emerged decades or centuries ago and underwent slight variations only, or adapted changing styles from the high-fashion sector for more popular markets. The creation of a vast new market by rigid mechanization and standardization of production[43] and a new system of

[41] See Möhle, *Exporthandel*, p. 31.

[42] According to the monthly special periodical *Messer und Feile*, edited in Ludwigshafen and financed by the employers of the German small-iron industry, the US import duties in 1913 were as follows: for spring-knives 77.6 percent by value, for scissors 53.7 percent, for knives and forks 41 percent, for razors even over 100 percent on average (*Messer und Feile*, June 15, 1913, p. 463).

[43] Still a good account of the development of American cutlery industry is M. van Hoesen and A. Taber, *History of the Cutlery Industry in the Connecticut Valley* (Northampton MA, 1955). See also N. Rosenberg, "Economic Development and the Transfer of Technology: Some Historical Perspectives," in his *Perspectives on Technology* (Cambridge, 1978).

distribution, fostered by the growing demand of the poor East European immigrants into the US for basic and cheap household equipment and stimulated by a standardization of taste, reaching the stratum of skilled American workers well before World War I, became a new, alarming experience for Solingen exporters.

The Solingen cutlery industry remained restricted to the higher quality segments of the US market, which the American cutlery industry still was not able to supply. Though the cutlery industry of the US and other countries, building up their domestic production, were not seen as a challenge of Solingen's dominance of the world market in those years before the war,[44] the average Solingen entrepreneur became convinced that he had to react to the slow but constantly growing production of cheap, standardized cutlery.[45] It was now widely believed that – at least in the long run – the competitive power of the Solingen cutlery industry could only be maintained by abolishing the unions' right to co-determine the quality standard of the products, which had impeded a change to the production of simple cutlery in the past.[46] Because the extended production of simple and cheap cutlery would necessarily carry with it a substantial reduction of the average wage level as well, the new attitude of the entrepreneurs meant in consequence the total abolition of the historical "social contract" between capital and labor in Solingen.

The "hard-liners" with the Arbeitgeberverband tried to use this change of mentality of the average entrepreneur for a great confrontation with the local unions on the question of the introduction of factory-based piece-work, which was seen as a decisive step towards a transformation of production and a first step towards the introduction of "special-purpose" machinery in the grinding and

[44] See, e.g. the programatic article "Die heutigen Verhältnisse in der Solinger Stahlwaren – Industrie und ihre Zukunft," in September 15, 1912 *Messer und Feile*, p. 742 or the article "Das Angstgeschrei der amerikanischen Feilen- und Messerfabrikanten", September 15, 1913, p. 709. This article very competently analyzed the American cutlery production, coming to the conclusion: "First of all it clearly follows from this that the American superiority is sheer nonsense."

[45] The employers' periodical *Messer und Feile* began to propagate an enlarged specialized mass production instead of the traditional "flexible specialization." In an article on July 15, 1911 entitled "Suum cuique" it concluded: "The stream washing away the putting-out system can't be stopped. Wouldn't it be better, if the small entrepreneurs began to specialize, as the so-called bear-pits began to do during the last years? If they, instead of a large number of models, produced only some types of scissors and knives and developed the products on this line in every direction?" See also a more detailed proposal for specialized production in *Messer und Feile*, September 15, 1912, p. 742.

[46] See, e.g. *Messer und Feile*, June 1, 1910, p. 19: "No objection can be made to wage agreements between workers and factory owners in the present stage of industry, but we must reject the character of the collective agreement in the Solingen district, which holds the whole industry in bonds, hindering a thriving development." *Messer und Feile*, July 15, 1911, p. 18: "In particular it is such an abuse, if a workers' organization commands the employers: 'Only this way you are allowed to produce, in such a quality and using such a division of labor.' Or amounting to the same results: 'If you produce a different quality, you have to pay a higher price to the worker, if you manage to introduce a better division of labor, you still have to pay the workers the old wages.'"

hafting of cutlery. Though member firms of the Arbeitgeberverband and the local unions clashed several times, the majority of the entrepreneurs in the old *Fabrikantenvereine* remained neutral and all labor disputes ended in a tie. This was due on the one hand to the onset of a renewed boom from 1911 onwards, enabling the average Solingen entrepreneur once again to increase export profits. On the other hand, too many entrepreneurs – a large number still mere "putting-out" merchants – still profited from the "independent" social status of the grinders and hafters and their ability for "flexible specialization."

All they wanted was cheaper and socially unrestricted production, not a basic change of the structure and techniques. Therefore they avoided a great confrontation with the still powerful and financially strong local unions. They preferred to fight small wars against the social position and the "old" rights of the artisan workers, thus adopting a course leading towards the gradual dissolution of the corporate compromise of the 1870s. The course taken was indeed successful: from 1908 onwards the wages of the outworking grinders and hafters as well as the fatory-based forgers stagnated. This development was helped by a deep split within the Solingen labor movement, which made it unable to act effectively against the entrepreneurs' policies.

The reaction of organized labor to the emerging "sense of crisis"[47]

At the time of this challenge to the social and industrial "status quo" the old cutlery unions founded in the 1870s were no longer the only trade unions in the district. The German Metal Workers Union (Deutscher Metallarbeiterverband= DMV)[48] – a nationwide union, trying to organize the workers of the various fast growing German metal industries independently of their trade and qualifications – had established a local branch in the Solingen area. This "new model" union, aiming for mass membership, was first of all successful in newly industrialized areas and in the south of Germany. In Solingen as well as in other old industrial centers of the lower Rhineland it played an insignificant part, at least until the beginning of the twentieth century. By 1899 the local branch of the DMV in Solingen counted only 109 members, all of them working in metal trades outside the cutlery industry. By 1905, however, the local branch had organized nearly 2,500 workers, half of them working in the fast growing metal factories which were not involved in the traditional manufacture of cutlery. Nevertheless, the

[47] This section is based on Boch, *Handwerker-Sozialisten*, pp. 149–180.

[48] The DMV has been the topic of much recent historical research in Germany. A good collection of contemporary sources, but with a rather conventional interpretation: K. Schönhoven, *Expansion und Konzentration. Studien zur Entwicklung der Freien Gewerkschaften im Wilhelminischen Deutschland 1890–1914* (Stuttgart, 1980), esp. p. 91 and p. 306. In contrast, a very critical assessment of the organization model and the impact of the DMV on industrial policy in Germany is given in E. Domansky, "Arbeitskämpfe und Arbeitskampfstrategien des Deutschen Metallarbeiterverbandes 1891–1914," unpublished PhD, University of Bochum, 1981.

other half consisted of cutlery workers, most of them forgers working as factory workers in the mechanical forges.

The DMV regarded the precisely defined wage rates and the old local unions' right of co-determination with respect to the types of product produced as the main obstacle to the introduction of factory work. Therefore the DMV declared these rights and agreements to be against the interests of the factory workers, its own membership group. Whereas the artisan workers represented sectional interests of a guild-like type, the interests of factory workers were identical with the "general interest of the proletariat." The DMV defined "factory work as such" – regardless if it was in reality skilled piece-work – as representing "technical progress," as "a step in the natural technical and economic development."[49] "Technical progress" on the other hand was naively and optimistically seen as the natural ally in the preparation of the coming victory of the working class.[50] Due to this vulgar-marxist attitude the DMV did not support the old unions' fierce struggle. Rather, in industrial conflicts, it frequently took the side of the entrepreneurs, accusing the old unions of causing technological stagnation and of hindering the "necessary adaptation" to the "demand for cheap cutlery on the world market." "Technical progress" – the DMV argued – could not gain ground without freedom of decisions for the entrepreneurs in these matters.[51]

In Solingen the leaders of the DMV, representing a new political type of Social Democrat in pre-war Solingen as well, were the real protagonists of an "alternative" model of mass production in the cutlery industry. While – before the war – even the entrepreneurs within the Arbeitgeberverband took specialized mass production as a mere point of reference, a regulative idea predicting that only as a final – though undoubted – result of a long process the outworking grinders and hafters would die out,[52] the DMV officials wanted this "necessary" change speeded up. They dreamt of huge factories, producing in large series for an international mass market; and this process – in their view – was not to be obstructed by the organized labor movement.

The answer of the local unions to the positions taken up by the entrepreneurs and the DMV can be reconstructed from a series of articles in the unions' weekly journal:[53] The old unions believed in the possibility of influencing certain

[49] DMV-Hauptvorstand (i.e. general committee) (ed.), *Was bietet der Deutsche Metallarbeiter-Verband den Stahl- und Schneidwarenarbeitern des Kreises Solingen?* (Stuttgart, 1906).

[50] For instance, the minutes of the general meeting of the DMV in 1905 read: "The faster the technical development, the earlier the capitalist mode of production will have reached the point . . . at which it will have to be replaced by a higher stage of production." See *Protokoll des DMV-Verbandstages 1905* (Stuttgart, 1905), p. 138.

[51] In a speech at a mass meeting in 1905 the local representative of the DMV in Solingen, Sendler, went so far as to produce the following statement, which was to become a "discussion point" within the old local unions: "If I were an entrepreneur, I would not permit workers to tell me what to do or not to do either" (*Bergische Arbeiterstimme* – the local daily newspaper of the Social Democratic Party – June 5, 1905).

[52] See, e.g. *Messer und Feile*, December 15, 1910, p. 17; September 15, 1912, p. 742.

[53] *Der Stahlwaren – Arbeiter*, esp. June 15, 22, 29, 1906, "Was bietet der Deutsche Metallarbeiter – Verband . . ." and July 27, and August 10, 1906, "Qualitäts – oder Schundwarenindustrie?"

conditions in the world market. They did not credit the DMV and the entrepreneurs' argument that the world market demanded cheaper cutlery and that the Solingen industry had to produce such goods before international competitors beat them to it. Instead, they were convinced: "Not the world market demands cheaper goods, but the capitalists greed for profit."[54] In their judgment, the tendency towards the production of cheap and simple cutlery was based on competition among entrepreneurs, not on an existing demand. It was the function of unions to protect the workers from the laws of capitalist competition, and they did not see this function restricted to the protection of wages. It had been their experience for more than thirty years that by taking part in the determination of the types of products, competition among entrepreneurs could be reduced and influenced by the workers.

Even if there existed a certain demand for cheap cutlery in the world market they did not accept that supply had to follow demand, but they believed that the demand for high-quality goods could be produced by a supply of such goods which shaped the tastes and habits of the buyers. In continuing to influence the type of demand for cutlery, the local unions saw their only opportunity to keep up the cutlery grinders' relatively high standard of living and rights of co-determination. To the local unions, enforced production of cheap cutlery marked the beginning of the destruction of the Solingen cutlery industry. Such innovation would slowly destroy the basis of Solingen's position on the world market, namely the skills of its highly trained workers, and in the end, competition for the cheapest wages would be the sole criterion for survival. Thus they accused the entrepreneurs of the Arbeitgeberverband and the local officials of the DMV of having forgotten or having never understood the historical lessons of the past century and of ignorance of the specific conditions of the industry.[55] Only a common interest in a high standard of quality could – in their view – avoid a real crisis in the future.[56]

The role of mechanization by means of "single-purpose" machinery

Curiously, during the lively debate about the structure of industry in pre-war Solingen the question of the mechanization of grinding – still the stronghold of artisanal work based on the traditional "multi-purpose" grindstone – was of secondary or even less importance.

In contrast to the ongoing entrepreneurial plans for the division of the artisanal grinding process into skilled piece-work – performed under the roof and the supervision of the factory owner – mechanization was seen as a challenge to the industry in a distant future.

Though a "single purpose" grinding machine for the so-called hollow-grinding of razors was successfully introduced in the 1890s,[57] new types of grinding

[54] *Ibid.*, August 10, 1906. [55] *Ibid.*, December 3, 1909. [56] *Ibid.*, October 2, 1908.
[57] See Boch, *Handwerker-Sozialisten*, p. 135.

machines for knives[58] invented at the turn of the century showed poor results and lacked economic efficiency, because they could execute only some working processes on a restricted number of models. As late as 1911 a technically adequate grinding machine for a wider range of knives was invented, but its productivity per hour rarely reached that of a skilled grinder working on a "multi-purpose" grindstone.[59] Even this level of productivity was only achievable, if the machine could be used in the manufacture of a large series of the same type of knives. Every adjustment to the shape of another model needed a lot of additional work by a skilled mechanic, thus lowering the productivity substantially.

Thus, though mechanization of the grinding process had become possible from a purely technical point of view, there was no economic reason to follow this line[60] as long as no decisive change in the market structure and in the traditional structure of production by outworking grinders had taken place.

Even those firms most engaged in changing the "status quo" saw the introduction of factory-based piece-work grinding and the employment of unskilled workers as a necessary first step towards mechanization. Less expensive than experiments with mechanical grinding, piece-work could alter – in the long run – the type of products offered by the Solingen industry in the direction of cheaper and more standardized goods.

Only the combination of piece-work and the centralization of grinding were seen as creating the market conditions for the introduction of advanced "single purpose" grinding machines. Under these conditions alone, mechanical grinding could become an alternative to the established system of production, because large series of fairly high-quality cutlery could be produced this way.[61]

But these were thoughts about future developments. In reality, production was much more changed by the unforeseen effects of World War I, which brought these debates and labor conflicts to a sudden end.

The impact of World War I on the Solingen cutlery industry and the rise of US competition

The first blow to the "artisanal backbone" of industry, i.e. the constant reproduction of skills needed for the established form of "flexible specialization" came with the disorganization of apprenticeship training during the war. Though

[58] A reasonable "single purpose" grinding machine for scissors was only invented as late as 1928. See Putsch, *Heimarbeit*, p. 181 (n3).

[59] *Messer und Feile*, December 15, 1911: "Die neue Messerschleifmaschine und die Solinger Schleifer."

[60] Though many entrepreneurs saw the possibility of introducing a functioning grinding machine as a welcome means to threaten the strong knife-grinder's union, which had rarely lost a strike in the past thirty years. An article in *Messer und Feile*, December 15, 1911, p. 17, expressed this clearly.

[61] This possible future development was proposed in *Messer und Feile*, September 15, 1912: "The invention of a [highly productive, R. B.] grinding machine, designed only for the grinding of very few types of blades, could be achieved much more easily than that of a universal purpose machine, aiming to grind all types of knives and scissors."

this institution was reorganized until the early 1920s, it never regained the importance, strength and strictness it possessed in the years of the *Kaiserreich*.

Since the greater part of the cutlery industry was not declared essential for the wartime economy,[62] a large number of apprentices and skilled artisans were drafted, and a considerable number died. Many older "independent" outworkers lost work and had to look for a job in local factories or the important military industries of the nearby Ruhr area – joined by the apprentices too young for recruitment.

When, in late 1916, the bulk of the Solingen cutlery industry was included in a new wartime scheme – aimed at mobilizing the productive capacities of the economy – mass production of standardized razors and knives for army use as well as equipment for field-kitchens was introduced. This made the introduction of machines and employment of unskilled labor possible and economical, at least for some larger firms profiting from the program. For these firms World War I initiated an unprecedented opportunity to experiment with new types of industrial production without the restrictions of the "normal" market structure and international competition. In addition, due to the conditions of the wartime economy, the local unions could not strike against the introduction of piece-work and new brands of cutlery or the employment of boys or women to work on the new machines.[63]

But, while rationalization through standardization affected only a rather small sector of industry, the war had another, more disastrous effect. For almost five years Solingen was cut off from its international buyers. Its ability – seen by the old unions as a central strategy for the perpetuation of "flexible specialization" – to shape more or less, due to its dominant position on the world market, the general taste in cutlery according to the types of products Solingen supplied, was severely shaken. Some smaller countries were forced to build up their own cutlery production, while others, like Spain, Sweden and Japan took the opportunity to export their highly specialized cutlery products, enlarging their insignificant, tariff-protected pre-war capacities. But the main winner of the war was the American cutlery industry, becoming the most important supplier of many overseas markets, e.g. South America, formerly dominated by Solingen.[64] For the US producers, World War I was a necessary precondition or at least an accelerator for establishing themselves as the main

[62] The production of weaponry – in the late eighteenth century the most important part of Solingen's industry – held only a share of 10 percent of the overall cutlery productions in the district in 1913. Its enlargement and administrative protection did not alter the result that the "flexibly specialized" cutlery industry suffered severely from the war.

[63] For a detailed account of the economic and social changes in wartime Solingen, see M. Krause, "Die Gewerkschaftsbewegung und die sozialistische Arbeiterbewegung in Solingen 1914–1925," unpublished manuscript, University of Bochum, 1981, and Putsch, *Heimarbeit*, p. 90.

[64] See F. Fesenmeyer, *Ursachen und Gründe der Wandlungen im Export deutscher Eisen- und Stahlwaren nach Südamerika* (Cologne, 1933), pp. 27, 49, 62.

competitor of the Solingen cutlery industry on the world market, leaving Sheffield far behind.

Due to a constant shortage of skilled labor – despite renewed immigration of European cutlery workers from Sheffield and Solingen – the American firms had experimented with mechanical devices from a very early stage.[65] The large homogeneous domestic market, new methods of selling and retailing as well as the "standardization of taste" in the lower segments of the market, fostered the production of large series of cheap cutlery which in turn favored the employment of "single purpose" machinery. In determining patterns of design and levels of quality, the requirements of mechanized production therefore increasingly became the dominant factor. As a result, two-thirds of the cutlery produced in the USA – though mainly of lower quality – was machine-ground already in 1918.[66]

The American production strategy is precisely described in the following contemporary quotation. "What has been the American response? Briefly that where mechanical devices cannot be adjusted to the production of the traditional product, the product must be modified to the demands of the machine. Hence, the standard American tableknife is a rigid, metal-shape, handle and blade forged in one piece, the whole being finished by electroplating."[67] The inherent tendency towards the development of a totally new product in order to overcome the "natural" advantages of a "flexibly specialized" competitor can most clearly be demonstrated by the example of the "safety razor" invented by the American Gillette company. The "safety razor" can be seen as the perfection of the American logic of production, hand forging and hand grinding having been practically eliminated. While at the turn of the century the US production of razors had to be heavily protected against imports from Germany, the Gillette company conquered the world market with its disposable blades during the war and its aftermath. The success of the company was apparently fostered by a new consumers' attitude: the readiness to throw away instead of resharpen partially used products.

The 1920s, the international economic crisis, and World War II

With the end of the war economy and the return to peace it became apparent that during the war the productive capacities of the international cutlery industries had been multiplied – compared to pre-war conditions – through the enlargement of existing and the building of new factories.

In Solingen, it was mainly the production capacity of cheap and simple cutlery in those larger firms having produced exclusively for the army, that had risen

[65] For the European immigrant workers and for the technological development of the American cutlery industry, see van Hoesen and Taber, *A History*. [66] *Ibid.*, p. 46.

[67] Lloyd, *Cutlery Trades*, p. 394.

substantially. Despite postponed consumer demand, prices fell after the war, due to the now fierce competition and over-capacity.[68]

In addition, the structure of the world market was now partially shaped by the mass production introduced by Solingen's US competitor. US exports to Latin America, and also Europe,[69] were growing rapidly. However, the new role of the American competitor must not be overestimated: as late as 1928 over 80 percent of the American export of cutlery products was made up of "safety razors" and changeable blades. Table and kitchen knives made up only 4.3 percent and scissors only 1.1 percent of the total export figures.[70]

Thus the only sub-market that was radically changed by the American methods of production was the production of razors. For the European firms and workers in this sector, this change had dramatic consequences. But, mass production using "single purpose" machinery during this time was thus economically superior to the traditional European system of cutlery production by means of "flexible specialization" only if accompanied by the introduction of a totally new product, i.e. in its most radical form.

The Solingen cutlery industry met these new challenges in the world market by reactivating the full potential of the system of "flexible specialization" at all levels of the production process. Yet, as will be described below, the destructive forces that seem to be inherent in this system of production if not combined with a "special contract" between labor and capital were unleashed at the same time. In a surprisingly short time the Solingen industry was able to rebuild its network of foreign business connections and to profit from the incipient post-war boom. The policy that enabled Solingen to do so was two-fold. Wherever possible, the Solingen industry aimed at the market for the highest quality products which could not be served by the new national industries. Solingen pursued this policy not only in European countries with relatively young cutlery industries but also in the US and Asian market. In the latter, the Japanese cutlery industry had started to dominate the market for cheap cutlery. In all other markets, by contrast, including those of England and France, the other traditional producers of cutlery, Solingen competed also for a share of the low and lowest-quality products.[71]

In contrast to the pre-war period, however, no limit for the down-grading of the standard of quality seemed to exist any longer: whereas in 1913 the difference between the lowest and the highest value of cutlery exported by Solingen had been 100 percent, the difference was 500 percent in 1920.[72]

The astonishing recovery of the Solingen industry in the immediate post-war

[68] Large segments of this chapter are based on the recently published work of J. Putsch. Stimulated by my book *Handwerker-Sozialisten*, he analyzed the further development of the Solingen cutlery industry from the 1920s onwards from a similar perspective.

[69] See Putsch, *Heimarbeit*, p. 360. [70] *Ibid.*, pp. 130, 131.

[71] *Ibid.*, ch. 3.3 "Dominanz unter veränderten Vorzeichen," p. 108. Partially due to this flexibility the cutlery industry was one of the few German industries which already in 1925 exported more in volume and value than before the war.

[72] Figures quoted from: *Enquete-Ausschuß zur Untersuchung der Erzeugungs- und Absatzbedingungen der deutschen Wirtschaft. Die Deutsche Eisen- und Stahlwarenindustrie* (Berlin, 1930), p. 374.

period was based not only on the advantages of "flexible specialization" but to a considerable extent on the high German post-war inflation. With workers' wages constantly lagging behind the inflationary rise in prices the industry was able to offer even high-quality products at "dumping prices." In addition, the export of Solingen cutlery was favored by the fact that the German exchange rate depreciated more rapidly than even the rapid rise of German prices. This was an ideal situation for manufacturers of cutlery who exported goods with a low import and a high labor content.[73]

What proved to be good for the performance of industry was disastrous for the cutlery workers and the old social compromise in the district:[74] the lowering of wage levels introduced in the final years before the war continued during the inflation and came to a bizarre climax during the hyper-inflation of 1923. For several months the latter caused a complete breakdown of the wage system. For a short while even payment in kind instead of cash money became a common feature again. It was not only the shortlived revival of the "truck system" reminiscent of the times of crisis in the first half of the nineteenth century. In tendency the whole local industry – due to external changes in the market situation and internal changes in employers' attitudes – was again running against the interests and the economic substance of the "direct producers."

During the whole Weimar period, wage levels remained below those of 1913. Furthermore, for the first time in many decades, long-term unemployment became a characteristic feature of the working life of the skilled artisan worker. It first hit the forgers, grinders and hafters in the large, traditional manufacture of razors which was severely shaken by American competition.[75] It became more widespread when the post-war boom, backed by inflation, came to an end in the mid-twenties, revealing Solingen's structural overcapacity in the production of simple – and average – quality cutlery. Long-term unemployment became common, when the World Depression after 1929 destroyed the last chances for a readjustment of the industry to the demands of the international market. From their pre-war status as a labor aristocracy the Solingen cutlery workers were thrown back into the day-to-day struggle of proletarian existence as were their predecessors a hundred years ago after the abolition of the "industrial code" in 1811 and the substantial changes in the world market after 1815.

The crucial weakness of the old local unions after World War I was due not only to these economic developments but also to internal changes within their most important membership groups, the "independent" outworking grinders and hafters. The "renaissance" of the "putting-out" system – the number of outworkers soon equalled and after 1920 exceeded the pre-war level – was only seemingly a connecting link to the "golden years" of the cutlery industry in the times of the Kaiserreich.

[73] This analysis was offered by: H. Townsend, "World Trade in Cutlery 1920 to 1951," in *International Cutlers*, 2, 4 (1952), 10–16.

[74] For the following see Boch, *Handwerker-Sozialisten*, p. 294.

[75] Several hundred razor grinders, forgers and hafters lost their work due to the conquest of the market by the American Gillette Company. See my article, "Changing Patterns."

The increase of outworkers continued until 1925. However, in contrast to the traditional outworkers of the Solingen system of production, these new outworkers were not properly trained and, like those grinders and hafters employed for piece-work on the premises of the larger firms, frequently produced cheap standardized goods. Since the training of young artisans had been seriously interrupted during the war the growing number of independent outworkers were recruited from the ranks of factory workers. Through the fast quantitative growth of half-trained outworking grinders and hafters quality standards dropped considerably. And, in addition, the economic conditions of the established artisan workers deteriorated. The percentage of unionists decreased, and did not keep up with the increase in outworkers. Most importantly, though, the old local unions totally lost the control over the everyday working life they had previously possessed.[76]

The outworking grinders and hafters started to lose the common set of ethical and professional standards which had unified all members of the profession. These written and unwritten rules had been followed during the time of the guilds, had formed the content of an artisanal counter-culture from 1811 to 1870 and had been reinstituted by the workers' movement in the last third of the nineteenth century. Now they were being lost. Not only did the number of artisan workers producing "shoddy wares" increase, but also the number of workers prepared to work at wages distinctly lower than those defined in the price-lists of the wage-agreements. Thus, in the late twenties, the workers' movement no longer provided a social basis for the control of the (old-established) "social contract." While the entrepreneurs were not interested to secure the observance of the legally still binding tariff wages, the old local unions had lost their ability to do so.

The discussion about the future of the Solingen industry became lively again with the end of the post-war boom in the middle of the twenties. At first sight, the issue was discussed even more abstractly than before the war: in the absence of the real changes of the previous decade, i.e. the many small technological improvements in several individual sectors of production[77] and the change in the number and status of the outworkers. Before 1914 the debate centered around the introduction of the factory system as the dominant form of production in

[76] This was partially due to the spread of electricity as a new and comparably cheap source of energy for the driving of the traditional "multi-purpose" grindstones. It led to a sharp rise in the numbers of small grinders' workshops in the backyard of their family homes during the last years of the pre-war period and in the first half of the 1920s. At the same time, many grinding halls – organized under the "paying for steam" system – in which up to 100 still "independent" grinders worked together had to close down. This development fostered the unlimited expansion and the "individualization" of the grinders trade, making union control over work hours and quality standards very difficult. The spread of electricity, however, did not lead by necessity to individualization and the weakening of the labor movement. In Saint-Etienne (France) at the turn of the century, for instance, the ribbon weavers, acting through a Socialist municipal government, controlled the length of the working day simply by regulating the hours at which the local power station supplied current to the looms. See C. Sabel and J. Zeitlin, "Historical Alternatives to Mass Production," *Past and Present*, 108 (1985), 133–176. [77] See Putsch, *Heimarbeit*, p. 169.

order to create first of all the necessary precondition for standardized production of large series. Now, the organization of the process of production within this new form of production became the focal point. Much more explicitly than before the war, "Taylorism" and "Fordism" as a system of production were being discussed.[78] But, as before 1914, the American cutlery industry played no part, as a model, in the debate. J. Putsch, in his study of the twenties, found not a single article in the technical journals and special periodicals of the small-iron and cutlery industries which referred directly to the production of cutlery in the US.

Earlier, I observed that at first sight the debate about the future of Solingen's industries had a somewhat remote character, not touching on recent social and economic changes. In hindsight, however, the debate assumes a different social function: it finally made the new paradigm of industrial development acceptable to the majority of Solingen entrepreneurs and to the bourgeois public of the district. Furthermore, the owners of leading firms could now mentally link this paradigm to their own experience of wartime production.

The debate about mass production with "single purpose" machines, however remote it was at the time from the conditions prevailing in Solingen, thus became an important factor in its final realization. Whether this function of the debate was consciously known to the participants I cannot say. In contrast to the debate before 1914, after the war the old production system was no longer defended. Resistance was very weak, on the organized workers' part, compared to the pre-war days. From 1926 onwards, the "German Metal-Workers' Union" (DMV) was the only union representing the workers' interests, and the position of this organization on the issue of rationalization was very uncritical indeed. Whereas before the war the widely read weekly of the old local unions, the *Stahlwaren–Arbeiter* had functioned as a forum for precise arguments against the introduction of piece-work and "single purpose" machines, only one booklet continued this tradition after the war. Significantly it was written by a former artisan worker, now a functionary of the DMV.[79]

Rationalization, though widely and publicly discussed, was in fact introduced

[78] *Ibid.*, p. 160; For example the – in the local commercial world – intensively discussed article by Hermann Bick, owner of the important and old established cutlery firm Herder: "Henry Ford und sein System," *Zeitschrift des Vereins für Technik und Industrie*, 7 (1924), 199. See also – besides many articles on "Taylorism" – the programmatic contribution in the trade journal *Messer und Feile*, February 15, 1926: "Wichtige Vorbedingungen bei der Umstellung der Solinger Industrie auf das 'amerikanisch-Fordsche-Arbeitssystem.'"

[79] J. Kretzen, *Die Krise in der Solinger Stahlwarenindustrie und ihre Lage auf dem Weltmarkt* (Solingen, 1926). After the war the local unions only managed to publish a monthly bulletin. Only some numbers have survived in the city archive. The right-wing Social Democratic leadership of the local DMV remained close to the naive point of view on "technical progress" this union had propagated before the war. In the now communist daily political newspaper *Bergische Arbeiterstimme* the word "rationalization" – appearing in many articles in the second half of the 1920s – became a more and more meaningless label, used to "analyze" and describe nearly every conflict within the industry. The Communists backed workers' resistance to the introduction of piece-work on grinding machines for political reasons – enforced radicalization – only. The idea of an alternative model of future industrialization based on the old artisanal "brains-trust" – as proposed by the local unions before the war – was alien to the young functionaries of the Communist Party in Solingen.

by very few, large firms only.[80] Grinding machines were continuously developed and improved in Solingen and several new larger firms even produced grinding machinery. However, these machines were either exported or sold to the formerly small – now growing – factory-based cutlery districts of Thuringia and Württemberg. Ironically, the machines for the international mass production, so frequently deplored by the Solingen entrepreneurs for its competition with Solingen, were, to a great extent, produced in Solingen itself![81] In large parts of the Solingen cutlery industry itself, the installation of such grinding machines was still not profitable: the Solingen system still functioned according to a totally different logic of production, the means of production – the traditional "multi-purpose" grindstones – still being owned by the majority of the "direct producers," the artisan workers. Despite all technical improvements of grinding machines there was still no "multi-purpose" machine on the market. All current types were still inferior to the work performance of a trained grinder in terms of speed and precision. Only if used as "single or near single-purpose" machines in factory-based production of large series designed for certain segments of the market, were grinding machines economically superior. Considering the given market structure – shaken though still dominant – and the production structure of the Solingen industry, this kind of superiority was at that time relevant only to a limited sector of the Solingen industry.

In addition, in 1925, the climax of the debate about the future industrialization of Solingen, a central element of the traditional "social contract" between capital and labor was formally[82] abolished: the local unions' right to co-determine the products, i.e. their type and their process of production. When several firms introduced piece-work (typically *not* on grinding machines) by semi-trained boys in the grinding of scissors without asking the unions, the latter made a last attempt to co-determine the methods of production through industrial action. The strike, beginning in the autumn of 1925, ended in total defeat. At the end of January 1926 the employers' association locked out all workers, approximately 3,500, engaged in the production of scissors all over the district. In April 1926, after many weeks of industrial action and severely weakened by high unemployment due to the end of the post-war boom, the local unions were forced to sign an agreement in which they promised not to oppose any future changes of the process of production, whether introduced through piece-work or machinery (!). From now on, only the employers were to have the right to decide all matters concerning the process of production and the types of product. As to the introduction of grinding machines it was rather a matter of principle – like the whole debate

[80] As Putsch, *Heimarbeit*, p. 335, points out very clearly. But – furthermore – these few, larger factories for different sorts of cutlery were not reorganized according to "Fordist principles" either. The only sectors of the industry which were organized in this way were the production of the newly invented hair-cutting machines, and the production of "safety razors" in some former traditional razor factories, most of them having been taken over by the American Gillette Company. *Ibid.*, p. 197. [81] *Ibid.*, p. 175.

[82] For the following see Boch, *Handwerker-Sozialisten*, pp. 293, 358, and "Changing patterns."

on the future of the industry – fought out *a priori*, in the absence of any economic necessity.

After that crucial defeat the *raison d'être* of the local unions as independent unions became obsolete. In July 1926 they amalgamated with the "German Metal Workers' Union" (DMV). A number of important concessions on the part of the DMV had eased the decision for the local unions. The outworking grinders and hafters were to man, for instance, half of the local administrative body of the DMV.[83]

But neither the defeat of the local unions nor the rationalization measures implemented during the following two decades caused the direct and immediate destruction of the system of "flexible specialization." Rather the system finally collapsed through the sudden breakdown of the already-shaken apprenticeship system and the effects of the unforeseeable international economic crisis, which hit the mainly outworking producers of cutlery of better qualities even harder than the rest of the crisis-ridden industry.

Since pre-war times dedicated followers of factory-based mass production had euphorically propagated the necessary decline of the "putting-out" system based on artisanal skill. Suddenly the prophecy produced the first real – unforeseen and unwished for – effects: the youth of Solingen lost interest in the cutlery industry and hope in its future. From 1925 onwards more apprenticeship positions were offered than could be filled. Whereas in a survey of 1925 the Solingen labor office found forty-eight pupils of the last forms of the local elementary schools who wished to become grinders, this number suddenly dropped to thirteen pupils in 1927. For professional hafters, the respective numbers were nineteen in 1925 and three in 1927.[84] The high unemployment rate in 1926, the enforced public debate – reaching the level of the daily local newspapers again – but first of all the defeat of the local unions seemed to confirm the impression among the working populace that an immense, unfavorable restructuring of the cutlery industry was coming or already taking place. A growing number of artisan workers expressed this pessimism on the occasion of bringing their sons to the Solingen labor office for occupational counseling: "For God's sake, no occupation in the Solingen industry, it will go to pieces for sure."[85]

A new attitude of denial spread across the Solingen industrial district: the sons of the artisan workers became unwilling to suffer the uncertainties of the export-dependent industry. They were tired of seeing their "independent" position used by large firms as a mere "buffer" to soften the effects of the business cycles and of being forced to bargain continuously for their wages and to undersell their colleagues. Thus, these sons of the artisan workers of Solingen –

[83] By the time of amalgamation, the DMV's position had weakened, too. It had lost several industrial conflicts as well and its membership had fallen from around 15,000 in 1921 to 6,940 by the end of 1925. The local unions during the same time-span had also lost half of their former 8,500 members. See Boch, *Handwerker-Sozialisten*, p. 359.

[84] See Putsch, *Heimarbeit*, p. 225.

[85] *Bergische Zeitung*, September 27, 1927, quoted from Putsch, *Heimarbeit*, p. 226.

confronted with the fate of their fathers – fled the system of "flexible specialization." Without social bonds, responsibilities and regulations jointly respected by capital and labor alike, this system proved particularly anarchical and detrimental for the dependent "independent" producers. Like the out-workers after 1815, they had experienced only the negative side of the system. The advantages of the system experienced by the artisan workers during the Kaiserreich had all but disappeared: self-determination of work, rights of co-determination and comparatively high wages on the basis of high quality production.

However, in contrast to the artisan workers after 1815, the young generation in the 1920s did have an alternative: the rise of a new metal industry out of the cutlery industry since the 1890s[86] created attractive training opportunities in the Solingen district. These industries, less dependent on export and, in the case of the machine-tool industry growing at high speed, were regarded less crisis-ridden and more diverse. To train as a turner or fitter, for example, opened up much more diverse chances for employment than the professions exclusively adapted to the specific Solingen industries which could not be practiced elsewhere. Also, due to the larger size of the firms in the new metal industry, they offered new chances for occupational mobility. The choice of a different occupation than one's father and in many instances the whole family had held over generations was eased by the increasing individualization of working class life. In Solingen, this general development was accompanied – in the early years of the Weimar Republic – by the breakdown of the traditional occupational conditions and of the guild – like local unions which had shaped social life for more than fifty (!) years.

From 1925 there were more exits due to age than entries into the cutlery industry. This trend was speeded up during the Great Depression: there was not only a lack of apprentices but, also, of masters fit to train, either because they were unemployed at the time or because they had left the cutlery industry to seek gainful employment elsewhere. From 1928 to 1936 the number of cutlery workers dropped by 34.7 percent, the number of outworkers – most of them skilled artisan workers – even by 58.2 percent. Whereas in 1925 50 percent of the cutlery workers were outworkers, by 1936 the percentage had dropped to 23 percent. Of the total number of ca 4,600 outworkers merely sixty-five (!) were younger than twenty-five years old in September 1937.[87]

From 1932 onwards the average value of the exported cutlery wares dropped continuously: whereas in 1931 the value per two hundredweight of cutlery was 944 Reichsmark, its lowest value was reached at 573 Reichsmark in 1935, to be

[86] For the emergence of these new industries see Boch, *Handwerker-Sozialisten*, pp. 149, 181.

[87] All figures are taken from Putsch, *Heimarbeit*, pp. 227 and 391. During the Great Depression Solingen belonged – together with Plauen and Chemnitz – to the German cities most severely hit by unemployment. In the winter of 1933 nearly 40 percent (!) of the whole resident population had to be supported by public relief. In 1936 – while the Ruhr area was already in the midst of the rearmament boom – the Solingen district still counted more than 10,000 unemployed workers. See Boch, *Handwerker-Sozialisten*, p. 360.

exceeded only minimally during the following four years. The decline was not solely a result of a number of price reductions. In addition, the structure of the Solingen products had changed: the greater part of the exported goods – and apparently also of those produced for the domestic market – had been shifted to a simpler quality level.[88] As a result, the skilled outworking grinders who made their living through the grinding and sharpening of higher quality cutlery were most severely hit by the lack of work. The dramatic decline in the number of outworkers during the economic crisis supports this interpretation. The motto for managing the crisis had already been announced on August 17, 1930, by the periodical *Messer und Schere* ("Knife and Scissor"), namely: "by changing the type of the Solingen cutlery products from durable goods to wear-and-tear products."

In Solingen the effects of the international economic crisis lasted longer than elsewhere and seems to have removed the last obstacles against abolishing the traditional standards of quality. Thus the Great Depression finally accelerated the production of cheap cutlery and, therefore, factory-based grinding on piece-work. Induced by consumer impoverishment and regulations and orders by the German state, simple, even very simple and cheap cutlery remained the bulk of the Solingen products through the Second World War and the post-war years well into the fifties.

The breakthrough of mass production in the absence of an alternative: Solingen 1948–1960

The destruction of the apprenticeship system beginning in 1925 and the halving of the number of outworking, "independent" artisan workers during the Great Depression can be seen as the decisive caesura between the system of "flexible specialization" and the system of specialized "mass production," though in Solingen the latter never appeared in a pure form even after 1960. Neither the training system nor the number of "independent" artisan workers ever recovered. Both developments proved lasting. World War II, similar in its impact on the "independent" grinders and hafters to World War I, brought the numbers of outworkers down to 3,215 in 1948; less than a quarter in comparison to the 13,000 in 1925.[89]

When around 1950 production and exports began to normalize and a demand for higher qualities reappeared – especially in some of the traditional foreign markets for Solingen cutlery – the shortage of skilled artisan workers was badly felt. This shortage of skilled labor remained a problem despite considerable efforts by the Solingen employers' association and the local labor office. Since the late 1940s both organizations had sought to raise the attractiveness of the apprenticeship system and to anchor it among the larger firms rather than the outworking master-artisans. Skilled labor again proved a crucial element in the

[88] *Ibid.*, p. 232. [89] *Ibid.*, pp. 305 and 391.

Solingen system of production: it was neither technically obsolete nor, given the organizational structure of the average Solingen firm, could it easily be replaced. In the early fifties the collapse of the training system, i.e. the constant lack of apprentices since the twenties, culminated in a structural crisis: despite favorable business conditions, the number of outworking grinders had not increased beyond a mere 3,600 in 1952. Of this number, 80 percent were older than forty-four years old and a remarkable 42.4 percent older than fifty-four.[90]

After it became obvious that the reorganization of the apprenticeship system was doomed to failure a wave of mechanization and rationalization swept the Solingen cutlery industry. Significantly, the Solingen entrepreneurs introduced mechanization and rationalization on a large scale only after the disappearance of an alternative, i.e. the system of "flexible specialization."

In view of the imminent danger of having to stop production before long for lack of skilled labor trained on "multi-purpose" grindstones, mechanization by "single purpose" machines became unavoidable. Many entrepreneurs were conscious of the fact that this change made them a competitor of the many new cutlery industries developed on that line. However, for the entrepreneurs mechanization became all the more urgent, the more the remaining outworkers were able to raise their wages, taking advantage of their growing bargaining power during the boom of the Korean War. Indeed, the skills of the artisan workers had a high price once again. In the mid-fifties Solingen's productive capacities were fully exhausted. Without basic changes in the methods of production no further growth was possible. This was the situation in small and large firms alike. Now, with a timelag of thirty years between its announcement and its realization, the grinding and hafting of cutlery on "special-purpose" machinery in huge series was introduced. It was possible, however, to rely on technological research done in the 1920s.

In 1960 artisan workers still lived and worked in Solingen; the process of dissolution of the "social formation" they represented had not yet reached its end.[91] Yet from their former position of a social power co-determining the structure of the system of production, having been able in the last third of the nineteenth century to shape the development of mechanical forging according to their own needs and logic, they had been reduced to a position on the very margin of a factory-based industry.

The training of highly-skilled artisan workers stopped with the introduction of mechanization or, rather, was shaped to the needs of the new logic of production. Despite the fact that the greater part of specific artisanal skills could not be replaced by mechanized work processes, these skills died out in the following years. The Solingen cutlery industry, once a stronghold of skilled artisan workers, was now dominated by semi-skilled workers, themselves soon to be replaced by female workers. Nothing distinguished Solingen from its interna-

[90] *Ibid.*, p. 305.

[91] In 1961 the Solingen outworkers (mainly grinders) still numbered 2,522, holding a share of 16.5 percent of the whole labor force of the cutlery industry.

tional competitors any longer. The number of resident firms, the rate and volume of exports all decreased.[92]

Though after World War II the decrease of skilled outworkers was often deplored by Solingen employers, the incapacity of the Solingen cutlery industry to secure the training of new generations of skilled workers became a constant feature. Engulfed in mass production, Solingen was unable to maintain its traditional potential for the production of quality products, a potential which may well have been a source of innovation for its industry. The new model of industrialization which had become the dominant vision of future development by the 1920s was still too powerful during the following decades, when the euphoric hopes gave way to disenchantment about its technological possibilities.[93]

Today mass production and the highest quality are no longer incompatible. It has been possible since then – due to further technological progress – to overcome the previous contradiction. Today, however, the Solingen cutlery industry faces severe difficulties in keeping up with the revived demand for artisanal quality products and for special designs of cutlery produced in small series.[94]

[92] Putsch, *Heimarbeit*, statistics pp. 346, 356, 384.

[93] Already in the 1930s the attitude of many entrepreneurs and of the technical journal *Messer und Schere* towards the future rationalization of the industry had changed. The euphoric presentation of the supposed technical possibilities for the mechanization of grinding and hafting (even in the higher quality sector) of the 1920s – which tended to deny the existing sharp contradiction between quality production and the desired mass production, had given way to a more realistic judgment. In 1934 *Messer und Feile* wrote that with respect to the future of the industry "nobody can take comfort with technical progress anymore, for, especially in the production of high-quality cutlery, the trained worker, doing his job with love and pride, cannot be replaced by machines in a foreseeable period of time." (*ibid.*, p. 507) See also Putsch, *Heimarbeit*, p. 259. Despite the continuity of this rather realistic approach to mechanization in post-war Solingen and despite the experience of the early 1950s that "normal" times created a substantial demand for fine or very fine cutlery, nothing – aside from proclamations – was done to reconstruct or assist the comparatively small, but lively sector of artisanal "flexible specialization." This task, however, would have needed stronger efforts than the intensified search for more (and cheap) apprentices undertaken in Solingen. Only if the entrepreneurs had declared their readiness to pay high wages and to grant certain amount of self-determination and co-determination would the life of an outworking grinder have become attractive again.

[94] According to Putsch, the French cutlery town of Thiers – having undergone the same destruction of its capacity for "flexible specialization" as Solingen – is now trying to reconstruct an artisanal sector of industry through municipal aid. A training program for traditionally specialized artisans as well as a retail cooperative are under way. See J. Putsch, *Konzept für ein Industriemuseum* (Solingen, 1984), pp. 139–140.

4 Manufacturing flexibility in nineteenth-century Switzerland: social and institutional foundations of decline and revival in calico-printing and watchmaking

Béatrice Veyrassat

Introduction: the Swiss experience of industrialization – a flexible craft alternative

Technical challenges and response strategies

There were two particular phases in the nineteenth-century Western industrialization process when the force of international competition and the impact of new technologies threw Swiss manufacturing milieux into disarray. The first phase was the surge of British industrialization at the beginning of the nineteenth century: the Manchester effect[1] or the challenge of relatively standardized mechanized production of intermediate products like cotton yarn or certain finished products like printed fabrics. The second was the emergence in the last quarter of the nineteenth century of the "American system of manufactures" whose originality lay in perfecting the standardization of the semi-product, or progressing towards the interchangeability of components so as to make possible standardized fabrication of the finished product.

These important changes in production techniques and the forms of industrial organization linked to them met with a variety of responses in Switzerland: some saw in them the key to progress while others saw these foreign models as destructive of craft traditions and incompatible with the specific requirements of Helvetic industrial production.

Responses to these challenges were ambivalent. In the cotton industry at the start of the nineteenth century the Swiss spinning sections – which became an arena for intense experimentation with advanced technologies – did not imitate

[1] When he visited Manchester in 1814 the Zurich industrialist Caspar Escher, founder of one of the first modern spinning mills in Switzerland (*Escher Wyss & Co*, 1805), observed "that in a single street in Manchester there were more cotton spindles than in the whole of Switzerland," quoted by W. O. Henderson, *J. C. Fischer and his Diary of Industrial England, 1814–1851*, (London, 1966), p. 57.

the English model blindly. While adopting those new techniques from abroad that had proved their effectiveness elsewhere, the country's spinners adapted English prototypes (or machines of French construction that were easier to import) to the specific needs of the domestic market – i.e. to the extremely diverse needs of sectors further on in the manufacturing chain (indigenous weaving of numerous specialities). Consequently, in Switzerland the mechanization of spinning was not synonymous with the standardization of production but was accompanied by a marked differentiation in the quality of the threads. In 1875 an observer noted:

What is particularly characteristic of Swiss spinning, clearly distinguishing it from English spinning, is the extraordinary mobility and diversity of its production. We do not really possess any spinning mill that produces the same quantities from year to year, but each establishment of a certain size is equipped with a system of machines that allows it to adapt rapidly to a constantly shifting demand.[2]

Generally speaking industrialists and workers reacted to external challenges by seeking solutions in keeping with the particular features of external outlets or of the domestic labor market. The latter – a workforce that was overwhelmingly engaged in textile activity at the end of the eighteenth century[3] – was significantly destablized by the mechanization of spinning after 1800 and especially after the end of the continental blockade. However the contraction of the job market in the technologically dynamic sectors was advantageous for the traditional labor-intensive sectors. In the first decades of the nineteenth century a massive transfer of production took place away from spinning towards sectors that followed it in the production process: this involved either the redistribution of activity to highly diversified specialized sectors (hand weaving, hand embroidery, dyeing and printing of fabrics in the context of non-mechanized protofactories) or conversion for work on other textile fibers (manual weaving of silk). This redeployment of the workforce in a plurality of secondary industries (finished products) that were not as yet mechanized, guaranteed the survival of cottage industry well beyond the first half of the nineteenth century.[4]

Thus, once technological unemployment was overcome the advance of innovative techniques in spinning created conditions elsewhere that were favorable to the regeneration of traditional forms of production and industrial organization.

[2] Hermann Wartmann, *Industrie und Handel des Kantons St. Gallen auf Ende 1866. In geschichtlicher Darstellung*, ed. Kaufmännischer Direktorium (St. Gallen, 1875), pp. 494–495.

[3] In about 1787 the cotton industry employed around 70,000 male and female spinners, 28,000 weavers, 30,000 in embroidery (Eastern Switzerland and peripheral areas), all working at home: Peter Dudzik, *Innovation und Investition. Technische Entwicklung und Unternehmerentscheide in der schweizerischen Baumwollspinnerei 1800 bis 1916* (Zurich, 1987), pp. 64, 406–407.

[4] Béatrice Veyrassat, *Négociants et fabricants dans l'industrie cotonnière suisse, 1760–1840. Aux origines financières de l'industrialisation* (Lausanne, 1982), pp. 50ff.

This interdependence of radically different production and organizational forms was also a feature of Swiss watchmaking. In the last quarter of the eighteenth century attempts at mechanization were made in watchmaking, tending towards standardization and serial fabrication of parts. The inventor of this new procedure may well have been a Swiss artisan, Jeanneret-Gris, but the real innovative advance took place on the other side of the frontier, at Beaucourt in Franche-Comté, thanks to a French watchmaker called Japy – whom some regard as the actual inventor. In a few years Japy moved from artisanal production to the mass production of standard parts (*ébauches*), flooding the Swiss market at prices challenging those for hand-produced goods.[5] The imperfections of mechanized production meant that Swiss watchmakers went on for a long time making movements by hand in home workshops, and only two or three small manufactures took on board Japy's innovation.[6] Like the spinners, the watchmakers in Switzerland preferred the enrichment of their ranges to standardization. Thus the Fontainemelon factory, which was one of the few Swiss companies to have mechanized the manufacture of parts at this time, very quickly diversified its production range and between 1825 and 1870 built up a stock of about 1,000 different specialized calibers (types of movement or models), a move that was entirely customer-driven.[7] This was a striking example of the fact that the mechanization of component parts was in no way a hindrance to the availability of an infinite variety of finished products.

Jura watchmaking also greeted with scepticism the plans of the Swiss engineer Ingold (1787–1878) who strove in the 1830s to pioneer the manufacture of watches "by means of powerful and ingenious auxiliary machines." This was the end result of lengthy researches into the unification of tooling and into the methods appropriate to the various stages of watch manufacture, in order to make possible the serial mechanical production of individual parts (plates and wheels) that were uniform and interchangeable. Some saw Ingold as the "initiator and invisible director of the huge American factories."[8] He received no encouragement at home and set off on a journey of a highly symbolic nature: first to London, where in the early 1840s he managed to perfect his prototypes definitively, but was again defeated by the "stubborn conservatism" of the

[5] Pierre Lamard, *Histoire d'un capital familial au XIXe siècle. Le capital Japy de 1777 à 1910*, thèse de IIIe cycle (roneotyped), Université de Franche-Comté, 1984, t. I, pp. 99ff., esp. 145, 149, 266; David S. Landes, *Revolution in Time. Clocks and the Making of the Modern World* (Cambridge/London, 1983), pp. 261ff.

[6] Philippe de Coulon, *Les Ebauches – Deux siècles d'histoire horlogère* (Neuchâtel, 1951).

[7] In addition to ordinary pieces this firm produced "fines excentriques," "fines à mobiles noyés," "superfines à l'anglaise," pieces "à la Berthoud," "à l'anabaptiste," "à la française"! (Paul Robert, *La fabrique d'horlogerie de Fontainemelon. Une usine plus que centenaire; 1793, 1825–1925* (Neuchâtel, 1925), pp. 14, 23.

[8] Jules-Fu.-U. Jürgensen, "De l'emploi des machines en horlogerie, spécialement dans la fabrication des montres de poche," *Musée Neuchâtelois*, 14 (1877), 160–161.

watchmaking corporation as well as by the refusal of parliament to legalize the establishment of his company, the *British Watch and Clockmakers' Company*; then in 1845 to the United States where he sought to arouse interest and obtain financial backing for his plans.[9]

This rejection by Swiss manufacturers of the standardization of production was repeated later in the nineteenth century, following the "American challenge" of mass production: a form of division and organization of labor of which the principles were discovered by European industrialists and engineers at the Philadelphia Universal Exhibition of 1876.[10] In the following decades the watchmaking sector and the machine industry had to abandon technologies that had hitherto been essentially manual. But the unhurried pace of their evolution towards the machining of interchangeable individual parts was beneficial in terms of the greater diversity and quality of the finished product.[11] We can say the same of the cotton industry which in its phase of technical modernization in the second half of the nineteenth century, managed to combine mechanization and flexibilization of product lines.[12] During this period, marked as it was by the development of international mass markets, a Swiss textile magnate affirmed in 1874 that it was necessary "to continue in the sphere of specialities . . . to rely on excellence of quality and not on cheapness: this is our best policy . . . [for]

[9] R. F. and R. W. Carrington, "Pierre-Frédéric Ingold and the British Watch and Clockmaking Company," *Antiquarian Horology*, 10 (Spring 1978), 698–714; Landes, *Revolution in Time*, pp. 283–5.

[10] This enormously impressed Swiss visitors to the New World where they did in fact discover a whole new world: "the particular character of American industry has been strikingly evident to us from the start. In America the aim pursued in industrial manufacture is much more elevated and rational than at home." In this respect the famous pamphlet from whence the above quotation comes, entitled *L'industrie dans les Etats-Unis de l'Amérique du Nord. "Garde à vous!" adressée aux industriels suisses à propos de l'Exposition universelle de Philadelphie*, by Edouard Bally (Geneva/Neuchâtel, 1877), p. 6, is highly indicative of the astonishment and shock felt by the Swiss.

[11] The Swiss machine industry was distinguished up to modern times by its "very varied manufacturing programs," aided by "simple and consequently universal tools and machine tools," by "a constant renewal of products" and by the specialization of certain businesses in the construction of individual models for very specific orders: Hannes Hofmann, *Die Anfänge der Maschinenindustrie in der deutschen Schweiz, 1800–1875* (Zürich, 1962), pp. 169, 150. This author writes: "By contrast with the American machinery industry, for example, that makes its customers choose the most appropriate model from a series of standard types, in Switzerland they have continued the nineteenth-century tradition of building machines that could handle the most sophisticated requirements of customers" (*ibid.*, pp. 169–170). On the Swiss machinery industry, "whose articles of predilection did not generally lend themselves to mass production," see also B. Lincke, *Die schweizerische Maschinenindustries und ihre Entwicklung in wirtschaftlicher Beziehung* (Frauenfeld, 1911), pp. 68ff., 194ff, 207. The Swiss automobile industry was similarly an example of flexible specializaton: David Asseo, "La place de la Suisse dans l'industrie automobile mondiale d'avant 1914," in Paul Bairoch and Martin Korner (eds.), *La Suisse dans l'économie mondiale (15e–20e siècles)* (Zurich, 1990).

[12] For an idea of the immense variety of cottons produced in Switzerland in the nineteenth century, the reader is referred to the catalogues and reports of national and universal exhibitions.

everywhere that we encounter English mass production there is no room left for Swiss business."[13]

Thus, whether we are talking about cotton spinning, the embroideries and figured stuffs sector (Saint-Gall, Appenzell), printing on cotton (Neuchâtel and Zurich cantons), Basle ribbons or Zurich silks, the building of machines (eastern Switzerland) or watchmaking (western Switzerland), the technological options chosen were not such as to lead to a manufacturing system thriving on mass market conditions, on the limitation of product ranges to a few standard goods or on recourse to unskilled labor.

Thus Swiss experience in the age of industrialization was similar to that of those famous manufacturing centers whose technological development was conceptualized by Sabel and Zeitlin as a model of "flexible specialization"[14] – i.e. as a production system able to respond to changing markets by an innovative diversification of products, thanks to the extension of flexible and multivalent technologies that put a high priority on professional qualifications and autonomy of labor. The way Swiss industry came to technological maturity runs counter to the thesis of Landes, for whom "The logic of the machine pushes it in the direction of uniformity and standardization . . . the logic of industrialization has pushed all economies towards mass production, some faster than others."[15]

In short, the Swiss technological model may be summed up by this fundamental characteristic: far from destroying product diversity, mechanization – first in sectors earlier in the chain (semi-finished), then in end sectors (finished products) – actually extended the range of products and human skills.

Markets, production and technological choices

It seems that the structure of the markets available to Swiss industry provides a fundamental explanation of the rejection of mass production: around 1845 two-thirds of Swiss exports, especially of cotton and silks, went outside a Europe that was bristling with protectionist barriers.[16] Internationally, the commercial advance of the British and their textiles for everyday needs was such that the Swiss had to pursue a niche strategy involving the continual diversification of production ranges, the promotion of product innovation, and an emphasis on

[13] Jb. Steiger-Meyer, "Bericht über Gruppe V, Sectionen, I, II, III, V, VI: Schafwollwaaren, Baumwollwaaren, Leinenwaaren und Strohflechterei, Stickereien, Bekleidungsgegenstände; Gruppe XXI: Hausindustrie," in *Wiener Weltausstellung 1873. Schweiz. Bericht über Gruppe V: Textil-Industrie [. . .], Truppe XXI: Hausindustrie* (Schaffhausen, 1874), pp. 60–61, 63.

[14] Charles Sabel and Jonathan Zeitlin, "Historical Alternatives to Mass Production: Politics, Markets and Technology in Nineteenth-Century Industrialisation," in *Past and Present*, 108 (1985), 142ff.

[15] David S. Landes, "Small is Beautiful. Small is Beautiful?" in *Piccola e grande impresa, un problema storico*, Fondazione ASSI, Instituto per la storia dell'Umbria contemporanea (Milan, 1987), p. 27.

[16] B. Veyrassat, "La Suisse sur les marchés du monde. Exportations globales et répartition géographique au XIXe siècle. Essai de reconstitution," in Bairoch and Korner (eds.), *La Suisse dans l'économie mondiale.*

labor skills. The best way to counter the dumping methods employed by the English or the large-scale textile production of the Americans was to offer not only quality but also specialities at affordable prices in areas with limited buying-power (Latin America, Asia, Africa) and top-of-the-range articles for countries with high customs tariffs (Europe, United States).

The need to react quickly and flexibly to these highly unstable market conditions discouraged the immobilization of capital in heavy dedicated machinery and encouraged skill-intensive technologies. The typical example here is that of textile printing where the Swiss manufacturers did not follow the example of the Manchester calico-printers in aiming for continuous production in long runs, but persisted with manual printing methods that left more scope for technical opportunities and allowed for greater flexibility in dyeing.[17]

Even if such options did not *necessarily* involve the implementation of flexible technologies in the service of the diversification of products, they nonetheless tended to encourage it. Thus Swiss textile (and also watchmaking) centers produced a constant flow of technical progress "on craft lines." Numerous parallel mechanical inventions and improvements came from indigenous artisans. All were perfectly integrated into a decentralized system of family craft production.

Having said this, we must admit that the flexibility of Swiss companies had its ups and downs – in some sectors this meant temporary alterations and in others more substantial ones. Some manufacturing centers evolved towards a less flexible utilization of technology and labor, while in others there was a coexistence or alternation of antagonistic forms of production: high-quality short-run production and/or the manufacture of cheaper articles in much longer runs.

[17] As someone well-placed to observe the state of international competition a Swiss businessman in Brazil was somewhat contemptuous of the mechanical printing methods used by the English who, he said in 1837, "in spite of their efforts did not manage with their machines to produce the same articles as us – and our work is done by hand" (Federal Archives, Berne: *E2200/Recife-Pernambuco, 8, fol. 91)*. This is how a chronicler justified the rejection of the machine alternative by the Swiss calico-printers:"Mechanical printing is only worthwhile when the quantity of pieces of a single pattern is large enough. In the absence of a market big enough for a new pattern, manufacturing in rolls poses *problems*" *(Bericht über die dritte schweiz. Industrie-Ausstellung in Bern 1857*, by Dr. [P.-Alexander] Bolley *et al* (Bern 1858), p. 105). See also Veyrassat, *Négociants et fabricants dans l'industrie cotonnière suisse*, pp. 36ff., 47–50; Pierre Caspard, *La Fabrique-Neuve de Cortaillod. Entreprise et profit pendant la Révolution industrielle 1752–1854* (Paris/Fribourg, 1979); the very precise profit calculations of the Cortaillod calico-printers for each of their articles (pp. 146ff.) show that it was indeed in their interest to favor the manual printing of fabrics in order to incorporate in them a maximum of handwork (p. 158), rather than to have recourse to the mechanical roller – to such an extent that even after the adoption of a printing machine in 1810, the technology employed in this company, as elsewhere in the Canton of Neuchâtel or indeed in Switzerland, was still essentially based on plate work (p. 141). Caspard notes in this connection that "the most insignificant event" for the activity of the Fabrique-Neuve was "the one that theoretically ought to have had the most revolutionary consequences: the introduction of mechanisation in 1810" (p. 116).

The changing fate of manufacturing flexibility in Switzerland

The purpose of this study is to discover the reason for these contrasting types of development by focussing on two examples: the watch manufacture in the Canton of Neuchâtel, sometimes *craft-oriented*, sometimes *mass-oriented*, and the cotton industry in Glarus, whose initial *craft* potential was destroyed in the course of the nineteenth century by large-scale production strategies.

In both these industrial centers – areas of mountain ranges and pastureland, one in the Jura and the other in the Alps – the origin of manufacturing activity went back to the seventeenth century. From the valleys of the Canton of Neuchâtel it spread through the parts of the Jura in the Cantons of Vaud and Berne[18] and watchmaking soon became second nature to the Jura peasant, the peasant-watchmaker of the eighteenth and even of the nineteenth century. In spite of all the revolutions it has experienced down to modern times in production and the organization of labor, and in spite of dramatic structural crises – as recently as the period between 1975 and 1985 – watchmaking is currently one of the dynamic sectors of Swiss industry. As for the Canton of Glarus, it developed in the nineteenth century towards an industrial monoculture based essentially on cotton fabrics: printing of fabrics for everyday use at moderate prices became the key sector, the most important from the point of view of employment, and dynamic enough to give an impulse to regional development and the growth of the sectors that fed it (spinning and weaving). But only up to the 1860s, when textiles in the canton went into decline (1865–1900).

Calico-printing and the manufacture of watches have in common the fact that they are activities orientated towards export, working almost exclusively for supra-regional markets, in the nineteenth century strongly integrated into the global market and therefore both equally dependent on international trade. Moreover these were labor-intensive non-mechanized industries, both based on technologies that were still largely manual (except at the stage of intermediate products, where the manufacturing process was mechanized at the end of the eighteenth century for spinning and watch pieces and after 1840 for the weaving of plain cottons). But the similarities end there.

An exception to flexible production: calico-printing in Glarus Glarus was in fact the only manufacturing center to distance itself decisively from the model of flexible specialization. From the 1840s industrialization there was increasingly marked by the stigmata of mass production which transformed this little Alpine canton into a factory valley.

It is worth noting at this point that this mass production involved some rather curious modalities: it was simultaneously based on a *non-mechanized* sector

[18] I am omitting from this study any reference to Genevan watchmaking, which in opting for the luxury end of the market was too different from the more typical and more modestly-priced watchmaking of the Jura.

dominated almost up to the end of the nineteenth century by a manual technology – calico-printing – and on mechanization of the work that fed it – spinning and weaving of the necessary semi-products. Nonetheless, this division of labor – mechanical factory production of semi-finished articles; hand printing of fabrics by plates in the context of impressive proto-factories[19] – may be described in terms of "mass production." Contemporaries themselves referred to printed fabrics made in Glarus as "cheap mass-produced articles," or "articles for mass consumption" or "cheap mass production."[20] This concept seems all the more pertinent to me in that it is to observers like these that we can trace back the catalogue of key elements that historians today associate with serial production: the massive production[21] of a limited number of varieties of printed fabric and even of relatively "standardized" articles – "a single pattern in the same colour might be turned out day after day for more than a year by the same printer using the same plate all the time and at the same table";[22] a division of labor taken to extremes and work that was routine, hasty and careless,[23] a progressive deskilling, and a trend towards poorer quality merchandise and the erosion of the original innovativeness of local artisan manufacturers.[24]

19 Of course this is all relative: Glarus had 157 calico-printers for each company in 1837 while in the Haut-Rhin there were 350 in 1839 (Gret Heer and Urs Kern, "Industrialisierung und Fabrikarbeiterschaft am Beispiel der Glarner Tuchdruckerei im 19. Jahrhundert," unpublished manuscript, Zurich University, 1978, p. 316; Robert Levy, *Histoire économique de l'industrie cotonnière en Alsace* (Paris, 1912), p. 95).

20 Fridolin Schuler and A. E. Burckhardt, *Untersuchungen über die Gesundheitsverhältnisse der Fabrikbev Ikerung in der Schweiz mit besonderer Berücksichtigung des Krankenkassenwesens* (Aarau, 1889), p. 115; *Bericht über die dritte schweiz. Industrie-Ausstellung in Bern 1857*, p. 104.

21 According to a local historian, a manufacturer-printer in one of the big Glarus companies, "there were periods when week after week in the biggest factories 1,800 to 2,000 pieces of 75 ells (162–180 km!) were put through washing and preparation processes, which took the annual volume to 100,000 pieces (9,000 km.)": Adolf Jenny-Trümpy, *Handel und Industrie des Kantons Glarus*, Offprint from the *Jahrbuch des Historischen Vereins des Kantons Glarus*, 34 (1902) (Glarus, 1903), p. 637.

22 Heer and Kern, *Industrialisierung und Fabrikarbeiterschaft*, p. 112.1

23 Certain observers were struck by "this movement, always the same, repeated one hundred times a day, from the printing table to the color frame, from that to the printing table, by the mechanical regularity of the mallet blows struck on the engraved plate, by the rapidity and hastiness of the gesture," (*Bericht über die sanitarische Bedeutung und Tragweite des Zeugdruckes in den glarnerischen Fabriken. Dem h. Rath des Kts. Glarus vorgelegt von den Herren Prof. E. Kopp in Zürich, Dr. J. Göttisheim in Basel und Direktor Ph. Imbach in Lörrach* [Glarus, 1872], pp. 8–9).

24 As was noted by one of the Swiss chroniclers of the Universal Exhibition in Vienna in 1873, certain factories "are getting into more and more difficulty" and the single solitary Glarus exhibitor "showed how trivial things had become; it was only because the official programme required that articles also be exhibited that were completely ordinary and cheap, that this exhibitor received an honorable mention"; Jb. Steiger-Meyer, "Bericht über Gruppe V," p. 61. On the cheapness, the defects and the far-from-fast colors of certain articles from Glarus, see also: *Berichte über die internationale industrie- und Kunstausstellung in London 1862* [Technischer Bericht über die Betheiligung der Schweiz . . .," by P. A. Bolley] (Zurich, 1863), p. 67.

The ups and downs of flexibility: the case of the watchmaking industry

Such slippages were not restricted to Glarus. Other sectors of activity in Switzerland were subject to similar lapses – the forcing of production, the bodging of work and the qualitative deterioration of merchandise – or to be more precise, to alternating phases of tendencies towards mass production and returns to flexibility.[25] This was the case with the watch industry in the parts of the Jura in the cantons of Neuchâtel and Berne. On several occasions the watchmaking "art," manual and domiciliary as it was throughout the nineteenth century, was contaminated by the first advance viruses of mass production and succumbed to a cheapening of quality, talent and craft ethics. This phenomenon was sometimes hidden and sometimes of crisis proportions; the first signs of it appeared right at the start of the nineteenth century and it reached a peak in the 1840s and 1850s as a consequence of a massive growth in markets and in particular (an interesting detail) of the focussing of watchmaking exports on the North American market.[26] These boom times led to an increased division of labor,[27] but in specializing tasks this division of labor shortened apprenticeships or even made them superfluous. Worse, it turned the workman into "a machine hardly aware of what he is doing." Increased productivity and lower prices then provoked "ferocious competition" between manufacturers together with its more dangerous corollary – unfair competition. This inevitably gave rise to a fatal spiral – a fall in profits and wages, a "lower grade workforce" and ultimately the hasty introduction of the "junk watch."[28]

[25] E.g. silk-weaving in the canton of Zurich. As old as the watchmaking industry and also dominated by an essentially manual technology right up to the last quarter of the nineteenth century, it kept going successfully – though not without its periodic crises – right up to the modern era. At certain periods it was orientated towards mass production, specializing in a small number of standardized mass-consumption articles (especially from the 1840s on when the United States became its main outlet), at other periods it was geared to production of a wider range, flexibilization of products with top-of-the-range markets, and organized to respond more quickly to fluctuations in demand. At the present time the silk industry is comparable for flexibility to that of the eighteenth century: see Michael Bernegger and Hans Georg Rhonheimer, "La soierie zurichoise, de la révolution industrielle à nos jours," in *La soie. Recueil d'articles sur l'art et l'histoire de la soie*, ed. Jean-Pierre Jelmini and Caroline Clerc-Junier and Roland Kaehr, Musée d'art et d'Histoire (Neuchâtel, 1986), pp. 141–161. Other non-mechanized or not very mechanized branches of the textile industry went through similar phases: cotton-weaving in the St. Gall canton, for example, where commercial success in the 1840s led to rushed production, decline in quality and fraudulent practices (H. Wartmann, *Industrie und Handel des Kantons St. Gallen*, pp. 514–515). For this phenomenon elsewhere in the nineteenth century, see Rudolf Boch, "The Rise and Decline of 'Flexible Production': The Cutlery Industry of Solingen since the Eighteenth Century," in the present volume.

[26] Between the late 1830s and 1872 the USA was the main outlet for Neuchâtel watches.

[27] The watchmaker made all the watch parts himself up to the end of the eighteenth century, but the work was progressively subdivided into ever more numerous "separate parts" (54 around 1830; 100 or so around 1870), involving small specialized crafts practiced in workshops by artisans or at home by workmen: Marius Fallet-Scheurer, *Le travail à domicile dans l'horlogerie suisse et ses industries annexes. Rapport final publié au nom du Comité d'organisation des expositions de Zurich et de Bâle de l'industrie à domicile (1909)* (Berne, 1912), p. 300.

[28] On this evolution towards overproduction of poor quality merchandise see the documents cited by Auguste Bachelin, *L'horlogerie neuchâteloise* (Neuchâtel, 1888), pp. 113–114, 91.

Unlike what happened in Glarus, this evolution never left the watchmaking milieu indifferent; from the moment of the first signs of decay the milieu systematically developed a remarkable aptitude for producing effective antidotes with the aim of producing "fewer but better watches,"[29] "keeping up with technical progress and consistently outgunning the competition by quality manufacturing" – and not by dumping prices – and finally encouraging "a moral understanding between manufacturers and workmen" to prevent unfair price-cutting practices.[30]

The end of the Glarus "economic miracle"[31] and the triumph of the Swiss watch: some questions

The present chapter, devoted to the divergent development of these two regions, will try to show what were the reasons for the decline of a "flexibly specialized" industry in Glarus; it will also seek to highlight the causes and conditions of the persistence of the flexible watchmaking system throughout the whole of the nineteenth century.

Glarus: a conflictual system

We saw above how an exogenous element – the markets created by the Swiss – provides a substantial explanation for the appearance of flexible features both in the area of the variation of products and in that of capital equipment and the adaptability of the workforce.

Nonetheless, despite its continuous commercial dynamism up to the end of the nineteenth century and its consistent sensitivity to stimuli from outlets, the Glarus manufacturing community took the mass production model as its guide. So we need to consider other types of explanation in addition to external conditions such as market structures and look at the peculiarities of the local context. Likewise we shall have to reject the classic interpretation and its technological reductionism: my analysis will show that the transformation of Glarus into a center for mass production cannot be explained in terms of the intrinsic advantages of large-scale production or the allegedly superior efficiency of standardized production, but that it was the result of conscious socio-economic choices, choices determined by local conditions in respect of the distribution of resources and social power and molded by a specific political context.

In the third section we shall see that the emergence of a class of powerful industrial leaders and integrated companies dominated by short-term commer-

[29] Robert Jaccard, *Sainte-Croix et ses industries. Notice historique publiée sous les auspices de la Société industrielle et commerciale de Sainte-Croix* (Lausanne, 1932), p. 122.

[30] According to contemporary testimonies cited by Frédéric Scheurer, *Les crises de l'industrie horlogère dans le canton de Neuchâtel* (La Neuveville, 1914), pp. 40, 47–48.

[31] The expression was coined by the Swiss historian Walter Bodmer, in *Das glarnerische Wirtschaftswunder* (Glarus, 1952).

cial interests, was to have a crucial impact on the disappearance of a local "craft entrepreneurship" and on the impoverishment of craft know-how, work skills and quality standards – defined as key elements in the system of *flexible specialization* and indispensable for the perpetuation of the latter. Furthermore this tension in the production system was not the only one: there were also tensions between industrial leaders and the state, which found itself at odds with the freedom of the employers – the sacrosanct freedom of trade and industry – as it strove to provide legal protection for labor. There were tensions too between employers and employees. This evolution eventually destroyed the political and social consensus necessary for the development and good functioning of industrial regulation mechanisms, thus paradoxically making the Glarus system of direct democracy a conflictual one.

This explanation is clearly very different from the traditional explanations given up to now, which attributed the decay in Glarus printing and the industrial decline of the canton after the 1870s to the external market and in particular to competition from other countries that were getting involved in mass production in ever-increasing numbers. Such *dependentist* conceptions are very common in Switzerland and developed because of the great dependence of Swiss industry on foreign markets, but I shall instead propose a contrasting interpretation that is more *autonomist* and takes into account domestic forces encouraging either stagnation or dynamism. After all the Swiss watchmaker surely had to face up to the same foreign demons as the manufacturer-printers of Glarus and he too had to cope with decreasing competitiveness on the international market – and at the same period too.

Diverging paths of development

It was in fact in the 1870s – a time of great technological challenges when the triumphant industrialism of certain nations (especially the US) was rampant – that the contrast between the two manufacturing centers became clear: both facing the same kind of changes, they reacted very differently to the question of technological shifts and the rationalization of production methods, questions that were at the time to the forefront in an intensive debate in both sectors.

Let us first summarize these changes and the distinctive nature of the respective regional responses to them.

Technological challenges In the area of fabric printing, German and Austrian companies moved over to the mechanical production of a particular category of printed cottons, *handkerchiefs*, which had for technical reasons depended for a long time on manual technologies. Now these hand-made articles had in the nineteenth century become the *pièce de résistance* of the Glarus manufacturers who successfully occupied this niche in the international market up to the 1870s. Would local industrialists continue to progress, modifying the allocation of tasks between man and machine, and acquiring modern equipment at the expense of a reduction of their labor force? This question was all the more urgent for the workforce in that another innovation was spreading at the time in the area of the manufacturing process, an innovation that had actually come from the local employers. One of the biggest companies in the area, specializing in *Yasmas* or "Turkish bonnets" (light printed materials for veils and turbans), managed to develop a revolutionary method of printing that enabled it to double or triple production without additional investment, simply by superimposing two, three or even four layers of semi-fine or fine fabrics (a technique called "double printing").[32] This form of rationalization, accompanied by a substantial fall in remuneration for labor and in the sale price, did however involve a risk, a risk that was soon noticed by the calico-printing workers: that of overproduction. Imitated in the late sixties by all the local firms specializing in oriental articles, it soon demonstrated its perverse features: the eruption of a monetary crisis in the East caused a substantial fall in demand for these articles in 1870/1. Reductions in work and redundancies brought about a sudden dramatization of relations between bosses and workers and the principle of "double printing" became a matter of urgent public debate.

A similar technological challenge and market crisis shook the watchmakers' world after 1872 when the value of watch exports to the United States fell by two-thirds between 1872 and 1876[33] and the threat of a new type of competition appeared: the 1876 Philadelphia Universal Exhibition made abundantly obvious to visiting Swiss experts that the Americans had made tremendous progress on the path of mass production between 1850 and 1870s. Firms like Aaron Dennison (The Waltham Watch Co. or American Watch Co., MA), pioneers in the mechanization experiment, soon followed by imitators like the National Watch Co. of Elgin, IL managed to set up serial production of complete watches on the basis of standardized interchangeable factory-made components.[34] While the Swiss watchmaking milieu had found its special niche and prosperity in the continuous manufacture of an infinite variety of sometimes very sophisticated and sometimes highly specialized products, for which only manual technology had up to now proved effective,[35] the Americans were offering a watch that was "of standard construction on a few standard patterns," wholly manufactured by machine and in the factory.[36]

From then on, in the eyes of the Swiss watchmakers, standardization of

32 For details, see Jenny-Trümpy, *Handel und Industrie des Kantons Glarus*, pp. 429, 541ff.

33 Still 18.3 million Swiss francs or 366,000 watches and movements in 1872, but no more than 65,000 pieces in 1876 or 3.5 to 4 million francs in 1877/8: Landes, *Revolution in Time*, p. 383.

34 Landes, *Revolution in Time*, ch. 19; Charles W. Moore, *Timing a Century. History of the Waltham Watch Company* (Cambridge MA, 1945).

35 Based on "inheritance of skills," on "the transmission by fathers to their sons of a delicate touch, a secret know-how, and artistic skill with the file and the burin": J.-F.-U. Jürgensen, *L'horlogerie neuchâteloise et suisse en 1881* (La Chaux-de-Fonds, 1881), p. xlii.

36 *Ibid.*

manufacture and large-scale production[37] seemed like an alternative to artisanal forms of production, dominated by manual techniques, by equipment that although it may have been perfected and indeed mechanized for the fabrication of certain pieces, was utilized in the traditional framework of the domestic workroom or small workshop. Here was an opportunity to discuss what was at stake in such innovations in the manufacturing process: whether to move to the new mode of production centralized in large units and adopt the cause of mechanizations, or whether to maintain the traditional organization of decentralized manufacture and safeguard the virtues of the workman's know-how and the watchmaker's craft.

The debates In both regions there was a mobilization of a real "Estates General of industry" in the face of unavoidable production choices. Tensions emerged that were to feed public debate and lead in both cases to a politicization of innovation. The canton of Glarus like the Jura watchmaking area became the forum for a discussion where although the cause was the same – technical innovation – the issues at stake were fundamentally different.

In Glarus the debate sounded like a social conflict and revealed the disfunctioning of the system. It set the employers against the mass of workers, with argument focussing on the jobs threatened by the rationalization measures of local industrial chiefs. Here the working population was set on defending its right to work – a sort of social pact with a long history and to which I shall return – and took particular exception to "double printing" which was creating more victims than the advance of mechanization. Invoking the insalubriousness of the working conditions associated with it (this process required greater physical force and gave rise to uncomfortable heat in the factories), the democratic assembly of citizens – the supreme legislative authority in this canton that was a direct democracy – passed a law suspending the process in 1872.

In the Jura and in particular in the canton of Neuchâtel the debate was at an entirely different level: the central issue was less the fate of the workmen than the future of Swiss watchmaking. Key questions for the watchmakers were the way forward for technological development, the possible scenarios for a restructuring of production and the organization of labor and problems of industrial ethics. In addition – very importantly – they tended to advocate the advantages of a strategy of collaboration in the face of the industrialist threat.[38]

[37] At the Waltham Company a workman produced 150 watches a year around 1880 – in Switzerland a workman produced 40 a year in the mid-1870s: Landes, *Revolution in Time*, p. 318.

[38] On this debate, see Jean-Marc Barrelet, "Les résistances à l'innovation dans l'industrie horlogère des montagnes neuchâteloises à la fin du XIXe siècle," in *Revue Suisse d'Histoire*, 37 (1987).

Regional responses What strategic response developed in this critical period, seemingly a time of rupture? The strategies in the two regions were diametrically opposite.

In Glarus the policy of workers' self-defense was a failure and the employers not only re-established the much-contested process of "double printing" after having remedied the sanitary problems it was causing, but advanced still further on the path to mechanization.[39] A block on dialogue between the social partners, technological unemployment and emigration, a rapid fall in numbers of jobs in industry and – in spite of the rationalization measures adopted – a loss of competitiveness in external markets leading to the collapse of calico-printing:[40] such was the outcome of this story.

The watchmakers did not try to imitate the Americans blindly even though there were convinced advocates of the new mode of production in their ranks. The watch industry was to set out on the path to a modernization that did not break with the indigenous manufacturing model. Some thought that it was possible to go down the path to mechanization without casting aside the resource of a skilled workforce and the polyvalence of an industry that had so far managed to satisfy all tastes and respond to the most varied demands.[41] Others raised questions about the social consequences of a reorganization of labor in integrated establishments and considered that the improved efficiency allowed by the American methods was not sufficient to make it worthwhile sacrificing the existing order and a cultural identity that had developed in the context of putting-out and "work localized in the family."[42] Finally the transformation of the watchmaking sector would take place in a dualist fashion. Decentralized homeworking, the foundation of artisanal manufacturing in "separate parts" combined with concentrated mechanical fabrication (small- and medium-sized units) and gave way to the latter only slowly.[43] For the execution of certain

[39] Jenny-Trümpy, *Handel und Industrie des Kantons Glarus*, pp. 510–513, 606.

[40] Between 1865 and 1901, 2/3 of the jobs in calico-printing were lost (according to statistics in Heer and Kern, *Industrialisierung und Fabrikarbeioterschaft*, p. 48; see also p. 21. On emigration, *ibid.*, pp. 280–281, 379–380). Exports of printed fabrics fell from 25 million francs in the 1860s to 16 million in 1885 and to an average of 7.4 million in the years 1898–1900 (Jenny-Trümpy, *Handel und Industrie des Kantons Glarus*, pp. 367, 535, 695; Jenny-Trümpy, "Textilindustrie: Baumwollindustrie," in Handwörterbuch der schweizerischen Volkswirtschaft, Sozialpolitik und Verwaltung, ed. N. Reichesberg, III [Berne, 1911], p. 915).

[41] Thus Jürgensen, "De l'emploi des machines en horlogerie," pp. 154–156; Golay, "La vallée de Joux," (1979 edn), pp. 39ff.

[42] See, for example, *Rapport présenté au Conseil d'Etat [de la République et Canton de Neuchâtel] par les experts horlogers délégués à l'Exposition universelle de 1878 [à Paris]* (Neuchâtel, 1879), pp. 129–130.

[43] In 1870 out of 39,348 persons employed in watchmaking, about 35,000 (89 percent) worked at home. By 1905, the proportion of homeworkers had dropped to 23.5 percent (12,000 out of 50,938: Fallet-Scheurer, *Le travail à domicile*, pp. 307–308, 311). On these two sides to the working pattern and on the so-called "mixed" system of manufacturing, combining decentralized work with centralized manufacturing production: *ibid.*, pp. 17–18, 286–289, 302, 315. The average number of employees per company around 1900/1 was 38 in Switzerland and 529 in the United States (Landes, *Revolution in Time*, p. 323).

specialities (complicated mechanisms, precision instruments or fantasy or luxury articles), handwork remained indispensable (especially at the finishing stage), while mechanization advanced in the area of components (semi-products such as movements and covers) and the trend was towards the interchangeability of separate parts, even while an immense variety of models went on being made. Moreover the modalities of the change were governed by regional affinities: the centers that were the most attached to the craft traditions cultivated the skilled, delicate and relatively dear end of the market, while the new factories, more profitable for the production of cheap watches or the serial production of expensive watches, were set up on the periphery (cantons of Berne and Soleure).[44]

The reaction to the challenge from America was thus hybrid: the adoption of new production processes but without abandoning skilled labor: mechanization in the domain of intermediate products, but without standardization or impoverishment of product ranges. This modernization was achieved *in spite* – so classic historiography tells us – of a strong "resistance to innovation," *in spite* of "the aversion to change" of numerous watchmakers. I am inclined to say on the contrary that the success of this modernization was really due to the latters' resistance to the deskilling of labor and the mechanization of man, together with their opposition to the American principle of standardization of production. At the end of the nineteenth century, Swiss watchmaking recovered its lost ground, doubling the value of its exports between 1885 and the eve of the First World War, and was probably responsible by then for more than one half of global production. Even better, it kept the upper hand in the domain of technical and mechanical progress – even American watchmaking eventually came to be dependent on the Swiss manufacturers again.[45]

The watch industry in the Swiss Jura: a system of solidarity?

Why did the industrialization of watchmaking in the nineteenth century not follow the "Glarus" path of expansion, where the class of sub-contractors – small workshop heads, artisan-manufacturers, piece-workers – saw themselves progressively eliminated in the course of a movement of concentration in which a deep socio-political divide emerged between industrial chiefs on the one hand and workmen on the other? Why did watchmaking, which had been based on the small business and homeworking throughout almost the whole of the nineteenth century, not slip towards forms of the *sweating system* like other collective manufactures in Europe? And how was it that the watch industry employers, contrary to the Glarus textile bosses, preferred rather to preserve

[44] Henri Borle, "Les transformations industrielles dans l'horlogerie suisse," *Revue suisse pour l'Enseignement commercial*, 4 (1910), 303ff;(Landes, *Revolution in Time*, pp. 323–324).

[45] Landes, *Revolution in Time*, pp. 324–326. Export of watches went from 82 million Swiss francs in 1885 to nearly 183 million in 1913: Scheurer, *Les Crises de l'industrie horlogère*, pp. 136–137.

"craft" structures in the traditional framework of decentralized production? Inertia, innate fear of the new, resistance to innovation, corporatist reactions in the face of any prospect of change? This is the type of explanation that is central to the majority of the replies given up to now either by recent authorities in the history of watchmaking or by those involved in it or, more precisely, by those of them who lived through the "shock" of the Philadelphia Universal Exhibition of 1876. Becoming fully aware at that moment of the extent of the differences between American industrial organization and their own, they concluded that Switzerland was "backward." It was they who were behind an interpretative tradition based on the myth of watchmaking employers who were not innovative – an interpretation that has been perpetuated up to modern times and reinforced once again by recent events.[46]

It took a historian with a fresh way of looking at things to cast doubt on this negative vision: paradoxically it was Landes, convinced of the superior technical efficiency of mass technologies and of the inevitability of movement towards this type of industrial organization, who discerned most clearly the original flexible qualities of watchmaking in the Swiss Jura, and who first gave full weight to – or restored full weight to, with respect to a rather dismissive traditional vision – the inner dynamism of this example of *a collective manufacture*. Landes explains the Swiss watchmaking phenomenon, "almost without parallel in the annals of commerce," in part by its responsiveness to market opportunity, but especially as a model of cohesion, a "model of *collective* effort and performance," one of the sources of which in his opinion was cultural: the community of language and customs (favoring the communication of knowledge and talent through personal contacts) and religion, the influence of Protestantism on the intellectual and cultural level of the population, its contribution to the development of education and of a widespread ability to read, write and count. But in my opinion this explanation of the success of Swiss watchmaking is inadequate. The Glarus manufacturing center was also a homogeneous linguistic and cultural community and was also predominantly Protestant. In Glarus too, as in the Jura, "they could all read and write."[47]

I am much less convinced by the religion/culture explanation than by the interpretations of Y. Schwartz (on Mulhouse calico-printing), who sees an "essential link between the requirements of the work process . . . and the creation of a certain type of collective relationship between the manufacturer and his workforce."[48] In all the chain of operations leading to a final product like a watch

[46] The double challenge from America and Japan after the Second World War: the competition from Timex and then particularly from the quartz watch – a new product with new technology – which aroused a lot of opposition in Swiss watchmaking circles.

[47] Landes, *Revolution in Time*, pp. 303–307. On the prevalence and level of literacy and education in Glarus, see Jürg K. Siegenthaler, *Konflikte der Industrialisierung. Industrie, Raum und Gesellschaft in Glarus und Scranton 1840–1914* (Bern/Frankfurt/New York, 1985), pp. 125–134.

[48] Yves Schwartz, "Pratiques paternalistes et travail industriel à Mulhouse au XIX siècle," *Technologies–Idéologies–Pratiques*, 1, (1979), no. 4, p. 36 and section III.

with operations split between different categories of executors, none is technically dissociable from the one before or the one after and the creation of the necessary collaborative and cooperation links involves a solidarity based both on technical and on social values. But even if watchmaking, by the nature of the work it involves, is an activity with strong human cohesion (a premise for social cohesion), this is still not an adequate explanation of the preservation and successive reproduction of the system. For after all at the beginning of the nineteenth century Glarus calico-printing was equally demanding from the point of view of technical constraints and dyeing processes, from the point of view of skills and the coordination of tasks. Besides, as has been said, watch manufacturing was subjected to repeated shocks as it slipped like Glarus manufacturing towards mass production. At certain moments in its development it too abandoned its quality ideal and moved towards forms of untrammelled competition, and it too had experience of the unscrupulous entrepreneur and of unfair competition. So we need to consider – and this will be the subject of the fourth section – what kind of system of internal regulation it was that was able to guarantee the perpetuation of the flexible features of watchmaking.

Reasons for the decline of Glarus

There seem to me to be essentially two explanations for the way that Glarus cotton industry lost its initial flexibility: one is *social* and is to do with the evolution of employer structures, and in particular with the rise of a dominant employer class with increasing economic and political influence. From the 1830s/40s on this class took over the financial, organizational and technical leadership in manufacturing, forcing through the technological choices that fitted in with *its* conception of production. This conception was increasingly profit-obsessed and tyrannical – though for a long time accepted by the workforce – to the detriment of what was still at the beginning of the nineteenth century a "craft system."

While the first explanation involves consideration of the way that local production declined in quality, the second explanation is of an *institutional* order. If things evolved in this direction it was because there was no authority capable of putting an effective brake on the omnipotence of the business *bourgeoisie* and on the emergence of an employer ideology focussed on profit more than on product excellence. Even the growing interventionism of the state in the area of the regulation of labor did not succeed in providing such a brake. The result was that freedom of trade and industry as embedded in the liberal constitution of 1836/7 tended in this particular case towards freedom to exploit the calico-printing worker – at least up until 1860/70, when the ever-increasing polarization of social relations of production led to a rupture between the industrial chiefs and their workforce.

Moreover there was nothing in the local institutional structure to encourage training or promote quality of production or cooperative practices between firms.

I will now consider each of these explanatory elements in turn at greater length.

Vertical concentration and the disappearance of craft entrepreneurship

At the turn of the eighteenth and nineteenth centuries, there was still no real obvious difference between the Glarus system and watch manufacturing. Both were home-based and hybrid and involved a combination of wage-earners (homeworkers earning piece rates), small independent workshops or bigger-scale entrepreneurs (merchant-manufacturers in the cotton industry; assemblers in watchmaking) and an intermediate regime where work was done to order and remunerated according to the amount of labor involved, but where each artisan could in principle retain a measure of autonomy by working on his own account at the same time.

Each system was however subjected to separate influences. The development of the one was a reflection of the constitutive complexity of the watch; it was towards forms of production that were increasingly differentiated technically and professionally in the direction of fragmented production broken up into a plethora of tasks carried out in a multitude of small independent workshops (especially through subcontracting); the development of the other was towards the destruction of a class of artisan workers – subcontractors or piece-workers – and towards a concentration of the whole production cycle in the hands of a few generalist firms.

In the canton of Glarus, this integrative activity was manifest from the late 1820s and especially in the two following decades; by seizing financial and organizational control, local merchant capital managed to extend its influence over an ever-longer chain of production; certain export houses integrated some of the operations prior to fabric printing while a few prosperous weaving employers established their own printing manufactures or invested their profits in the construction of spining and weaving mills. Consequently the piece-worker – enjoying an intermediate status between the wage-earner and the employer and dominant in Glarus calico-printing at the start of the nineteenth century – was on his way out by mid century. The small manufacturers of the early days, seeing that they were losing contracts, since their former sources of orders were setting up their own printing works, were obliged to organize markets for their printed fabrics themselves. Those who had the nerve joined the ranks of the independent entrepreneurs, sometimes going as far as to take control of international outlets, while the more vulnerable quickly found themselves driven into liquidation (1834–45).[49]

Unless we bear this social dimension in mind, we will not be able to understand

[49] On this movement of concentration and integration: Jenny-Trümpy, *Handel und Industrie des Kantons Glarus*, p. 365; Heer and Kern, *Industrialisierung und Fabrikarbeiterschaft*, pp. 29, 69–71, 74–77, 313–316; Silvia Oberhänsli, *Die Glarner Unternehmer im 19. Jahrhundert* (Zurich, 1982), pp. 24–25, 42–44, 280; Veyrassat, *Négociants et fabricants dans l'industrie cotonnière suisse*, pp. 253–262.

the printing sector's loss of its initial flexibility. For the elimination of these small master-artisans or artisan-workmen meant that a "craft" dimension of production, incarnated by the early nineteenth-century Glarus calico-printers – disappeared at the same time. While the roller printing machine was spreading more or less all over Europe, the latter actually distanced themselves rather contemptuously from mechanical printing, whose potential was still rather limited. As craftsmen (and sometimes experienced technicians) trained in various continental calico-printing centers, they had looked for inspiration to "craft" centers in Alsace, Geneva or Neuchâtel – i.e. to places where as one of them put it, "you see a bit of everything; a place like Mulhouse where new removable colours are used . . . rather than places where only one type of article is produced [all allusion to the first roller-made articles manufactured in series to the detriment of the patterns, which were excessively simple].[50] At this first crucial turn in the technological context of the period the mechanical alternative was then clearly rejected in favor of skill-intensive activities. Without launching into the top of the range like their rivals in Alsace and Neuchâtel, the Glarus cotton men nonetheless remained attached to quality standards, to articles whose complex design required the utilization of several plates and colors, to the careful manipulation of mordants and dyestuffs, and to the employment of demanding dyeing techniques that were at that time incompatible with any mechanical process. A cheap labor force allowed them to sell quality at tolerable prices, and this was a further incitement to focus on specialization in types unsuitable for machine treatment. Thus in the 1820s and 1830s they diversified in ranges of article that required an important input from workers and a certain dexterity, forcing local calico-printing into the mould of a skill-intensive industry.

But with the advance of industrial concentration and the emergence of the new entrepreneur with his large-scale printing plants things could no longer be the same. Calico-printing would also come within the sphere of attraction of factory concentration and although still employing traditional unmechanized modes of production up to the 1870s, it was to bow to the same productivist demands as spinning and weaving, with their machine-driven rhythms. And the destruction of craft know-how at the entrepreneurial level was to spread to the level of the workforce: Glarus calico-printing, based on a kind of compromise between a manual technology and a way of organizing work conceived so as to be as efficient as a mechanized system of production, was transformed into a *sweated industry*.

[50] According to passages in the correspondence between Glarus calico-printers (1809) cited by A. Jenny and F. Luchsinger, *Die industrielle und gewerbliche Benutzung der Wasserläufe in Glarus und Umgebung* (Glarus, 1935), p. 122. On geographical mobility and professional training, see Heer and Kern, *Industrialisierung und Fabrikarbeiterschaft*, pp. 69ff. On the multiplicity of printing techniques and types of printed fabrics in the first half of the nineteenth century: Jenny-Trümpy, *Handel und Industrie des Kantons Glarus*, pp. 189–192 and *passim*; Veyrassat, *Négociants et fabricants dans l'industrie cotonnière suisse*, pp. 47–50.

1830/40: the end of a craft system . . . and of a patriarchal ethic

A crucial moment was the turn of the 1830s and 1840s, which saw the end of the old proto-industrial equilibrium in the region and the birth of a *new logic of strategic choices*, accompanied by a substantial growth in manufacturing activity leading to a rapid growth in production between 1840 and 1860/65.[51] Glarus export manufacturers signally failed to exploit the relatively skilled artisans and well-motivated workforce available to them or the large variety of techniques which could have increased the variety of articles and the flexibility of employment.

What precisely was this new logic?

Price competition versus quality competition: the evolution towards mass production Was it the growing grip on local manufacturing of commercial leaders represented on all the markets in the world, which gave pricing policy the predominance over quality criteria? We do know that the Glarus houses had an unfortunate tendency to undersell one another.[52] At any rate the trend was towards the *progressive adoption of large-scale production strategies* which, without meeting the needs of a mass market, was governed by a logic of economies of scale.

Now there are two ways of reducing costs when (at least in the opinion of industrial leaders) rationalization through the implementation of new technologies does not represent a solution suitable for the local situation:[53] either an improvement of the work-rate, the key to increased productivity, or a deskilling strategy to lower labor costs. It was in this direction – *maximal increase in the workmen's efficiency at minimal cost* – and so towards the mass production of cheap handmade articles, that most of the Glarus calico-printers opted to go. Profiting from the abundance of local labor and deciding not to mechanize printing, they nonetheless organized it in the most efficient manner possible. The means? Organizational improvements above all: a refined division of labor combined with the system of piece-work (which encouraged over-work) made possible an increase in productivity. This strategy meant the abandonment of articles whose

[51] Unlike other textile regions in Switzerland where structures of a proto-industrial type held firm, Glarus saw manual weaving collapse and with it the regime of dispersed industry. From the late 1830s to the 1860s, the number of homeworkers in the textile industry fell by 2/3, or from 2,900 active persons (2,545 of them in cotton) to 900 or 1,000, while the number of factory workers (printing, spinning and weaving) tripled at the same time – from 3,417 in 1837 to about 10,000: Heer and Kern, *Industrialisierung und Fabrikarbeiterschaft*, pp. 29, 32–33.

[52] Jenny-Trümpy, *Handel und Industrie des Kantons Glarus*, pp. 622, 643.

[53] It needs to be remembered that the first Glarus calico-printers gambled from the beginning of the century on techniques that were labor-intensive and had specialized between 1820 and the 1840s in types of printed fabrics at the time incompatible with roller printing (handkerchiefs and batiks). The industrialists of the next generation doubtless thought it pointless to make the financial sacrifice that would have been required for the technical modernization of printing – or preferred to invest in the integration of mechanical spinning and weaving.

production process was technically complicated, slow and difficult, in favor of a simplification of the work process[54] which eventually led to each firm having a single technical plant suitable for a single type of printed fabric (handkerchiefs or *Yasmas*). Each company thus specialized in a particular type and sometimes even in a single segment of the market, limiting its room for maneuver solely to the creation of novelties in patterns and coloring.

The result was the *abandonment of an innovative product strategy* which for three or four decades – a period of intercontinental diversification of Glarus outlets – had favored the differentiation of articles in terms of new and changing demand. It also meant the *standardization of work* – routine and monotonous processes, hasty and repetitive gestures that were often very rough – and in the long term the disappearance of know-how.[55] Moreover instead of investing in the long term by training and skilling the workforce, calico-printing bosses sought to reduce labor costs by resorting to the feminization of employment.[56]

By thus lowering production quality,[57] they transformed the character of local calico-printing: they progressively removed the skilled dimension from it, turning what had been an artistic industry requiring competence and talent at the beginning of the nineteenth century into an efficiency-driven industry.

Certain contemporary chroniclers were fully aware of this slippage – but that did not lead to any collective reaction as in the case of the watchmaking industry. Around the middle of the nineteenth century, two observers deplored the stagnation of Glarus manufacturing, particularly on the level of technical contributions, since the late 1820s:

The lack of local institutions able to provide a training suited to the industrial class . . . Progress that does not always match current requirements in the area of engraving and printing etc. . . . with respect to the precision and skill of the work, we are surely

[54] Thus the Glarus industry almost entirely abandoned the printing techniques of removable print on a Turkey Red background (which continued to be practiced in other centers in Switzerland), as well as batik processes that required highly skilled work and a great number of precautions in handling. Between 1855 and 1875, these two specialities represented only 16 percent of total output, while their value was slightly above that of the other articles (handkerchiefs and *Yasmas*, or, in volume, 84 percent of local production): Jenny-Trümpy, *Handel und Industrie des Kantons Glarus*, pp. 11, 38, 164ff., 401–402, 536, 628–629; on the preference of the Glarus cotton men for simple processes, easier to handle and cheaper: *ibid.*, pp. 56–61 and P. A. Bolley, *Die Betheiligung der Schweiz an der internationalen Ausstellung von 1867. Technischer Bericht zu Handen der Eidgenössischen Ausstellungskommission* (1868), p. 225.

[55] On all this, see Heer and Kern, *Industrialisierung und Fabrikarbeiterschaft*, ch. 3 and G. Heer and U. Kern, "Alltag der Glarner Tuchdruckereiarbeiter im 19. Jahrhundert," in *Arbeitsalltag und Betriebsleben. Zur Geschichte industrieller Arbeits- und Lebensverhältnisse in der Schweiz* (Diessenhofen, 1981), pp. 78–116.

[56] From the late 1820s to the 1840s, about one third of the workers were women; in the 1850s, their proportion rose to one half of the printing personnel: Jenny-Trümpy, *Handel und Industrie des Kantons Glarus*, pp. 397–398.

[57] On carelessness about quality, especially in the area of "double printing" see Jenny-Trümpy, *Handel und Industrie des Kantons Glarus*, pp. 544, 640, and in the area of handkerchiefs and shawls, see Bolley, *Technischer Bericht über die Betheiligung der Schweiz an der allgemeinen Industrie-Ausstellung in London im Jahre 1862*, p. 67.

no further on today than we were 20 years ago . . . Pointless migrations of workmen from one establishment to another, very damaging for the perfecting of tasks . . . The disappearance of the cordial and even paternal relations between factory chiefs and their workmen . . .[58]

Thus at the same time as what had in the opening decades of the nineteenth century been a craft industry was deteriorating, a patriarchal conception of work and of relations between employers and workmen was collapsing.[59]

The ambiguities of a mode of industrial regulation

Despite a precocious institutionalization of labor relations in favor of the employees, the existing institutional apparatus showed itself quite unable – at least up to 1860 – to guarantee any effective control of employer strategies, and equally incapable of preventing excesses in their employment of labor. Before looking at the reasons for this, let us see what this apparatus consisted of.

Direct democracy and labor legislation A product of the "Regeneration," the liberal movement of the Thirties, the Constitution of 1836/7, "the most perfect of all the Swiss democracies,"[60] gave each citizen aged eighteen and over both the right to vote and that of formulating proposals for laws. After being examined by *ad hoc* political and administrative authorities, such proposals were put to the "*Landsgemeinde*" (the annual assembly of citizens on the public square to discuss the affairs of the country, to elect representatives and pass laws – it was the supreme legislative authority in the canton). Thus for example the new constitution established the formal basis "for an integration of workmen into the state":[61] by attributing legislative competencies to the people on the one hand and by creating a legal procedure for the promotion of individual claims on the other, this political instrument had the potential to become a tool for pressure and negotiation in the hands of workmen or their defenders (pastors, doctors,

[58] Oswald Heer and Johann Jakob Blumer-Heer, *Der Kanton Glarus, historische, geographisch, statistisch geschildert von den ältesten Zeiten bis auf die Gegenwart. Ein Handund Hausbuch für Jedermann (Historisch – geographisch – statistisches Gemälde der Schweiz)* (St. Gallen/Bern, 1846), pp. 460–461.

[59] On the artisanal character of the first printing workshops, their *ambiance* of harmonious cooperation and the conviviality between master artisans and artisan workmen, still so close in their level of skill, see Oberhänsli, *Die Glarner Unternehmer*, pp. 56–57; Veyrassat, *Négociants et fabricants dans l'industrie cotonnière suisse*, pp. 181ff, 194ff.

[60] According to L. Snell, a German jurist who moved to Switzerland and became one of the intellectuals leaders of the liberals, cited by Otto Bartel and Adolf Jenny, *Glarner Geschichte in Daten*, vol. 1 (Glarus, 1926), p. 216. On the history and the authors of the constitutional revision of 1836/7: *ibid.*, pp. 198–218; also Jakob Winteler, *Geschichte des Landes Glarus*, vol. 2 (Glarus, 1954), pp. 405–446.

[61] Heer and Kern, *Industrialisierung und Fabrikarbeiterschaft*, p. 234. A similar view was taken by W. Grieder-Tschudi, who saw in the democratic machinery of this canton and especially in the *Landsgemeinde* an open road for each to encourage social progress and a useful political asset for the working class (*Wandlungen in den Beziehungen zwischen Arbeitgebern und Arbeitnehmern im Kanton Glarus*, Offprint, Glarus, 1953).

educational or ecclesiastical representatives, whose moral authority undergirded the first measures of industrial legislation) and to introduce forms of control over employers' strategies.

The canton of Glarus, one of the first in the Helvetic Confederation to have adopted measures to protect workmen in 1824,[62] stood out in Switzerland for its commitment to social legislation, even being considered a model in the area of the defense of workers' interests. However, up to now historians have not paid much attention to some of the "vices" of this industrial legislation. In Glarus – and this is a point that must be stressed – the legislative dispositions adopted up to 1856/64 were discriminatory.

Calico-printing: the preserve of the local employers Up to 1860 in fact, these measures applied to only a proportion of industrial establishments (mechanical spinning and weaving mills, where moreover they were often evaded by the employers) and not to the manufacture of printed fabrics which employed more than half of the canton's factory workers in the 1840s.[63] Thus among other examples the prohibition of work by children under twelve did not affect fabric printing works, where their involvement was nonetheless far from negligible (nearly 27 percent in 1837). And even the first *general* law on factory working, and of 1864, applicable from then on to *all* establishments (spinning, weaving and printing), did not affect the unskilled workers (about 1/4 of the fabric printing workers around 1860): another gift for the calico bosses, who were quite happy to lengthen the working day of this latter category as it suited them.[64]

Thus under the appearance of being very progressive the interventionism of the state had spared all the calico bosses, who had total freedom of maneuver in the area of printing up to 1856/64. This they exploited to the utmost, launching into the productivist marathon that marked the whole period between 1840 and 1865 and imposing "slave labor"[65] on the whole working class.

If the industrial chiefs of the area were successful in this strategy, it was because the worker-printer had broadly accepted it. And all the more so in that in this

[62] 1824: prohibition of night work in spinning mills (because of the danger of fire), thus putting an end to their round-the-clock use; a prohibition repeated in the 1845 ordinance but with certain concessions to the spinning bosses. The 1845 ordinance also prohibited the employment of school-age children there. 1848: adoption by the sovereign people (*Landsgemeinde*) of the first law on workers' protection, fixing 14/15 hours as the maximum working day in spinning establishments. 1856: a law extending the prohibition of night work and of work by children under 12 to all factories (only spinning mills had been affected by the preceding regulations), supplemented in 1858 by the prohibition of Sunday working (Emil Hobi, *Die Entwicklung der Fabrikgesetzgebung im Kanton Glarus* [Bern, 1920]; Erich Gruner, *Die Arbeiter in der Schweiz im 19 Jahrhundert. Soziale Lage, Organisation, Verhältnis zu Arbeitgeber und Staat* [Bern, 1968], pp. 227–237.

[63] Around 73 percent in 1841; still 63 percent in 1864 (calculated according to figures given by Heer and Kern, *Industrialisierung und Fabrikarbeiterschaft*, pp. 21, 48).

[64] *Ibid.*, pp. 100–101, 129ff., 173–174, 245.

[65] Jürg Davatz, *Glarner Heimatbuch. Geschichte* (Glarus, 1980), p. 205.

mono-industrial system no other industry paid better wages than calico-printing, which conferred on it a social status superior to that of the spinners.[66] Although vulnerable to the despotism of the bosses and to their utilitarian strategy, it had for a long time internalized factory discipline with its rigid hierarchy and its apparatus of coercive control. Moreover the calico-printer considered overtime as a way of maximizing wages. In fact on several occasions employees manifested opposition to restrictions on the length of their working day or on that of their children.[67] A Glarus pastor – one of the ardent defenders of workers' interests at that – denounced "the often fearful greed of the workmen and their desire for huge wages."[68]

Ultimately, although relations between employers and employees had become tense, it cannot be denied that a certain community of interests persisted for a long time in this industrious valley. A sort of tacit agreement had existed from the time of the harsh economic crisis of the beginning of the century (1799–1817) between manufacturers who were providers of jobs and a population that had been in the grip of terrible unemployment (a crisis in manual spinning, war and famine). This agreement drew together a small group of employers who were generally respected and who were sincerely convinced they were contributing to the material and moral progress of the workmen and of the collectivity, and a workforce that saw in them salvation from poverty. The multiplication of printing establishments after 1820 and even the birth of the first little mechanical spinning mills combined with a generally positive collective perception of the transition to the manufacturing system and an uncontroversial vision of the future prospects.[69] The philanthropic discourse of the times, and especially that of the Swiss Society for Public Unity, of which certain Glarus entrepreneurs were members, contributed greatly to spread this community and solidarity-oriented state of mind, a paternalist ethic (promotion of workers' education and of associative institutions – savings banks and health insurance), and the idea that industrial development implied a potential for social harmony.[70]

However this consensus and the "labor peace" that resulted from it did not

[66] Heer and Kern, "Alltag der Glarner Tuchdruckereiarbeiter," p. 101.

[67] Thus after the passing of a law in 1872 shortening the working day from 12 hours to 11, several workmen got together to propose to the 1873 *Landsgemeinde* a return to the 12-hour regime . . . in order to raise their wages (Hobi, *Die Entwicklung der Fabrikgesetzgebung*, pp. 86–87).

[68] Bernhard Becker, *Ein Wort über die Fabrikindustrie. Mit besonderer Hinsicht auf den Canton Glarus* (Basle, 1858), pp. 37, 38–40, 90.

[69] "At the time we wanted industry," explained a Glarus man in 1873, remembering his childhood, a time when "rich and poor," "the masters of the trade and the people walked hand in hand": [Salomon Feldmann], *O's Gmeivolch und d'Heere! Eine Betrachtung bei Anlass der Arbeiterbewegung im Kanton Glarus. Von einem glarnerischen Arbeiter* (Glarus, 1873), pp. 13, 15. On all this see also: S. Oberhänsli, *Die Glarner Unternehmer*, pp. 63–65, 74–75; Siegenthaler, *Konflikte der Industrialisierung*, pp. 58ff.

[70] Christoph, Trümpi, "Bericht über die im Fache des Armenwesens ausgeschriebenen Fragen," in *Neue Verhandlungen der schweizerischen gemeinnützigen Gesellschaft, 28. Bericht 1843* (Glarus, 1844), pp. 91–131.

persist beyond the middle of the century in the face of the disintegration of the collective consciousness into separate antagonistic identities,[71] resulting from the emergence of an absolutist and despotic conception of employer power and by the deterioration in working conditions – in complete contrast to the entrepreneurial successes of the boom years between 1848 and the 60s, when Glarus industry was at the (quantitative) peak of its development.

1860/70: the end of a socio-political consensus

After the middle of the century, when the social question was at its most acute and new laws were projected on working conditions, the head of the Glarus government quite rightly noted that "at bottom, the relations that need to be regulated are analogous to those that derive from a contract" (1864). On the other hand he was wrong to prophesy that state intervention on behalf of the factory workers, perfectly justified in his opinion, could not "have any bad consequences, since relations between bosses and workers have so far been free from animosity."[72]

If the eighteen-sixties and seventies were marked by a sudden dramatization of relations between bosses and employees, if there were key moments at the time of the popular consultation of 1864 (passing of the first general law on factory working) and even more so in 1872, when the question of "double printing" became acute, it was because both camps simultaneously broke the "pact" that had cemented working relations thus far.

On the one hand the working population exploited its civic rights to attack the local manufacturers' "closed shop," with the introduction of a bill intended to improve working conditions in all factories and not just in spinning mills.[73] This bill also included measures for the creation of a factory inspectorate – a supreme obstacle to employers' freedom[74] – and was submitted to the Landsgemeinde of 1864, when the factory workers, who then represented more than one half of the active population of the canton,[75] won a majority.

On the other hand calico-printing bosses broke with their employees by adopting certain labor-saving rationalization measures. The first steps towards machine printing of handkerchiefs and "double printing" raised worrying

[71] "The common people and the masters of the trade are no longer at one": on the divisions between manufacturers and workmen, see *O's Gmeivolch und d'Herre!*, pp. 21–22.

[72] Quoted by Th. Curti, "Classes ouvrières. Le socialisme," in *La Suisse au dix-neuvième siècle*, ed. Paul Seippel, vol. 3 (Lausanne/Berne, 1901), p. 202.

[73] Twelve-hour day and prohibition of night work: regulation of sanitary conditions of production, etc. On the beginnings of local workers' organisation in relation to the drawing up of the 1864 and 1872 laws, see Heer and Kern, *Industrialisierung und Fabrikarbeiterschaft*, pp. 234–268.

[74] Nomination of official experts having the right to inspect the internal life of all the industrial establishments in the canton and having the responsibility of enforcing labor laws.

[75] 57 percent in 1864 – as against about 1/4 in 1837: Heer and Kern, *Industrialisierung und Fabrikarbeiterschaft*, pp. 237 and 245.

prospects of unemployment and put an end to the kind of tacit convention that had bound the workforce so closely to the manufacturer and the factory.

From then on it was open war and dialogue between the social "partners" was blocked. At the time of the 1872 Landsgemeinde a new law was accepted by the people, further reducing the working day (11 hours) and, even worse, prohibiting "double printing" a matter close to the heart of the material interests of the calico-printing bosses and on which they were not ready to negotiate: indeed they managed to reintroduce the process not long after it had been suspended, though making some concessions to the workers on the issue of the amelioration of the sanitary conditions linked to this particular technique.

These two laws of 1864 and 1872 marked a critical moment in the socio-political context of the canton: a rigidifying of social relations and a loss of political cohesion. Thus the regime of direct democracy, attributing legislative competencies to the sovereign people, and giving to the working population (predominant in this canton) the possibility of an institutional normalization of strategies of resistance to employers' prerogatives, was eventually transformed into a conflictual system. A fruit of the democratic power of the citizen-workers, the legislative measures adopted in 1864 and 1872 were felt by industrial leaders to be police interference in their private sphere and the official inspection of factories was viewed as an intolerable strangulation of the free enterprise regime, an act of "terrorism."[76]

The private versus the public sphere[77]

So Glarus was no longer a cooperative economy, as it had been at the beginning of the century: there was no longer an ethic of collaboration between an industrial elite now turned in on itself and the calico-printers, nor was there even any solidarity between the employers themselves. Competition between local manufacturers and reciprocal mistrust seemed to be hallmarks of the system.[78] Competition that was often unfair, aiming to undercut prices (skillfully hidden in employers' rhetoric, which acknowledged only the pressure of *foreign competition* on the prices of Glarus printed fabrics), and not a healthy rivalry in promoting the whole sector.

Neither the local authorities nor the representatives of the textile employers nor the workers, who only began to organize in about 1860 and then only very timidly, managed to establish institutions able to maintain – or restore – the initial flexibility of local industry.

76 In the words of a local liberal publication: quoted by E. Gruner, *Die Arbeiter in der Schweiz*, p. 233. See also Oberhänsli, *Die Glarner Unternehmer*, pp. 120–121.

77 "Public sphere": a network of institutions for the regulation of industry encouraging cooperation between rival firms, created by private enterprise or by the public authorities. See Paul Hirst and Jonathan Zeitlin, "Flexible Specialization and the Competitive Failure of UK Manufacturing," in *Political Quarterly*, 60 (1989), 164–178.

78 Jenny-Trümpy, *Handel und Industrie des Kantons Glarus*, pp. 560, 622, 643.

In Glarus there was no place for an exchange of technical experience linked to production, and no place for discussion of innovations as in Alsace, where the Société Industrielle de Mulhouse provided a kind of platform from 1826 on for the development of a technical culture and for the mobilization of a scientific capital serving the needs of industry.[79] In Glarus there were none of those industrial societies, often so appositely named societies "for industrial emulation," that flourished elsewhere in Switzerland, especially in watchmaking Switzerland, and that contributed so much to the advancement of science as applied to industry, to professional education and to the diffusion of technical knowledge and inventions.[80] On the contrary, the Glarus manufacturers kept their manufacturing secrets jealously to themselves[81] and their opposition to factory inspection was partly inspired by fear of "industrial espionage."

Another contrast with Mulhouse calico-printing: the lacunae in professional training. Already in 1840 or thereabouts contemporary observers were deploring the lack of educative institutions for training the workforce, and the absence of a quality ethic and even of a work ethic. Nor did anything come along to fill the gaps later: no private or public investment in science or technical advance, no prefessional school, nor classes for technical apprenticeship. While in the watchmaking sphere or in the eastern Swiss textile industry there was a multiplication of institutions specializing in the promotion of technical know-how, in Glarus the attitude was rather reticent and Malthusian.[82]

It was also in the logic of the system that any attempt at cooperation between firms, rare as it was, was bound to fail,[83] while such employers' associations as

[79] The *Technischer Verein*, established in the 1840s by chemists, pharmacists, dyers and manufacturers in the canton, was in fact only an informal reading club prior to its reorganization in 1864: Oberhänsli, *Die Glarner Unternehmer*, pp. 120–121. A Glarus section of the Swiss Society for Public Utility was not set up until 1844, after many difficulties. This devoted itself above all to the struggle against pauperism: in the second half of the nineteenth century, however, some of its publications did deal with professional training: see E. Hafter, *Hundert Jahre Gemeinnützige Gesellschaft des Cantons Glarus 1844–1944* (Glarus, 1944) (esp. pp. 36ff.).

[80] As in Geneva, the *Société des Arts* (Classe d'Industrie) or in Neuchâtel, the *Société d'émulation patriotique*; there were endless societies of this kind even in the remotest parts of the Jura valleys (*Société industrielle et commerciale de la Vallée de Joux, Société d'émulation du Val-de-Travers, Société industrielle et commerciale de Sainte-Croix, etc.*).

[81] Jenny-Trümpy, *Handel und Industrie des Kantons Glarus*, p. 542.

[82] Siegenthaler, *Konflikte der Industrialisierung*, pp. 125–134 (esp. 131–133). The Glarus folk who went to industrial schools had to do so in the neighboring cantons: this option was available only to the sons of businessmen: Oberhansli, *Die Glarner Unternehmer*, pp. 84–85. We should also note the absence of any formal apprenticeship in calico-making – apart from that for engravers (2–4 years): Herr and Kern, *Industrialisierung und Fabrikarbeiterschaft*, pp. 123, 160. On developments in industrial teaching in Switzerland: O. Hunziker, "Beiträge zur Entwicklungsgeschichte des gewerblichen Bildungsschulwesens in der Schweiz," *Zeitschrift für Schweizerische Statistik*, 28 (1892), 42–74.

[83] For example, to prevent turnover of workers (Heer and Kern, *Industrialisierung und Fabrikarbeiterschaft*, pp. 124, 168) or to organize the communal acquisition of coal (Jenny-Trümpy, *Handel und Industrie des Kantons Glarus*, pp. 560ff.). On the failure of several attempts for each agreement between manufacturers or to make "conventions," types of cartel arrangements: Jenny-Trümpy, *ibid.*, pp. 641–643, 544 and Oberhänsli, *Die Glarner Unternehener*, pp. 122, 240–242.

were created took on the character of organizations for the defense of group interests, against the state or in reaction to the workers' movement.[84] Finally, the "management" of calico-printers depended on a very hierarchical division of labor which generated very ineffective relations between employers and employees, and not (as in Alsace for example) on a strategy of social integration.[85]

Conclusion

To conclude this section, we may affirm that the Glarus model provides eminent proof of the idea that evolution towards large-scale production was not conditioned by some logic inherent in technical progress, but by a certain type of power relations.[86] This proof is all the more impressive in that technical progress was precisely the great absentee from the local scene in calico-printing, even though it was indeed dominated by mass production.

This is a kind of negative proof: no massive technological investments in the main branch, printing, but a modest capital input; not technicist engineering, but simple tools; no machines – or very few – but the repetitiveness of manual operations, the mechanization of the human gesture. An example of industrial "archaism" by its "backwardness" in the technological domain, Glarus was nonetheless a perfect example of the mass production system. The social origin of this development was the formation of a powerful authoritarian group of employers over against a docile workforce which had no alternative but to put up with the move to factory production or else to emigrate.

The reasons for the decline of textiles in Glarus at the end of the nineteenth century? They are not to be sought exclusively, as with the classic explanations, in the conditions of the international market or in the "technical superiority" of rivals – we have seen that the watchmaking branch managed to survive in spite of external competition and American "superiority." The first enemy to be fought was of course domestic competition: the Glarus manufacturers exhausted themselves in internal battles and did not as the watchmakers did choose to institutionalize "the spirit of emulation,"[87] i.e. create the conditions for balanced competition, or "cooperation in competition."[88] As a historian of the local

[84] The *Börsenverein*, for example a militant organization in the confrontation between employers and employees founded in 1864 (a critical date if ever there was one for the employers) – included almost all the local firms (Oberhänsli, *Die Glarner Unternehener*, pp. 240–241, cf. also 231). We might also refer here to the *Technischer Verein* which sought to boycott the laws on factory work (*ibid.* p. 123).

[85] Y. Schwartz, "Pratiques paternalistes"; Sandrine Kott, "Enjeux et significations d'une politique sociale: la Société Industrielle de Mulhouse (1827–1870)," in *Revue d'histoire moderne et contemporaine*, 34 (1987), 640–659. In Glarus there was a turn towards the end of the nineteenth century back towards a paternalist policy – but especially in the domain of spinning, to try and tie down a workforce that was rather slimmer than before: Oberhänsli, *Die Glarner Unternehmer*, pp. 124ff.

[86] Sabel and Zeitlin, "Historical Alternatives to Mass Production." pp. 134, 161.

[87] A concept that never appears in the internal historiography of Glarus while it was an integral part of the watchmakers' rhetoric.

[88] Sabel and Zeitlin, "Historical Alternatives to Mass Production." p. 144.

employers put it, these "entrepreneurs did not prove capable of rising above the preoccupations of their own companies . . . nor of uniting to promote the industry as a whole."[89]

As for the regulatory mechanisms produced by this democracy in the domain of labor legislation, they proved to be counter-productive, for they upset employers who felt themselves more and more discriminated against by the law and were beginning, by the last quarter of the nineteenth century, to lose interest in the industrial future of the area and to move capital and production away to Italy.

Key factors in the success of watchmaking

I now propose to demonstrate (as I have done for Glarus) the relations between the socio-institutional context of Jura watchmaking and the strategic orientations of regional industry. More particularly, I propose to seek the elements of an explanation for the successful reproduction of a manufacturing system that (unlike the Glarus system) managed – with the occasional period of "slippage" – to safeguard the characteristics of flexible specialization.

The aim of the following pages will be first to highlight the fundamental features of the watchmaking milieu; not (as in the case of Glarus) dominated by an industrial caste aware of its prerogatives, turned in on itself and trapped in a profit-obsessed ideology, but a multipolar society which had remained a veritable reservoir of skills and marvelously exemplified this "pattern of diffused authority, autonomy and responsibility that flexible specialization demands."[90]

This *pattern of diffused authority* might be thought to have been the result of the political organization peculiar to the canton of Neuchâtel. I shall in fact go on to ask whether the aristocratic structure of power in the first half of the nineteenth century, which kept the bourgeoisie away from the political and administrative decision-making center, was not of such a kind as to call for a *parallel power*, in which that bourgeoisie – which included in its ranks entrepreneurs, traders, manufacturers and artisan watchmakers – could apply its political will. It seems that the energy that was forced away from high political spheres sought an alternative outlet in *associative force* on the local or communal level. This energy voluntarily structured the public space at that level and laid the informal foundations for the self-regulation of the regional manufacturing order (which had remained more or less at the margins of ruling-class concerns). The formation of the new republican political class that would overthrow the aristocratic regime in 1848 and show an extraordinary creativity on the level of associations of public unity and institutions for collaboration in the industrial domain – a key point in the argument that follows – seems to have developed through an informal sociability, evolving gradually towards more organized forms. In the case of Neuchâtel, we unfortunately know too little about this

[89] Oberhänsli, *Die Glarner Unternehmer*, pp. 242–243.
[90] Hirst and Zeitlin, "Flexible Specialization."

subject.[91] But the emergence in the Neuchâtel mountains (watchmaking districts) of a public spirit,[92] the product *par excellence* of the liberal republican movement of the first half of the nineteenth century, seems undoubted.

So in conclusion I will try to capture the influence of this public spirit on the birth of a series of institutions for industrial regulation and to examine the role of such institutions in the internal organization of the watchmaking economy.

The watchmaking milieu: productive decentralization and the continuity of skills

One important initial point: the watchmaking milieu was not invaded like the Glarus system by capitalist influences strong enough to dissolve its proto-industrial structures: the production equilibrium remained unchanged as late as 1880, and was based on the putting-out system and on the autonomy of labor.[93] It is even difficult to say here where the employees ended and where the employers began. It is difficult to decipher the social morphology of this complex edifice of the "collective manufacture" type that the watch industry became and remained in the nineteenth century,[94] fragmented into a multitude of workshops

[91] For a survey of research in Switzerland: *Sociétés et sociabilité au XIXe siècle. Colloque à l'Université de Lausanne, 13–14 Juin 1986* (Lausanne, 1986) (Etudes et mémoires de la section d'histoire de l'Université de Lausanne, publiés sous la direction du Prof. H. U. Jost, 5). See also François Kohler, "La vie associative dans le Jura au XIXe siècle – première approche," H. U. Jost, A. Tanner, *Sociabilité et faits associatifs* (Zurich, 1991) (Société suisse d'histoire économique et sociale, cahier no. 9), pp. 159–177; the author demonstrates the strong associative density of the industrial communes in the present Jura canton, which is next door to the canton of Neuchâtel.

[92] Alphonse Petitpierre, *Un demi-siècle de l'histoire économique de Neuchâtel, 1791–1848* (Neuchâtel, 1871), p. 436. Still on the subject of Neuchâtel, another nineteenth-century author wrote: "The communal life of the workshops and factories has stimulated their native gaiety and sociability. There are few towns and countries where there is a greater number and variety of societies as in the Neuchâtel mountains": Bachelin, *L'horlogerie neuchâteloise*, p. 210.

[93] In 1857 3/4 of the watchmaking workers worked in their homes; the others were waged workers, employed in specialized workshops: Elie Wartmann, "Horlogerie," in *Bericht über die dritte schweiz. Industrie-Ausstellung in Bern*, p. 226. And in 1873, a report on the Universal Exhibition in Vienna noted that in Neuchâtel watchmaking "the domestic establishment is still more common than the workshop, and the workshop itself still tends for the most part to be a family matter": Charles Kopp, "Rapport sur le développement de l'industrie horlogère dans le canton de Neuchâtel," in *Exposition universelle de Vienne 1873. Suisse. Rapport sur le groupe XIV* (Schaffhouse, 1874), p. 12.

[94] Robert Pinot – a representative of the sociological school of Frédéric Le Play – was almost certainly the first person to compare Jura watchmaking to a collective manufacture, following an on-the-spot inquiry in 1885, the results of which were published first in *La Science Sociale* (Paris, 1887), then republished with the title *Paysans et horlogers jurassiens* in 1979. In fact this term was not current in Switzerland where the usual concept was that of putting-out. On this system of productive, technical and geographical decentralization: Fallet-Scheurer, *Le travail à domicile*, pp. 292ff.; Jean-Frédéric Gerber, "Le syndicalisme ouvrier dans l'industrie suisse de la montre de 1880 à 1915," in Erich Gruner, *Arbeiterschaft und Wirtschaft in der Schweiz 1880–1914. Soziale Lage, Organisation und Kämpfe von Arbeitern und Unternehmern, politische Organisation und Sozialpolitik*, II/1 (Zurich, 1988), pp. 482–485.

and statuses: from the jobbing workman to the semi-independent piece-worker bound by labor rates imposed by someone else; from the small supplier, more autonomous, capable of dictating his prices, to the putter-out, that central entrepreneur, manufacturing chief, distributing and coordinating the work on a scale that was sometimes huge. In this aggregate of small employers working for themselves, skilled artisans and piece-workers (half-waged, half entrepreneurs, employer-workmen or employed small businessmen?), what was the degree of autonomy of each? What was the degree of satellization by the putter-out/wholesaler who orchestrated their work and dictated his conditions – but was himself also of course subject to those laid down by his suppliers?

What is quite clear is that these professional structures, so diversified and open, formed an ideal seedbed for a profusion of mini-bosses – a factor in the dynamization of the whole system – and that, between the substantial manufacturer and the workman, intermediate figures were not (as in the Glarus textile industry) absorbed and eliminated at the behest of a predatory capitalism determined to stifle small businesses. Thus even if the reality of watchmaking activity was that of an integrated/integratory production from the point of view of the technical structure of manufacturing and the collaboration of specialized workshops, for a large part of the nineteenth century it did not get translated as such onto the level of employer power and capital. There was fragmented production and there were dispersed small businesses; isolated portions of capital and disconnected innovations. Each person remained free to work at technical advance for himself within his own domain and according to his own needs.

Decentralization and the aptitude for innovation In a system like this innovative capabilities, so important for flexible specialization, were broadly safeguarded. As Landes rightly pointed out, implicitly recognizing the limits of mass production, it was "excessive industrialization [which], suppressing the multiplicity of place for potential innovation, lowering the general level of workmen's skills, reducing them to the role of docile executives more and more cut off from global knowledge of the craft . . . dried up the springs of invention and broke the previous chain of personal transmission of skills and ideas."[95]

There was nothing of all this in the old Jura manufacturing, where innovation flowed up continuously from the base. It was among the independent workshops that watchmaking tools were constantly perfected: a slow technological moulting comprising partial subterranean improvements that produced a gentle technical maturation of the sector. These multiple individual achievements by ingenious workmen and artisans brought about the gradual replacement in one specialism after another – and there were tens (or indeed hundreds) of them in watchmaking – of more or less simple tools by an unbelievable variety of

[95] "Histoire des montres suisses," in *L'Histoire*, 23 (May 1980), 34.

machine-tools:[96] a smooth technological transition which undoubtedly contributed to the accumulation of a collective capital of craft experience and knowledge, to the continuity of skills, and also to the perpetuation of a tradition of "high watchmaking."[97]

A high technology system: perpetuum mobile Even if a large section of the watchmaking population was neither particularly skilled nor trained in the more refined elements of the craft, there were numerous specialized artisans who lived the techniques daily, facing each day the challenges of precision work that required the investment of an incalculable amount of assiduity in the domain of the infinitesimal. We should not underestimate the effects of social training that may well have been the result. And Bachelin, the historian of the "men with watchmaking in their blood," was not wrong to write: "When the industrial genius of a people has awoken, it develops rapidly in several directions: one inventor stimulates another and emulation and rivalry themselves give birth to new inventions."[98]

In this respect the old Jura watchmaking industry did not really differ from the high technology cottage industry born in the 1960s and 1970s in central and north-east Italy.[99] Like that industry, watchmaking had the look of a decentralized system of creativity and innovations that produced a multiplicity of specialisms, processes, and tools, a system of synergy of talents, cooperation and exchanges between manufacturers,[100] a system without the hierarchical rigidities of an industry like the Glarus one.

If, at the moment of the great public debate after the "American challenge," a rather diffuse conflictuality emerged in the watchmaking milieu (unlike in Glarus where two clearly defined camps linked up against each other – employers against employees), it was because the specialized crafts, small workshop heads, skilled workers and a multitude of artisans including "men with watchmaking in their blood," did not disappear from the scene. Since the founding myth of Jura manufacturing, built on the personality of Daniel Jean Richard, the *genius inventor*, and since the eighteenth-century innovators, the technological vitality of watchmaking was for ever being fertilized by a professional elite: indefatigable researchers, skilled inventors and artisans, "artists,"

96 Jürgensen, "De l'emploi des machines en horlogerie," p. 155; Bachelin, *L'horlogerie neuchâteloise*, chs. 20 and 22.

97 Petitpierre, *Un demi-siècle de l'histoire économique de Neuchâtel*, *passim*, esp. pp. 260, 262.

98 Bachelin, *L'horlogerie neuchâteloise*, p. 21.

99 Charles F. Sabel, *Work and Politics. The Division of Labor in Industry* (Cambridge, 1982), pp. 220ff.

100 Cooperation in the domain of research, communication of discoveries and improvements: see Petitpierre, *Un demi-siècle de l'histoire économique de Neuchâtel*, p. 262; other examples: A. Bachelin, *L'horlogerie neuchâteloise*, pp. 63, 93; *Rapports au Conseil fédéral sur la participation de l'industrie et de l'agriculture suisses à l'Exposition universelle de Paris en 1855*; a first-class medal was won by a Neuchâtel watch manufacturer "for the generous manner in which he communicates his improvements to all his colleagues" (p. 40).

who often devoted more time to research than to their own business.[101] True precursors of hi-tech, these fervent defenders of a specific know-how and of a tradition of quality manufacturing, who consistently reacted with alarm signals when there was a tendency to sacrifice quality production, were probably the soul and conscience of the milieu, the inspiration for a cooperational management conceived so as to make the very best use of what industrialization had to offer. Their ability to mobilize the watchmaking milieu was crucially important in the sector's phases of decline – decline in the quality of the watches and decline in the quality of the work. Unlike in Glarus, the signs were always clearly perceived in certain areas of watchmaking society and were always the object of self-criticism or public warnings – a phenomenon that was all the more remarkable in that such critical moments (and especially the 1840s and 1850s) corresponded to phases of growth in production and in general prosperity.

The political constitution and the structuring of the industrial public sphere

The political form of the canton of Neuchâtel probably favored the emergence of the kind of pattern of diffused responsibility referred to above. There was a fundamental difference on the level of the organization of power between this canton – in fact a principality under Prussian sovereignty, but accepted in 1814 into the Helvetic Confederation – and the canton of Glarus. Up to the radical republican revolution of 1848 (which took Neuchâtel away from the King of Prussia and gave it an unequivocal status in the federal Swiss state),[102] the aristocratic regime in place did not give the local bourgeoisie the chance to hold any power and, consequently, it did not allow the entrepreneurs to play a part on the political stage. Throughout the first half of the nineteenth century, they could not therefore like their fellows in Glarus[103] integrate into the politico-administrative mechanisms of the state (either at the level of the parliament or at that of the government) and influence the economic organization of the canton. While Glarus respected the role of the manufacturer, not only by including freedom of trade and industry in the liberal Constitution of 1836/7, but also by acknowledging him as a benefactor of the public community and guarantor of employment in a poor and overpopulated area, the urban oligarchy of Neuchâtel kept its distance from the watchmakers of the mountains, after all, they were still just mountain folk even though they might occasionally be geniuses. While in Glarus the liberal debate and the Regeneration doubtless facilitated the rise of a "conquering"

[101] One example among many: Antoine Lecoultre, a symbol of this pursuit of excellence, detached from financial worries. On this artisan of genius, more passionate about mechanical improvements than about the accumulation of a fortune, see, François Jequier, *De la forge à la manufacture horlogère (XVIIIe–XXe siècles). Cinq générations d'entrepreneurs de la vallée de Joux au coeur d'une mutation industrielle* (Lausanne, 1983), pp. 134–183.

[102] On the political regime in the Principality of Neuchâtel, see Jean Courvoisier, *Panorama de l'histoire neuchâteloise* (Neuchâtel, 1972), pp. 118–123.

[103] Oberhänsli, *Die Glarner Unternehmer*, pp. 69ff., 99ff., 235ff.

industrial bourgeoisie, the "failed regeneration"[104] in Neuchâtel was translated into the repression of the liberal movement, closing off access to power at the highest level. There remained only the communal level as a place where this bourgeoisie could exert its will to socio-economic organization. And as the reigning oligarchy, having no relation with industry, remained "closed to the aspirations of modern life and to the approaches of the bourgeoisie,"[105] this was seemingly an incitement for the latter to take in hand *collectively*, beyond the management of day-to-day affairs that were traditionally down to the administration of the local community – forest, pasture, etc. – that of manufacturers' specifically local interests.

The dense network of associations even of a most ordinary kind and the multiplication of institutions with economic aims that (in total contrast with Glarus) appeared after the change of political regime seems to encourage such a hypothesis. In the middle of the nineteenth century there was created an institutional structure of a corporative type based on the defense of the general interests of the watch industry – and on a communal collective search for solutions to the problems posed by the way that markets evolved. Contrary to the situation in Glarus, labor relations in watchmaking prefigured the model of "labor peace" that prevailed in Swiss industry after 1937. On this point I think we can say with J. Zeitlin that, comparing the two regions, "differences in industrial relations can best be explained not by variations in social and economic structure, but rather by historical divergences in institutional development resulting from the organization and strategies of trade unions, employers and the state."[106]

It should be noted that in the period in question, the degree of unionization in watchmaking remained small. It was only in the 1880s that the labor movement woke up, and right up to the eve of the First World War its development was marked by discontinuity, absence of cohesion and social and mental resistances. Nonetheless, this movement, dominated though it was by a collaborationist ideology and rather refractory to confrontations with employers, did produce some positive though ephemeral results on the level of the regulation of competition between firms, the labor market and collective conflicts. I shall come back to this. In my opinion the main actors involved in the construction of an institutional framework appropriate to a flexible economy were recruited especially from the communal and municipal authorities – and after 1848 the cantonal authorities.[107] Their objective was mainly to improve production

[104] Rémy Scheurer, Louis-Edouard Roulet and Jean Courvoisier, *Histoire du Conseil d'Etat neuchâtelois. Des origines à 1945* (Neuchâtel, 1987), ch. 4.

[105] Courvoisier, *Panorama*, p. 118.

[106] "From Labour History to the History of Industrial Relations," in *Economic History Review*, 2nd ser. 40 (1987), 178.

[107] Unfortunately we do not know the exact nature of the links between the politicians of the republican generation and the watch industry, its employers and its artisans, for the study of the social components of the institutional network born after the 1848 revolution is still to be done.

quality and to promote a wider diffusion of knowledge. In the following pages I shall examine the development of the institutional apparatus proper to Neuchâtel watchmaking and also its role, especially in the industrial restructuring after the "American challenge."

The recurrent threat of mass production and institutional solutions to the "watchmaking question"

Specialization of Swiss watchmaking in the most varied types in the middle of the range,[108] which did not prevent certain regions or businesses from aiming for the upper end of the market (complicated pieces or de luxe watches), equally failed to prevent an opposite tendency: a slippage towards the bottom end of the range, towards facilities for hasty and cheap production of a "mass of watches," the specimens of which were catalogued, in the vocabulary of the trade, as junk watches or, even worse, watches so useless that they "did not work and their value consisted entirely in their low-grade silver cases."[109] Such watches were the very embodiment of *mass production*. They were the products of prosperity, when "everyone got used to letting the communal vessel drift,"[110] until there arose one of those recurrent crises, so much feared in the milieu, that were more and more identified as crisis of "overproduction" internal in origin rather than as temporary reverses down to fluctuations in the market.

So while Jura watchmaking excelled in the middle of the range, in its worst moments it also enjoyed the global monopoly of certain forms of mass production. It was characterized then by a certain ambivalence – by variable relations between flexibility and mass production. But it is not easy, since the sources are insufficiently informative (or insufficiently exploited), to distinguish between periods of "decline," i.e. of a trend towards the expansion of junk production, and periods of reaction, revitalizing manufacturing flexibility. Nonetheless, I am going to propose a periodization and try to formulate the phases of development of a substratum of collective services intended to correct the course of industrialization.

1790–1810/15: the decline of manufacturing Around 1790 there came to a close a period that was particularly dynamic and marked since the mid eighteenth century not only by the extensive development of which manufacturing in the Jura, but also by *flexible specialization*: product innovations (the automatic watch for example[111]), innovation in strategic compon-

[108] The good quality watch for the middle classes and the ordinary low-priced watch marked with the triple seal of first-rate quality, satisfactory precision, and a reasonable price.

[109] *Rapports des délégués neuchâtelois à l'Exposition universelle de 1867, à Paris, présentés au Conseil d'Etat de la République et Canton de Neuchâtel* (La Chaux-de-Fonds 1868 (rapport Hirsch)), p. 32. [110] Golay, "La vallée de Joux," (1979 edn), p. 37.

[111] Bachelin, *L'horlogerie neuchâteloise*, ch. 7; Landes, *Revolution in Time*, p. 269.

ents;[112] innovation in the manufacturing process (mechanization of movements[113]), innovation in traditional equipment[114] and innovation in marketing, with the risk being taken for the first time of crossing the Atlantic.[115]

However at the end of the century things began to go badly: mass production was tending to take over and Neuchâtel watches were losing their reputation for solidity and quality. A significant indicator: in 1790, thirty manufacturers complained to the government about frauds and abuses of gold and silver hallmarks and about the lack of respect for official quality control (established in 1775 in Neuchâtel).[116] Bad workmanship, the simplification of manufacturing processes, carelessness and a lack of mathematical exactitude on the level of the assembly and regulation of the watch, contempt for apprenticeship, degradation of the labor force, undercutting of prices and the truck system – all this spelt disorder in the manufacture: "For many people competition meant only that one manufactured as badly or as cheaply as possible. Europe was flooded with watches made to last for 24 hours with flimsy cases and misleading hallmarks and selling *by the gross*."[117]

The chief cause of this decadence – and contemporaries were perfectly well aware of it – lay in the revolution of the "machines."[118] The serial production of watch mechanisms by Japy (Haut-Rhin, France), which reached mass proportions after 1795,[119] gave an enormous impetus to Swiss manufacturing, which from now on could import them cheaply. The result was a sudden acceleration of production, involving in turn a sharper division of tasks (still manual). But other results were also a lowering of rates of pay for the work and a marked trend towards over-hasty production.

[112] The introduction of various types of escapements (the escapement is the most important part of the watch), allowing a greater variation in calibers and so in the final product: Fallett-Scheurer, *Le Travail à domicile*, pp. 199, 34–35/6.

[113] A French or a Swiss invention? The question remains an open one. Exploited by Japy (France), it went on to revolutionize Swiss watchmaking.

[114] A source of industrial diversification, the manufacture of tools became a separate branch in the second half of the eighteenth century and was localized particularly in the Val-de-Travers: Bachelin, *L'horlogerie neuchâteloise*, ch. 22; Landes, *Revolution in Time*, p. 261.

[115] The Neuchâtel watchmakers were among the first Swiss artisans and businessmen to emigrate to America towards the end of the eighteenth century and to check out its markets: B. Veyrassat, *Réseaux d'affaires internationaux, émigrations et exportations en Amérique latine au XIXe siècle. Le commerce suisse aux Amériques* (Geneva, 1993), chs. 2, 5. 9.

[116] Bachelin, *L'horlogerie neuchâteloise*, pp. 184–185. On the development of mass production to the detriment of quality at this period, see Fallet-Scheurer, *Le travail à domicile*, pp. 199–201, 219–220, 226–227; Scheurer, *Les crises de l'industrie horlogère*, pp. 15–16; de Coulon, *Les Ebauches*, pp. 37–38.

[117] Petitpierre, *Un demi-siècle de l'histoire économique de Neuchâtel*, pp. 248–249; see also pp. 247–251.

[118] See the minutes of the Commission for Industry (1811), La Chaux-de-Fonds Committee, whose discussions are related by Scheurer, *Les crises de l'industrie horlogère*, pp. 14–16.

[119] Lamard, *Histoire d'un capital familial*, pp. 145, 149.

It was then that the "watchmaking question"[120] surfaced for the first time: the public discovery of the phenomenon provoked a crisis of conscience, with debates on the mechanical manufacture of watch mechanisms, on the benefits or disadvantages of the division of labor, and on the means of stopping competition from junk watches. In this sort of discussion the camps at odds were always the same: on the one hand the partisans of government intervention and of the establishment of guilds and on the other, the supporters of liberty – or rather of an enlightened liberalism, under the auspices of which would be seen "the formation of industrial circles, prizes, incentives, and above all of special schools for watchmaking." The second group moreover, just like their adversaries, "did not imagine there were other ways of opposing competition from junk watches than providing all the markets with good ones."[121]

So it seems that a consensus was established at this time, if not on means, at least on ends.

1810/15–1830/40: a revolution of quality and precision

In reaction to this "slippage" phase, a resolve to cooperate took shape among local actors, embodied in the desire to organize the provision of collective public or semi-public services to remedy the situation.

Efforts were directed especially in the direction of *research*, in the framework of an active collaboration between the Neuchâtel Société d'émulation patriotique and a few watchmakers in the mountains. Around 1815/16 in fact, the center of concern for this institution for "public utility," founded in 1791, shifted very obviously away from the rural economy towards industry, and its activity in this new sphere received wide publicity.[122] This collaborative institution – to an extent the equivalent of the Société Industrielle de Mulhouse – occupied a key position in the general movement towards the revival of watchmaking.[123]

[120] The expression was coined by Jules Gfeller who, a century later, in similar circumstances (competition between mechanized manufacturing and the old putting-out system), reflected on the decay of production: "L'horlogerie suisse en 1886," in *Journal de statistique suisse*, 2 (1886), p. 75.

[121] Petitpierre, *Un demi-siècle de l'histoire économique de Neuchâtel*, p. 252. See also Scheurer, *Les crises de l'industrie horlogère*, pp. 17–19.

[122] See A. Petitpierre, *ibid.*, who follows the evolution of the watchmaking sector as presented by this institution's sources – in the studies and memoirs it encouraged or published (pp. 13–14, 8–27, 253–266). F. Scheurer also used the minutes of the *Société d'émulation patriotique* for his work: *Les crises de l'industrie horlogère*, pp. 17–18, 23, 31–33.

[123] A place for the exchange of practical experiences and theoretical reflections, the *Société d'émulation patriotique* made a great contribution to the diffusion of the spirit of research both in the domain of complicated pieces and in that of simple watches or indeed in that of working procedures (for example, it encouraged research on the gilding of watch parts – mercury gilding having shown itself to be very harmful to workers' health). In 1817 contacts were established between watchmakers, traders and this institution, which had then taken as its task the revival of watchmaking. It established several public competitions and advertised in the press; commissioned studies of certain technical subjects, giving awards for the most useful memoirs and publishing them, criticizing and encouraging watchmakers who

Worried by the serious recession of the period 1806–12, the state attempted to encourage the industry, though without affecting the administrative autonomy of the communes that were affected by the crisis: in 1811 the government set up an *Industry Commission*, with the task of reviving (or studying the means of reviving) watchmaking in the mountains. But this project was financed collectively by a share subscription in the whole country and the *Industry Commission* immediately ramified into local committees, each one free to suggest or take supportive measures.[124] From this resulted a few initiatives aimed at the amelioration of the quality of watches and the encouragement of new specialisms. Thus, on the proposal of one or other of these committees, the government worked for the reinforcement of quality control of gold and silver work by successive ordinances and decrees on hallmarks and hallmarking offices (1819, 1821, 1822), up to the unified regulation of 1839.[125] Between 1812 and 1814, these decentralized committees gave a further impetus to *industrial diversification*: the manufacture of instruments for physics, mathematics and surgery (navigational and mathematical compasses, alidads, microscopes, etc.) was introduced in various places.[126]

Finally, attention was given to *training* with an attempt to organize maths and physics courses (1823), and courses in geometry and mechanics as applied to the arts (1828), financed again by local public subscriptions and by a government gift. The experiment did not last. Two watchmaking schools were set up, one at La Chaux-de-Fonds (1831), the other at Le Locle (1826 or 1827), but their existence was precarious, and they were "sponsored" by individuals.[127]

On the level of formal institutions, except for the Société d'émulation patriotique there is not much to be said, since government measures remained limited to the supervision of watch cases, while the results of the private initiative in the area of technical teaching proved to be ephemeral. But although pragmatic, the forms of this cooperation between local actors took on a crucial importance in sustaining trust. The collective effort called for by subscriptions

submitted their inventions and work for the judgment of its experts; distributed prizes either for the making of watches which met certain criteria, or for the establishment of watch parts (spiral springs, for example, which were still not made in Switzerland) or for the provision of tools capable of bringing a solution to the technical problems of the day (for example, a machine to cut and round off gear wheels, in order to remedy imperfections in their teeth, one of the great weaknesses in Swiss manufacturing at the period). On all this: Petitpierre, *Un demi-siècle de l'histoire économique de Neuchâtel*, pp. 253–266, 13ff.

124 On this *Industry Commission* and the activity of the subcommissions: Jequier, "L'action des comités neuchâtelois d'industrie durant la crise horlogère de la fin de l'époque napoléonienne," *Musée neuchâtelois* (1975) and Scheurer, *Les crises de l'industrie horlogère*, pp. 21ff.

125 Bachelin, *L'horlogerie neuchâteloise*, pp. 184–185.

126 Petitpierre, *Un demi-siècle de l'histoire économique de Neuchâtel*, p. 253.

127 The Le Locle establishment, "founded and maintained by voluntary subscription" numbered "8 to 10 poor pupils" in 1836: according to a document quoted by J.-M. Barrelet, "La situation économique dans les Montagnes neuchâteloises vers 1836," in *Musée neuchâtelois*, 24 (1987), 246; it remained open for impoverished children up to 1879. On all this: A. Petitpierre, *Un demi-siècle de l'histoire économique de Neuchâtel*, pp. 256–257, 261; Scheurer, *Les crises de l'industrie horlogère*, p. 33; *Les écoles suisses d'horlogerie*, published by the Association des directeurs des Ecoles suisses d'horlogerie (Zurich, 1949), pp. 15–16.

for financial support for the revival of manufacturing, the psychic dimension to this mobilization, as well as the attention given by the press to the problems of the "industrial arts": all this was to contribute to promote various kinds of progress from the 1830s on in the area of quality and precision.

For example product diversification began to increase; although communal watchmaking remained the warhorse of Jura manufacturing, there was a trend towards technically refined watches (high-precision watches, marine chronometers, gold watches).[128] A ladies' watch was launched – this was a totally new market[129] – and the *Lépine* watch became common.[130] Improvements were moreover under way in the area of escapements.[131] In the area of equipment (machine tools operated by hand), important modifications occurred in the 1820s,[132] making possible much greater precision in the manufacture of watches, which consequently functioned more reliably: from now on the Swiss were in a position to compete with the English on level terms. Finally, where markets were concerned, Neuchâtel companies not only opened new outlets (Near East, China, Far East, Latin America), but also – and this was new – developed direct relations by progressively eliminating intermediaries.[133]

Prosperity returned and became established.[134]

A new era of decline: the era of shoddy goods and discredit (from 1830/40 to the "American" crises of 1857 and 1876)

This trend was developing by the 1830s: in 1828 the Société d'émulation patriotique was told that the reputation of Swiss watches was declining in the United States; a report presented to this same society in 1834 and another report to the government in 1836 criticized both the over-production of watches and their bad quality; in 1836 fresh complaints were heard about the quality control

[128] Fallet-Scheurer, *Le travail à domicile*, pp. 234–235; F. Scheurer, *Les crises de l'industrie horlogère* p. 142. [129] Landes, *Revolution in Time*, p. 270–271.

[130] Requiring mathematical rigor and great precision in the manufacture: it was the "revolution" of the flat watch, invented in France (in the second part of the eighteenth century) but adapted and perfected by the Swiss: Landes, *Revolution in Time*; Petitpierre, *Un demi-siècle de l'histoire économique de Neuchâtel*, p. 259.

[131] Fallet-Scheurer, *Le travail à domicile*, p. 232; Bachelin, *L'horlogerie neuchâteloise*, p. 112.

[132] Especially the innovation of the fixed burin, a machine with many uses: Bachelin, *L'horlogerie neuchâteloise*, pp. 146–147; see also Adolph Pfleghart, *Die schweizerische Uhrenindustrie, ihre geschichtliche Entwicklung und Organisation* (Leipzig, 1908), pp. 40–42.

[133] B. Veyrassat, "Présence du Jura horloger au Brésil pendant la première moitié du XIXe siècle," *Musée neuchâtelois*, 23 (1986); B. Veyrassat, "De Sainte-Croix à Rio de Janeiro: fromages et absinthe, dentelles et musiques contre café (1820–1840). Entreprise, région et marché mondial," in *Mélanges d'histoire économique offerts au Professeur Anne-Maria Piuz*. Etudes réunies par Liliane Mottu-Weber et Dominique Zumkeller (Geneva, 1989); Fallet-Scheurer, *Le travail à domicile*, pp. 227–228; Scheurer, *Les crises de l'industrie horlogère* pp. 29–30, 33, 38; F. Jequier, *Une entreprise horlogère du Val-de-Travers: Fleurier Watch Co S.A. De l'atelier familial du XIXe aux concentrations du XXe siècle* (Neuchâtel, 1972), chs. 1, 2.

[134] Barrelet, "La situation économique dans les Montagnes neuchâteloises vers 1836," p. 242; Scheurer, *Les crises de l'industrie horlogère, passim.*

of gold and silver materials – the regulations and supervisory offices for the guarantee of hallmarks were being evaded.[135] This was the beginning of the "great crisis of the craft"[136] which was to mark Jura manufacturing for two or three decades. Towards the middle of the nineteenth century, while orders poured in, quality fell off increasingly while unfair competition spread.[137] Among all the technical reports published after the great universal exhibitions of the second half of the nineteenth century, there was not one before 1880 or thereabouts in which the official Swiss experts did not express their alarm at the damage caused to good watchmaking by junk watchmaking.[138]

This new slippage towards mass production should not surprise anyone: we observed the phenomenon at the same time for Glarus and for other sectors too. No doubt it should be related to the huge extension of outlets for Switzerland, and especially to the Americanization of its exports: the growing attraction of the North American mass market. Between the years 1840 and 1872, the peak year for watch exports to the United States, that country was in fact the chief outlet for Swiss watchmaking.[139]

But as has already been said, the market does not explain everything. We should see another cause of this *laisser-aller* phase of the period in the general *laisser-faire* mood, the liberal trend of the 1830s, that time of "the apotheosis of the businessman," of "limitless competition"[140] and of rampant individualism. A time apparently less favorable to collaborative relationships. For example around 1839/41, the attempt to create a *Société industrielle* at La Chaux-de-Fonds for the defense of technical progress and good workmanship was a failure; while after 1841 the *Société d'émulation patriotique* – which had since 1816 put the interests of industry at the top of its concerns – turned towards history and the natural sciences.[141] The intervention of the state in the area of industry remained more or less a dead letter: in 1836, the government opposed all protection of inventions, designs and patterns and for this period we have only the adoption in

135 Scheurer, *Les crises de l'industrie horlogère*, p. 31, note 1, p. 62, note 1; Robert, *La fabrique d'horlogerie de Fontainemelon*, p. 23.

136 Scheurer, *Les crises de l'industrie horlogère*, p. 39. For Scheurer the crisis lasted for almost the whole of the second half of the nineteenth century: for Fallet-Scheurer, the "decay" of Swiss watchmaking occurred between 1848 and the (symbolic) date 1876 (*Le travail à domicile*, p. 251); for others, between 1850 and 1870 (Golay, "La vallée de Joux," p. 37).

137 See for example the reaction of Val-de-Travers manufacturers to unfair practices: Scheurer, *Les crises de l'industrie horlogère*, pp. 47–48.

138 Rapports au Conseil fédéral sur la participation de l'industrie et de l'agriculture suisses à l'Exposition universelle de Paris en 1855, op.cit., 1857, p. 33. See also Barrelet, "Les résistances à l'innovation," pp. 397–401.

139 Already around 1836, according to a contemporary witness, the United States was the main outlet for Swiss watches (Barrelet, "La situation économique dans les Montagnes neuchâteloises vers 1836," p. 243); this was still the situation towards the middle of the nineteenth century (*Bericht über die dritte schweiz. Industrie-Ausstellung in Bern 1857*, p. 226) and up to 1872; (on the decline of watch exports to the United States after this date, see the statistical table published in the *Feuille officielle suisse du commerce*, [1900], p. 46).

140 Gfeller, "L'horlogerie suisse." p. 25; Scheurer, *Les crises de l'industrie horlogère*, p. 34.

141 Petitpierre, *Un demi-siècle de l'histoire économique de Neuchâtel*, pp. 264, 15.

1852 of a law on gold and silver hallmarks, making the supervision of precious metals obligatory, but it was very often violated and finally replaced in 1865 by an extremely liberal law (freedom of hallmarks, official checks only optional).[142] Finally, between the early 1830s and 1856, no new industry school was founded.

Sales at low prices and overproduction after the 1850s resulted in the same procession of evils and the same complaints as in the years 1790/1815:[143] badly-made goods, a stationary level of quality – or indeed a decline, hallmark fraud, the truck system, a mediocre workforce, inadequate apprenticeships. Where models were concerned, creativity seemed diminished: everyone specialized in a particular type of watch:[144] there was a trend towards a simplification of calibers.[145] On the technological level it was a period of relative stagnation, with very few new improvements.[146] Even in the area of marketing this phase seems to have been marked by decline and by a recrudescence of trade through intermediaries.[147]

The "watchmaking question" emerged again with the first signs of crisis. For it was at times of depression – which affected junk watches much more than good watches – that public opinion was most aroused by a development of this kind. The two "American" crises in particular, strongly felt in the watchmaking population, led to self-criticism, as the regional press reveals (1857/59):[148]

> we are wrong to think that America is sick if trade with that country is going badly. It is we who are sick, we who have so much difficulty in ridding ourselves of our wretched business fever, so unfortunately contracted in the prosperous years between 1850 and 1857. America no longer asks us for three hundred thousand watches a year, now it's no more than one hundred and twenty-five thousand at the most. But we carry on manufacturing and sending three hundred thousand . . .

And again in 1861:

> In the time of prosperity there were abuses. Defective movements were put in gold cases. We could see gold cases . . . plated metal sold as solid gold. This shamelessness was responsible for the greater part of our evils. Not only did we make too much, but our production was tainted by bad faith and lack of loyalty. The universe is stuffed full of our watches, and everything is wrong with them: defective movements without escapements or with broken stones, pinions without pivots, and cases without hallmarks or with false ones.

[142] Scheurer, *Les crises de l'industrie horlogère*, pp. 32, note 4; 53, note 3 and 61–63.

[143] Fallet-Scheurer, *Le travail à domicile*, pp. 251–253; Scheurer, *Les crises de l'industrie horlogère* pp. 45, 47, 53, 57, 61ff., 77.

[144] Scheurer, *ibid.*, p. 52, note 4. This seems however to be contradicted by the exhibition catalogues, which often contain a wide variety of models made by certain exhibitors . . . but perhaps only the most "flexible" among the manufacturers exhibited – those who held a variety of stocks.

[145] Fallet-Scheurer, *Le travail à domicile*, p. 302.

[146] *Ibid.*, p. 251; Scheurer, *Les crises de l'industrie horlogère*, p. 78.

[147] Fallet-Scheurer, *Le travail à domicile*, p. 251; Scheurer, *Les crises de l'industrie horlogère*, pp. 65–73.

[148] Passages cited by Scheurer, *Les crises de l'industrie horlogère*, pp. 63–64; 57–58, 78.

A few years later when exports of Swiss watches to the United States were continuing to fall (between 1872 and 1877/8), the "American challenge" was the opportunity for a new bout of self-examination (1876):

At times when business was going well, we thought we would never be able to make enough watches, and we looked for the speediest and least costly ways to saturate the market with products of an inferior quality. For this type of product it made no difference whether the workman was good or bad . . . provided that he produced a few dozen pieces every day.

Swiss watchmakers would draw a lesson from this period of crisis (1850–76), which was also "a period of education. Employers and employees realized that quality production is the only way to make a name for the industry and create outlets for it . . ."[149]

From the years 1860/70 to the First World War: the institutionalization of cooperation and the restoration of flexible production The institutional response was amazing. It materialized on two levels. First of all, that of the *provision of collective services*, where it was for the most part the work of public authorities – of the new political regime put in place after the republican revolution of 1848. After 1857, there were increasingly numerous initiatives by the authorities – and also by individuals – to put at the service of the manufacturing collectivity institutions for the correction of a development considered to be fatal. Then, with the awakening of the labor movement (late 1880s), it was in the domain of the regulation of the labor market and of collective struggles that systematic efforts were made, especially by a few workers' unions, to have the principle of "limited free competition" (the standardization of prices and wages) accepted and to create *conciliation and arbitration instruments* to regulate disagreements between employers and workmen.

Let us begin with this last point. Numerous attempts were made to restrain and regulate competition between producers on the domestic market from the last third of the nineteenth century, although they rarely met with any durable success. They did, however, contribute – and this is undoubtedly their great merit – to spread and establish the practice of negotiation on a basis of equal representation.[150] The whole period between the great depression of 1874/9 and the 1914–18 War was marked by a *concerted* quest for institutional solutions to the problem of the bitter competition between the watchmaking entrepreneurs – a rivalry that had disastrous effects on wage levels.

Negotiations were carried on both on the terrain of trades union organization

[149] Fallet-Scheurer, *Le travail à domicile*, pp. 259–260.

[150] On this strategy of associationism on the basis of equal representation, very characteristic of the watchmaking sector: Gerber, "Le syndicalisme ouvrier." For a comparison with the cotton industry for example, where power relations clearly favored the employers, and fixed price lists were unilateral and not very common, see Andreas Balthasar, "Die Geschichte der Tetilarbeiterbewegung in der Schweiz," in Gruner (ed.), *Arbeiterschaft und Wirtschaft in der Schweiz 1880–1914*, II/1, esp. pp. 473–474.

(small craft corporations, unions of specialized workers, craft unions and federations) and in the framework of organizations with equal representation (mixed syndical chambers). Trade union strategy was essentially based on a traditional closed shop policy: defense of the craft and of craft skills, a quota system for the workforce and a limitation of apprenticeships, the exclusion of women, and the standardization of salaries and the regulation of the working day. A network of more or less loose agreements was thus created and conventions concluded with the employers, in particular fixing the prices for labor and minimum wage rates – known as "*tarif* agreements" or "collective agreements," then as "collective labor contracts" (the official term after 1911).[151] Mixed unions bringing together employers and employees in bodies where there was equal representation had the task of regulating product prices from the employers' side and wages and overall working conditions on the workers' side. The development of these conventions necessarily encouraged the creation of supervisory commissions and private conciliation and arbitration organs. Responsible for settling differences, these contractual institutions, rudimentary enough at the start, were to become a fundamental element in the collective contract, at least as important as the *tarif* itself.[152]

However, as a historian of industrial relations in the watchmaking sector has pointed out, these coalitions for most of the time had only a temporary character, for lack of discipline on the part of those concerned, and the contracts concluded had only a limited scope, without real influence on industrial reorientation, for they did not include the totality of watchmaking professions.[153] All the same it should be noted that this readiness to collaborate on the part of employers and their workers laid the foundation for a *collective regulation* of labor relations and labor conflicts and that the main bases for a *contractual strategy* (*tarifs* and institutions for arbitration) were in place on the eve of the First World War. Even if the beneficial effects for production are difficult to evaluate concretely, the progressive institutionalization of procedures for the regulation of collective conflicts[154] seems to have contributed to the relaxation of labor relations:[155] collective contracts, with their *tarifs* and their pacification clauses established voluntarily and on a private basis, gradually made the employers aware of the advantages that could come from labor peace and the non-intervention of the state, while the workers profited – at least temporarily – from the *tarifs* negotiated.

151 On the history of *tarif* practices in Switzerland, also in a comparative international perspective, the fundamental work is that of E. Gruner, cited above, II/2, pp. 1,213–1,297.

152 See Hans Born, *Einigungsämte und Schiedsgerichte in der Schweiz* (Berne, 1925), pp. 10ff. For reflections on the wider significance of such institutions: Gruner, *Arbeiterschaft und Wirtschaft in der Schweiz 1880–1924*, II/2, pp. 1,279–1,297.

153 Gerber, "Le syndicalisme ouvrier," p. 525.

154 The organizational model for this was borrowed from that of the *Conseils de Prud'hommes* (equal representation; two-stage procedure; conciliation and where necessary arbitral judgment).

155 Gerber notes the small number of strikes in the history of watchmaking between 1880 and 1914 ("Le syndicalisme ouvrier," p. 519); other authors speak of their great frequency in the watchmaking districts (see Gruner, *Arbeiterschaft und Wirtschaft in der Schweiz 1880–1914*, II/2, pp. 860, 864).

More than that, the contractual policy made the workers *responsible*: they became reliable interlocutors for the entrepreneurial class and also became more aware of their own role in the organization of production.

However the main innovation of this practice of collective contracts seems to me to have been the substitution of a regime of flexible dynamic regulation for the old system of rigid static regulation – that of individual contracts. Unlike the individual contract, which sets out the conditions of work between the master, his workers and his apprentices on a case-by-case basis, the collective contract – which is an arrangement negotiated between an employer or association of employers and a workers' organization – not only manages collective interests, but also influences the future evolution of labor relations and makes it possible to fix *strategic options for the future* (the collective regulation of apprenticeships, which acts directly on the labor market, especially on the supply of skilled labor, is the most spectacular example). The instruments that derived from this contractual strategy thus seemed like the institutional corollary of the new production and social structures that were being established in the watch industry: at the turn of the nineteenth and twentieth centuries, this industry abandoned a mode of operation based on relations between individual and individual in favor of industrial concentration, with company bosses on the one side and employees on the other. New mechanisms for the regulation of disputes thus became necessary in order to handle conflicts that were from now on essentially collective (strike, lockout) – something that was outside the competence of the traditional tribunals (tribunals for industrial arbitration or CONSEILS DE PRUD'HOMMES). The competence of the latter was in fact limited to individual disputes of a juridical nature (which excluded *economic* conflicts or conflicts of general interests), to the interpretation of existing labor contracts (which excluded any influence on the agreement of new contracts) and to cases whose litigatory value was below a certain sum fixed by cantonal laws. From this point of view we may regard the late introduction of tribunals for industrial arbitration in the canton of Neuchâtel (law of August 13, 1874), transformed in 1885 into CONSEILS DE PRUD'HOMMES in accordance with the French model, as an *anachronism* – which also explains why they never played the role in the watch industry that they did in the nineteenth-century Lyons silk-manufacture (see the article by Cottereau in the present volume).[156] One thing that must be remembered however: the CONSEILS DE PRUD'HOMMES as a model –

[156] Their introduction was in any case optional and only three communes in the canton of Neuchâtel established CONSEILS DE PRUD'HOMMES between 1886 and 1892, the others keeping their industrial arbitration tribunals (individual disputes having a litigatory value lower than 500 francs). Let us note however that before 1914 Neuchâtel was one of the few Swiss cantons to have established a cantonal conciliation and arbitration office to settle differences of a collective order between employers and workers, especially in cases where a strike or a lockout were threatened (law of 23.11.1899): Born, *Einigungsämte und Schiedsgerichte*, p. 57 (just as the state of Geneva gave competence to the Geneva Industrial Tribunal to act as conciliators in collective conflicts). The federal law of June 18, 1914 on labor in factories decreed the obligatory introduction in all the cantons of permanent conciliation and arbitration offices.

but solely on the organizational level – in the modernization of procedures for the settlement of conflicts, both in the private sector (norms fixed by professional associations in the framework of collective contracts) and in the domain of norms decreed by the state (cantonal conciliation and arbitration offices).[157]

We now come to the positive institutional apparatus, the provision of collective services. In the area of *technical professional training*, there were no fewer than eight schools of watchmaking, design or art applied to industry, opened in the canton of Neuchâtel alone between 1856 and the late 1870s – the first law on industrial schools having been passed in 1853.[158]

On the level of *production*, a series of public services was set up to promote and support centers of excellence: precision watchmaking (marine and pocket chronometers). Thus in 1858 the Neuchâtel Great Council decreed the establishment of a cantonal astronomical and chronometrical Observatory for the scientific supervision of the measurement of time.[159] Later this institution was the location for the famous chronometric competitions, annual competitions established by the government which did a lot to help improve regulatory techniques and the overall quality of pocket chronometers.[160] Among other initiatives also aiming at the improvement of watchmaking and at the promotion of competition in the area of quality, we might also note the foundation of new local "emulation" societies,[161] the establishment of collections of "industrial art" and special-

157 The CONSEILS DE PRUD'HOMMES also served as a model for the English system of Mundella and Kettle (1860s): Born, *ibid.*, pp. 1–3.

158 Courvoisier, *Panorama de l'histoire neuchâteloise*, p. 147; Bachelin, *L'horlogerie neuchâteloise*, chs. 27 and 28; Scheurer, *Les crises de l'industrie horlogère*, pp. 54, 71, 87; Hunziker, "Beiträge zur Entwicklungsgeschichte des gewerblichen Bildungsschulwesens," p. 46. This promotion of technical teaching also occurred in the other watchmaking cantons: see *Les écoles suisses d'horlogerie*, pp. 5–28.

159 After this, apparatus for the daily transmission of the exact time – via telegraph lines – communicated with the main watchmaking centers, making possible the setting of finished watches. This establishment also verified the working of marine or pocket chronometers sent to it by manufacturers. Thus Switzerland had two of the five main observatories of the period (Geneva, Greenwich, Hamburg and Paris): Bachelin, *L'horlogerie neuchâteloise*, ch. 25; Fallet-Scheurer, *Le travail à domicile*, pp. 234–235, 259.

160 From 1866 the Neuchâtel government organized official regulating competitions with prizes for the best chronometers and publication of results – excellent publicity for the winners. In fact the first such competitions had been started back in the early 1790s in Geneva and 1826/8 in Neuchâtel: at that time use was made of the movements of the sun for controlling the precision of watches! Later on experiments were carried out on the basis of less rudimentary procedures and they were progressively perfected until the Observatories set up at Geneva (1829) and Neuchâtel (1858) gave a more systematic and scientific character to the control tests. Up to 1968, when they were suspended, these chronometric competitions, which had been internationalized in the meantime, were an arena for the technological supremacy of the Swiss. On all this: Eugène Jaquet and Alfred Chapuis, *Histoire et technique de la montre suisse de ses origines à nos jours* (Basle/Olten, 1945), pp. 78–79, 189–196, 205–217; Charles Thomann, *Les dignitaires de l'horlogerie. La merveilleuse et tragique épopée des derniers régleurs de précision qui participaient aux concours de l'Observatoire chronométrique de Neuchâtel, 1923–1967* (Neuchâtel, 1981).

161 In 1858 the Comité neuchâtelois pour le perfectionnement de l'horlogerie, replacing the Société d'émulation patriotique, dissolved in 1850: in 1864, the Société industrielle et commerciale de Sainte-Croix; in 1878, the Société industrielle et commerciale de la Vallée

ized libraries,[162] as well as the creation of a press for professional information.[163]

Alongside such institutions there developed various forms of collaboration. *Collaboration between employers*: price or production cartels to limit the effects of domestic competition,[164] groupings on the level of marketing or commercial credit for example,[165] groupings for the promotion of the general interests of the watchmaking economy (the creation of the local industrial societies mentioned above). Interregional collaboration, with the foundation in 1876 of the *Société intercantonale des industries du Jura*, an organ for the defense of watchmakers' interests, bringing together several local manufacturers' associations. Created at the time of the "American challenge," this grouping served as their mouthpiece to the public authorities (especially in questions having to do with custom policy and industrial legislation) and played a far from negligible role in the subsequent reorientation of the watchmaking industry, by reacting "to the excessive individualization that has prevailed for too long in our watchmaking industry," by working in favor of a "regime of good and intelligent solidarity"[166] as well as the qualitative amelioration of the Swiss watch. It was for instance to play an active role in the preparation of two federal laws: that of 1880 on the supervision and guarantee of gold and silverwork and that of 1886 on patents.

In the 1880s, there was in fact a general consolidation of that institutional infrastructure, not only in the cantonal framework, but also on the federal level,[167] as well as a vigorous development of institutes for technical teaching, which from then on benefited from federal subsidies.

What were the concrete results of this institutional proliferation? They are to be seen first of all in the improvement of the quality of products. Already in 1873,

de Joux and the Société d'émulation industrielle in La Chaux-de-Fonds (which organized annual competitions as well as a National Watchmaking Exhibition in 1881); in 1891, the Société industrielle et commerciale du Val-de-Travers, etc. (Fallet-Scheurer, *Le travail à domicile*, pp. 419–423; Scheurer, *Les crises de l'industrie horlogère*, pp. 54, 87).

162 Bachelin, *L'horlogerie neuchâteloise*, ch. 28.

163 The *Journal suisse d'horlogerie*, a technical and scientific periodical that was to perform important services, was set up in 1876.

164 But these were often ephemeral: Gerber, "Le syndicalisme ouvrier," pp. 494–497.

165 Founded in time of crisis were the foundation of a Société d'exportation at Le Locle (1858), and the Union horlogère at La Chaux-de-Fonds (also in 1858). The latter, a sort of bank founded by 57 shareholders, especially watch manufacturers, to assist industrialists and facilitate credit, turned into an export syndicate: J.-M. Barrelet, "Diplomatie, commerce et ethnographie. Le voyage d'Aimé Humbert au Japon, 1862–1864," *Musée neuchâtelois*, 23 (1986), 149–150; Max Hintermann, *Vertriebsorganisationen in der schweizerischen Uhrenindustrie* (Weinfelden, 1928). On other initiatives in the area of credit (the creation of banks giving credit on merchandise), see Scheurer, *Les crises de l'industrie horlogère*, pp. 61, 65, 73, 83.

166 These are the words of one of its founders, cited by J.-M. Barrelet, "Les résistances à l'innovation," p. 407; see also the same author's "Diplomatie, commerce," p. 163.

167 Thanks to the revision of the Federal Constitution in 1874, the federal state had increased powers of intervention in the domain of the economy. Federal laws were passed on the protection of trade marks (1879), the supervision and guarantee of hallmarks (1880), the right of obligations (1881), making it easier to combat unfair competition, trade in gold and silver offcuts (1886), patents and the protection of designs and models (1888).

in his report on Swiss watchmaking at the Vienna Universal Exhibition – where there was no more sign of "those 15-franc 'silver' watches, that are neither watches nor silver" – the director of the Neuchâtel Observatory noted with satisfaction that "the whole watchmaking industry has made a big step forward."[168] Although they had not entirely disappeared, junk products had greatly diminished in number. The tendency to simplify calibers was no longer interpreted as a proof of decline, for it had concrete benefits in terms of the rational construction of the principal organs of the watch.

Moreover it needs to be emphasized that in the long term a certain number of product innovations had a positive effect, the most notorious of these being the application of electricity to watchmaking,[169] the launching in 1866 of the Roskopf watch, and that of the bracelet watch around 1880. The Roskopf – from the name of the manufacturer who perfected it – accessible to the poor and for that reason called "the workman's watch," became a big seller around 1895. Solid and reliable, it was the very symbol of the victorious battle against poor-quality goods.[170] As for the fashion for bracelet watches – they were a great hit in 1910/20.

As we have already seen, it was in this new phase of flexibility, between the early 1880s and the First World War, that the competitiveness of Swiss watches was restored. By evolving towards modernization without sacrificing a manufacturing identity that was in serious difficulties after 1876, Jura manufacturing managed to pick up the technological gauntlet of mass production. At the end of the century, on the occasion of his visit to the National Watch Co. in Elgin (Illinois), Houriet could write with that craft pride typical of centers of flexible specialization:

> We are "watchmakers" and the Americans are "mechanics." [. . .] I was disappointed to see the techniques employed by this factory, and I can say that thanks to the immense variety of our own and to the improvements introduced since the warning we got from our delegates at Philadelphia, we have overtaken them, not in a general manner [. . .] but, thanks to the impetus given, our upward march is continuing while over there manufacturing is stuck where it was then.[171]

Conclusion

All the internal historiography of Jura manufacturing – a reflection of watchmakers' rhetoric – shows that the major dilemma that this industrial district had

[168] *Exposition universelle de Vienne 1873. Suisse. Rapport sur le groupe xiv: instruments de précision et de l'art médical.* "Rapport de Mr. Hirsch sur l'horlogerie," (Schaffhouse, 1874), p. 38.

[169] Bachelin, *L'horlogerie neuchâteloise*, ch. 29.

[170] Although it had a simplified movement, requiring less labor input, the "workman's watch" was not a standardized article, for it was made in various qualities, depending on the taste of the customer (*Exposition universelle de Vienne 1873*; C. Kopp, "Rapport sur le développement de l'industrie horlogère dans le canton de Neuchâtel," p. 22; Fallet-Scheurer, *Le travail à domicile*, pp. 36, 272).

[171] Charles Houriet, *Les fabriques américaines d'horlogerie et leur outillage* (Geneva, 1895), pp. 9, 11.

to face in the nineteenth century was that it was caught between two antagonistic tendencies: manufacturing that was faithful to its international reputation for quality or production that was excessive and careless; between a craft identity and tendencies to cut-throat competition, "domestic competition, the only type that was really to be feared":[172] the kind of competition that led to a slump in prices and wages, to a lowering of the quality of products, and to discouragement of innovation.

One initial point to be emphasized in the functioning of the watchmaking system was the elaboration from the start of the nineteenth century of *horizontal control strategies*: before 1848, in a political era marked by the aristocratic and virtually anti-industrial organization of power, an informal network of collaborative relations was woven within the watchmaking population (why have so many Swiss historians stressed the individualism of watchmakers' behavior?). Among the examples given in this article: that of the Société d'émulation patriotique (promotion of research and innovation) or that of collective financing by public subscription of campaigns of support for industrial activity.[173] Since then appeal has always been made to local or regional soldarity in moments of crisis for the craft, each time that the "watchmaking question" arose.[174]

But to safeguard an ethic of quality manufacturing and maintain the spirit of competition and aptitude for innovation in the longer term, it was necessary to formalize such practices. The installation of a new political regime after the republican revolution of 1848, in which the watchmakers in the mountains in fact took an active part,[175] gave them the political means for an institutionalization of cooperation.

The latter initially took the form of cooperation between the state and the manufacturing milieux, then, towards the end of the nineteenth century, between professional partners.

The intervention of the public authorities after the 1850s and their continuous action up to the First World War played a role of major importance in the strategic reorientation of an activity that was in crisis in the mid nineteenth century and in the restoration of its international competitiveness after the "American challenge." It was towards the *organization of a wide panoply of educative services and technical assistance* that collective efforts converged. Among

172 J. Gfeller, winner of a competition for memoirs on "the watchmaking question": "L'horlogerie suisse," p. 31.

173 Practices that were perpetuated into the second half of the nineteenth century, when the creation of certain schools was financed by popular subscriptions.

174 "We must appeal to solidarity to put an end to the mindless unbridled competition that is causing a general lowering in prices and wages . . . everyone must be involved, bosses and workers . . .," – one example among many others (J. Gfeller, "L'horlogerie suisse," pp. 30–31); see also Barrelet, "Le développement d'une ville industrielle au XIXe siècle. Le cas de La Chaux-de-Fonds, 1850–1914," in *Les passages des économies traditionnelles européennes aux Sociétés industrielles. Quatrième rencontre franco-suisse d'histoire économique et sociale, Genève, mai 1982*. Etudes édités par Paul Bairoch et Anne-Marie Piux (Geneva, 1985), pp. 34, 38.

175 Fritz Jung, "Comment on fait une Révolution. Artisans Loclois de la République de 1848," in *Annales Locloises*, Cahier II 175 (Le Locle, June 1948).

the institutions created to provide the manufacturing community with the means the better to defend quality niches, a major part were intended to act *indirectly* for the amelioration of manufacturing, by encouragement of training (the creation of numerous vocational and technical schools and art and craft schools; the creation of libraries and museums); while others were conceived in such a way as to influence the maintenance or improvement of quality standards *directly*: thus the supervision and guarantee of hallmarks or the network of services installed by the Neuchâtel Astronomic Observatory for the chronometric regulation of finished watches – not to mention its annual chronometry competitions. These institutions were sometimes interdependent – we should note in passing – and stimulated each other.[176]

The rehabilitation of the watchmaking industry also involved a quest for adequate regulatory instruments for the labor market, for structural transformations at the end of the nineteenth century destroyed its juridical framework. The old system of individual negotiation, which favored the masters above all, was replaced by a system of *collective regulation*, whose slow and difficult gestation occurred between 1880 and 1914. Slow and difficult, for the success of collective negotiations depends on a solid organization not only on the employers' side but on the workers' side too. But before the First World War, the degree of unionization remained weak in the watchmaking industry. Nonetheless, the will to interprofessional collaboration was one of the noteworthy features of the sector: watchmakers in Switzerland played a pioneer role (after the typographers) in the area of associational strategy.[177] The issue was the collective agreement – the "*tarif*" – a peace agreement that could regularize domestic competition by stabilizing wages and calm collective conflicts. The content of it was variable, but what distinguished watchmaking conventions from those in other sectors was that corpratist spirit typical of artisan classes at the quality end trying to defend their craft, to defend a craft capital.

The entry at the end of the nineteenth century of the Jura watchmakers into the era of associationism based on equal representation and *tarif* agreements may thus be seen as the natural extension of that strategy of horizontal control referred to above; a strategy that was a feature of the first half of the century. As I have already said, the results of contractual policy were variable, the leveling of conditions of competition far from being achieved, and social dumping far from

[176] Thus for example at La Chaux-de-Fonds the Office for the Supervision of precious metals was behind the foundation of a local business school; or there were vocational schools that organized competitions for their pupils with a view to the great industrial exhibitions, in order to give the younger generation experience of international competition.

[177] Gruner, *Arbeiterschaft und Wirtschaft in der Schweiz 1880–1914*, II/2, pp. 1,254–1,255. This author notes that with the exception of watchmaking, Swiss industry was behind in the area of the collective regulation of labor: *tarif* practices and agreements were much less widespread there than in England and Germany – and it was essentially the German model that inspired sectors of activity other than watchmaking. In contrast, Switzerland was the first country in Europe to sanction by law (1911) the practice of official *tarifs* between professional partners (*ibid.*, pp. 1,159, 1,240–1,241, 1,272, 1,277).

having been eliminated. But these results were clearly more positive than in the textile industry for example, where with a few exceptions the outcome was a vertical control that served the interests of the employers. For power relations in watchmaking, a production system mostly based on the small business, were much better balanced, despite bitter struggles for influence, than in the textile industry – Glarus provides a significant illustration of this.

Part II

The battle of the systems

5 Between flexibility and mass production: strategic ambiguity and selective adaptation in the British engineering industry, 1830–1914

Jonathan Zeitlin

Few cases are more difficult to categorize as *either* flexible *or* mass production than that of the British engineering industry between 1830 and 1914. Through most of the nineteenth century, Britain's continental competitors saw her dominance of world trade in machinery – like that in cotton textiles – as the result of mass production in the contemporary sense of the term: the use of powered machinery and the subdivision of labor to manufacture significant quantities of relatively standardized products at low prices. So great was Britain's advantage in textile machinery, for example, that emergent manufacturers in Germany were forced from the outset to concentrate on "specialized products which complemented British spinning equipment or looms or produced yarn or cloth that could not be made with standard machines."[1] Yet in the 1850s, when British engineers inspected American techniques for manufacturing products such as locks and small arms, the latter's use of jigs, fixtures, gauges and special machines to produce interchangeable parts seemed distinctive enough from British methods to be termed "the American system of manufactures."[2] By the end of the nineteenth century, when American firms manufacturing large volumes of standardized equipment had begun to penetrate world markets, the contrast with their British counterparts' use of craft methods to manufacture customized products in small batches had become a commonplace of engineering comment on both sides of the Atlantic. These conflicting perceptions of British engineering reflected real ambiguities in domestic practice, whose significance would only become apparent as new competitive pressures from both mass and flexible producers forced the industry to redefine its identity through a process of strategic debate and selective adaptation.

The plan of this essay is as follows. The first section examines the evolution of British engineering practice during the industry's nineteenth-century commercial heyday. The second section analyzes the debate touched off by the onset of American competition from the mid-1890s, while section three explores the

1 C. F. Sabel, G. B. Herrigel, R. Deeg and R. Kazis, "Regional Prosperities Compared: Massachusetts and Baden-Württemberg in the 1980s," *Economy and Society*, 18 (1989), 386.

2 N. Rosenberg (ed.), *The American System of Manufactures* (Edinburgh, 1969); S. B. Saul (ed.), *Technological Change*, 1972; M. J. Piore and C. F. Sabel, *The Second Industrial Divide* (New York, 1984), p. 42.

practical responses of the British engineering industry up to the First World War. The conclusion considers the relationship between these pre-war choices and the industry's subsequent decline.

Engineering practice in the "workshop of the world"

There can be little doubt that the founding fathers of British engineering shared many of the characteristic aspirations of mass production as it was later to develop: uniformity, standardization and the substitution of machinery for manual skill. The blockmaking machinery devised by Bentham, Brunel and Maudslay for the Portsmouth dockyard at the beginning of the nineteenth century has long been recognized as a pioneering use of special machine tools to manufacture interchangeable parts (sponsored, like American armory practice, by the military authorities). But as Musson and others have demonstrated, Maudslay, Roberts, Nasmyth, Whitworth and other early British machinery manufacturers also developed the first "self-acting" or semi-automatic machine tools such as the slide-rest lathe, the planer, the slotter and the shaper, as well as such crucial instruments of precision engineering as bench micrometers, standard gauges, true planes and standard screw threads. By the 1830s and 1840s, their firms were producing textile machinery and locomotives in substantial batches to gauge and template, while advertising their own range of standard gear-wheels, machine tools and other engineering products in printed catalogues. On the shop floor, these policies led naturally to an extensive subdivision of labor, the employment of non-apprenticed men and boys on machinery, and the introduction of other cost-reducing measures such as piece-work and systematic overtime.[3] Such practices were vigorously opposed by the nascent societies of journeymen engineers, but the abundant supply of unskilled labor in the industry enabled a combination of London and Lancashire masters to defeat the newly formed Amalgamated Society of Engineers in a lockout over these issues in 1852 and impose the non-union "document" as a condition of re-employment.[4]

Despite these precocious steps towards standardization and deskilling, the developing pattern of demand for British engineering products placed stark limits to further progress in the direction of mass production. An emblematic case in point is that of James Nasmyth and the Bridgewater Foundry, recently studied by Cantrell. In his large purpose-built engineering works opened near

[3] A. E. Musson, "British Origins," in O. Mayr and R. C. Post (eds.), *Yankee Enterprise: The Rise of the American System of Manufactures* (Washington, D.C., 1981); and "Joseph Whitworth and the Growth of Mass-Production Engineering," *Business History*, (*BH*) 17 (1975); and E. Robinson, *Science and Technology in the Industrial Revolution* (Manchester, 1969), chs. 13–15.

[4] K. Burgess, "Technological Change and the 1852 Lockout in the British Engineering Industry," *International Review of Social History*, 14 (1969); and "Trade Union Policy and the 1852 Lockout in the British Engineering Industry," *ibid.* (1972); J. B. Jefferys, *The Story of the Engineers, 1800–1945* (1946), pt. I; W. McLaine, "The Engineers Union," PhD thesis, University of London, 1939.

Manchester in 1836, Nasmyth set out to build standard machinery for stock using "self-acting" machine tools arranged on the "straight-line system" to minimize movement of work-in-progress from shop to shop. While Nasmyth did turn down orders for non-standard patterns wherever possible, the commercial realities of his enterprise soon necessitated modifications of the original project. After an unsuccessful attempt to manufacture standard locomotives in advance of orders, Nasmyth discovered that the purchasers of such large capital equipment insisted on designs tailored to their specific requirements. Customers for one type of machinery wanted to obtain orders from the same supplier as well, and Nasmyth's catalogue of 1843 offered 126 different products, not including stationary steam engines and locomotives. Even his own patented machine-tools such as the steam hammer were too costly and came in too many sizes to be produced in advance of orders, and only 13 percent of total turnover between 1836 and 1856 appears to have been supplied from stock. While Nasmyth developed special machines for particular operations, making wide use of laborers and "intelligent lads" on machine tools which "did not require a skilled workman to guide or watch them," the assembly of his products seems to have "required a certain amount of fitting and filing" and he was never able to dispense with the services of refractory skilled mechanics. Despite successive victories in industrial disputes, Nasmyth was likewise unable to eliminate trade unionism among his workforce, and Nasmyth attributed his own premature retirement – whether sincerely or not – to the impending threat of future labor disturbances.[5]

Nasmyth's experience encapsulated in microcosm the development path of British engineering during the second half of the nineteenth century. As the surge of demand for textile machinery and railway equipment slackened after 1850, British engineering firms turned towards an ever greater diversity of markets at home and abroad. At the same time, moreover, their products became increasingly specialized, reflecting the growing sophistication of machinery itself and its progressive adaptation to the requirements of particular tasks.[6] By the 1890s, therefore, the British engineering industry had become a congeries of distinct but overlapping sectors linked by a common set of metalworking processes and the manual skills associated with them. Textile, railway and marine engineering were the largest single sectors, though significant clusters of firms could also be found in the manufacture of prime movers, boilers, armaments and machine tools, as well as agricultural, mining, sugar and printing machinery. But a large section of

5 J. A. Cantrell, *James Nasmyth and the Bridgewater Foundry* (Manchester, 1984), esp. chs. 4 and 8; cf. also S. Smiles (ed.), *James Nasmyth, Engineer: An Autobiography* (1889).

6 For overviews of the economic history of British engineering during the second half of the nineteenth century, see A. E. Musson, "The Engineering Industry," in R. A. Church (ed.), *The Dynamics of Victorian Business: Problems and Perspectives to the 1870s* (1980); S. B. Saul, "The Market and the Development of Mechanical Engineering, 1860–1914," *Economic History Review*, (*EcHR*), 2nd series, 27 (1967); and "Engineering," in D. H. Aldcroft (ed.), *The Development of British Industry and Foreign Competition, 1875–1914* (Glasgow, 1968), pp. 191–195.

the industry was also committed to general engineering, while many more specialized firms manufactured basic products such as machine tools, engines and boilers for their own use or as a means of adapting to the trade cycle.[7]

Individual sectors were typically concentrated in particular districts close to the end-users of their products: textile machinery in Lancashire, Yorkshire and the East Midlands; marine engineering on Clydeside, the North East Coast and the Thames; agricultural equipment in Lincolnshire and East Anglia; hardware, cycles and later motor vehicles in the West Midlands. The characteristic unit of production was the small to medium-sized family firm employing less than 500 workers, while most larger enterprises remained collections of craft workshops dealing with complex, bulky and expensive products rather than integrated factories engaged in high-volume manufacture. Vertical integration, where it occurred, was usually undertaken less to secure economies of scale than to offer a full product line to certain types of customer: thus, for example, it was the capacity to supply complete battleships to governments at home and abroad which lay behind the formation at the turn of the century of giant steel, machinery, armament and shipbuilding combines such as Armstrong-Whitworth, Vickers-Maxim and Cammell-Laird.[8] Outside of the railway company workshops, government arsenals and naval dockyards whose geographical isolation and distinctive employment conditions made them worlds of their own, even the largest engineering establishments were normally integrated into the surrounding regional economy through subcontracting relationships, while their workers often moved from firm to firm during the course of the trade cycle.

Within this pattern of sectoral and regional specialization, the heavy capital equipment which dominated British engineering output was generally produced to customer specifications. Each export market had its own distinctive requirements for products such as locomotives, marine engines or machine tools, while many domestic buyers, as the chairman of the Manchester Association of Engineers observed in 1896, insisted "upon makers constructing machines, specially designed according to the purchasers' ideas for every little operation, instead of accepting the standard pattern."[9] Hence even quite specialized enterprises were forced to maintain a wide product range in practice: the substantial Leeds machine tool firm of Greenwood and Batley, for example, manufactured 793 distinct products between 1856 and 1900, 457 of which were one-off jobs.[10] Even where customers could be induced to accept standard designs, sharp cyclical fluctuations in demand for capital goods and the high

[7] See tables 5.1–5.2 for a breakdown of output by sector in 1907, by which time newer products such as cycles, motor vehicles and electrical goods had come to form a much more significant component of British engineering production.

[8] C. Trebilcock, *The Vickers Brothers* (1977); S. Pollard and P. Robertson, *The British Shipbuilding Industry, 1870–1914* (Cambridge MA, 1979); K. Warren, *Armstrongs of Elswick* (1990).

[9] *Colliery Guardian*, April 2, 1896 quoted in Cantrell, *Nasmyth*, p. 71.

[10] R. Floud, "Changes in the Productivity of Labour in the British Machine Tool Industry, 1856–1900," in D. N. McCloskey (ed.), *Essays on a Mature Economy* (1970), p. 321.

Table 5.1 *Sectoral structure of British engineering, 1907*
Gross Output (£m.)

Sector		Value
Textile machinery		13.0
Railway locomotives: Private builders	4.5	
Railway companies	7.9	12.4
Railway carriages and wagons: Private builders	8.2	
Railway companies	7.6	15.8
Marine engineering		10.0
Steam engines (excluding locomotives and agricultural steam engines)		6.9
Cycles, motor cycles and parts		5.7
Motor vehicles and parts		5.2
Boilers		4.1
Machine tools		2.8
Agricultural machinery and steam engines		2.4
Internal combustion engines (except motor vehicles)		2.1
Hydraulic machinery		1.4
Mining machinery		1.3
Other machinery		11.7
Electrical engineering		14.1

Source: Final Report of the First Census of Production 1907 (1913).

Table 5.2 *International competition in mechanical engineering exports, 1913 (£m.)*

Product	UK	USA	Germany
Agricultural machinery	3.0	6.7	2.5
Boilers	1.8	1.9 (Boilers and Prime movers combined)	0.8
Prime movers	5.2		4.7
Machine tools	1.0	2.9	4.0
Locomotives (rail and road)	3.4	1.2	3.9
Sewing machines	2.4	2.4	2.8
Textile machinery	8.3	0.3	2.8
Miscellaneous	9.7	14.2	15.7
Total	34.8	29.6	37.2

Note: Table excludes cars, cycles, firearms, and railway carriages and wagons. Miscellaneous section largely consists of cranes, pumps, mining, hydraulic, sugar, grain milling, paper making and woodworking machinery.
Source: Balfour Committee, *Survey of the Metal Industries*, pp. 138–206, analyzed in Saul, "Engineering," p. 227.

costs of maintaining inventories discouraged firms from stockpiling much beyond small components and inexpensive equipment.[11]

Most British engineering products were thus manufactured in relatively small batches using flexible methods of production. Despite the development of the counting house and the drawing office, cost accounting remained rudimentary and production planning limited, as employers sought to keep the overheads associated with white-collar staff confined to a minimum. Detailed responsibility for the direction of production was largely devolved on craft-trained foremen, who also retained broad powers of hiring and firing.[12] Although great care was taken to minimize the movement of heavy components, British engineering establishments were typically laid out on the "grouping system," with general-purpose machine tools arranged by the class of work – such as milling, turning or boring – rather than by the sequence of operations for individual products.[13] Where "special" machines were used, these too were often capable of handling a range of different components, as in the case of steam hammers for large marine forgings, outsize lathes for railway work or boring machines for heavy guns.[14]

These product and production strategies made heavy demands on the skills of the labor force. While templates and gauges might be employed to meet customers' demands for interchangeability, jigs and fixtures were rarely used and workmen were often left to mark out the job and settle the details of machining operations themselves (under the supervision of foremen and chargehands), as

[11] The influence of cyclical fluctuations on the organization of production has been examined most fully for shipbuilding and marine engineering: see Pollard and Robertson, *The British Shipbuilding Industry, 1870–1914* (Cambridge MA, 1979); K. McClelland and A. Reid, "Wood, Iron and Steel: Technology, Labour and Trade Union Organization in the Shipbuilding Industry, 1840–1914," in R. Harrison and J. Zeitlin (eds.), *Divisions of Labour: Skilled Workers and Technological Change in Nineteenth-Century Britain* (Brighton, 1985); A. Reid, "'Employers' Strategies and Craft Production: The British Shipbuilding Industry, 1870–1914'," in S. Tolliday and J. Zeitlin (eds.), *The Power to Manage? Employers and Industrial Relations in Comparative-Historical Perspective* (1991).

[12] For discussions of foremanship, see J. Melling, "'Non-Commissioned Officers': British Employers and their Supervisory Workers, 1880–1920," *Social History*, 5 (1980), and "Leading Hands and Industrial Techniques: Workplace Supervision and Management Practice in the British Engineering Trades, 1870–1914," in Associazione di Storia e Studi Sull'Impresa (ASSI), *Annali di storia dell'impresa*, 1 (Milan, 1985). For a survey of the development of cost accounting in British industry, including engineering, see P. Armstrong, "The Rise of Accounting Controls in British Capitalist Enterprises," *Accounting, Organizations and Society*, 12 (1987).

[13] For discussions of the respective merits of the two systems, see T. Mudd, "Plan and Construction of Marine Engine Works," *Transactions of the North East Coast Institution of Shipbuilding and Engineering* (*TNECIES*), 1886–87; T. Westgarth, "Works Organisation," *ibid.*, (1899); "Some Aspects of Workshop Management: The Machine Shop," *The Engineer*, September 12, 1902; "The Laying Out of New Works," *ibid.*, January 18–February 1, 1907; J. Horner, "The Arrangement of Machine Shops," *Engineering*, July 28, 1916–December 28, 1917; H. C. Armitage, "The Grouping of Machine Tools: A Comparative Analysis of Grouping and Sectioning," *Engineering and Industrial Management* (*EIM*), March 4, 1920.

[14] See, for example, "Special Machine Tools for Locomotive Shops," *Cassier's Magazine*, (*CM*) (May 1904) and "Arrangement of Machine Shops."

well as to select their own feeds and speeds.[15] Manual skills and long experience were also needed to produce reliable castings from wooden patterns, and hand labor played an extensive part in fitting and erecting the final product. Enhanced skills were required even on the "self-acting" machine tools introduced during the first half of the nineteenth century. For as J. B. Jefferys observed in his history of engineering trade unionism,

> It was true that the jobs which had required a high degree of skill and experience to produce on a hand lathe or with a chisel and file, were usually comfortably within the ability of a youth or less skilled man on a "go-cart" or planer. And as long as the demands on the engineering industry were similar, except in number, with those made in the "handicraft" period, there was some justification for the employers' high hopes. But industry was not to remain satisfied for long with the hand tool standards of quality and speed of production. The new machines were capable of an entirely different range of work, and with the demand for larger jobs, greater accuracy, intricacy and speed, the operator, youth or skilled man, had to develop a range of skill to match the capabilities of the machine or be replaced by the man who was so able.[16]

By the end of the nineteenth century, therefore, some 50 or 60 percent of British engineering workers were formally classified as skilled craftsmen, though proportions could vary significantly from one firm to another. While some workers "picked up" a trade by moving from job to job, craft training remained the central route to skilled employment and most engineering firms of any size maintained an ongoing commitment to the production of apprentices. A final indication of the skill-intensive pattern of development in the industry can be seen in the resurgence and expansion of engineering trade unionism after its mid-century defeats: membership of the ASE, which catered primarily for skilled fitters and turners, grew from 9,737 in 1852 to 72,221 in 1891, and parallel increases were registered by craft societies catering for other metalworking occupations such as boilermakers, patternmakers, moulders and smiths.[17]

But British engineering firms were deeply concerned with price competitive-

[15] In 1885, for example, a leading trade journal felt obliged to explain to its readers the meaning of the terms "jigs and fixtures" in the course of a detailed account of the Singer Manufacturing Works near Glasgow: see *Engineering*, August 18, 1985, p. 211. "Templating," as Joseph Horner observed in 1917, which was the "crude prototype" of jigs and fixtures and "is still employed for pieces of large dimensions, leaves the actual methods of tooling, and also of holding, usually undecided and open to the choice of the workmen." *Ibid.*, December 28, 1917, p. 680. [16] Jefferys, *Engineers*, p. 16.

[17] *Ibid.*, pp. 51–67 and 291; M. and J. B. Jefferys, "The Wages, Hours and Trade Customs of the Skilled Engineer in 1861," *EcHR*, 1st series, 17 (1947); K. Burgess, *The Origins of British Industrial Relations* (1975), pp. 25–28; C. More, *Skill and the English Working Class, 1870–1914* (1980), pp. 27–38, 181–197. Based on an analysis of government wages censuses. More estimates the proportion of skilled workers in engineering as 53.0 percent in 1886 and 49.5 percent in 1906 (*Skill and the English Working Class*, p. 186); while the Engineering Employers' Federation reported that 60 percent of workers in member firms were formally classified as skilled in 1914: see M. L. Yates, *Wages and Labour in British Engineering* (1937), pp. 31–33.

ness as well as flexibility, and the second half of the nineteenth century saw a constant struggle to cheapen and intensify skilled labor within the framework of craft production. Semi-skilled "handymen" and boys were employed wherever possible on the simpler types of machine tools, and increasing numbers of apprentices were confined to repetition work at the expense of their technical training. Fitting itself was sub-divided into distinct classes of work – engine or machine fitting, bench-fitting or erecting – and the rougher tasks assigned to specialized laborers known as "fitter's assistants." Piece-work, intensive supervision and systematic overtime were also widely used to undercut standard wage rates and squeeze increased output from skilled workers on existing equipment. Given the centrality of skilled workers in production, these policies in turn set the stage for recurrent conflicts with the craft societies over wages, hours and working conditions over the course of the trade cycle.[18]

The ambiguities of engineering practice in the "workshop of the world" can be appreciated more fully by examining those sectors in which standardization, specialization and systematic management had progressed furthest by the 1880s and 90s. Foremost among these was textile engineering, where six Lancashire firms accounted for more than three-quarters of the 40,000 employed in 1913; the largest, Platt Bros. of Oldham, had 12,000 employees and an output equal to that of the entire American industry on the eve of the First World War. In textile machinery, unlike other sectors, British manufacturers imposed a measure of standardization on foreign buyers and pursued interchangeability of components as much to facilitate the supply of spare parts to distant customers as to reduce costs of production. With batch sizes of up to 100,000 for small parts such as spindles, extensive use could be made of special-purpose equipment, much of it designed by the firms themselves. Thus a detailed survey of Platt Bros. in 1894 remarked that "as there is so vast an amount of repetition work here . . . a special machine tool can be made for each similar part. It is this fact which gives large establishments such an enormous advantage in competition . . ." But firms like Platt's were obliged to maintain a full product line to secure contracts from foreign clients for the outfitting of entire mills, and they also appear to have been willing to manufacture machines in special sizes with modifications to suit customers' individual requirements. Rapid technical progress in the industry meant that firms were reluctant to fill their warehouses with potentially obsolete equipment, and production for stock was largely confined to spare parts for existing machinery. Few components, finally, were truly interchangeable without a certain amount of fitting, and American firms such as the Pettee Machine Works which sought to copy British carding equipment with machine methods were long unable to match the quality of work turned out by British fitters. British textile machinery firms thus continued to employ a significant proportion of skilled workers, as Paul de Rousiers observed on a visit to Platt Bros. in 1892, but

[18] J. Zeitlin, "The Labour Strategies of British Engineering Employers, 1890–1922," in H. Gospel and C. Littler (eds.), *Managerial Strategies and Industrial Relations* (1983); and "Engineers and Compositors: A Comparison," in Harrison and Zeitlin, *Divisions of Labour.*

many of these were confined to piece-work on repetitive tasks and would have experienced great difficulty, as he also noted, "in disposing of their special skill elsewhere on equally favorable terms."[19]

The manufacture of agricultural equipment presented important similarities to that of textile machinery. Many of the firms were large, employing between 2,000 and 5,000 workers, standard products were marketed through printed catalogues, and interchangeability of components was vital to service the demand for spare parts from distant farmers. Like the textile machinery manufacturers, agricultural engineers made extensive use of machine moulding and special tools of their own design, such as milling and drilling machines with multiple heads; "intelligent laborers" and boys were employed on simpler machines, and piece-work was widespread. But agricultural machinery firms also manufactured a much wider range of products, from steam threshers, plows and mowers to transportation equipment, portable steam and later oil engines, many of which were intended for non-farm uses. Foreign agents expected manufacturers to offer a full line of equipment and each export market had different requirements for the major types of machinery; even in Britain, there was considerable variation in soil types and the same plow design could not be sold to farmers in different parts of the country. Here, too, product improvements and seasonal demand discouraged firms from manufacturing for stock, and the heavy machines such as steam threshers on which British firms specialized were produced in batches no larger than twenty-five per week. Hand labor was still required in fitting and erecting more complex products and skilled workers continued to be employed in significant numbers on direct production, even if their tasks were often defined more narrowly than in other sectors of engineering.[20]

In railway engineering, as already noted, standardization and interchangeabil-

[19] Saul, "Mechanical Engineering," pp. 112–114; and "Engineering," pp. 191–195; R. M. Kirk, "The Economic Development of the British Textile Machinery Industry, c. 1850–1939," PhD thesis, University of Salford, 1983, pp. 185–213; T. R. Navin, *The Whitin Machine Works since 1831* (Cambridge MA, 1950), p. 325; G. S. Gibb, *The Saco-Lowell Shops: Textile Machinery Building in New England, 1813–1949* (Cambridge MA, 1950), pp. 345–346; *Engineering*, July 27, 1994; P. de Rousiers, *The Labour Question in Great Britain* (1896), pp. 254–266.

[20] Saul, "Mechanical Engineering," pp. 118–120; and "Engineering," pp. 207–208, 211–212; R. Munting, "Agricultural Engineering and European Exports before 1914," *BH*, 27 (1985); C. J. Lines, "The Development and Location of the Specialist Agricultural Engineering Industry," MSc. thesis, University of London, 1961; *Engineering*, August 7, 1885 (Marshall and Sons), July 31, 1885 (Ruston, Proctor & Co., Clayton and Shuttleworth), August 21, 1885 (Robey & Co.), June 21, 1907 (Clayton and Shuttleworth), July 18, 1913 (Ransomes and Rapier); *The Engineer*, August 1, 1890 (Ruston, Proctor & Co.), March 6, 1908 (Hornsby); *Report of the Tariff Commission, vol. IV: The Engineering Industries*, (1909), paras. 638–677 (Harrison, McGregor & Co.; Midland Engine Works; Ransome, Sims and Jefferies); EEF Archives (MRC, University of Warwick): P(11)5 and P(11)7 (refusal of agricultural engineering firms in Lincoln and East Anglia to pay overtime on piece-work, 1907); evidence of H. Marshall (Marshall and Sons), P. W. Robson and S. W. Patteshall (Clayton and Shuttleworth), J. Bamford (Leighton Iron Works) and E. C. Ransome (Ransome, Sims and Jefferies) to the Board of Trade Committee on the Engineering Trades After the War, 22/9 and 3/11/16, PRO BT 55/23.

ity were of crucial importance almost from the outset. In Britain, however, the position was complicated by an institutional division between the railway company workshops manufacturing for their own use and the private builders producing primarily for export. The railway workshops were huge, autarkic enterprises – the Great Western works at Swindon employed 14,000 in 1914 and five others between 4,000 and 7,000 – which manufactured as much as possible in-house (including steel in some cases) and only ordered outside equipment to meet unexpected surges of demand. Each company pursued standardization and interchangeability of components across its own equipment, and individual locomotive types might be constructed in large numbers: thus the London and North Western works at Crewe built 943 nearly identical goods engines between 1858 and 1872. In 1876, however, the railway companies were forbidden by the courts from producing locomotives for commercial sale as an activity beyond the remit of their parliamentary charters, while accelerating product innovation, strict safety requirements and the combination of new and repair work on the same equipment further limited the scope for specialization in manufacturing processes. The railway workshops thus represented a curious amalgam of craft and mass production methods. Special tools were developed for machining standard parts such as wheels, tyres, axles, cranks and cylinders; templates, gauges and jigs were systematically used to obtain interchangeability; erecting work was simplified through extensive subdivision of labor and the use of sub-assemblies; and large numbers of specialized laborers were recruited from the surrounding agricultural areas for employment on repetitive tasks. But as an expert visitor to Crewe remarked as late as 1917, batch sizes remained small for many components, general-purpose machine tools were widely employed for complex jobs, and hand fitting continued to play a major part in ensuring high quality standards. Reliance on skilled labor and general-purpose equipment was even more pronounced among the private locomotive builders, whose effective exclusion from the home market obliged them to cater for the diverse requirements of a wide range of foreign railway systems.[21]

[21] Saul, "Mechanical Engineering," pp. 114–117; and "Engineering," pp. 195–205; M. W. Kirby, "Product Proliferation in the British Locomotive Building Industry, 1850–1914: An Engineer's Paradise?" *BH*, 30 (1988); and "Technological Innovation and Structural Division in the UK Locomotive Building Industry, 1850–1914," in C. Holmes and A. Booth (eds.), *Economy and Society: European Industrialisation and Its Social Consequences. Essays Presented to Sidney Pollard* (Leicester, 1991); D. K. Drummond, "Building a Locomotive: Skill and the Work Force in Crewe Locomotive Works, 1843–1914," *Journal of Transport History (JTH)*, 3rd ser., 8 (1987); and "'Specifically Designed'? Employers' Labour Strategies and Workers Responses in British Railway Workshops, 1838–1914," *BH*, 31 (1989); R. J. Irving, *The North Eastern Railway Company, 1870–1914* (Leicester, 1976), pp. 87–99, 107–112; *Tariff Commission*, paras. 570–589 (North British Locomotive Co.); Webb Collection E.A. XVII, ff. 3 and 6 (Crewe and Derby, 1892); *The Engineer*, October 29, 1886 (Beyer, Peacock and Co.), June 12, 1903 (Great Eastern Co., Stratford), July 31, 1908 (Atlas Locomotive Works), August 14, 1908 (Great Western Railway, Swindon); *Engineering*, July 22, 1898 (Midland Railway Co., Derby), January 4, 1907 (Beyer, Peacock); J. Horner, "Special Machine Tools for Locomotive Shops," and "The Arrangement of Machine Shops," pts. VI (London and South Western Railway, Eastleigh) and XIII (London and North Western Railway, Crewe), *Engineering*, November 3, 1916 and May 4, 1917.

A final illustration of the hybrid character of British engineering practice during the late nineteenth century can be drawn from the armaments sector. There the giant Tyneside firm of Sir W. G. Armstrong, Mitchell & Co. employed some 15,000 men in its naval shipyard, ordnance and engine works in the 1890s. Large orders for ammunition, shells and machine guns offered considerable scope for repetition production, so the firm made extensive use of semi-skilled machinists and its managing director privately boasted that, "I take as many apprentices as I dare." More original were Armstrong's methods of labor discipline and supervision. By 1890 the firm had already developed an elaborate system of timekeeping, using individual boards to monitor attendance and movements from day to piece rates. In shops engaged in repetition work, piece payment coupled with frequent rate cutting prevailed. Where the nature of the work precluded piece rates or where the men were thought to be limiting output to prevent rate cuts, Armstrong's introduced a special class of supervisors known as "feed and speed men" whose duty was "to keep moving through the shops in order to see that each machine is being kept at its proper speed and is producing the amount of work which it is known to be capable of turning out." But the firm's most lucrative products such as battleships and heavy guns were complex, custom-built items requiring highly skilled labor, and Armstrong's remained a stronghold of craft unionism up to and beyond the First World War.[22]

By the 1890s, therefore, even those sectors of British engineering which had advanced furthest towards mass production retained a substantial commitment to craft methods. But it was not until American competition began to make itself felt in domestic and Imperial markets that the ambiguities of British practice would become manifest, spurring the industry to redefine its identity through a process of strategic debate and selective adaptation.

Background/Competition

The American challenge

Through most of the nineteenth century, few American engineering products were competitive in international trade, and *ad valorem* tariffs of up to 45 percent were necessary to protect domestic machinery builders from foreign and particularly British imports. Between the 1850s and the 1880s, however, as Hounshell and others have demonstrated, American industrialists successfully adapted the techniques of "armory practice" from the manufacture of firearms with interchangeable parts to high-volume production of light, standardized

[22] W. G. Gordon, *Foundry, Forge and Factory* (1890), 34; Webb Collection, E.A. XXI, f. 18; Amalgamated Society of Engineers (ASE) *Quarterly Report* (June 1894); Committee on the Organization and Administration of the Manufacturing Departments of the Army (Morley Committee), *Report*, PP 1,887, XIV, qq. 8,998–9,016; Royal Commission on Labor, *Third Report*, PP 1,893–4, XXXII, qq. 23, 157; Select Committee on Government Contracts (Fair Wages Resolution), *Report*, PP 1,897, X, qq. 23, 157; B. Taylor, "The Machine Question and Eight Hours," *CM* (1897); R. J. Irving, "New industries for Old? Some Investment Decisions of Sir W. G. Armstrong, Whitworth and Co. Ltd., 1900–1914," *BH*, 17 (1975); Trebilcock, *Vickers Brothers*; Warren, *Armstrongs of Elswick*. For a fuller discussion, see Zeitlin, "Labour Strategies."

goods such as sewing machines, typewriters, agricultural implements and bicycles. New types of machine tools such as millers, grinders and turret lathes figured prominently in the "American System of Manufactures," and rapidly growing demand for their products stimulated the development of domestic machine builders. From 1879 to 1894, prices of American engineering goods fell steeply, and US machinery exports surged into world markets when domestic demand collapsed during the mid-1890s.[23]

American engineering products had begun to appear in Britain and her colonies from the 1860s and 70s. Charles Churchill opened the first agency for the import of American machine tools in 1865, Australian farmers turned towards US harvesting machines from the late 1870s, and Singer established a major branch plant near Glasgow for the assembly of its sewing machines in 1881. British engineers and managers were also familiar with American machinery and methods through international exhibitions, articles in technical journals, factory visits and sojourns abroad in what remained a closely integrated "Atlantic economy."

But it was only with the home boom of the mid-1890s that American engineering equipment began to make a significant impact on British practice, above all in bicycle manufacture. The bicycle was the first product manufactured with interchangeable parts to tap a mass market in Britain, and domestic production expanded massively between 1895 and 1897. Soaring bicycle sales stimulated demand for the new generation of machine tools required to produce them, and other booming sectors of the industry likewise turned to US suppliers able to offer quick delivery of capital equipment. Imports of American machine tools consequently mushroomed, and British machinery firms making the new products also prospered.[24]

[23] D. A. Hounshell, *From the American System to Mass Production, 1800–1932* (Baltimore MD, 1984); O. Mayr and R. C. Post (eds.), *Yankee Enterprise: The Rise of the American System of Manufactures* (Washington DC, 1981); N. Rosenberg, "Technological Convergence in the American Machine Tool Industry, 1840–1910," *Journal of Economic History*, 23 (1963); R. Floud, "The Adolescence of American Engineering Competition, 1860–1900," *EcHR*, 2nd ser, 27 (1974); W. H. Becker, *The Dynamics of Government-Business Relations: Industry and Exports, 1893–1921*, (Chicago, 1982), pp. 1–47. For the history of American tariff policy, which never fell below 35 percent on imported machinery before 1914, see F. W. Taussig, *The Tariff History of the United States* (8th edn, New York, 1931; reprinted, New York, 1967); G. R. Hawke, "The United States Tariff and Industrial Protection in the Late Nineteenth Century," *EcHR*, 2nd series., 28 (1975); B. D. Baack and E. J. Ray, "The Political Economy of Tariff Policy: A Cast Study of the United States," *Explorations in Economic History*, 20 (1983); and D. A. Lake, *Power, Protection, and Free Trade: International Sources of US Commercial Strategy, 1887–1939* (Ithaca, 1988).

[24] S. B. Saul, "The American Impact on British Industry, 1895–1914," *BH*, 3 (1960); and "Mechanical Engineering"; and "Engineering"; and "The Machine-Tool Industry in Britain to 1914," *BH*, 10 (1968); Floud, "Adolescence of American Engineering Competition"; and *The British Machine Tool Industry, 1850–1914* (Oxford, 1976); I. W. McLean, "Anglo-American Engineering Competition, 1870–1914: Some Third Market Evidence," *EcHR*, 2nd series, 29 (1976); S. J. Nicholas, "The American Export Invasion of Britain: The Case of the Engineering Industry, 1870–1914," *Technology and Culture*, 21 (1980); J. Blackman and E. M. Sigsworth, "The Home Boom of the 1890s," *Yorkshire Bulletin of Economic and Social Research* 17 (1965); A. E. Harrison, "The Competitiveness of the British Cycle Industry,

At their peak in 1899, imports of American machinery represented a small fraction of British production, and they did not regain this level until the pre-war boom of 1912–13. But the rapid growth of American exports made a powerful impression on British engineering opinion, and domestic alarm was heightened when US firms were awarded contracts in 1898–99 to supply locomotives for British railway companies and construct bridges for colonial governments – largely on the strength of rapid delivery dates rather than price or design.[25] From that point onwards, unfavorable contrasts of British with American practice – often written by US-trained engineers – began to appear in technical society proceedings and the trade press, particularly *Engineering Magazine* and *Cassier's Magazine*, which published editions on both sides of the Atlantic. A peculiar feature of this literature was its exclusive concern with the United States: Germany, Britain's major competitor in many markets, rarely appeared outside discussions of low wages, tariffs, cartels and technical education.[26]

Both sides in the ensuing debate shared a common understanding of the defining characteristics of the American model, however much this may have exaggerated the uniformity of contemporary practice across the Atlantic.[27] Perhaps the clearest and most provocative exponent of American methods was H. F. L. Orcutt, an engineer who had served an apprenticeship at Pratt and Whitney, a pioneering "American system" machine shop, and worked for the noted Berlin machine tool firm of Ludwig Loewe & Co. before establishing himself in Britain as a highly successful management consultant during the 1890s.[28] American engineers, Orcutt wrote in 1899, were generally amazed by

1890–1914," *EcHR*, 2nd series, 22 (1969); Becker, *Dynamics of Government-Business Relations*, pp. 1–47. Imports of machine tools were also stimulated by the national lockout of 1897–98 which paralyzed many British engineering firms for more than six months.

25 Floud, "Adolescence of American Engineering Competition," pp. 59–63; Saul, "American Impact," pp. 19–20.

26 For contemporary discussions of German competition, see E. E. Williams, *Made in Germany* (1896); "An American View of German Competition," *The Engineer*, November 18, 1898; A. Shadwell, *Industrial Efficiency* (1906); *Tariff Commission*; R. J. S. Hoffman, *Great Britain and the German Trade Rivalry, 1875–1914* (Philadelphia, 1933); C. Buchheim, "Aspects of XIXth Century Anglo-German Trade Rivalry Reconsidered," *Journal of European Economic History*, 10 (1981); R. C. Floud, "Technical Education, 1850–1914: Speculations on Human Capital Formation," in L. Jörberg and N. Rosenberg (eds.), *Technical Change, Employment and Investment* (Lund, 1982); S. Pollard, *Britain's Prime and Britain's Decline: The British Economy, 1870–1914* (1989), ch. 3.

27 For critical discussions of the realities of workshop management in American metalworking during this period, see D. Nelson, *Managers and Workers* (Madison WI, 1975); and *Frederick W. Taylor and the Rise of Scientific Management* (Madison WI, 1980); H. Harris, "Employers Collective Action in the Open Shop Era: The Philadelphia Metal Manufacturers' Association, *c.* 1903–1933," in Tolliday and Zeitlin, *The Power to Manage?*; P. Scranton, "Diversity in Diversity: Flexible Production and American Industrialization, 1880–1930," *Business History Review*, 65 (1991), esp. pp. 62–73.

28 For Orcutt's career, see his testimony to the BoT Committee on the Engineering Trades, 2/11/16, BT 55/23. During the 1900s Orcutt served as managing director of Clayton and Shuttleworth, a large manufacturer of agricultural equipment in Lincoln, and then of the British Small Arms Co. (which also made bicycles, sewing machines and motor vehicles) in Birmingham, before establishing his own precision grinding business using wheels supplied by Norton in the US.

the variety of products turned out by even the smallest European shops. "The very opposite policy," he continued, "is followed by most American makers of machinery. Many concerns could be mentioned which devote themselves entirely to one line of machinery, and have no hesitation whatever in refusing business or orders for anything which they are not specially equipped for producing, and of which they have not made a special study." US firms were thus able to manufacture machine tools such as lathes and drilling machines in batches of 100 or more, while standard parts such as hand wheels, change gears and cone pulleys were made in larger quantities and kept in stock. Such policies in turn facilitated the employment of semi-skilled operatives on special machines and turret lathes set up by foremen or toolsetters; rough work could be finished by grinding machines, fitting eliminated as far as possible, and the remaining skilled workers concentrated in a separate toolroom.[29] As he observed in a paper presented in 1902 to Britain's premier technical society, the Institution of Mechanical Engineers,

> Probably in no department of mechanical work are the contrasts between old and new methods more striking than in erecting or assembling. In the new methods, machining is done accurately to dimensions; in the old machinery and tools are mainly used for removing metal, and reliance is placed on the fitter for proper working fits. In the new, accuracy and interchangeability of dimensions are maintained by a suitable equipment of gauges and the establishment of limits; in the old there is a variety of sizes depending upon the skill and judgement of individuals . . . In the new, the use of fitters is strikingly small compared with the abundance of this class of helpers necessary in works running on old lines. The model erecting department, therefore, is one in which fitters' work is reduced to a minimum, and where the least possible time is occupied in assembling. These results are, of course, only possible when the best methods of machining are practiced, when jigs and fixtures are used as much as possible, when a proper equipment of gauges is installed, and a system of inspection is amplified.

Orcutt went on to castigate the backwardness of British managers in adopting "modern machine shop methods," singling out their training policies for special criticism: "It is wasteful and foolish,' he argued,

> for all to insist that a man must serve from four to five years before he can become a proficient turner, when we know that in a few weeks a laborer can learn to operate two grinding machines and produce cylindrical surfaces that it is impossible for the most skillful turner to duplicate, either in regard to quality or to cheapness. This is not saying that it is unnecessary to train men to become good turners; it is, however, unnecessary to train them to perform many operations which are superseded by new methods of working . . .

[29] "Machine Shop Management in Europe and America," *Engineering Magazine* (*EM*) (January–March 1902), pp. 551–554 and 703–710; for similar criticisms, see R. Grimshaw, "American vs. European Shop Practice," (*CM*) (March 1896), and the special issue on "American Competition in the World's Engineering Trades: From an American Point of View," *ibid.* (March 1901).

Under a "modern system of apprenticeship," Orcutt concluded, all-round training would henceforth be confined to "men who are necessary for supervising and keeping plant in a high state of efficiency."[30]

Some British engineers likewise urged the adoption of American methods, albeit in more measured tones. Thus Benjamin C. Browne of Hawthorn, Leslie & Co., a well-known Newcastle firm of locomotive and marine-engine builders, attributed American success in winning British contracts for heavy equipment to their use of standard types, and commended "specialization in design, manufacture and sale of products" as an extension of Adam Smith's principle of the division of labor.[31] Speaking before the IME in 1899, Ewart C. Amos underlined the cost reductions obtainable on products such as machine tools through greater specialization. "Where the American makers gain so much," he observed,

> is in their method of manufacture. Their practice is to put in hand at least a dozen machines at the same time, stocking what is not made to order, while British makers build, as a rule, to order only. Their system enables them to use elaborate jigs and templates, which secure accuracy, economy and rapidity in manufacture, but does not pay in the case of one machine. Having set a tool to mill, plane or drill a certain piece of work, it is evidently economical to be able to machine, say a dozen or fifty of the same kind before removing the jigs and resetting the machine for other work.[32]

For other British commentators such as Henry Webb, the president of the Manchester Association of Engineers, reductions in manufacturing costs through the adoption of American methods offered a possible means of defending British markets against low-wage rivals abroad without constant downward pressure on workers' living standards.

> There was an impression in the minds of a few employers of labour that the only hope of successful competition with foreign countries, where labour was comparatively cheap, and hours long, was to be found in the reduction of the wage scale, or increases of hours to approximate the low standard prevailing in those countries. Might not the true secret of success be in other directions?

"The majority of employers," Webb maintained,

> preferred to pay high wages, if they could have freedom of action, so that they could use a combination of improved labour saving machines, operated by intelligent mechanics, whose services would be worth a high price. Improvements could be made by adopting certain fundamental principles in the methods of manufacturing, such as interchangeability of parts and the designing of special machines to perform

30 "Modern Machine Shop Methods," *Proceedings of the Institution of Mechanical Engineers* (*PIME*) (1902); cf. also his series "The Training of Our Industrial Forces," *The Engineer*, August 24, September 7 and July 14, 1917.

31 "Standardising in Engineering Construction," *EM* (October–December 1899). Browne argued that standardization in design need not lead to stagnation in product innovation provided that the standards themselves were constantly updated. See also his comments on Westgarth, "Works Organisation," pp. 121–123.

32 "American Machine Tools and Methods: A British Opinion," reprinted in *CM* (November 1899).

special operations only, whereby the output was increased, division of labour systematised, the costly work of finishing and adjusting minimised, and the highest development of skill, accuracy and dispatch acquired.[33]

Similar views were also expressed by younger engineers such as William Weir and William Rowan Thompson who developed their own versions of American incentive wage or "premium bonus" systems which they popularized through presentations to technical societies during the early 1900s.[34]

The bulk of British engineering opinion, however, remained highly skeptical of American methods and their applicability to domestic conditions. Despite some favorable comments, for example, Orcutt's 1902 presentation to the IME aroused a storm of criticism from leading British engineers. As J. R. Richardson, a Glasgow marine engineer, observed,

It was not that English engineers did not understand American methods, but that Americans did not as a rule understand the conditions which obtained in large engineering works in England having a big general practice. There must be a large run of work. Even the most enthusiastic Americans had told him that a large quantity was not needed, that it could be done perfectly well with a dozen, but very often a dozen was a large quantity. Not only was it necessary to have on his catalogue 500 different types and sizes of steam engines, but an infinite variety of mining and general machinery; and in addition his firm was expected to do anything required, and had to do it even if it only had to be done once.[35]

Other speakers called attention to the difficulty and expense of re-arranging old works along Orcutt's suggested lines, demanding "ocular demonstration" that "it would pay to pull down workshops and build anew, in order to introduce new systems." Even a sympathetic discussant such as William Weir underlined the paper's "evident restriction to . . . conditions applying chiefly to engineering works doing fairly light manufacturing work," a conclusion Orcutt himself was obliged to accept. Not only did shops doing heavy and general work form "probably the largest portion of the engineering trade of this country," Weir continued,

but the general shop was . . . the one which had done and would do the original pioneer work in new lines of progress. By progress he did not mean progress in

[33] "English and American Methods in the Engineering and Iron Trade," *The Engineer*, Jan. 21, 1898. This article was written at the conclusion of the successful engineering lockout of 1897–98, in which the "machine question" and working hours both figured prominently among employers' grievances: for a fuller discussion, see Zeitlin, "Labour Strategies."

[34] W. Weir and J. R. Richmond, "Workshop Methods: Some Efficiency Factors in an Engineering Business" and W. Thompson, "The Manufacture of Marine Engines: Some Factors Affecting the Economical Manufacture of Marine Engines," papers presented to the International Engineering Congress, Glasgow 1901 and reprinted in *Engineering*, September 13, 1901; J. Rowan, "A Premium System Applied to Engineering Workshops," *PIME* (1903).

[35] *PIME* (Jan. 1902), 70–74; cf. the comments of J. Thorneycroft, p. 62. Similar objections were raised in a discussion of "Tool Rooms, Their Advantages and Disadvantages" at the Manchester Association of Engineers. See the report by Hans Renold in *The Engineer*, December 16, 1898.

methods, but progress in research, in design, and in invention; and he trusted that in specialising, and in the desire of being up-to-date in ways of doing work, engineers would not allow those considerations to blind them to the necessity of always keeping in view improvements in the design of their manufacture.[36]

The connection between market conditions and manufacturing methods was also a favorite theme of technical journalists such as Joseph Horner of *Cassier's Magazine*. Special machines, Horner observed in a review of modern machine shop practice at the turn of the century, "are immense factors in rapidity of execution and economy of results, because, though performing many times the quantity of work done by a single machine tool, they cost no more for supervision." But the employment of such machines, he cautioned,

> is generally justifiable only when there is sufficient work to keep it employed during the whole or a large proportion of the time. Firms, therefore, who do general work chiefly, fall into error when they purchase highly specialised machines with the idea that they will conduce to economic output.[37]

While acknowledging the cost economies obtainable through American methods, British trade journals were equally sensitive to the negative effects of standardization and specialization on product innovation and productive flexibility. "Specialisation in manufactures," *Engineering* observed in 1906, "can only succeed if the article made is of a standard type, and is likely to be in continued demand for a sufficient period of time to give a fair return on the plant before it is scrapped." But such products might become obsolete before the capital equipment had been amortized, while "times of trade depression may bring complete collapse rather than a partial cessation of work, as would be the case if they were making many products. There is truth in the old adage of having two strings to one's bow."[38] *The Engineer*, Britain's other major trade weekly, went even further in a resounding editorial on "The Danger Element in Standardisation" in 1907. "Standardisation is inevitable," the journal argued,

[36] *Ibid.*, pp. 58–62 (H. F. Donaldson), 66–70 (Weir). Contemporary German investigations of American engineering practice likewise emphasized its limited applicability to the market conditions facing domestic machinery manufacturers. Thus a report on "Die amerikanische Maschinenindustrie und die Ursachen ihrer Erfolge" (The American Machinery Industry and the Causes of its Success) prepared by Paul Moeller for the *Verein Deutscher Ingenieure* (the Association of German Engineers) in 1903 concluded that American techniques were designed for mass production of standardized goods, whereas "adaptability has opened many foreign markets to German manufacturers." See H. Homburg, "Anfänge des Taylorsystems in Deutschland vor dem Ersten Weltkrieg," *Geschichte und Gesellschaft*, 4, 2 (1978), esp. p. 174; and H.-L. Dienel, "German Opposition Against the American System of Production," unpublished paper presented to the conference of the Society for the History of Technology, Uppsala, Sweden, August 16–20, 1992, pp. 4–5.

[37] "The Modern Machine Shop," *CM* (March 1900), 399–400; cf. also "The Equipment of Machine Shops," *ibid.* (June 1903), 149–50.

[38] "Concentration in Manufacturing," *Engineering*, February 16, 1906; cf. also "Automatic Machinery and Standardisation," *ibid.*, October 25, 1912. But for more favorable treatments, see "Standardisation in the Drawing Office" and "Cost in Engineering Design," *ibid.*, May 17, 1907 and July 8, 1910.

"but a standardisation which admits of no elasticity is a very grave menace to engineering progress," citing examples from shipbuilding, locomotive construction and electrical engineering. "For the sake of an ever expanding industry, subject to modification almost yearly by the advances attained through engineering research," it concluded,

> such standardisation must be subject to frequent revision . . . and must possess a wide elasticity, otherwise the engineering historian of the future, musing by the relics of some of our most splendid steel structures, will date the decline and fall of British engineering supremacy from that period in which . . . the engineering community of Great Britain said, "Here let progress and invention cease, let us standardise."[39]

From 1911 onwards, the British response to American engineering practice became inextricably bound up with the controversy over scientific management raging on the other side of the Atlantic.[40] The most celebrated confrontation pitted J. A. Hobson, the radical economist, and Edward Cadbury, a Quaker chocolate manufacturer and pioneer of personnel management, against Frederick W. Taylor and his Harvard disciple C. Bertrand Thompson, with the British commentators emphasizing the negative effects of specialization and subdivision of labor on the character and initiative of the workforce.[41] Trade journalists and technical writers likewise doubted that British workers would accept "route cards" or detailed instructions from "motion masters" during a period of rising union organization and labor unrest.[42] But the central thrust of

[39] *Engineering*, Jan. 25, 1907. Cf. also "English and American Machine Tools," "American Competition," "Specialisation in Manufacture" and "Standardisation as a Principle," *ibid.*, November 21, 1898, August 31, 1900, February 10, 1905 and October 29, 1909; and P. Balard, "Scientific Management and Science," *CM* (May 1912). For evidence of the cumulative impact of incremental design improvements on the performance of British steam locomotives before 1914, see Kirby, "Innovation and Division in the UK Locomotive Building Industry," pp. 28–33; and for a modern analysis of the barriers to product innovation created by excessive standardization and specialization of production in the United States, see W. J. Abernathy, *The Productivity Dilemma: Roadblock to Innovation in the Automobile Industry* (Baltimore, 1978); and W. J. Abernathy, A. M. Kantrow and K. B. Clark, *Industrial Renaissance: Producing A Competitive Future for America* (New York, 1983).

[40] For the American controversy, see Nelson, *Taylor*, pp. 154–193; and for general discussions of the British reception of scientific management, see C. Littler, *The Development of the Labour Process in Capitalist Societies* (1982), pp. 94–96, 99–115; W. Lewchuck, *American Technology and the British Motor Vehicle Industry* (Cambridge, 1987), pp. 89–97; B. C. M. Weekes, "The Amalgamated Society of Engineers, 1880–1914," PhD thesis, University of Warwick, 1970, ch. 5.

[41] *Sociological Review* (1913–14). Other participants in the debate included G. D. H. Cole and Walter Hazell, a printing manufacturer, who were hostile to Taylorism, and Hans Renold, the director of a Manchester precision chain works, who drew attention to the upgrading of skilled workers to more responsible tasks as inspectors, toolsetters and rate fixers under systems of scientific management. Cf. also Renold's critical defense of Taylorism in his "Engineering Workshop Organisation," *Transactions of the Manchester Association of Engineers* (*TMAE*) (1913–14).

[42] See "Workshop Management" and "Taylorism," *The Engineer*, November 14, 1913 and May 19, 1911; "Automatic Machinery and Standardisation," *Engineering*, October 25, 1912; and D. Smith and P. C. N. Pickworth, *Engineers' Costs and Economical Workshop Production*

their critique was commercial rather than sociological, focussing on the high overhead costs associated with American management systems, particularly for shops doing a varied range of general work. Beyond a certain point, *Engineering* argued, economic loss ensued through the multiplication of supervision and expert skill: the new systems represented "pretty much what every wide-awake shop manager has been doing for years past, but the main ideas are carried beyond the limits of practical application."[43] *The Engineer* was even more sweeping in its denunciation of attempts to apply American management methods in British firms:

> The American system of management is bureaucratic. It requires about one "superior person" to every four and half workmen or so. Here the ratio is far greater. We deem one black coat sufficient for the needs of nine to a dozen producers. The amount of unremunerative clerical work involved in the American system is astounding and it is very difficult to believe that it is profitable. At any rate its success has been rare.

"The proof of factory management," the journal concluded,

> is to be found in the sales department. Unless scientific management enables us to produce more cheaply or more quickly, then it is of little avail. We have yet to learn that British works managed on American lines have paid higher dividends than British works managed on British lines.[44]

From strategic debate to selective adaptation

In no sector were the respective merits of American and British practice more hotly debated than in railway engineering. Between 1898 and 1901, a sharp upturn in railway traffic caused several British lines to place substantial locomotive orders with American companies, largely because the domestic builders were flooded with work and could not promise delivery in less than two years; colonial railway authorities in India and elsewhere likewise turned to American and German locomotive builders for similar reasons. Whatever the differences in design philosophy between British and American locomotives (American engines were cheaper and more powerful but less robust and more

(1914), pp. 91–94. For a discussion of the effectiveness of craft militancy in engineering during the years preceding the First World War, see Zeitlin, "Labour Strategies."

43 "Compromise in Engineering," *Engineering*, December 1, 1911; cf. also *ibid.*, February 1, 1907, June 2, 1911, September 15, 1911, March 1, 1912 and June 27, 1913.

44 *The Engineer*, November 14, 1913; cf. also *ibid.*, December 14, 1912 and April 25, 1913. For similar scepticism from industrialists about the value of bureaucratic systems of cost accounting and bonus payment, see the discussions of D. Cowan, "Workshop Administration," *Transactions of the Institution of Engineers and Shipbuilders in Scotland* (*TIESS*) (1900–1); and Rowan, "A Premium System Applied to Engineering Workshops." German discussions of Taylorist methods before the First World War likewise criticized the excessive costs of the detailed time studies necessary for their consistent application: see Homburg, "Die Anfänge des Taylorsystems in Deutschland," pp. 174–180; and Dienel, "German Opposition Against the American System of Production," p. 6.

expensive to run), contemporaries attributed this debacle primarily to variations in their mode of construction. American firms such as the Baldwin Locomotive Works, it was claimed (not without some exaggeration), tendered only for their standard types, while British makers designed to individual specifications, often imposed by an outside consultant engineer; the Americans built engines in batches of 100 or more and manufactured parts for stock, while the British produced each order from scratch even when standard patterns and templates could be used. American practice, its proponents underlined, permitted more rapid delivery dates, production economies and interchangeability of parts across models; but as British commentators were quick to point out, it also imposed costs on the user of equipment not designed for local conditions and discouraged innovation in locomotive design.[45]

The advantages and disadvantages of the rival approaches came sharply into focus with the attempt to standardize locomotive designs on the Indian railways after 1900. The Indian standardization scheme was sponsored jointly by British locomotive builders, who hoped it would enable them to match the lead times and production costs of their foreign rivals, and by the Indian government, which wanted interchangeability of equipment across the various railway lines to assist with troop mobilizations. A special subsection of the newly established British Engineering Standards Committee was assigned to draw up the specifications, and by 1910 all locomotives ordered by Indian railways, whether privately or publicly operated, were officially required to conform to six standard types for broad gauge tracks and four for metre gauge. But the extensive discussion which followed the introduction of this scheme soon made clear that the imposition of standard designs had largely failed to secure the desired results. Geographical conditions varied widely in India, and bridges and the permanent way were insufficiently strong on many sections to carry the standard engines; a number of lines refused to order the standard types or modified them once they arrived; Indian railway engineers pressed for the incorporation of design improvements in revised specifications, and British builders lacked the confidence to manufacture components for stock in the expectation that they could be put into the next batch ordered without modification. The result was the worst of both worlds: standardization remained incomplete and no reductions were obtained in

[45] Saul, "American Impact," pp. 20–21; and "Mechanical Engineering," pp. 115–117; and "Engineering," pp. 201–220; T. Yuzawa, "The Transfer of Railway Technologies from Britain to Japan, with Special Reference to Locomotive Manufacture," in D. J. Jeremy (ed.), *International Technology Transfer; Europe, Japan and the USA, 1700–1914* (Aldershot, 1991), pp. 199–201; C. Rous-Marten, "English and American Locomotive Building," *EM* (Jul.–Sept. 1899); "American v. English Locomotives" and "Locomotive Building in Great Britain," *The Engineer*, January 10, 1890 and April 7, 1899; *American Engineering Competition: Being a Series of Articles Resulting from an Investigation Made by "The Times" London* (New York, 1901), pp. 67–77. For a meticulously documented modern study which demonstrates that turn-of-the-century British observers overestimated both Baldwin's ability to impose standard designs on the market and its departure from craft production methods, see J. K. Brown, "The Baldwin Locomotive Works, 1831–1915: A Case Study in the Capital Equipment Sector," PhD thesis, University of Virginia, 1992.

production or maintenance costs, while the standard types were not fully up-to-date and the introduction of design improvements was slowed down. Many participants in the debate concluded accordingly that standardization should be confined to component details rather than extended to whole types of locomotive.[46]

Despite the amalgamation of three Glasgow firms employing 8,000 workers to form the North British Locomotive Co. in 1903, there proved little scope for rationalization and British locomotive builders continued to produce a wide range of models for fragmented export markets up to and beyond the First World War. But by 1914 UK imports of American locomotives had ceased entirely and British builders regained control of Imperial markets, with the partial exceptions of Canada and New Zealand. Even in the United States itself, an NBLC spokesman told the Tariff Commission in 1904, railway engineers now insisted on their own specifications unless the engines were wanted in a great hurry, so that firms such as Baldwin's had been forced to abandon the practice of working only to their own designs. The real challenge to British locomotive builders in many markets, particularly those of continental Europe, came from German manufacturers, who remained equally committed to product innovation and skill-intensive methods, while reaping the benefits of longer runs through close ties with domestic railway companies which unlike in Britain did not make their own rolling stock.[47]

In other sectors of British engineering, however, experiments with the

[46] C. Hitchcock, "The Standardization of Locomotives in India, 1910," *PIME* (1910), 1,409–1,521; "Standardisation of Locomotives," *The Engineer*, October 18, 1910. On the development of the Engineering Standards Committee, see Sir J. Wolfe Barry, "Standardisation in British Engineering Practice," *Engineering*, August 10, 1906; and "Standardisation and Its Relation to the Trade of the Country," *TIESS* (1908–9); and "The Standardization of Engineering Materials, and its Influence on the Prosperity of the Country," *Proceedings of the Institution of Civil Engineers* (1916–17); D. Hulme, "The Work of the Standards Committee," *Transactions of the Institute of Marine Engineers* (1905–6). Standardization of components across different types and sizes of engines was an important feature of Baldwin's practice: see Brown, "Baldwin Locomotive Works," pp. 32, 121, 151–153, 162–163, 264–286.

[47] Saul, "Engineering," pp. 201–203 and 228; R. H. Campbell, "The North British Locomotive Company between the Wars," *BH*, 20 (1978); Kirby, "Innovation and Division in the UK Locomotive Building Industry," pp. 37–39; *Tariff Commission*, par. 581 (testimony of J. F. Robinson); *Committee on Industry and Trade (Balfour Committee), vol. IV: Survey of the Metal Industries* (1928), pp. 168–177; R. A. Brady, *The Rationalization Movement in German Industry* (Berkeley, 1933), pp. 156–160, 165–166. From the 1850s onwards, as Brown shows, master mechanics on technologically progressive US railroads insisted on custom manufacture of locomotives to their own design specifications. Although "there is no doubt that the market for standard engines was Baldwin's mainstay," he concludes, these were largely adaptations of the more successful custom designs for "that majority of mainline railroads which believed that pioneering didn't pay," and the company's product line increased steadily from 15 different engines in 1846 to 72 in 1881, 316 in 1890, and 492 in 1915. See Brown, "Baldwin Locomotive Works," esp. pp. 93–132, 361–362; quotations from pp. 117, 120. For a case study of the Japanese market, where British locomotive exporters remained dominant before 1914, but American competition was more important than that of Germany, see Yuzawa, "Transfer of Railway Technologies from Britain to Japan."

introduction of American methods were often carried further. Such efforts were most characteristic of firms producing new products such as bicycles or electrical equipment. After a brief surge of imports in the 1890s, British cycle manufacturers standardized their products, installed American machine tools equipped with jigs and fixtures for interchangeability, and subdivided tasks to reduce their dependence on skilled labor. Since British firms continued to produce higher quality cycles with more innovative design features than their American or German competitors, the resulting fall in costs enabled them to reconquer the home market and regain their position as the world's largest exporter.[48]

In electrical engineering, by contrast, restrictive legislation and the prior development of other forms of power, lighting and transport slowed the growth of the British market and hindered the emergence of domestic producers, as did their initial commitment to direct rather than alternating-current technologies. The British industry was therefore dominated from the outset by the large-scale American and German electrical manufacturers such as General Electric, Westinghouse, Siemens and AEG, most of whom established substantial branch plants in Britain before the First World War. In some areas such as electric tradition, however, British firms took up American methods as the domestic market expanded and successfully used them to repel foreign competition. A case in point was Dick, Kerr and Co., which established itself as the leading British manufacturer of tramway rolling stock by copying American designs, standardizing parts and introducing special machine tools; its Preston factory was described in 1900 as "an experiment of bodily planting an American shop in England, arranging it on American lines, and working it according to the American system with British workmen." In cable making, similarly, where Britain had played a leading role in laying the international submarine telegraph network, the home market was controlled by four large firms (including the British branch of Siemens). Here the Cable Makers' Association was able to impose standard specifications on their customers, enabling its members to manufacture for stock using semi-skilled labor on specialized machinery, and compete with the Germans on equal terms in export markets. But other efforts to transplant foreign practice in the British electrical industry proved distinctly less successful. In 1902, for example, George Westinghouse opened an enormous manufacturing plant at Trafford Park, near Manchester, with machine tools, foremen and even some workers imported from the United States. But the plant proved far too big for the British market, and British Westinghouse paid no dividends until 1913, when its American founder had been deposed as chairman and effective control vested in the British management.[49]

48 Harrison, "Competitiveness of the Cycle Industry"; Lewchuck, *American Technology*, pp. 120–122.

49 I. Byatt, *The Electrical Industry in Britain, 1875–1914* (Oxford, 1979); R. E. Caterall, "Electrical Engineering," in N. K. Buxton and D. H. Aldcroft (eds.), *British Industry Between the Wars* (1979), pp. 246–249; L. Hannah, *Electricity Before Nationalization* (1979), pp. 1–53; T. P.

In the older sectors, too, British manufacturers sought to emulate the American example, particularly where their products were already quite standardized, as in the case of agricultural equipment. In some areas like reaping and binding machines, where the American giant International Harvester dominated export markets, British firms believed they had little choice but to adopt their rivals' methods whatever the cost. As Harrison, McGregor & Co. told the Tariff Commission,

> To adopt the policy of an increased production in the face of the severe American competition required a certain determination on the part of my firm. We had to considerably increase out-buildings and shops to put down special tools and plant and to standardise our machines as far as possible, with the result that we were able to reduce the cost of production.

Even firms producing more specialized equipment like threshing machines or stationary steam engines felt similar pressures to revise their manufacturing methods. Clayton and Shuttleworth, for example, brought in H. F. L. Orcutt as managing director, who reorganized their works on American lines, installed large numbers of light machine tools, and introduced standard systems of tooling, gauging and inspection; by 1907, the trade press reported,

> the room occupied by fitters is gradually being encroached upon by automatic or semi-automatic machine tools which turn out parts to standard dimensions with as near an approach to accuracy as was obtained by hand methods, and, of course, at a far cheaper rate, to say nothing of the advantages of interchangeability due to the elimination of the human factor.

Other manufacturers such as Hornsby of Grantham likewise tried to cut costs and extend their market by systematizing the flow of production, machining to jigs and templates, and breaking down the fitting process into a series of sections. But as marketing arrangements remained unchanged, none of these firms was able to make much progress in reducing the range of products and patterns

Hughes, *Networks of Power: Electrification in Western Society, 1880–1930* (Baltimore, 1983), esp. ch. 9; J. Wilson, "A Strategy of Expansion and Combination: Dick, Kerr & Co., 1897–1914," *BH*, 27 (1985), esp. p. 34; Saul, "American Impact," pp. 29–36; R. Jones and O. Marriott, *Anatomy of a Merger: A History of GEC, AEI and English Electric* (1970), pp. 43–67; *Tariff Commission*, pars. 412–496 (H. Hirst, General Electric Co.; W. L. Magden, Brush Electrical Co. Ltd; T. O. Callender, Callender's Cable and Construction Co. Ltd.); *Balfour Committee*, pp. 300–312; *idem, Minutes of Evidence*, (1924–27), II, pp. 834–848. Paul David argues that British efforts to develop high-voltage direct-current electricity supply (which was better suited to the domestic pattern of dense urbanization and decentralized industry than the polyphase alternating-current systems which eventually triumphed) might have proved more successful had not Thomas Edison, the leading sponsor of DC technology, prematurely abandoned for reasons of his own the struggle against George Westinghouse's AC system. See his "The Hero and the Herd in Technological History: Reflections on Thomas Edison and the Battle of the Systems," in P. Higonnet, D. S. Landes and H. Rosovsky (eds.), *Favorites of Fortune: Technology, Growth and Economic Development since the Industrial Revolution* (Cambridge MA, 1991), esp. pp. 99–105.

manufactured; where competition revolved primarily around price, therefore, American mass producers with their vast internal market retained a continuing edge.[50]

The adoption of American methods in British engineering was often facilitated by expanding demand for standardized equipment from the newer sectors themselves. The growth of the electrical industry, for example, stimulated manufacturers of steam engines for marine or agricultural use to redesign their products for dynamo work. Technologically distinctive products and the influence of American practice in electricity supply enabled firms such as Willans and Robinson, W. H. Allen & Sons, and Reavell & Co. to standardize their engines into a definite set of sizes and manufacture for stock using subassemblies of modular components machined in substantial batches. Reavell's, for instance, obtained forty-four different capacities of compressor from nine main castings and cylinders, allowing the firm "quickly to supply machines to meet any specified conditions . . . while carrying a comparatively small number of different classes of article in the finished stock." There, too, it was the practice "to work the finished stock on a minimum quantity basis; a certain definite and predetermined batch of new articles of any type being ordered as soon as the stock has fallen to the minimum"; and these quantities were large enough for standard components such as valves to keep capstan lathes permanently occupied on their production.[51]

In machine tools, similarly, expanding demand from new sectors like cycles and motor vehicles allowed some British firms to apply American methods to the manufacture of standard machines such as capstan and turret lathes. Foremost among these was Alfred Herbert, whose initial success as an importer of American machine tools led him to establish his own manufacturing operation during the 1890s. By 1910, Herbert employed 1,200 men in its Coventry works, confining itself, according to one report,

> to producing its own standard patterns of machines. The construction of single machines of special types is not undertaken except in the case of those of an experimental nature, for use in the company's own works, in the development of machines suited to market requirements.

With many machines built for stock in significant quantities, special tools and fixtures were widely used, a rigorous system of process planning and inspection

[50] *Tariff Commission*, pars. 638–677 (Harrison, McGregor & Co; Midland Engine Works; Ransome, Sims & Jefferies); *Engineering*, June 21, 1907 (Clayton and Shuttleworth); *The Engineer*, March 6, 1908 (Hornsby), May 23, 1913 (Ruston and Proctor); BoT Committee on the Engineering Trades, September 22, November 17 and November 21, 1916, BT 55/22–23 (Marshall & Sons Ltd.; Clayton and Shuttleworth; Ransome, Sims & Jefferies; H. F. L. Orcutt); McLean, "Anglo-American Engineering Competition."

[51] A. Lazenby, "Advanced Methods in a British Engineering Workshop" (Willans and Robinson), *EM* (December 1901); *Engineering*, February 27, 1903 (W. H. Allen & Sons), January 3, 1908 (Willans-Parsons turbines), July 27, 1917 (Reavell & Co. Ltd.); Saul, "Mechanical Engineering," p. 118 (Willans).

adopted, and piece-work generalized throughout the shops. Unlike American makers, however, Herbert also manufactured a wide range of products, seeking to stimulate the market by introducing new types of machine, consulting with users on design improvements and supplying special tools and fixtures to customer requirements. As Herbert himself later observed,

> It has often been suggested to me that we should have done better if we had specialised more intensively on a limited range of machines. There is no doubt much force in this contention, but in the early days I doubt very much whether there was scope for a growing business, which confined itself to the production of only one or two machines. Rightly or wrongly, I was more attracted by the idea of covering a fairly wide field. The cycle business, which was our principal customer, required in those days a variety of machines and not many of them of one kind.[52]

Perhaps the most interesting experiments with the adaptation of American practice to British conditions came in marine engineering. Nearly all ships were built to order for specific customers, and the heterogeneity of the final product naturally imposed demands for customization on manufacturers of marine engines and related equipment as well. But firms such as the North East Marine Engine Co. in Sunderland and David Rowan in Glasgow who manufactured engines for cargo steamers, the most uniform type of ship, sought to standardize their products as far as possible, even at the cost of losing a certain number of orders. At Rowan, for example, the entire range of engines was redesigned on modular lines at the turn of the century, facilitating the use of jigs for machining common components; where this proved impractical, similar pieces were grouped in graded series to economize on draughtsmen's time and smooth the progress of work through the shops. This firm also pioneered its own version of the premium bonus system, in which a maximum time was fixed for each task when it was assigned; if the worker completed the job in less than the time allowed, the proceeds were divided proportionately with the employer. William Rowan Thompson, the system's originator, attributed the company's standardization programme itself to "the discrepancies in the performance of similar tools on the same work" thrown up by the introduction of the premium bonus. While accurate records could be maintained on past jobs, however, not even Rowan could avoid the need to undertake a variety of work, and the firm's rate-fixing methods on new tasks retained a distinctly rough and ready air. As Thompson

[52] Saul, "American Impact," pp. 24–25; and "Machine-Tool Industry," pp. 31–33; *Engineering*, December 23, 1898, June 12, 1908 and July 22, 1910; G. C. Carden, *The Machine Tool Trade in Germany, France, Switzerland, Italy and the United Kingdom* (US Bureau of Manufactures, Special Agents Reports no. 26, Washington DC, 1909), pp. 225–229; Herbert's manuscript autobiography (n.d.), quoted in Floud, *British Machine Tool Industry*, p. 53; cf. his evidence to the BoT Committee on the Engineering Trades, July 27, 1916, BT 55/22. For other examples of standardization and specialization among British producers of machine tools and related equipment, see Carden, *Machine Tool Trade*, pp. 213–218 and *Engineering*, June 6, 1913 (John Lang & Co., heavy lathes); *The Engineer*, July 14, 1905 and *Engineering*, December 28, 1917 (David Brown, gear cutting); *The Engineer*, October 31, 1913 and *Engineering*, May 22, 1908 and July 21, 1913 (Hans Renold Ltd., precision chains).

himself commented, despite the limited amount of repetition, most marine engine work

> is sufficiently alike for me, when a new job presents itself, not to have to go into elaborate investigations with regard to time: we have a range of tables, and with an experienced man who has been working this thing for some time it is not very difficult, and it does not matter whether it is an hour or two one way or the other, it is quite good enough to be going on with; it is satisfactory for that job, and it saves the time of nearly everybody and gives the results that we desire.

But as contemporary observers were quick to note, the Rowan system had been carefully designed to fix a lower bonus ceiling than in other incentive wage plans, reducing the costs of unavoidable errors in the establishment of basic times for the long jobs characteristic of marine engine work.[53]

No British marine engineering firm made greater efforts to transplant American methods than G. & J. Weir, a Glasgow manufacturer of ships' pumps and other auxiliary equipment. Weir developed an innovative type of feed pump for naval boilers which found special favor with the Admiralty, and its patents and reputation for technical excellence enabled the firm largely to impose its designs on its customers. William Weir, the chief executive, had traveled extensively in the United States and American practice served as the model for the reorganization of the works in 1899. All components were standardized with the exception of the main castings; new shops were laid out with machinery grouped by "departments in which each distinct part of a pump could be completed"; automatic and semi-automatic tools were installed wherever possible; and Weir too introduced his own form of the premium bonus, a modified version of the American Halsey system. But the sophistication of the firm's management methods should not be exaggerated: while foremen were relieved of responsibility for inter-departmental work flows and the design of tooling, the bonus system for up to 1,000 men was "personally controlled by William Weir with the assistance of five men and two boys," 135,000 time cards were entrusted to a "solitary clerkess," and bonus times were based mainly on past practice rather than independent studies. Similarly, although there was more repetition than in marine engine building, "a large amount of special work" still remained, particularly in the foundry, where pump carcasses had to be cast individually to customer specifications. Hence as Alan McKinlay observes, "Weir narrowed but did not eliminate the power base of skilled engineering and foundry workers within the production process," and the factory became a major

[53] On the North East Marine Engine Co., see *Engineering*, July 3, 1914. On Rowan's and the Rowan system, see *ibid.*, September 13, 1901, May 9, 1902; *PIME*, 1902–3; testimony by W. Rowan Thompson to the BoT Committee on the Engineering Trades, September 21, 1916, BT 55/23; and *The Premium Bonus System* (1917); *The Premium System of Paying Wages* (5th edn, 1917; reprinted from *The Engineer*, 1902). A draft report prepared by the EEF on "Systems of Payment by Results" in 1918 [P(4)5], described the Rowan system as "most suitable for, say, Marine Engine works, where many of the jobs extend over a considerable period, and an error in time fixing may be considerable on account of the length of the contract."

focus of shop steward organization and craft militancy on the Clyde both before and during the First World War.[54]

Thus even the most committed exponents of American practice were rarely able to break completely with indigenous patterns of product customization, devolved management and dependence on craft skills in crucial phases of production. As the public debate made clear, moreover, few British engineering firms had either the power or the desire to recast their works on American lines, since productive flexibility remained a strategic priority in meeting the demands of diverse and volatile markets. In the bulk of the industry, therefore, piecemeal innovation rather than wholesale imitation constituted the main response to American competition.

A key example was the introduction of new machine tools developed across the Atlantic, which proved flexible enough to be widely used on the less standardized work characteristic of British engineering shops. As Joseph Horner noted,

> Although turret work lends itself best to repetition work, yet it is not to be despised in the work of the general shop, as witness the chucking turret lathe; for since it will hold drills, reamers, turning and boring tools and threading tools, its employment saves a deal of trouble in setting tools in the slide rest, or in changing work from lathe to drilling machine.

Orcutt himself observed that "the leading characteristic of the modern screw machine or turret lathe is its adaptability for comparatively short runs"; and similar comments were made about other techniques such as die forging or machine moulding. When interchangeability and precision were central considerations, jigs and fixtures might also be used to manufacture relatively small batches, as in the case of automobile engines, where products varied frequently and no British firm produced more than 5,000 vehicles before the First World War. British purchases of American-style machine tools shot up dramatically after 1900, and the employment of "handymen" on semi-automatic equipment capable of a wide range of work became a central grievance of craft unionists in British engineering during the following decade and a half.[55]

Another striking case in point was the adoption of new payment systems. British engineering employers had long viewed piece-work as a key method of

54 *Engineering*, January 26, September 13,1901, June 13, 1913; J. R. Richmond to *Tariff Commission*, pars. 635–638 and BoT Committee on the Engineering Trades, October 19, 1916, BT 55/23; W. J. Reader, *The Weir Group* (1971), chs. 1–2; A. McKinlay, "Employers and Skilled Workers in the Interwar Depression: Clydeside Engineering and Shipbuilding, 1919–39," PhD thesis, University of Oxford, 1986, pp. 43–52.

55 Horner, "The Modern Machine Shop," p. 391, "The Equipment of Modern Machine Shops" and "Modern Machine Shop Requirements"; Orcutt, "Modern Machine Shop Methods." For estimates of British machine tool purchases, see Saul, "Machine-Tool Industry"; and Floud; "Adolescence of American Engineering Competition." For the case of the automobile industry, see Saul, "The Motor Industry in Britain to 1914," *BH*, 5 (1962); Lewchuk, *American Technology*, pp. 114–130; and S. Tolliday, "Management and Labour in Britain, 1896–1914," in Tolliday and J. Zeitlin (eds.), *Between Fordism and Flexibility: The Automobile Industry and Its Workers* (Oxford, 1992; 1st edn, Cambridge, 1986), pp. 40–43. For a fuller discussion of conflicts over machine manning, see Zeitlin, "Labour Strategies"; and "Engineers and Compositors."

increasing output per man hour and lowering unit labor costs; with the defeat of the unions in the 1897–98 lockout, they accordingly pressed forward the rapid extension of piece payment. But a growing number of firms also began to experiment with American-inspired incentive wage systems such as the various types of premium bonus. The advocates of these systems argued that they would eliminate the tendency towards restriction of output built into piece-work because "no 'critical rate' of earnings looms in front of the [worker] to dampen his activity and lead him to slacken his energies." But as those subjected to the premium bonus soon realized, the division of time saved between worker and employer shifted to the advantage of the latter as output increased, building an "automatic price breaker" into the payment system itself. In practice, moreover, the premium bonus often degenerated into a more old-fashioned form of rate-cutting, since even under the Carlisle Agreement signed by the ASE and the Engineering Employers' Federation in 1902, a minor change in working practice could serve as the pretext for a sharp reduction in the time allowances. Similarly, the associated systems of work measurement meant that each man's performance could be ascertained and those not up to the maximum standard of efficiency weeded out. Even enthusiastic advocates of scientific management were obliged to concede that time study and incentive payment in British engineering firms were more often an extension of traditional strategies for cost cutting and work intensification rather than a breakthrough into a new Taylorist mode of organization. As A. Robert Stelling ruefully commented in 1919:

> The British manufacturer, observing the results of scientific management in America, made a somewhat superficial examination of the methods employed; he then tried such of these methods as appealed to his temperament particularly, or which seemed to him to promise the most immediate results.[56]

On the eve of the First World War, therefore, British engineering practice spanned a complex spectrum of hybrid forms between craft and mass production. Few firms had fully embraced the American model, but some had gone a long way towards routinizing production, and many more had been able to increase productivity and reduce costs by selective adaptation of American methods within a flexible framework.

If the impact of British debate over American manufacturing methods on domestic engineering practice remained equivocal before the First World War, so too were the results of international competition in world export markets. British engineering manufacturers had successfully defended the home market

[56] J. Slater Lewis, "The Labour Factor in the Intensity of Output," *EM* (November 1899); J. Holloway, "Remuneration of Labour and its Relation to Output," *CM* (July 1918); Trades Union Congress Joint Committee, *Report on the Premium Bonus System* (1910); A. R. Stelling, "Taylor's Principles in Modern British Management," *EIM*, October 9, 1919; Renold, "Engineering Workshop Organisation"; Weekes, "Amalgamated Society of Engineers," ch. 5; Zeitlin, "Labour Strategies"; Lewchuck, *American Technology*, pp. 72–77. For parallel modifications in the implementation of American incentive payment systems by German engineering firms, see Homburg, "Die Anfänge des Taylorsystems in Deutschland."

against imported equipment and continued to hold sway over most of the Empire as well. In 1913, as table 5.2 shows, Britain remained a more important exporter of mechanical engineering products than the United States, with unchallenged supremacy in established areas of specialization such as textile machinery. Where opportunities for standardization and volume production were greater, however, as in agricultural equipment, American manufacturers had outdistanced their British counterparts. But even more strikingly, Germany had overtaken Britain as the world's largest exporter of engineering goods, with particular strengths in heavy, complex and technologically sophisticated capital equipment such as locomotives and machine tools.[57]

From the turn of the century to the First World War, the "American challenge" dominated British debates about the reform of domestic engineering practice. Yet in many important markets – above all continental Europe which accounted for more than 50 percent of world demand – the most dangerous threat to Britain's competitive position came paradoxically not from American mass producers but rather from German manufacturers equally if not more committed to specialized products and skill-intensive methods. As Herrigel and others have shown, Britain's early dominance of world demand for more standardized types of machinery forced German manufacturers from the mid nineteenth century onwards to seek out new market niches through short-series production of specialized equipment to customer specifications. Even large, autarkic engineering firms such as Krupp, MAN, Hanomag and Borsig manufactured a vast array of complex capital goods from machine tools, steam engines and locomotives to chemical and steel-making plant, construction equipment, diesel motors, textile machinery and printing presses among many others. Such firms therefore depended heavily on skilled workers using general-purpose machinery to shift continuously from one relatively unique type of product to another.[58] Giant electrical manufacturers like Siemens and AEG likewise turned

[57] For a more detailed examination of the competitive position of British, American and German engineering in 1914, see Saul, "Engineering," pp. 227–237; F. Fröhlich, *Die Stellung der deutschen Maschinenindustrie im deutschen Wirtschaftsleben und auf dem Weltmarkte* (Berlin: Springer, 1914), esp. pp. 22–28; and J. Zeitlin, *Between Flexibility and Mass Production: Strategic Debate and Industrial Reorganization in British Engineering, 1830–1990* (Oxford, forthcoming), ch. 2.

[58] This picture of specialization and skill-intensiveness in the German machinery industry draws extensively on the work of Gary Herrigel, which synthesizes a vast body of contemporary literature: *Industrial Constructions: The Sources of German Industrial Power* (Cambridge, 1996), pp. 41–46, 100–106; "Industrial Order and the Politics of Industrial Change: Mechanical Engineering," in P. J. Katzenstein (ed.), *Industry and Politics in West Germany* (Ithaca, 1989); "Industry as a Form of Order: A Comparison of the Historical Development of the Machine Tool Industries in the United States and Germany," in J. R. Hollingsworth, P. C. Schmitter and W. Streeck (eds.), *Comparing Capitalist Economies: Variation in the Governance of Sectors* (Oxford, 1994). For other studies which support this picture, see Fröhlich, *Die Stellung der deutschen Maschinenindustrie*; Brady, *Rationalization Movement*, ch. 7; J. J. Pastor, "Die Ausfuhr des deutschen Maschinenbaus und ihre volkswirtschaftliche Bedeutung," dissertation, University of Cologne, 1937; and B. Dornseifer and J. Kocka, "The Impact of the Preindustrial Heritage. Reconsiderations on the German Pattern of Corporate Development in the Late 19th and Early 20th Centuries," *Industrial and Corporate Change*, 2 (1993).

out an increasingly diversified and rapidly changing range of products from complete power stations to a wide variety of high and low-voltage equipment for both industrial and consumer markets. Thus in these firms, too, as Homburg observes, "innovation, continuous change and experimentation with new products required – at least in the initial stages – an experienced, highly flexible, supportive, and motivated work-force," above all men with traditional metal-working skills.[59]

Why did the pre-war British debate focus so exclusively on the "American challenge" when German competition was often a greater commercial threat? A number of plausible explanations may be adduced. First, the real breakthrough of German machinery exports in world markets took place at the end of this period, above all during the years 1910–13 when foreign demand for British engineering products was also booming; at the turn of the century, when the debate over American engineering practice began in Britain, Germany was not yet so important a competitive rival.[60] Second, the overall productive achievement of the American metalworking industry was more impressive than its export performance, especially to visiting engineers and businessmen: thus according to one contemporary estimate, the United States accounted for 50 percent of world machinery production in 1913, but only for 26.8 percent of world exports.[61] Above and beyond such objective considerations, finally, many British engineers were fascinated by the American model even when they remained skeptical about its applicability to domestic conditions, at once attracted and repelled by the systematic development of principles such as standardization, specialization and mechanization which had deep roots in their own technological heritage. But despite Germany's remarkable success as a machinery exporter before the First World War, her engineering practice presented no such technological novelties, and with the important exception of

[59] H. Homburg, "The 'Human Factor' and the Limits of Rationalization: Personnel-Management Strategies and the Rationalization Movement in German Industry Between the Wars," in Tolliday and Zeitlin, *The Power to Manage?*, esp. p. 152; cf. also Dornseifer and Kocka, "Impact of the Preindustrial Heritage," esp. pp. 234, 239.

[60] For the growth of German machinery exports between 1899 and 1913, see Fröhlich, *Die Stellung der deutschen Maschinenindustrie*, p. 18; and H. Tyszynski, "World Trade in Manufactured Commodities, 1899–1950," *Manchester School of Economic and Social Studies*, 19 (1951), tables I–II. Before 1900, German manufactured exports consisted primarily of light consumer products such as woolen and silk cloth; clothing and hosiery; leather goods; musical instruments; books, maps and other printed articles; gold and silver objects; toys; and cutlery. See C. Buchheim, "Germany on the World Market at the End of the 19th Century: Successful Supplier of Consumer-Related Manufactures," *German Yearbook on Business History 1985* (Berlin); and "Aspects of XIXth Century Anglo-German Trade Rivalry."

[61] W. Schwade, "Untersuchung der vom Binnen- und Weltmarkt abhängigen Produktions- und Absatzbedingungen in der deutschen Maschinenindustrie," dissertation, Ludwig-Maximilians-Universität, Munich, 1934, pp. 36–38. In the same year, Germany accounted for 20.7 percent of world machinery production and 29 percent of world exports, and Britain for 11.8 percent of world production and 28.4 percent of world exports. Behind these divergences obviously lay the very different export ratios prevailing in the three countries at this time.

industrial training,[62] British engineers appear to have seen little on the other side of the North Sea which could be used to improve their indigeneous mix of flexible and mass production methods. Thus in responding to foreign competition, the ambiguities of British engineering practice functioned both as a source of insight and of blindness, alerting managers and technicians to certain possibilities for reform of domestic production methods while at the same time obscuring others.

Conclusion

The strategic debate examined in this essay forms the first act of a larger drama which preoccupied the British engineering industry through most of this century. From the First World War through the 1970s, as I hope to show elsewhere, a variety of actors from businessmen and engineers to trade unionists and government officials continued to weigh the competing claims of flexible and mass production as guiding principles for the revitalization of British metalworking manufacture. While the balance of the debate shifted markedly according to the economic and political context, tipping sharply in favor of mass production during the two world wars only to swing back in the other direction with the return to peacetime conditions, the terms in which it was conducted remained surprisingly constant for more than half a century. On one side of the scales lay the manufacturing economies theoretically obtainable through a full-scale application of the American model; on the other lay the practical advantages of more flexible methods in catering for the varied and fluctuating markets characteristic of British engineering.

Although many British metalworking firms experimented with American methods between the 1920s and the 1950s, particularly in sectors where amalgamation and cartellization made possible some measure of market control, selective adaptation rather than wholesale imitation thus continued to predominate as before 1914, giving rise to a range of hybrid forms combining elements of craft and mass production. It was only in the 1960s and 1970s, as a massive wave of mergers and takeovers transformed the structure of the industry, that British engineering companies decisively abandoned their indigenous model of productive organization in favor of imported management techniques such as multidivisionalization, work study and measured day work. Far from reviving its competitive fortunes, however, this putative Americanization of British engineering was associated instead with a rapid loss of market share both abroad and at home, resulting in a steep decline of domestic production and employment. By the 1980s, ironically, the competitive difficulties of British engineering firms, like those of the Americans themselves, were frequently attributed to their inability to

[62] For an analysis of the British debate on skills, training and technical education, in which the German example figured prominently, see J. Zeitlin, *The Sinews of Flexibility: Skills, Training and Technical Education in British Engineering, 1850–1980* (forthcoming), ch. 6.

match the standards of product innovation and productive flexibility set by the Germans and the Japanese in meeting the demands of increasingly diverse and volatile international markets.[63]

Why the British engineering industry ultimately failed to revitalize its indigenous model of productive organization along flexible lines, given the limited applicability of American methods under domestic competitive conditions, is a complex and thorny question which cannot be resolved here. No doubt the industry's inability to develop an appropriate institutional solution to key regulatory problems such as the training of skilled labor played a crucial part, as too did the growing intellectual hegemony of mass production in circumscribing the range of alternative strategies considered.[64] Yet whatever their later disorientation and decline, it is the resourcefulness and commercial success of British engineers in appropriating selective elements of American practice without sacrificing productive flexibility which appears most striking from the vantage point of 1914. Unlike in a Greek tragedy, therefore, the *dénouement* of this drama cannot be read back into its opening act, but would depend instead on the strategic choices of the actors themselves as they subsequently unfolded in historical time.

[63] This sketch of strategic debates and practical responses in the British engineering industry from the First World War through the 1980s will be developed more fully in Zeitlin, *Between Flexibility and Mass Production*. For preliminary accounts of the transformation of British industrial structure and managerial strategy since the 1950s and contemporary contrasts with German and Japanese practice, see Tolliday and Zeitlin, *The Power to Manage?*; P. Hirst and J. Zeitlin (eds.), *Reversing Industrial Decline? Industrial Structure and Policy in Britain and her Competitors* (Oxford, 1989); and "Flexible Specialization and the Competitive Failure of UK Manufacturing," *Political Quarterly*, 60 (1989).

[64] Among the vast literature on the failure of training reform in the British engineering industry after 1914, see especially A. McKinlay, "'A Certain Short-Sightedness:' Metalworking, Innovation and Apprenticeship, 1897–1939," in H. F. Gospel (ed.), *Industrial Training and Technological Innovation: Historical and Comparative Perspectives* (1991); and S. J. Prais *et al.*, *Productivity and Industrial Structure* (Cambridge, 1982), ch. 14.

6 The lost paradigm: an Italian metalworking empire between competing models of production, 1900–1920

Alain Dewerpe
Translated by Cyprian P. Blamires

At the beginning of the twentieth century a new model emerged for the organization of industrial production, characterized by the manufacture in long series of standardized products, by big firms using specialized tooling and a relatively unskilled labor force. It was a model that tended to oust more flexible forms of organization producing by the unit and based on a skilled workforce. This chapter is concerned with heavy engineering and shipyards in the strategic period before and during the First World War: it aims to study the operation of the different variables governing on the one hand the actual decision to move over from production by the unit or in small quantities to mass production and on the other the execution of that decision; and its focus will be on Ansaldo, a Genoese engineering and shipbuilding company owned by the Perrone Brothers.

Competition between antagonistic forms of organization of production was prevalent not just on the level of sectors or global economies but actually within each company. As two paradigms for the organization of production, mass production and flexible specialization need to be approached not just as explicit systems of representation but also as social strategies and social practices present in a diffused state in the minds of different agents of the business enterprise, either successful or doomed because of the contrasting or even contradictory constraints under which they have to operate. It is my intention to describe the modalities of this competition between flexible production and mass production and to survey the micro-economic factors operative in the social context within which companies operated. My observations suggest that the appeal to mass production initiated by the Great War, which might have led to a strategic reorganization of the firm and a rupture with earlier practices, in fact gave way to a return to the previous production system. What happened here needs to be understood in terms of the articulation of the totality of business variables – type of sector, type of products, relations with the market and the state, organization of labor and type of labor force employed. Over the relatively short period of two decades a flexible system and mass production competed in three complex and chronologically separated social contexts, within which the potentialities and limits of mass production strategy were exposed.

With Ansaldo[1] we see a move away from low-volume production of high-value units and into high-volume standardized production as a result of the First World War. But the development that was making such a major impact in the automobile sector was not consolidated here and there was a return to the traditional type of product. In the second half of the nineteenth century Ansaldo had specialized in heavy engineering, from locomotives to ships for the navy and the merchant fleet. The First World War led to diversification into the mass production of armaments and munitions. The speed and scope of this new development may be illustrated by a superficial inventory of war production, running to 10,099 guns, 4,168 gun carriages, 6,507 vehicles, 4,200,000 grenades, 1,400,000 shells, 200,000 bombs, and 2,170,000 shell cases.[2] But mass production failed to take: in 1925, after the bankruptcy of the Perrone brothers and the reorganization of the company, the La Vittoria factory (known as Electrotechnique), one of the biggest in the group, was producing by the unit or in very short runs, its output consisting of direct-current machines (propulsion engines, dynamos, engines), alternating-current generators (alternators, turbo-alternators), asynchronic engines (triphase engines), and various electrical machinery (cranes, traveling cranes) and electric traction (automotive, locomotive); this was a very diversified program of heavy electrical engineering. The shipyards meanwhile had reverted to production by the unit.[3]

Thus if Ansaldo went back to its traditional methods it was not for lack of having tested, tried and implemented the paradigm of mass production. The truth is that this pattern of development, which has parallels in numerous other metallurgical and engineering companies, was not solely or even principally due to the nature of the product supplied (production by the unit of ships or heavy machinery) or of the market (dependence on government orders); it is comprehensible only in relation to a more complex unfolding reality, whose successive logics imposed themselves on the company and governed it from the beginning of the century until the nineteen-twenties.

[1] On Ansaldo, cf. Emanuelle Gazzo, *I cento anni dell'Ansaldo 1853–1953* (Genoa, 1953); Paride Rugafiori, *Uomini, Macchine, Capitali. L'Ansaldo durante il fascismo 1922–1945* (Milan: Feltrinelli, 1981); "Agostino Rocca (1895–1978)" in A. Mortara (ed.), *I protagonisti dell'intervento pubblico in Italia* (Milan: Franco Angeli, 1984), pp. 383–403; Marco Doria, *Ansaldo. L'impresa e lo stato* (Milan: Franco Angeli, 1989).

[2] Archivio Storico Ansaldo (ASA), Gamba-4, Riassunto delle produzioni dei vari stabilimenti.

[3] ASA, Sarli-Biandra 6/2, Ansaldo S.A. Mechanical & Electrical Engineers Locomotives & Ship Builders (1925), 46 pp.

Pattern I

This emerged on the eve of the Great War; it involved a strategy of autonomization of the firm in the face of its competitors against the background of a contracting market and at the same time a crisis in the organization of labor, which was riddled with the idling and restriction of output typical of the situation where power was in the hands of the craft workers.

The state and the closed market

First variable: the product market. On the one hand there was the state's role as initiator and on the other the immediate extension of the company's industrial ambitions and productive capacities beyond the state's objectives. It is not surprising that at the beginning of the century market incentives to expansion into arms and iron and steel production found a ready response. This vertical integration, the "full-cycle vertical system" was the great strategic project of the Perrone brothers, Mario and Pio – who were at the helm of the company for the first two decades of the twentieth century – and it was at one and the same time both the means of their accelerated growth in wartime and the main cause of their downfall in 1921. By constructing a steel works at Cornigliano and factories for the production of armor plates and artillery materials, Ansaldo hoped to free itself from the control of its suppliers in cast iron, steel and naval war materials. In addition it wanted to set up a merchant fleet to transport raw materials for these factories, comprised of vessels built by the company's own yards. For the vigorous growth of the Italian iron and steel industry between the beginning of the century and the eve of the First World War – steel production rose from 115,000 tonnes in 1900 to 933,000 in 1913, production of cast iron from 24,000 to 426,000 tonnes – was accompanied by the creation of an oligopoly, a single supplier of armor plate and the special steels required for the production of battleships; while Ansaldo was excluded from the concentration of power by means of which the founder of Terni, Vincenzo Stefano Breda, then Attilio Odero and Giuseppe Orlando, competitors in naval shipbuilding, managed to corner the market. The iron- and steel-making "trust" they set up joined forces with the Banca Commerciale in the spring of 1902 – Terni linked up with Odero and Orlando in 1904, with Cantieri Navali Riuniti, then with Vickers – and provided the steels for naval equipment at below-market prices.[4]

Vertical integration thus corresponded to an extreme simplification of the market in which the state was the principal or indeed the sole buyer and seller and in which prices depended more on political choices than on the free play of the market. Hence the potential for big companies to exploit anticipated payments at

[4] On this whole story see: A. Confalonieri, *Banca e industria in Italia 1894–1906. III. L'esperienza della Banca Commerciale Italiana* (Milan, 1976), pp. 279–328; A. Confalonieri, *Banca e industria in Italia dalla crisi del 1907 all'Agosto 1914. II. Crisi e sviluppo dell'industria italiana* (Milan, 1982), pp. 48–135.

agreed fixed prices and organize their production plans.[5] According to R. Webster, the "vertical Ansaldo system" invented by Pio and Mario Perrone is to be interpreted as an alternative model for the development of the Italian economy,[6] whose logic arose from the conflict between Ansaldo and the ILVA-Terni "steel-making trust" linked to the Banca Commerciale Italiana. In order to counterbalance the power of the Comit group, Ansaldo set up an instrument, the *Banca Italiana di Sconto (BIS)*, which during the War opened up the generous line of credit essential both for its fixed capital investments and for its numerous purchases of companies. On the eve of the Great War the politico-financial group around Ansaldo and the BIS thus looked like the acme of a strategy of regulation and expansion of the Italian economic system.[7] Its ideology reflected the precocious decline of Italian liberalism.[8] Especially noteworthy was the aggressiveness

[5] Cf. E. Galli della Loggia, "La grande guerra e i nuovi equilibri del capitalismo internazionale ed italiano," in *Conflitti sociali ed accumulazione capitalistica da Giolitti alla guerra fascista* (Milan, 1972). For the firm's financial situation, cf. Ernesto Galli Della Loggia, "Problemi di sviluppo industriale e nuovi equilibri politici alla vigilia della prima guerra mondiale: la fondazione della Banca italiana di sconto," *Rivista storica italiana* (1970), p. 4; Giorgio Doria, *Investimenti e sviluppo economico a Genova alla vigilia della prima guerra mondiale*, vol. 2 (1883–1914) (Milan: Giuffré, 1973); Anna-Maria Falchero, "Banchieri e politici. Nitti e il gruppo Ansaldo-Banca Italiana di Sconto," *Italia contemporanea*, 146–147 (1982), 67–92; "La piramide effimera. Il sistema verticale Ansaldo dai primi passi alla distruzione," *Studi in memoria di Mario Abrate* (Università di Torino, Istituto di Storia economica 1986) pp. 379–409. For the fall of the Perrone brothers, cf. A. M. Falchero, "Il gruppo Ansaldo-Banca Italiana di Sconto e le vicende bancarie italiane del primo dopoguerra," in P. Hertner and G. Mori (eds.), *La transizione dell'economia di guerra all'economia di pace in Italia e in Germania dopo la prima guerra mondiale* (Bologna: Il Mulino, 1983), pp. 543–571; Franco Bonelli (ed.), *Acciaio per l'industrializzazione. Contributi allo studio del problema siderurgico italiano* (Turin: Fondazione Luigi Einaudi, 1982); Ludovica De Courten, "Marina mercantile e politica estera. L'Ansaldo di Pio Perrone nel primo dopoguerra," *Analisi storica*, 1/1 (July–December, 1983), 7–38; R. A. Webster, "La tecnocrazia italiana e i sistemi industriali verticali: il caso dell'Ansaldo (1914–1921)," *Storia Contemporanea*, 9, (1978), 205–239.

[6] Cf. Richard Webster, "La tecnocrazia italiana e i sistemi industriali verticali: il caso dell'Ansaldo (1914–1921)," *Storia Contemporanea*, 9 (1978). For the arms market on the eve of the Great War, cf. G. W. F. Hallgarten, *Storia della corsa agli armamenti* (Rome, 1972); Clive Trebilcock, "British Armaments and European Industrialization, 1890–1914," *Economic History Review*, 2nd series, 26 (1973), 254–272; Raymond Poidevin, "Fabricants d'armes et relations internationales au début du XXème siècle," *Relations Internationales*, 1 (1974); M. Pearton, *The Knowledgeable State. Diplomacy, War and Technology since 1830* (London, 1982); Luciano Segreto, "More Trouble than Profits: Vickers' Investments in Italy 1906–1939," *Business History*, 27 (November 3, 1985), 316–337.

[7] According to E. Galli della Loggia, "The Ansaldo-Banca di Sconto group represents the culmination of the development of Italian capitalism in its becoming aware of its own potential to influence both the political choices of the country and its general economic orientation," in "Problemi di sviluppo industriale e nuovi equilibri politici alla vigilia della prima guerra mondiale: la fondazione della Banca Italiana di Sconto," *Revista Storica Italiana*, 82 (1970), 825. For the expansion into steel-making, cf. Marco Doria, "Dal progetto di integrazione verticale alle ristrutturazioni dell'IRI: la siderurgia Ansaldo (1900–1935)," *Annali della Fondazione Luigi Einaudi*, 18 (1984), 411–453; "Strategie imprenditoriali e mercato mondiale degli armamenti: I rapporti tra l'Ansaldo e la siderurgia USA nel primo Novecento," *Società e Storia*, 38 (1987), 915–947.

[8] Cf. Silvio Lanaro, *Nazione e lavoro. Saggio sulla cultura borghese in Italia, 1870–1925* (Padua: Marsilio, 1979).

of the company's industrial policy, oriented towards "autarkic" withdrawal and the conquest of foreign markets. In theory, therefore, Ansaldo had the financial means to make the transition to mass production.

The policy of vertical integration became firmly entrenched. In the area of finished goods the *Società Nazionale di Navigazione* was established in 1915 with an initial capital of 3 million lire, going up to 15 million lire on October 25, 1916. In the area of raw materials the company bought mines in Cogne and built steel-making plants with the capacity to provide it with a permanent supply of steel, as well as an electric steel installation in the Val d'Aosta – *Alti Forni Elettrici ed Acciaierie Elettriche di Aosta* – and a hydroelectric plant – the *Impianti Idroelettrici*, also in the Val d'Aosta.

Vertical integration was linked to a policy of diversification into both military equipment and associated products. In 1915 a munitions factory (la Fiumara) and an internal combustion engine plant; in 1916 a factory for the construction of aero engines at Sampierdarena, a factory making artillery shell cases, yards and workshops at Cornigliano, a tube factory in Fegino, aeronautics plants at Borzoli, an iron foundry at Pegli and a heat-resistant materials factory at Serravalle Scrivia. In 1917 a fresh expansion of production capacity: for the most part, the new investment went into the aircraft workshops in Bolzanetto, La Spezia and Turin and into the expansion of the one at Borzoli, but three new units were set up, one for the production of oxygen and hydrogen, another a thermochemicals plant for tungsten and molybdenum, and in particular a new artillery factory at Cornigliano. In the same year a shipping line, the *Transatlantica Italiana*, was purchased, and also Fiat San Giorgio, producers of aero engines. This very broad strategy for control of the market was clearly incompatible with Fordian specialization.

Workers' control and the crisis in the organization of labor

The second variable was relations in the workplace, which were profoundly affected by a crisis in the organization of labor.[9] Here a rationalizing logic from pre-war days clashed with a very violent critique of the organization of labor. According to the accountant Raimondo, an attentive observer, the problem was much more evident in the shipyards – where "unbelievable chaos" reigned – than in the other factories and where "bad administration" was the order of the day. With more than a hint of understatement, he remarked that "the situation [is] not very satisfactory."[10]

[9] For the organization of labor, cf. Alain Dewerpe, *L'ouvrier, le travail et l'usine à l'Ansaldo (1900–1920)* (Genoa: Mémoire de l'Ecole française de Rome, typescript, 1984); "Modi di retribuzione e organizzazione produttiva all'Ansaldo (1900–1920)," *Studi Storici*, 26, 1 (January–March 1985), 29–52; "Miroirs d'usines. Photographie industrielle et organisation du travail à l'Ansaldo, 1900–1920," *Annales e.s.c.*, 5 (September–October 1987), 1,079–1,114; Paride Rugafiori, "Occupazione e composizione operaia all'Ansaldo," in Giovanna Procacci (ed.), *Stato e classe operaia durante la prima guerra modiale* (Milan: Franco Angeli, 1983), 244–267.

The first item in the dossier of obsolescent working practices was the situation at Meccanico, the great engineering factory in the group, located at Sampierdarena. Reporting on it in 1906, the English engineer Brackenbury commented on the old and poorly-maintained machines that were quite unsuitable for the work: "The majority of the work is done on machines and it is done slowly and without finesse," a view which hardly squares with the usual image of Ansaldo work as marked by quality and precision finishing. In addition to the loss of time there were expenses caused by modifications to the fittings. Finishing grinding was almost completely unknown. The way the pieces were set up was poor: for heavy machining they had to be supported and reinforced. High-speed steel was practically unknown: there were a few bars of it, according to the director Omati, but they were not used because it was thought – and especially by the manager – not to be suitable for old machine tools. Work was done slowly with files, calipers and hand or pocket rules, with each piece adjusted to its corresponding piece and interchangeability non-existent. Tools for measuring and leveling were of inadequate quality. Fitting was slow too and took up a good part of the manufacturing time. This slowness could be explained in terms of a sum of convergent factors: ageing machinery badly set up, the insufficient and mediocre and antiquated gauges, the lack of adequate gauging of the machine-finished pieces, and finally the absence of grinding machines. The workers sharpened their own tools, leaving the machine unused in the meantime. Quality control was seen to by the head foremen themselves, but it was "of a very elementary nature." These were classic weaknesses brutally exposed by the norms set by the new engineering emerging in the first decade of the century.[11]

But it was at Sestri Ponente, in the shipyards, that the crisis in the organization of labor was at its gravest. "In the course of the numerous visits that I was able to make to the yards, and in particular recently," wrote Mario Perrone to the director of the Soliani shipyards, "I have observed that though work is done with care, there is not enough of it for the labor force to apply themselves diligently and economically. You see inactive and idle-looking workers all over the place. It seems to me that the engineers and head foremen are failing to show the requisite zeal and vigilance."[12] The accountant Raimondo's long reports give detailed and well-argued confirmation of this laconic summary and reflect a crisis on the work floor that suggests an organization in the grip of growing contradictions.[13]

Insufficient loading of the machines is a leitmotif: in the specialized workshops for repair or tooling, machine-tools remain unused, while pneumatic drills for slipway riveting are left in the stores. This exposé of the chronic under-utilization of productive potential is paralleled by criticism of the uneven loading of their

[10] ASA SSR 404–26, Riservata, Note Sparse, 1913, pp. 8, 15, 20.

[11] ASA SSM 4–51. On Brackenbury, cf. H. I. Brackenbury, "Modern Machinery and its Future Development," *The Engineer*, August 30, 1907, 220–22; H. I. Brackenbury, W. J. Guthrie, "Workshop Methods in Heavy Engineering," *Transactions of the North-East Coast Institution of Engineers and Shipbuilders*, 43 (1926–27), 99–112.

[12] SSR 178–22, Mario Perrone to Soliani, April 11, 1912.

[13] SSR 404–26, Riservata, Note Sparse, Raimondo, 1913, 15.

machinery by the different workshops. Production accounting is maximally imprecise and requisitioning accounts not to be believed. The total amount of material recorded as being taken from the stores for the vessel the *Città di Bari* appears as 100 tonnes less than what was in fact employed on the hull alone. The reason is that the head foreman used materials he found on the walkways – material meant for other vessels under construction – and neglected to have the necessary accounting alterations made out by the offices. Workshops preferred to make their own tools rather than go to the toolroom, while the latter, which had a small machine park of lathes, drills, grinding wheels and milling machines, either left them unused or used them to make tools which it then left in the stores: this preliminary stocking involved pieces whose production was not a matter of urgency. The same phenomenon was evident with the machine repair service, also independent and apparently no more efficient. So there was a lack of balance in the loading of machines – in one place they did little or no work while in another they were overworked. There was a clear contradiction between the two principles of specialization (the creation of autonomous tooling and repair shops) and coordination. But the specialization was also inadequate since the works manager was at the same time yard manager, design office head and administrative services chief: "The fact of wanting to concentrate all the responsibilities means that it is not possible to devote all the necessary commitment to each of these functions, and so things proceed as they can." More generally the allocation of responsibilities was defective. While staff could sign requisition slips without consulting anyone and without being answerable to anyone after the event – and in certain cases without putting the material back after use, no one knew who had the competence to requisition materials for tool-making and material for the stores.

Wastefulness was rampant. A significant example: a boiler was used as a siren and this meant that the heating was kept on even in summer, despite the fact that the yard had an electricity supply, which would have made it easy to install an electric or compressed-air siren. Another symptom: supervision of materials was disorganized but the storekeeper's complaints about it fell on deaf ears. The stores were often kept out in the open and materials such as fine metal sheets or steel sections left exposed to the inclemencies of the weather and spoiled or made unusable. Stocks of wood went missing or rotted because they were left around all over the place instead of being collected together in one store. Wastefulness was so bad that top-quality wood like teak and mahogany was sometimes used to feed the boilers.

The image of the engineer had gone downhill. Proposing a new organization chart for the yards, to be supervised by a chief engineer for the yards and three engineers, the accountant Raimondo observed that such engineers did exist "but one has to admit that everything has not been done to give them the right kind of authority, whether in writing or in the day-to-day agendas or orally; neither the positions they are to occupy nor their functions nor their responsibilities are ever properly allocated or clearly defined and their prestige in the eyes of their inferiors cannot but suffer." Positions were in fact "vague and ill-defined." While head foremen had the authority without the responsibility, the engineers had responsibility without authority. Their pay was poor – after two years' experience at Sestri an engineer was paid by the day and given only 7 lire, considered appropriate for a good professional. His authority was all the weaker in that scales of remuneration did not reflect any obvious hierarchical ladder. The key person in the organization of work remained the head foreman, the *capo maestro*. He had great power and an equally great indifference to costs, paying scant attention to the expense occasioned by the way the work was organized. One of these men, criticized for his way of supervising wages, and "struck by the fact that his practices were not shrouded in a comfortable mystery," had recourse "to the most vulgar of threats." Another might remove the steel sheets from a vessel to use on the one he was in charge of without making any kind of declaration to the requisitioning office, for "he doesn't think these administrative matters have any importance." "Up to now the *capi maestri* have always had too much authority, which has led to abuses and irregularities – but absolutely no responsibility." Hence a strategy of progressive autonomization leading to "the formation of numerous little republics and, what is even worse, they are at war with one another." The relationship with the power structure in the yards was clear: "shipbuilding as things are is in the hands of the *capi maestri*, and they are still far from convinced that the construction of a merchant vessel can and must be just as carefully supervised as that of a battleship."[14]

The management saw the solution to the crisis in making precise distinctions between the different competencies, in the division of labor between engineer and head foreman, in the specialization of tasks, and in the transformation of wage systems – in short, a new hierarchical line to resolve the contradiction between the will to specialize and the necessity for coordination, based on a concentration of power at the top. All this corresponded to an administrative paradigm *à la Fayol* and, more discreetly, to a Taylorist paradigm of division and specialization.

The employers' strategy aimed less at creating the conditions for a transition to mass production than at the provision of politico-administrative means for controlling the work process: so the proposals formulated at the time were primarily of an administrative and bureaucratic order and concerned the concentration of powers and the allocation of tasks by the offices. In 1911 the management of the shipyards emphasized that: "To obtain good results, it is important to break the work up as much as possible and subdivide it according to its different elements, which is equivalent in other terms to dividing the different categories of workmen, while avoiding over-complex and over-long jobs, for which it is very difficult if not impossible to establish a definite price."[15] This strategy was based on a Taylorist acculturation, as was borne out in 1914 by the

[14] SSR 404–26, Riservata, Note Sparse, Raimondo, 1913, 15.

[15] SSR 322, Studio teorico-pratico sulla retribuzione delle prestazioni economiche del personale operaio.

engineer Pozzo's notes on the Taylorist system, which were edited and translated in two articles in the *Bulletin de la Société d'encouragement pour l'Industrie nationale*; these notes are descriptive and entirely devoid of commentary.[16]

One of the main elements in the "unbelievable chaos" of the organization of work at the shipyards on the eve of the war related to piece rates. The case of the construction of the vessel *Giulio Cesare* was an example. "Here I could write a whole volume. In my opinion one of the reasons for the construction work being behind and one of the reasons for the excessively high costs of vessels and the indiscipline of the workmen is the poorly-regulated and indisciplined output-based wage structure."[17] Left completely to the whims of the *capi maestri*, piece rates lost any sense: they gave no stimulus to the workman, they did not incite him to greater productivity and higher earnings, nor did they enable the company to save in time and costs. They became "a way for the head foreman to give his workers a bonus." The foreman for the sailors for example set the rate for tasks and put them in his own name: "a disgusting attitude" but one that apparently suited everyone, including the deputy manager, who took quite a while to wake up to the doubtful aspect of the procedure; it took him all of ten years to get the head foreman onto a fixed monthly salary.

Thus we find at Ansaldo a copious inventory of practices for the regulation of work and pay: assignments begun without a price being fixed and then paid for *a posteriori*, premature ending of assignments, commissioning delayed until after work was already under way, increases of rates under various pretexts when pay was low and crude calculations applied to production quotas to safeguard income. Far from having any incentive to speed up his rate of work, the workman would prefer to go slow "in order to end up with a higher labor cost, which meant bigger bonuses" (for he was sure of a percentage). Taking into account the losses involved in the greater time taken, such practices must have doubled labor costs for the construction of the *Giulio Cesare*.

There was no attempt to allow for differences in terms of productivity levels between the various possible techniques while rate changes seem to have been governed solely by the purely external social imperatives of control of the mass of pay distributed and of the provision of incentives through exploitation of bonuses. Initially the bonuses distributed for pneumatic drilling seemed too high, 320 percent at the maximum, 185 percent at the minimum with an average of 250 percent (drilling of bridges on the *Giulio Cesare*). Then the level was lowered by 40 percent since it was clear that there had been a massive slowdown: in a fortnight the workmen – we do not know how many – drilled 2,400 holes, which means 200 for each 10-hour day, a paltry amount. The reduction left bonuses at a maximum of 253 percent and an average of 136–138 percent. There was a change of policy and the head foreman suggested an increase of 10 percent; the bonuses leapt to 395–292 percent, an average of 150–128 percent, admitted-

[16] SSR 307–24, August 1, 1914, Organizzazione del lavoro, p. 24.; with graphs.
[17] SSR 404–26, Riservata, Note Sparse, Raimondo, 1913, 15.

ly for work on material that was difficult to handle. The example of two workmen is enlightening: in the second fortnight of August 1911, Erigerio received 204 lire, representing a bonus of 300 percent and Brezzo 194 lire, a bonus of 237 percent.[18]

The spread of piece rates was a direct response by the employers to the crisis in the organization of work. A classic circular of July 6, 1914 instructed the head foremen to have all the work done on a piece-work basis, apart from small jobs;[19] reference was made to an earlier injunction, whose repetition shows that it wasn't being observed. By the eve of the First War then the system was becoming the rule. From the workers' side there was no rejection of piece rates in principle, only a strategy of regulating its most harmful effects in the knowledge that it put pressure on the workers: it was "a system that enslaves, a feverish rhythm of work that sacrifices the workman to physical effort and puts years on him at the same time."[20] In spite of this absence of any systematic rejection of the new system of remuneration, its implementation was difficult and hazardous. The negative aspects of piece-work can be read between the lines of the workers' claims: wasted tasks subsequently subsumed under profitable ones, the non-payment of tasks carried out if the piece was later scrapped for reasons that had nothing to do with its actual manufacture – design errors or poor-quality metal; delays in the payment of sums owed as a result of often considerable gaps between manufacture and quality control; and hourly payment of laborers working together on the same terms with workmen on piece rates, which meant that the former were subjected to the accelerated work rhythms generated by the remuneration system without benefiting from any of its advantages.

Between the beginning of the century and the First World War everything was thus put in place for a broad transformation of systems of remuneration. With the establishment of mass and serial production in the firm's arms factories during the War, this cycle of transition from one type of pay to another may be considered complete; it had begun in the shipyards and was concluded in the serial production plants for munitions and armaments – the one at La Fiumara and the Sestri projectile factory. But in the absence of any calculation of timings such as would allow the setting of precise targets, the only stimulus to speedy work Ansaldo used was the addition to the rate of a supplement proportional to the savings of time achieved.[21]

The same administrative and technical strategy was employed at Meccanico.

[18] SSR 404–26, 30 August, 1913, Breve relazione sul lavore di foratura.

[19] ASA Ansaldo lotto 2, 270–271, no. 67, July 16, 1914.

[20] *Memoriale degli operai dello Stabilimento Meccanico Gio. Ansaldo Armstrong & C. di Sampierdarena, e relative osservazioni e decisioni della Società* (Genoa, Stab. Tip. del Giornale Il secolo XIX, 1907).

[21] For a comparison with Great Britain, cf. A. Reid, "The Division of Labor in the British Shipbuilding Industry, 1880–1920," Phd thesis University of Cambridge, 1980; E. Lorenz, The Labor Process and Industrial Relations in the British and French Shipbuilding Industries from 1880 to 1970; Two Patterns of Development, Phd thesis University of Cambridge, 1983.

In 1906 it was noted that the method of high-speed roughing-out with a grinder being used for the finish was not being employed; high-speed steel was hard to come by; the state of the machine tools was poor, they were poorly maintained and in bad repair and workmen sharpened their own tool, leaving their machines unused while they did so. The Englishman Brackenbury reported: "I found many head foremen ready and willing to adopt my suggestions and in a few cases excellent use was being made of the old machines under their control, but there seemed to be a lack of knowledge about modern methods and tools." Brackenbury indicated that everyone seemed to want "to adopt the new methods." So there was no *a priori* rejection of a transformation of working methods. Brackenbury had tools in high-speed steel made at once; he showed that a saving of four fifths could be made on manufacturing times. He particularly stressed interchangeability and suggested that employees specialize for this end independently of the head foreman. It was the emergence of the American system: "at the present time demand for interchangeable precision pieces may not be very substantial in Italy, but that will soon change," he stressed. Such transformations involved intensification and adaptation and therefore production methods. A tooling and gauging office was set up to deal with time-wasting. A special office depending on the "intelligence department" was to take particular responsibility for the quality of work with the aim of avoiding the need for finishing as much as possible. The workmen's ability to control the work through their mastery of the tooling was only gradually weakened. If the manager of the shipyards still felt the need to state in 1913 that moving about during the working day was no longer necessary and that installation and preparation of tools should be done at the start of each half-day of work, time-wasting must still have remained prevalent.

Hence the suggestion for a new department to be called the Intelligence Department whose brief would be as follows:

to set rates based on estimates produced by calculation and by experience of the productive potential of the machine tools.
to ensure that the machines were used to the maximum.
to ensure that head foremen were careful about the quality of pieces produced so as to cut down as far as possible on rectification and fitting work.
to gather and take account of all possible information on types of machine and on new methods and to visit innovative workshops.

Finally Brackenbury suggested that three head foremen go to Elswick, even though it was a factory that was very traditional in its methods, and more generally that visits should be made to English locomotive factories. They would be shown "the best of our methods" and the author was quite certain that "when they return to Sampierdarena they will be in a position to implement the suggestions I have made with great success."[22]

[22] SSM 4–51.

The strategy of this first approach was rationalistic, barely Taylorist, and very technicist: power relations in the organization of work as in that of the markets made the road to mass production extremely costly. In truth it had no internal logic and the resources to impose it were very scanty. In this first approach, serial production was not really at any moment an objective of the firm, for which on the contrary the political dominance of the market and a substantial increase in its productive potential seemed the only effective strategy.

Pattern II

The adoption of serial mass production, with a production system made up of old manufacturing units recycled and new ones created *ex nihilo* for the purpose, constituted a rupture. The history of the company shows that war production was a special opportunity for experimentation in rationalization, mass production and volume work, but the old and new paradigms clearly coexisted.

The war factory and standardized work

The change brought about by this shift needs to be understood differently depending on whether we are discussing new production – mass production by definition, the production of armaments and munitions – or the transformation whereby traditional products made by the unit could now be envisaged as produceable in volume. In the first case a production space, machines, and production processes were created *ex nihilo*, with a final product (munitions) on a massive scale. In the second case, an old production process was reinterpreted in the light of new requirements and adapted appropriately. The former was rapid but ephemeral, for the experiment was already over by 1918; the latter was much slower and less visible but its impact was more profound: it was to prove a lasting phenomenon.

Creations ex nihilo*: the new factory for the mass product*

The production unit typical of the first case was the Fiumara munitions factory. Though conceived according to the mass production paradigm it was subjected to very particular constraints. Its stock of machines was limited – 5 reverberating furnaces, 5 circular saws, and 4 fly presses – and its production was equally limited – 4,000 "102–5" pieces and 5,000 "75s" for each working day of twenty hours in two shifts. There was a significant degree of decentralization, with the ironworks divided up according to product and each section, rough forging and billets, having its own furnaces. The fuel supply was equally decentralized, the fly presses being supplied by two Fergusson oil furnaces, while the disc saws were in two groups, each one driven by a separate motor. The only element of centralization was the tool room serving the whole of the ironworks section of the factory.[23] It was a modestly sized establishment rationally planned on paper at least – "layout by product" determined the division of the workshops according to the calibers of the pieces. The machine shop was much more ambitious: it was in an old mill that had been reorganized and covered 8,126 square metres with six floors and three goods lifts capable of carrying two tonnes, 1,200 machines, a transformer supplying 1,000hp, and a daily production of 4,000 "75" shells, 3,000 "75" shrapnel shells, 4,000 "102–5" shells and 4,000 mortars.

A second example; the Sestri Ponente munitions factory, created out of nothing on a 20,400 sq m site, with 6,500 sq m in use from the beginning of production, plus the 7,500 sq m of the new ironworks for medium and large calibers. The production target was 1,000 grenades per day rising to an anticipated optimum of 3/4,000. A memo of October 1915 informs the management that "in setting up the new engineering plant you should adopt a type of ground plan that conforms to modern criteria, with practical low-cost machining by groups of machines, and you must take into consideration post war needs as of now."

The investment involved was therefore substantial – 8 million lire for the installation period – and the thinking ambitious from the start; the dominant conception was that of a factory for volume production and one that would last, so a new organization of work was called for. Thus the machines were all brand new: 800 machines were ordered, including a section with 210 semi-automatic Lo-Swing lathes specially designed for the machining of shells, which "made possible a very high level of productivity." The installation of machines by product confirms this fact, the heart of the effort focussing on a single series of "102" and "105" shells – 5,000 per day and shrapnel shells of the same caliber – 3,000 per day.[24] From then on all the tooling was planned *ex ante* in the context of a predetermined product.

A great deal of creativity went into the planning of the new factory, but the firm was of course subjected to a certain number of external influences and also took over certain pre-established models, and this led to the acculturation of mass production. One conduit for such influence was the volume published by the National Munitions Committee[25] together with related booklets on small caliber projectiles, munitions, "149" grenades and mines and "75 mod. 906" grenade-torpedo cases. One booklet was entirely devoted to the utilization of high-speed steel.[26] These works were fairly classic in their treatment of hot machining and the modeling of projectiles, but the most interesting thing was the stress put on the use of the milling machine and particularly the circular miller for threading, and on the use of the ordinary lathe adapted for the new machining. For thermic

23 SSR 542–13, Stabilimento della Fiumara per la produzione di munizioni da guerra.

24 SSR 542–13, Proiettificio di Sestri Ponente.

25 *Alcune notizie sulla fabbricazione dei proiettili* (Rome: stabilimento tipografico Riccardo Garroni, 1916), 32 pp., signed by the engineer Antonio Cortese.

26 *La fabbricazione dell'acciaio per utensili (acciaio rapido). Informazioni su alcune Case che construiscono le macchine necessarie per la fabbricazione delle munizioni* (Rome, December 1915).

treatment there was explicit reference to the works of Taylor. In this way the technical office of the sub-secretariat for Arms and Munitions diffused the leading schemas of mass production: one of these texts noted that "it is a question of constructing with all speed machines suited for use by very inexperienced turners."[27] These texts provided complete machining sequences, especially for "75" grenade-torpedoes, and gave "a detailed account of the successive operations to be carried out on the bars of metal to make the case and the conical point" and "the successive stages in the machining." Work done by women was naturally an abundant source of data: it made possible and indeed indispensable some consideration of intensive production techniques not requiring much know-how.[28]

The division of labor and deskilling

We know that munitions production requires an elaborate division of the production process which in turn makes possible the employment of an unskilled workforce. A great number of turning operations had to dispense with skilled labor, and the large-scale munitions production required by war raised the problem of the reorganization of work with particular clarity: a tendency to replace a male by a female workforce was a natural corollary of all this. In the first quarter of 1916 the programs of La Fiumara and the Sestri munitions factory were planned on the basis that 200 female workers would be engaged in April, the first of a legion. Female labor was thus explicitly assigned to the specific tasks of metallurgy or turning (including roughing-out) – drilling, threading, the application of brass rings, varnishing, packing and checking, an ensemble of operations that were primarily those required for the standardized production of munitions, excluding the work that involved excessively heavy loads. This connection was not fortuitous.[29]

The fast-growing popularity of bonus-linked wages was a direct consequence of the forms of mass and serial machining that were brought in as a result of the War. This "fast-growing" popularity had as its corollary a perceptible increase in bonus percentages. But variations in the bonuses were still enormous, ranging from 20 percent to 300 percent in 1916 at the Sestri *Proiettificio*. Again it was the team leader who set the level of the bonuses; if the job had been completed before he did so, he worked them out on the basis of the previous rate which he renewed or divided, and if the work was without a precedent, he made the assessment on the basis of his practical technical competence. In no case, let me repeat, was there any calculation of timings. The offices checked the head foreman's

[27] *Opuscolo sulla fabbricazione dei proiettili, pubblicato per cura del Comitato lombardo di preparazione per le munizioni*, by the engineer Camillo Arpesani, p. 31.

[28] Consult the little work by the national committee *La donna nelle officine di munizionamento*, Rome, topografia dell'Unione Editrice, May 1916 and the British example analyzed in *Il lavoro femminile nella lavorazione dei proiettili*, Rome, *ibid.*, April 1916.

[29] SSR 541–28, nota sul personale della Fiumara e di Sestri, March 1916.

estimates, made a record of them, and calculated payments: but in the absence of any real scope for initiatives on costs it was an *a posteriori* office. The employer's strategy was to extend the power of the offices at the expense of technical workshop managers by means of accounting precision and the multiplication of bureaucratic procedures: but what resulted in practice was a flow of paper – because of the increased number of modules, the number of bodies involved, and the multiplication of the number of signatures requisite to validate a task – while the subordination of office to workshop remained unaffected in practice. Paradoxically, the office thereby liberated the head foreman from the essentially bureaucratic part of his work and increased his autonomous power of decision-making over the establishment of production norms. This hardly represents a transfer of power away from the workshops on the basis of mastery of costs and of the real flows of merchandise.

So the establishment of the system of remuneration of the standardized factory brought about wider changes, and the connection between mass machining on specialized machine-tools, the introduction of unskilled labor, and piece rates could be regarded as automatic. This global connection did not however become immediately apparent. This was primarily because Ansaldo took on a fresh intake of workers without increasing its existing stock of machines. Hence the necessity, since piece rates were introduced only slowly, for a system of remuneration based on bonuses. Such a system was established in the Sampierdarena artillery factory in March 1914 and was intended to encourage both production and the upkeep and use of machine-tools (it worked "in favor of workers who performed particularly well with the machine tools and looked after them in exemplary fashion").

The resistance of the flexible production systems

The continued existence of the two paradigms side by side was not without considerable tensions. Consider the report of April 30, 1916 in which Raimondo painted a dismal picture of the establishment of the serial production factory at La Fiumara.[30] Nothing was really working well there and structural obstacles were blocking any rational organization of the work. The maximum level of case production possible without forcing the machines should have been 4,000 per day (over a 24-hour working day) but actual figures fell way short of this target. There is a damaging assessment of the high number of rejects: "According to the data collected from competent personnel, such a high percentage of rejects can be justified at the beginning of the work when the staff are not yet quite up to scratch, but the percentage ought not to exceed 5 percent when the work and the staff are properly organized, as seems to be the case now." The level of definitive rejections was running at 15 percent. This high percentage was only partially due to the hardness of the metal and to the inevitable experiments with different

[30] SSNB 1098–50, report by the engineer Scavia on La Fiumara, May 19, 1917.

tooling: in essence it was a question of resistance by the old-style organization, which had resurfaced in the new factories.

In practice, the new factory operated to a large extent according to the rules of the old one. Production was actually lower, and for four kinds of reason. The first kind: work on the machines. No machine had sufficient equipment and almost none had spare parts, "so that the workman had to stop the work either to ask for new parts in the case of a breakage or to sharpen or repair the ones in use." A circular saw, for example, remained unused for two hours while its operator sharpened it. The same went for machines for cutting bars or those for machining cases and their tips. It is true that in theory there was a specialist for sharpening but in practice not every workshop had one, and even when it did the workman stayed with him throughout the operation. Finally, indecision over the positioning of machines caused frequent moves, an additional reason for inaction. Another cause: interruptions to the electric current. In the drill workshop the motor would get overloaded and overheated and then had to be turned off.

A second kind of reason for poor production levels: supplies. There were sporadic lacunae in stocks of raw materials, (responsibility for which fell to the supplier, ILVA) and of tools for the machines. On one occasion a whole workshop for the threading of cases remained idle because there were no cases. Finally, certain machines would be not totally operational while awaiting their tooling and lay inactive in the meantime. The general picture was pretty dismal (though it may have been exaggerated) since reports from the workshops suggested to the author of this report that out of six working days four were productive and two unproductive or not very productive because of the lack of bars, tooling and raw materials.

A third kind of reason for poor productivity: the wages problem. Raimondo blamed poor supervision for all the rejects: inadequate supervision in the mornings, since reports by testers were ignored, lack of supervision of work on the machines and lack of analysis of the causes of rejects, and finally the lack of any preventive measures – here it was the team chief who was to blame. But rejects were not deducted from workmen's wages – at least not till April 15, 1916 – so management were not providing any incentive to quality production. There were insubstantial fines for bad workmanship, but a special rate penalizing rejects was not introduced till the second fortnight in April. A study of the accounts shows that half the work force of the factory was paid by the day. But it was clear that workers paid by the day were less productive and could give part of their work to their comrades who were on piece rates with a resultant loss of production.

Finally the number of pieces produced was paid for in accordance with a formula arrived at by the time-keeper on the basis of declarations by team leaders rather than by the tester for each workshop on the basis of individual calculations for the production of each workman; the management standard established right from the opening of the factory was ignored. Moreover for lack of time tasks were often paid for at the end of the month and not each fortnight, and this delay made it difficult to keep up sustained supervision of the production workshops and individuals. Finally the workers paid on piece rates could not cope, a "cause that was far from being secondary" of the lack of productivity, and in any case gaps in the provision of stock meant that workmen paid by the piece were in a hurry once they had received the pieces to catch up on lost time.

The fourth kind of reason for poor productivity: deficiencies both quantitative and qualitative in the labor force. There was an evident lack of military personnel for testing and therefore a bottleneck at the despatch end, with shells and grenades held up for lack of tests. A request by the lieutenant with the job of chief ester for thirty additional soldiers went without response for a long time. At the factory, not enough staff were allocated to deal with the supply of parts for the machines. The workmen, laborers and peasants, were particularly lacking in skills. Assembly of grenades was disorganized and labor in short supply. Only the women recently taken on gave some ground for satisfaction. On the management side, some team chiefs were poor. Moreover there were too many time-keepers, many of them were incapable, and their chief had little dynamism or skill.

The general picture then was pretty depressing and problems could not be attributed exclusively to the fact that this was a learning exercise in the new paradigm of mass production, which would come to an end after a running-in period; it was first and foremost a matter of the compatibility of the two paradigms.

The measures proposed in response by management were typical of a rationalizing strategy. For sharpening, Raimondo suggested that each group of similar machines doing the same work be given a stock of spare parts so that "while the blunted part is under repair by the specialist operator, the machine can continue to function and produce." This system has "moreover been adopted all over industry and is particularly appropriate for standardized work." All workmen were to be paid on a piece-work basis and the calculation of bonuses was to be done by the accounts department on the basis of daily reports from the tester. In order to ensure that power was transferred from the team head to the tester Raimondo proposed that the latter be the only one to establish what effective level of production had been attained, supplying targets per workman to the accounts office, so that actual production corresponded exactly to what was paid to the workmen; and that as a corollary rejects be set against workmen's pay. However, the proposals do not seem to have been acted upon, for in the second fortnight in April team heads were still signing for payments for tasks carrying a bonus.

Far from being a matter of remedying mere technical deficiencies then, the establishment of new kinds of production logically entailed a transformation of hierarchical relations within the workshop together with a transformation of modes of remuneration and technical organization (supply of raw materials, power and tooling). What emerged was essentially a new factory. A mass-production factory overseen *ex ante* and *ex post* by the toolmaker, governed by the law of piece-work, and with systematic supervision of all technical functions. The position of the tester was strengthened since it was his daily report that set the wage level.

Lastly, the workmen had to be convinced that it was in their interest to increase productivity: "it had to be made very clear to the workman that rates would not under any circumstances be reduced, whatever the profit being made by him." At the same time foremen and team leaders were to be galvanized by profit-sharing schemes, the former on the basis of surveys of completed production, the latter on that of production for each of the fifteen machines they were supervising.

One year later a series of reports on standardization confirmed the tensions between the two paradigms. While the organization of the offices was appropriate to a mass production model, a good part of the machine stock, with its universal machines, was actually suited to flexible specialization. At La Fiumara, the engineer E. Scavia[31] gave an account of the transformations effected. He laid stress on the work office (*Ufficio lavorazione*), "one of the most important management offices, much esteemed by the chief American companies," with its management personnel kept under the direction of an engineer [*capo della lavorazione*] and with a technical and moral authority over the workshop engineers that included the area of tooling. The activity of this office was "technical control and the allocation and registration of orders for the workshops," and the bureaucratic activity of the workshop itself had to be limited as much as possible so that the work office would take complete control of these tasks. Scavia made a distinction between the experimental operations, the big series of small-scale products, and the small series of large-scale products. For the big series, what emerged was the ideal of a product store that held standardized tooling and materials; the workshop could send finished pieces to this general store instead of making items specifically for particular given end products (vessels, locomotives, etc). While recognizing that the "type of organization I have been analyzing has been partially established over here," Scavia was critical of the widespread employment of universal machines ("I am convinced that the majority of Italian companies have persistently made the mistake of purchasing machines that are too universal, and that it is a mistake they are still making"). And he added "I am fully aware of the possible objections but I have not the space to discuss them. What I do know is that (provided the usual caveats are borne in mind) there are very good grounds for my belief. Therefore, particularly when confronted with the requirements of future standardization, in purchasing and planning for machine tools our firm must be very mindful that the basic qualities required are strength, power and simplicity; and that they are not to be universal."[32]

A neo-Taylorist strategy against mass production?

The radical transformation involved in the transfer of bureaucratic skills from the workshop to the offices was made difficult in the old production units by endemic

[31] *Ibid.* [32] *Ibid.*

inertia. The same was not true in respect of the new factories, created by and for the war, which played the role of midwife of projects. Distinguishing between a technical workshop office and a piece-work office the *Proposta di piano di organico per le officine di artiglieria* of March 7, 1916 marked a notable evolution in the reorganization of work: a new coherent network capable of guaranteeing mass production was perceptible behind the arrangement of the responsibilities attaching to the new jobs, and in the effort to define decision flows and relations of subordination.[33]

The rise of the technico-administrative function was evident in the attributions of the new technical workshop office (*Ufficio tecnico di officina*) under the direct control of the chief engineer. It took responsibility for the allocation of tasks between the different workshops, thus over-riding the executives *in situ* – engineers and head foremen and their deputies – but the collaboration of these latter remained an indispensable prerequisite for the smooth operation of the arrangement. The office was in charge of the overall production process and took control of it away from the workshops, henceforth reduced purely to an executive role. It was a centralizing planning organism that was really a design department and a tooling office at one and the same time. Before jobs were allocated it took responsibility for the study of the design of the tools for the manufacturing and supervised the establishment of an inventory of the materials required, taking account of deadlines and of the production capacity of each workshop. Although load charts for the machine tools and work sheets for the personnel available for finishing were done by the workshop engineers, responsibility for them automatically came down to this office. "Allocation of tasks to each workshop must be carried out in such a way that there cannot be any uncertainty about it." The workshops were supervised by *incarichi speciali* responsible for coordination, particularly in relation to manufacturing times (and hence delivery deadlines).

The task of preparation for execution, detailed analysis of each operation, equipping of the work stations and responsibility for the machine tools were all put in the hands of a single organism to which autonomous engineers and designers were assigned. In overseeing the study of improvements to ongoing work and the tooling of special pieces this technical office represented an embryonic methods office. Yet continuity was always more fundamental here than innovation and we are still a long way from the functional division of rationalized organization, and all the more so in that we are talking here about a big company. The real function of the technical office was in fact first of all – and in reality perhaps only – to centralize preparation for manufacturing. At no time did it concern itself with studying times – which would have been a logical consequence of the responsibilities of a methods office – or with prototypes or tests, which would have been a matter for a mature design department; or finally with the analysis of potential uses for the machines, which would have been the

[33] SSR 490–49, *Proposta di piano di organico per le officine di artiglieria*, March 7, 1916; report on the workshops, April 16, 1916.

task of a tooling office. The only rationalizing principle remained that of centralization.

The way responsibilities were shared between the workshop engineer and the head foreman made it possible to perceive the gap that had opened up between administration and production tasks. A significant pointer to the persistence of worker-based control was the supervision of piece and flat rates: it was the head foreman who "laid down the cost of production of pieces not yet covered by a rate." Here the engineer acted as a registration chamber, "responsible for informing the piece-work payments office about rates so that the latter could fix the rate and intervene in cases of conflict (with the workman)." To this bureaucratic administration of the administrative production authorities – labor vouchers, flat rates, fines and punishments, delivery slips for raw materials and tooling – the engineer contributed only his ability to make "concrete proposals" for the improvement of the workshops.[34]

The piece-work office (*Ufficio cottimi*) is equally disappointing for admirers of Ford. In fact it had responsibility for the calculation of rates only for pieces which were not the object of earlier estimates or which required the employment of special tooling. More significant still, it was not totally autonomous, for it had to get collaboration from the technical office on the one hand and the "workshop experts" on the other. Provision was even made for an appeal – optional it is true – to the workman himself, which implied the maintenance of negotiation on the rate, or at least retained the possibility of this. We have seen that the head foreman fixed rates on the basis of information supplied by this office, but the latter in turn relied on the collaboration of the head foreman to set its estimates. This vicious circle was the logical result of the nature of the rate, which was not imposed – as it was when the calculation of times became the norm – but could only be negotiated and contractual; the intermediary and place of the contract remained the workshop which, collectively or by the agency of the head foreman, was the sole locus of the experience necessary for the fixing of the pieces. Overseer and administrator of a relation of social force, the *Ufficio cottimi* modified the organization of workmen's remuneration only to a limited extent.

In 1916 a reply by Mario Perrone to the engineer Combes states: "To have taken away from the head foremen all the burden of the bureaucratic side represents great progress, and so there will no longer by any excuse for delays and negligence in production. The definition before work begins of each of the tasks and pieces required prevents any favoritism and any irregularity. The responsibilities and functions of each person have been clarified. The head foreman may at any moment familiarize himself with the state of play in the work and give an account of it and of the efficiency of the machines and the level of activity of the staff."[35] Here was a foretaste of changes, more theoretical than actually visible, and without the separation of the conception/direction functions on the one hand from the execution function on the other being really carried through. An

[34] *Ibid.* [35] SSR 490–47, April 16, 1916.

emphasis on the problem of respect for deadlines indicates that the question of costs remained secondary, for what mattered above all was that they produce and produce quickly.

The adaptation of the old factories

In most cases, organizational changes necessitated by the increased demand for war materials did not get beyond a straightforward increase in the endowment of tools. In March 1916 a memorandom from Electrotechnique responded to the problems raised by more and more ambitious specifications from the Navy and the draconian deadlines it was laying down for the delivery of the electric fittings for submarines and torpedo boats by calling for "greater intensification of the work." The term was ambiguous: specifically it involved fresh acquisitions of machine tools, the construction of new workshops, and the fuller utilization of under-used machine tools. All this was a long way from rationalization in the direction of Ford-type mass production. However the report adds that "new arrangements in the machine shop" were planned, which suggests a transformation of the production space and perhaps a transformation of the production process. But in the case of the small machines shop reorganization corresponded simply to a linear increase in production capacity: its area was doubled and new machines were purchased.[36]

Standardization and rationalization in the old factories remained very limited. At the steel-works[37] the engineer Pozzo criticized the shortcomings of draughtsmen and planners who did not have the knowledge and experience "necessary for the delicate work required to ensure that the form of a piece is suitable for speedy and economic production." He suggested that projects be discussed by the workshops and that each of them have a standardization office (*Ufficio di standardizzazione*) which would bring to bear on projects at the time of their conception its familiarity with the work and its knowledge of the machines required and put to the planners the modifications necessary for reductions in costs. The workshop manager would let the offices know his point of view, meticulously comparing, cataloguing and examining the pieces one by one, and identifying those of which only one sample needed to be cast. A dialogue between the standardization office and the workshop developed, leading to uniformization and product modification in the light of the machine stock and the technical processes involved. The planning office (*Ufficio progetti*) approved the decision: sanctioning the production order, it was the guarantee of the continuity of the standardization policy ("so that in the future they stick to the types chosen"). But standardized production itself was always based on the skill and abilities of the *capi-tecnici* in the workshop, and in particular on their ability to reduce costs ("from this point of view the value and skill of the *capi-tecnici* appropriately

[36] SSR 542–13, Electrotechnique, March 21, 1916.
[37] SSNB 1098–50, report by engineer Pozzo, May 19, 1917.

guided and directed by the workshop directors will have a very wide field of application").

With standardization decentralized to the workshop by the workshop/standardization office combination, a central organism became necessary to link the different production units: this role fell to the head standardization office (*Ufficio capo centrale*) which received information about production (form, quantity, destination) from each workshop: as the registry office and central distributor for the firm, it passed this information on to the other workshops which were expected to take note. To begin with at least it did not possess any particular authority: it did not give orders and contented itself with transmitting information, although its staff had to be highly qualified (in working hot materials, cold materials, and manufacture of steel, bronze and cast iron ingots). It gave advice, and whether this was followed was verified by teams of inspectors going round the workshops. Finally it also pronounced on the possibility of producing pieces ordered from outside and of managing existing stocks of products that were already standardizable: valves and taps, oilers, supports, joints, pulleys and transmission components, tubes and keys, screws and bolts, tooling for hot and cold work and supplementary tooling.

At Meccanico[38] the standardization principle was also accepted: "there is not much to say about the necessity of 'standardization', since it is universally accepted, especially where the postwar period is concerned, when domestic and foreign competition will make itself felt on our markets, so that the intelligent adoption of these principles could mean life or death for very many industries," it was observed. But its adoption was a long-term process, for it came into conflict with the planning offices, whose old "arbitrary criteria" had to give way to "precise criteria." Here the obstacle was a matter of the specialization of office skills over against workshop skills: "in general a person who begins a career in the technical offices does not interrupt it to begin another in the workshops and vice-versa. It is not possible to find persons with a competence in the area of machining and there are even fewer persons who know both the types and the features of the machines used and the conditions for the employment of these machines." There was an ideal of decentralization in the design department workshops: the study of pieces in series "must be carried on *exclusively* in the workshops" (italics in original). Special staff would be trained and taught workshop production and would have experience of real problems "that are only comprehensible to those who really know machining," and this staff would constitute a workshop design department (*Ufficio tecnico di officina*). The transformation was thus global: "a final difficulty to be considered and overcome is this one: the reorganization of the workshops on criteria completely opposed to those followed up to now, the retaining of the staff, and the change or transformation of a good part of the materials."

[38] SSNB 1098–50, report by the engineer Giovanelli, May 18, 1917.

The debate on Fordism

Competition between the two systems was very real in the minds of the group management, where reflection on American methods of organization began very early on. Pio Perrone followed the publications of *The Engineer* and paid particular attention to the article of February 4, 1916 on the standardization of shipbuilding.[39] He grasped the fact that "each class of vessel has to be built with perfectly identical interchangeable parts." The Ansaldo stores held pieces already assembled: if the method was adopted, the return would be 25 percent of the cost, not to mention the reduction in time taken, for a 6,000 tonne cargo vessel could be assembled in thirty days and only fifteen with yards in perfect working order. "It is an American thing! but there's no question that the Americans are astonishing at standardized work," concluded the entrepreneur admiringly. He could not see any real problem with the implementation of standardized work(!) on the one hand, or with the simplification of designs on the other – though this latter was a necessity he believed the technical offices had not really grasped.[40]

But resistance was fierce: commenting in October 1916 on how difficult standardization was Oglietti, the director of Delta, gave a negative account of American factories on the basis of experience gleaned on visits there since the early 1890s. He contrasted the example of the Westinghouse Factory – a workshop of more than 500 automatic machine tools just turning out cast iron sewing machine wheels, each machine having a store of materials, four machines per girl, for a production of millions of pieces, all interchangeable and identical – with the numerous cases of sizable factories that were not standardized because they could not be, such as the Cramp yards in Philadelphia or Baldwin. These factories he reckoned to be closer to what Ansaldo was and should be, manifesting a scepticism and a reserve that were equivalent to a defense of the benefits of flexibility.[41]

The debate broadened when in 1917 the printers of *Secolo* – the Ansaldo Group newspaper – published a twenty-nine-page brochure that was a fairly precise description of the Ford Factories ("Alcuni dati sugli Stabilimenti 'Ford'," Genoa, Tipografica Secolo XIX, 1917, 29pp.). The stress was on the vast scale of the plants – the 11,168 freehold hectares in Detroit, 1,922 of them covered, the turnover of 150 billion dollars, the 18,000 employees and the production of 500,000 vehicles per year – and the new modalities of production. In the huge single hall of 64,400 square meters (three-quarters of the total area of Meccanico) there were more than 6,000 machines organized into production lines:

[39] For the attempt at standardized production in American shipbuilding during the First World War, cf. David A. Hounshell, "Ford Eagle Boats and Mass Production during World War I" in Merritt Roe Smith (ed.), *Military Enterprise and Technological Change. Perspectives on the American Experience* (Cambridge MA: MIT Press, 1986), pp. 175–202.

[40] SSR 502–24, letter from Pio to the naval yards, April 6, 1916.

[41] SSR 596 bis–48, report by Oglietti to Mario Perrone, October 1916.

Visitors used to factories will soon note the special organization of this tooling. Machines of the same type are not set up in a single group or unit. Each department contains all the machines necessary to produce the piece it works on completely and in all its parts. For example, they begin the machining of a raw piece from the foundry or forge at a given point in the department and it goes through the necessary machines to leave this department completely finished and ready for assembly. This system requires that machines of different kinds be grouped together – including soldering furnaces, tempering furnaces, and other special tooling that is more generally grouped in special workshops. The work moves rapidly from one machine to the next in such a way that a workman can transport a piece from his machine to the next one as if by gravity.

This movement was kept going by the famous conveyor belts. For assembly, the pieces "are assembled progressively on belts in the following manner: each assembler or operator carries out a single operation and repeats this operation on each piece that passes through the workshop." The procedure is the same in the final assembly hall where "when the chassis reaches the end of the chain it is ready to move on its own." "The result is that each operator quickly becomes a specialist, and specialization is the chief principle of the Ford Company. The economic results of a system like this are incredible . . . saving on space and elimination of confusion, with the units being assembled freeing up stocks since the pieces are for different parts." With Ford as its reference point the mass-production paradigm had found its way into Ansaldo.

A new stage was reached in March 1918: Mario Perrone distributed copies of the book *Ford Methods and Ford Shops* to company executives, initiating a more or less widespread movement to plan the import of the methods of organization of standardized work. Twenty-three copies were distributed by the management to begin with, and funds were allocated for an Italian translation, while the factory managers were invited to get engineers and technicians to read it. This policy was an initiative coming from management, from the center rather than from the periphery, and although the importance of the imported system, its potential and its utility were explicitly acknowledged, this was less an open discussion of principles than an incitement to reflect on how to implement it. So the managers' minds seem to have been made up at this date: "this book contains a lot of useful information about technical organization and standardized work which I reckon can be applied advantageously to our plants too." Reactions were favorable on the score of the principle, but sceptical about the possibility of implementation. At the steel works Almaggià gave the pamphlet to his technicians to read. At Delta Oglietti emphasized the immediate and obvious importance of it – his first glance had convinced him of that. At the ship yards Segala gave strategic assent: "in my opinion, the adoption of systems for the scientific organization of work, of which Ford is one of the finest examples, constitutes the absolutely necessary goal for all our efforts."[42]

According to Mario Perrone the principles embodied in the new organization

[42] SSNB 619.

of work are perfectly applicable to Ansaldo; he rejected any idea that the particular features of the automobile industry limited their relevance or applicability, or that their success could really be attributed to the favorable commercial situation. The objective of the adaptation of American rationalization would be to "make savings in production costs, bring down the sale price of products, and have the chance to defeat the competition, even the competition from abroad." It was wrong to suppose that these production methods "were not applicable, wholly or partially, to other industries as well." A general rationalization program was theoretically conceivable in the steel mill, where extreme division of labor could limit the skilled labor force "to the minimum," ideally to 5 percent of the total labor force; it was conceivable in the area of research and design services at Meccanico, where expansion could reduce the time required for assembly and work with machine-tools and for general planning of tooling; and finally it was conceivable in the area of the transport of materials to the ship yards, where savings could be made. An explicit policy of cost reduction would be a first important stage, to be followed by a strategy of replacing the skilled labor force with an unskilled labor force by means of office-planned organization. In this way of looking at things the Fordist conception was wedded to Taylorist considerations and there was both a social and a technical approach; the workshops were to be stripped of their control of the work and it was to be handed over to the offices, which would standardize production, cut down production time and thus production costs, and plan production *en masse* in an expanding market.

This impetus from the center received hardly any encouragement from the factories. Agreement was stated in principle but subsequent statements frequently demonstrated considerable scepticism. The production process seemed incompatible with a healthy application of the new methods. For locomotives, "in truth the variety of types of wheeled vehicles required by the needs of rail traffic and the continuous improvements that had to be accepted unhesitatingly to raise the technical efficiency of the locomotives did not fit in with the ideal of the sort of factory that would produce just one single type." The arguments put forward by Electrotechnique were along the same lines and are worth quoting at length:

But we note that all the different arrangements adopted by the Ford Company require as their basis the presupposition of an absolute uniformity and constancy of production. In fact the Ford plants make one single product, the Ford car, and it is always the same. Consequently each workshop makes the same pieces repeatedly, each machine works over and over again on the same production and on the same pieces, while each worker goes on repeating the same operation. In these conditions it is perfectly understandable that the technical office is able to predict all the details of the work (even the minutest) precisely, to prescribe inflexibly the progression of materials from entry into the factory to exit, and even the times involved in the various movements, to study the best modalities of execution for each product, and to have not just the tooling but also the machine tools made and adapted solely for the production of a single piece. This characteristic of Ford production also leads to a special arrangement of the workshops in which the machines are set out on the materials line in the precise order in which the machining operations are carried out, so that the shortest trajectory is thereby obtained. This makes it possible to install the means of transport most suitable to ensure movement in the most ordered and most economic manner. However none of these systems is applicable to the kind of variable production typical of our establishment. In the course of production different operative machines are called on to carry out different operations on various materials and it is absolutely impossible to establish the succession of operations and the line for the materials in a mandatory and invariable manner. Consequently instead of the extreme specialization typical of the Ford factories, our machines and handling methods must exhibit the maximum universality in order to fit in with the various requirements of production. So it is not possible to apply the Ford principles to our establishment since we find ourselves in conditions diametrically opposite to those necessary for their application to be appropriate.

With its variety of types and its tooling of maximum universality, this is in fact a program of flexible specialization based on prior experience. While the offices could not keep effective control of prices,[43] Taylorist standardization and rationalization made awkward bedfellows at Ansaldo.[44] In October 1916 Oglietti, director of Delta, who remembered earlier failures, said he was convinced that standardization of machinery could not be generalized. Around 1905 a first attempt known as "single type" had led to a determined and persistent drive to interchangeability in taps, valves and tubes. It had been blocked by the equally obstinate underground hostility of the entire factory staff, from top to bottom, articulated in the introduction of successive modifications that finished off the "single type" in just a few months. Moreover, "the Royal Navy helped to frustrate this praiseworthy experiment: it never allowed the adoption of the single type and always wanted to impose its will in the smallest details." At this time the Schichau class torpedo boats were starting to be built and they were so different from the single Ansaldo type that the specifications of the latter were consigned to the archives and very quickly forgotten. As a producer of machines for the Navy and locomotives for the railways the firm was in the grip of its customers, who successfully blocked any standardization. "This first attempt to establish a rational system that was clearly of great utility was shipwrecked then not simply by the inertia of the whole mechanism, but also and above all because our production did not lend itself to a unification of types": The failure was total.[45]

Conclusion: the transformations that would have been required by a Fordist strategy in a group like Ansaldo were enormous and the obstacles many. The whole traditional way of operating had to be questioned but it offered an overwhelming resistance. The standardized production of munitions created more problems than it resolved, and as an exogenous factor therefore the War, far

[43] SSNB 124–21, January 29, 1917, Breve riassunto, movimento cottimi e lavori d'ufficio compiuti nell'ufficio per servizio cottimi.

[44] ASA Copialettere 247, Meccanico, July 25, 1917.

[45] SSR 596 bis–48, report by Oglietti to Mario Perrone, October 1916.

from being an opportunity to begin again from the beginning, was rather a cause of major disorganization.

Pattern III

The third pattern: post-war reconversion. It was a time for settling accounts. Firstly on the financial level – for the company could not take on artificial indebtedness to a credit institution which was its own creature, while the state, liberated from economic crisis by victory, fell disastrously behind in relation to payment for the military goods produced by Ansaldo, which was driven into bankruptcy. Secondly, settlement of accounts on the social level too, where an accumulation of tensions resulted in unprecedented agitation, a series of strikes and the occupation of factories in September 1920. Far from bolstering a mass production strategy, these extreme tensions actually stimulated a return to tradition. There had never been such a huge gap between programs and reality.

Imperialism and Fordism

The Fordism of the Perrone Brothers was closely bound up with an imperialist and nationalist strategy that linked market expansion to the utilization of a large (unskilled) labor force, seen as the chief resource of the Italian economy. This was a paradoxical program to the extent that the economic situation was fragile and ambiguous and the phase of adaptation and growth was followed by an anxious questioning about the options open to Ansaldo once the war was over. In a secret anonymous report of March 1917[46] the Perrone brothers, dramatically overestimating the size of the markets, opted for a labor-intensive strategy – the only one they thought appropriate to Italy's resources, given that though a poor country she had a population of 40 million. Their argument for market regulation and stable outlets was anchored in the imperialistic anti-liberal program they advocated. "Given the amount of capital invested in it, our great steel industry," they observed, "needs to have the quantity of its annual production constantly adjusted; it needs to create a market for itself which gives a constant return, and which is not subject to the politico-military variations of the nation, but which is based on a stable and secure customer base." This applied even more to the engineering industry and all of its "post-war problems perhaps represented the one that was the toughest to resolve." This naturally led to a study of the potential market for the steel and engineering industries. Further back up the chain, an undeveloped mining industry with old-fashioned techniques would not be able to help much: open-cast mines without much quarrying depth could not generate a strong demand for machinery. Equally, the relative lack of coal deposits compared with the massive quantities produced elsewhere made growth in the market difficult.

[46] SSNB 1098–30, Pensieri sui problemi del dopoguerra.

So the strategy was explicit: to utilize few raw materials, little capital, and a large labor force while tolerating a fairly hefty wages bill, if we are to take the 1922 Commission of Inquiry on the steel industry at face value: "Italy needs to generate wealth by labor and must therefore export manufactured products which have absorbed higher costs from the labor force that made them. This seems obvious, since our country is dependent on foreign countries for the greater part of its coal and raw materials, which means that it benefits from working a small amount of materials while employing a large number of workers." This was a strange argument, aiming to conquer the market by reducing investments in fixed capital and raw materials costs and based on the single factor of labor. The ambition was all the harder to put into practice in that it implied both means of pressure on labor costs themselves ("we shall have to fight with all our strength to prevent the eight-hour day from being introduced into Italy and keep the so-called 'English working day' at bay," said the management) in a situation of revolutionary crisis and a capacity for organization able to reduce expenditure by adopting a more rational manner of organization: but we have seen that the firm's capacities were very limited in these two connected domains and the bankruptcy of 1921–22 made the reconversion strategy pointless anyway. The firm was then reconstituted on a humbler basis, losing part of its financial power and some of its production units.

However, the strategy did find some reinforcement in the new allocation of production units. The heavy artillery factory, called Vittoria and then Electrotechnique, was to become a great electrical engineering unit whose production would find its outlet in the electrification program to make the peninsula independent of coal (as advocated by the Perrone brothers). In addition there was a very ambitious electrical steel-making program in the Val d'Aosta. In this context the factories established to meet wartime requirements for production were reorganized in the light of harsh experience. At La Fiumara the layout was set up appropriately for a single product machined in series: taps, bolts, helical points, millers. A workshop of 1,600 square meters divided into three departments with 208 machines produced more than 2.7 million bolts in the first semester of 1920 alone. But not everything worked together for the best. The helical points section covering 900 square meters and with 73 machines made only 500 per day instead of the 2,000 planned for. The milling workshop covering 600 square meters and with 36 machines did not even give an account of its production, while the tap shop, covering 900 square meters and with 62 machines made 500 pieces per day instead of the 1,500 planned.[47] With regard to methods, a committee for the standardization of installations and tooling was set up; it produced confidential minutes, encouraged "referenda" over the choice of standard tools and fixed the dimensions and characteristics of the steels necessary for overall tooling in the company and the global requirement for steel and stock levels.[48] These procedures were linked to the publication of standard

[47] SSNB 678–34, Situazione lavori e impianti, Fiumara, July 8, 1920.

tooling tables for each establishment and for a fairly large spread of operations (tools for cold machining, drilling, reduction cones, cutter spindles, helical points, special drilling tools, frontal millers).

From mass production to bureaucratic rationalization

There was an imperialist and Fordist coherence to the Perrone program, despite the fact that it was the product of the management of a firm in crisis, but the multiplication of plans, reports and projects relating to organization indicates the presence of deeply-rooted ambiguities and the practical results seem to have been limited. The tendency was to speak of a transition from disorganization to organization – a fetish word that covered a multitude of evils. In 1919 a report by the engineer Bruni, who managed the Campi steel foundry, to his manager in Pozzo strongly emphasized "there is no real industrial organization. During the war the exclusive concern was to produce, to produce and again to produce, without worrying about costs or even quality. Industrial accounting is non-existent, for in the [earlier] system it had no use, and in mid-1919 the accountants are still doing the accounts for the years 1914–1918 . . .!"[49] This lack of organization and industrial accounting was the dominant feature of management during the war: "In these conditions, the work of the company management was greatly facilitated." These "elementary" considerations were known to everyone.[50] The starting-point was thus very basic – an organization which had no real rules for the rational administration of its costs but which had nonetheless for over seventy years of existence embodied habits of organization that were a terrible drag on the rationalizatory potential that employer strategies might seek to tap. It was not until autumn 1920 that the Office for Statistics and Supervision of Shipbuilding suggested the adoption of unified accounting systems and homogeneous estimates to make possible detailed production and cost calculations: hitherto differences in the systems had not allowed for any accurate comparison of data.[51]

An Ansaldo publication put together by chief accountant Fossati (undated but probably from 1921) states that there was no "industrial organisation" – in the modern sense of the term, before the First World War.[52] "This," he adds, "was especially true for the engineering industry, where the infantile character of the administrative organism and methods made it impossible to profit properly from the fortunate circumstance of the availability of a labor force that was much cheaper than those of the great industrial nations." This was the dominant

48 SSNB 769–7, 1919, Fiumara. See the 18 standardization sessions from January to May 1919, in: SSNB 678–23.

49 SSNB 778, report by Bruni to Pozzo, steel works, August 28, 1919. 50 *Ibid.*

51 SSNB 564–13, cantiere, Ufficio statistica e controllo delle costruzioni navali, verbale della seduta del 20 ottobre 1920.

52 ASA, *Industria e amministrazione. La riorganizzazione aziendale delle società industriali del Gruppo Ansaldo nel dopo guerra con speciali accenni della parte amministrativa*, note compilate dal rag. Salvatore Fossati, n.d.

interpretation that the post-war rationalizers were to pick up on as they built a bipolar system combining the old and the new. The prevailing opinion was still that work was the job of the workmen, and many employers left the real management of the production process to them, because "it was experience alone that enabled them to regulate manufacturing and manage the labor force on the basis of traditional empirical criteria." Fossati's analysis derived logically from the crisis that befell the earlier work process. According to him "the work was very repetitive and labor costs and general expenses were limited, so that the productive life of establishments with an existing high profile – like those (that Ansaldo) possessed at the time – was more or less guaranteed." But the huge productive advances during the war, which were accompanied by increased competition and a rise in wages and general expenses, and the fall in the productivity of the labor force deriving from the reduction of the working day, suggested a need to reorganize and introduce "modern scientific organization." Taking its cue from the teaching of Taylor, Fayol, Paul Négrier, Warner, Evans, Gantt and Schwab, this publication put administrative skills in the first rank of the requirements of the modern company. The schema proposed had the advantage of simplicity; it contrasted "scientific organization" with the "empirical patriarchal" organization of the old-style Ansaldo, structured with its principal and its manager and the limited number of members of the inner circle – head foreman, accountant, stores manager – who went to report and receive instructions every morning. But in spite of the Perrone brothers the philosophy that won the day was neo-Taylorist and not Fordist.

The weight of realities

Mass production, rationalization, organization: at Ansaldo these terms remained more theoretical and verbal than practical, for the reality was very different from the programs. From 1918 to 1920 post-war reconversion planned for factories to be retooled for products for which outlets were to be guaranteed by a strategy for the conquest of new markets; but this diversification option ran counter to the whole ethos of Fordist production. Part of the Campi steelworks was kept ready for operation in hopes of an expansion of shipbuilding. Munitions factory no. 2 was organized for the production of related material and naval accessories, and workshops were set up there for building combustion engines; the aviation workshops at Borzoli were geared up to production of naval fittings and the La Spezia seaplane workshops were converted into yards for small craft and barques. In short, the accent was on flexibility and short runs (making allowances for the size of the market). The return to flexible production was particularly evident in shipbuilding. The completion of a series of 6,450 tonne cargo vessels for the transport of wood, all identical and with a 400hp diesel engine, represented the beginnings of serial production of standardized cargo vessels – "i.e. the same in all their dimensions, identical in all the parts of the hull, engine and accessories" – during the war, but the end of the war saw a stasis and a return to the one-off order from the shipowner; the industry was market-led, although the firm did retain standardization just in steel sections and sheets. But it was not only the market that decided: management were quick to emphasize the quality of the traditional product in the face of the standard cargo vessel and the tradition of quality, supposed to be consubstantial with flexible production, eventually prevailed again.[53]

The case of Electrotechnique was in this respect typical of the real mode of operation of the company. The factory was initially geared to a non-diversified production and set up solely for naval electrical machinery. Up to the last years of the war, it was still not very good at electrical engineering overall, especially big electric motors. Nonetheless, its situation was not without advantages, a "privileged situation, almost a monopoly one, in respect of the supplies necessary for our wartime navy." In this market that was limited but dynamic in terms of the profits it could bring and the technical impetus it imparted, the factory remained modest in size and employed only 300 to 400 workers before the war. After the victory there was a growth in the factory's industrial activity in accord with "the objective of expanding into the massive area of electrical engineering constructions *in their totality*, i.e. the manufacture of big machines, transformers, electric rail traction and tramways, small and medium-sized machines of all types, and general electrical equipment."

But by the late twenties prospects were far from bright, for "no-one has really weighed up the full significance of the decision to develop in this direction." This was a function of the technical complexity of the product: the great Italian electrical engineering industry was in its infancy and with nine-tenths of electrical plants built abroad before the War, "the race of engineers, technicians, and electrical engineering constructors was only just beginning to develop in Italy," according to one report. Though foreign companies – mainly American and Swiss – had factories in Italy in 1927, these worked on the basis of plans drawn up abroad by the parent companies. This slavish dependence on technical "wet-nursing" naturally had a paralytic effect and hindered any development or refinement of autonomous technical strengths.

A further constraint was the narrowness of the market. The major part of the heavy materials produced in the twenties amounted to three deliveries of electric locomotives for the state railways, an order that was based on an agreed allocation among the six Italian producers of electrical materials. Otherwise activity was sporadic and output of heavy materials "could be counted on the fingers of one hand." From 1920 to 1924 the factory produced only ten big and medium-sized machines and a few moderately-sized transformers. The financial situation was deplorable: there were disequilibria between costs incurred and what was invested in production, disequilibria between productive and unproductive labor, and between general costs and the value of the productive labor

[53] SSNB 768–27, Materiale navale della nostra marina mercantile, May 31, 1919.

force (more than 300 percent). Here were "clear symptoms that these establishments are massively underperforming and that there needs to be a radical but healthy revival of their productive rhythm."[54] The implementation of a more energetic policy of recruitment of office staff, a more enthusiastic quest for orders and a more vigorous reorganization gave the factory something of a boost in 1926, when the worrying disparity between general costs incurred and costs recuperated on the products actually completed decreased while the percentage relationship between productive and unproductive labor forces slowly fell from 103 percent to 63 percent.

Production was still slow: a first big alternator for the Cellina plant, ordered in April 1920, took over four years to complete before its delivery in August 1924. Locomotives – a product that presented no technical planning difficulty, since detailed plans and drawings were provided by the state railways – were produced by Ansaldo three or even four years late (depending on the series), which was way behind the five other companies. Here was the same old story of inadequacy, neglect, disorganization, and "fantastic delays in the work." To attract customers the company made promises on deadlines and prices for large-scale powerful machinery of considerable technical difficulty without having the means to fulfil its promises. A significant detail: promises were made regarding the supply of very powerful high tension transformers even though there were no test transformers or installations for seasoning materials in the workshops.

In 1930 the factory was still trying to reduce costs – an evident indication of earlier failure to do so.[55] There were gaps in the stock of machines, resulting in slowness, lack of precision and high costs. This lack of efficient machines appropriate for modern machining was due to the absence of any replacement policy, and the stock was more or less completely made up of machines bought just before the war or during the war which had not been specifically chosen for electrical engineering manufacture; it was out of date and unsuitable. Not until 1925 was the need recognized to acquire appropriate means of production and the decision taken to acquire the necessary machines. So even in the case of new sectors where there was an evident dynamism, the weight of traditional habits of production was such that it presented obstacles to the mass production way even when that way was accepted by management in principle.

The loss of the paradigm

In the last resort it was the combined effects of the social crisis, market imperatives, and power relations within the company which were responsible for the post-war arrangements and the failure of the Fordist rationalization programs. The bankruptcy that drastically reduced the size of the firm and its financial means also led to the dissolution of the mass production paradigm

[54] Sarli-Biandra, 53–5, pro-memoria, 1927, Electrotecnico, 6.
[55] Sarli-Biandra, 53–12, April 14 an VIII, pro-memoria.

under the convergent pressure of the constraints – both effects and causes of this dissolution – that hampered Ansaldo. Far from helping to encourage in-depth transformations of the organization of production, the endless internal conflicts, tensions, and contradictions actually ended up by favoring established positions. In fact, a number of factors came together.

Management style. There were two separate and constantly conflicting employer strategies: delegation and control. On the one hand the Perrone brothers wanted to be free from the day-to-day management, which they delegated to factory managers. On the other hand their preference for being kept in touch with relatively minor expenses and for holding on to the reins of production times led them to intervene constantly. This contradiction was ever-present and it was intensified by an even deeper hesitation between the principle of authority and the rule of negotiation. On the one hand, management policy was based on conviction: it thought that if everyone was convinced, and especially the top management in the different factories, the new organization would have sufficient prestige to face down the most recalcitrant and convince them of the usefulness of the reform. The reorganization was approached as a battle, a struggle between the old and the new, a major transformation leading to reflection on all the aspects of the company, but its success was thought to be dependent on sincere acceptance. On the other hand, according to a 1919 report by engineer F. Galeazzi,[56] "Unfortunately the present approach of the factory management rests almost universally on an absolute concentration of a wide variety of functions in the hands of a single manager, a kind of *Deus ex machina* who has to solve every problem." The staff below him had no autonomy and therefore no sense of responsibility and the general policy was to take the least possible amount of risk, since anyone who made a mistake would incur "a rocket from the management" whereas those who did do well got no thanks for it. "The administrative staff to which you [Perrone] allude and which is required for a healthy approach to organization will encounter a major obstacle in the mentality of (almost all) the factory managers, since for them decentralisation means loss of authority." So it was not a question of temperament but of management style, for all who had management functions had this same mentality, not just in the upper echelons of the hierarchy but the executives in general, which Galeazzi attributed to the mediocrity of their education.

Management style. There was a way of matching staff to posts that was based on the personal value of the individual at work and a very individualistic mode of reasoning which meant that the rules fluctuated with each different person holding a given post. This was not the rational-bureaucratic style; agents relied "on their particular skills and qualities and never on intelligent scientific training." Management thus became a "matter of art, purely subjective and variable with every individual's attributes and temperament." Those who had such "favorable" attitudes and temperament were precious, but there was a price

[56] SSNB 842–17, January 10, 1919.

to be paid for them, and it was often high. As for the others, they were not much use. A more rational system would reduce the power of those given prominence by the old one, for it would reduce their value by making them replaceable. So even the best – indeed especially the best – were against. Moreover mass production entailed the emergence of new engineers, executives and "advanced" head workmen and that was an irritation to chiefs who had come up through the ranks. These new actors, most often trained in foreign establishments and with a rapidity that contrasted with the slowness of earlier "fossilizing" career paths; these promotions were carried out in the teeth of a complete lack of understanding on the part of the old managers, head foremen and ex-principals who were distrustful of "new methods." Thus the experience of large-scale serial production in wartime had as its backcloth embittered internal conflict.

The style of production and the fascination with "sporting" achievement. In this same 1919 report which Pio Perrone sent round to the engineers and head technicians anonymously ("so that they did not just swallow it whole"),[57] engineer F. Galeazzi noted the existence of numerous machines that yielded a very poor return and caused heavy losses, especially in the big workshops, losses that never really registered. There were several reasons why they were ignored: on the one hand, an almost universal ignorance about the potential returns on the best modern machines meant that no one bothered to make comparisons; on the other hand there was a preoccupation with "a peculiar kind of almost 'sporting' industrial practice: people became fascinated with the production of an exceptionally fine machine, vessel or engine," while no one bothered to work out how to get a better return on capital, something that would have required "intensive production." The report also condemned the "entrenched vice" that underlay the desire to perfect the product in the course of production work. According to Galeazzi, if a product was improved while actually being made, the thing to do was first to sell the machine or series of machines being produced at a price based on their initial machining, and then to build a new product incorporating the improvements already developed. Instead what actually happened was that production was halted while the improvements were planned, and this of course disorganized the work; part of what had been done was jettisoned and the resulting product was poor because it was the result of "a coupling of different criteria" and because of the transformations in the course of production. Hence a financial mess resulting from pieces being thrown on the scrap heap and from wasted time, and opportunities to design properly perfected machines lost. With regard to the choice of plant, the head foreman drew up "proscription lists," which included good machines that only needed a little adjustment and machines that he simply did not like or that certain team leaders did not like. "As for the machines spared? . . . I am quite certain there will be some worthy lathe on the list of those to be kept, either because it is so old or because the person who works it is so old." This seemed perfectly normal because no one was acquainted

[57] *Ibid.*

with the scientific bases for the assessment of returns on machine tools. In 99 percent of cases there was no such thing as a list of the particular features of the machines, with technical data derived straightforwardly from empirical research.[58]

Paradoxically, in spite of the failure of the workers' offensive, the social crisis finished off hopes of a reduction of the power of the workshop professionals. Starting from the Electrotechnique strike in September 1915 and then the October strike at Meccanico, the cycle of unrest continued up to 1921; it focussed on the struggle "against the high cost of living," on a strategy of eliminating wage differentials between companies and factories through wage rises, and on an attempt to assert workers' control, in particular with the claim to recognition on the part of the elected Internal Commissions (*Commissioni interne*). In 1919 the Sestri no. 2 munitions factory went on strike over the sacking of a worker and the Martin steel furnace plant successfully followed suit. The week of February 16–21, 1920 was crucial, with an initial occupation of factories and in particular of the steelworks – the strike being caused by deductions being taken from advances paid to workers made technically redundant as a result of the electricians' strike. The unrest reached a climax with a series of conflicts within workshops or individual factories, as at the Multedo factory, where there was not a single day when some group or other was not protesting up to the final paroxysm of the great September occupation. Even after the failure of the movement, the return to work was fairly slow and the carpenters – who had been on strike for two months – persisted with their protest. Although the failure of the occupation movement led to an exodus of skilled workers, the *Biennio Rosso* nonetheless helped to make any transformation of the production system impossible: for the craft worker was still there, difficult to replace, and the bitterness of the conflict sharpened the difficulty of facing the kind of direct confrontation that would have resulted from the deskilling associated with standardized production. Employers' strategies were thus blocked by a workers' initiative that was all the more restricting in that it aimed beyond wage claims at workers' management.[59]

So there was a hysteresis that resulted in the crystallization of the status quo and the persistence of practices. "Those bad unhealthy attitudes that have persisted for too many years and that certainly cannot be eradicated in a few months, however good the intentions"[60] reflect a culture of entrenched factory ways, resistant to change and riddled with deficiencies. Largely ignored, the

[58] *Ibid.*

[59] Cf. SSR 593–4, October 6, 1915; *Lotta Operaia*, 8–5, February 28, 1920, "Lo svolgimento dell'agitazione operaia a Campi. La settimana 'rossa' nelle Acciaierie"; *Lotta Operaia*, December 16, 20, 1919, "Dalla fonderia Multedo"; SSNB 564–15, Maestranze, Cantiere, October 7, 1920; Sarli-Biandra 5–1, November 17, 1920, Vittoria; Sarli-Biandra, 26–27, October 4, 1921; Gianfranco Faina, *Lotte di classe in Liguria dal 1919 al 1922* (Genoa: Istituto storico della Resistenza in Liguria, 1965); G. Perillo, Antonio Gibelli, *Storia della Camera del Lavoro di Genova. Dalle origini alla seconda guerra mondiale* (Rome, Editrice Sindacale Italiana, 1980).

[60] SSR 404–26, Riservata, Note sparse, Raimondo, 1913, 17.

management abandoned its efforts to introduce change all the more readily in that it was kept in the dark about essential aspects of the firm's operations – and indeed the firm was close to bankruptcy.

Conclusion

In the light of M. Montagnana's[61] hypothesis that Italian capitalism gambled on the maximization of profit rates by exploiting absolute surplus-value while the other European countries relied on the expansion of the mass of profit on the basis of technological modernization by an increase in the relative endowment of fixed capital, we can see that the transformation effort demanded by Fordism was a much too costly investment for Ansaldo, once the collapse of the *Biennio Rosso* had blunted the threat from the workers. Technological constraints deriving from the heavy engineering market were not the only ones limiting the firm's choices. The quasi-permanent crisis in the organization of work and social relations in the company, the persistence of certain practices involved in the handling of goods, machines and men, the management's imperialist and nationalist culture – everything combined to make the Fordist path a difficult one. Nonetheless, that path was always open and in time of struggle was boldly and openly urged by the group's leadership. But although the Perrone's strategy was receptive to mass production, it still thought in terms of an ideal of a closed market, and there was no complementary Fordist appeal to high wages and mass consumption.

The organization of work was modified under the pressure of criticism focussed on the risks of accumulation: "organisation" was in fact a key word and a word that was replete with a fundamental ambiguity. The issue of rationalizing modernity as a synthesis of procedures already present in American production organization or being introduced into it at the turn of the century suggests some more subtle questions. Ansaldo could not Fordize in the way Ford itself could. The persistence of global schemes for flexible production was a powerful constraint; certain options were only open sporadically and were very far from being the product of rational planned decisions implemented in accord with anticipated or predictable deadlines. If the slowness of the rationalization of production disciplines at Ansaldo is obvious, it cannot be measured solely against the criterion of the pioneering companies, for which the eponymous example in Italy would be FIAT. The microhistorical operation of industrial transformation was thus very complex. Management strategy was certainly important but it needs to be understood within the context of the respective positions of the managements themselves, and within the context of the company

[61] Cf. "L'organizzazione scientifica del lavoro in Italia", *Stato operaio*, Sept.–Oct. 3, 1929, 609–15, taken up by P. Fiorentini, "Ristrutturazione capitalistica e sfruttamento in Italia negli anni '20'," *Rivista storica del socialismo*, 1967, n. 30.

structure. This was the basis for an arrangement of possibilities that were relatively open at any given moment, that were in permanent interaction, and that were involved in the perpetual dynamic of the procedures put into practice by production.

7 "Have a heart for the manufacturers!": production, distribution, and the decline of American textile manufacturing

Philip Scranton

In standard chronicles of American industrialization, the textile industry appears in two roles during widely separated epochs. In the early republic, *c.* 1790–1840, it serves as vehicle for discussion of technology transfer and the rise of the factory system. Samuel Slater's Rhode Island alliance with the Almy and Brown merchant house brings mechanized spinning to the new United States, setting the stage for Francis Lowell's copying of British power looms, the establishment of integrated, capital-intensive production at Waltham, Massachusetts, and with the addition of Paul Moody's inventiveness plus the marketing connections of Boston investors, the elaboration of staple cotton goods production at the falls of the Merrimack River. The political dimension of the account involves Alexander Hamilton's promotion of national economic independence through state policies fostering manufacturing, the stimulus that Jefferson's embargo and the War of 1812 provided to textile entrepreneurs, the protection tariff of 1816 which favored domestic cotton goods, widespread ruin in the late 1810s as foreign goods flooded markets, and the politics of incorporation (as legislative acts had to be passed to charter each of the first generation corporate enterprises). The faltering steps of early promoters and politicians lead toward an heroic climax in the erection of the Lowell System, a comprehensive net of interlocked corporations controlling water power, machinery building and fabric construction by the 1830s. Technologically and organizationally sophisticated for their time, the Lowell mills set out a template for others to adopt/adapt as the railroad phase of industrialization opened.

Once the thematic center of the mainstream narrative turns to railways, steel, agricultural equipment, and machine building, textiles cease to be of interest, surfacing only in the early 20th century as an example of interregional competition, the evils of state regulation, and backward managerial practice. At the close of the 19th century, entrepreneurs in and near the southern cotton states mobilize local capital and northern credit/marketing facilities to create a growing array of rural staple-goods mills. Catching a technological breakthrough on the rise and hiring displaced white farm labor at just above the area agricultural wage, they mount a sustained challenge to veteran Yankee corporations. The latter, unable to generate decisive scale economies due to throughput obstacles, and unable to integrate forward into market and price management, are driven to attack labor rates, efforts strenuously if not always successfully opposed by northern trade unions. Southerners are immune from the union plague, and as well have few restrictions on hours of work, negligible taxes, etc., creating powerful comparative advantages that work permanent harm on the northern complex. When persistent overproduction drives prices downward in the 1920s, New England mills collapse serially and southern profits shrink painfully, a process intensified by the Great Depression. This sad tale provides a vivid counterpoint to the saga of consolidation and triumph in "second industrial revolution" sectors where du Pont, General Motors and General Electric stand as the hallmarks of managerial capitalism, and in which the United Auto Workers and other industrial unions of the 1930s exemplify labor's progress while the United Textile Workers carry the burden of dreams unrealized.

As it takes into consideration technology, labor, marketing, government, interregional investment and competitive dynamics, this account appears to be both empirically and conceptually satisfying, for it places textiles' fate within the larger process of the technical and organizational evolution of "modern business enterprise."[1] Yet it has profound limitations. Sectorally, cotton goods stand for the industry as a whole, despite the fact that cotton production displayed a steadily declining proportion of textile capital (1880, 51 percent; 1904, 49 percent; 1929, 27 percent, in constant 1929 dollars).[2] Moreover, this textile saga considers only the staple segment even of the cotton manufacture. The above capital figures include specialty firms as well, in damasks, coverlets, lace, upholstery and the like. This double compression, of all textiles into cottons, and of all cottons into staples only, indicates that in order to generate a more satisfactory, inclusive image of the textile industry's transformation, it is essential to move beyond received formulations. Only when batch-oriented specialty trades in and beyond cottons, in both weaving and knitting, and in those vital districts between Massachusetts and the South (the missing middle term spatially) are treated in tandem with the bulk system can the contrapuntal form of the industry's expansion and decay be adequately portrayed.

In doing this it may be helpful to think first about the range of parameters along which firms exercised choices that inclined them toward routinization of flexibility in production and distribution, recognizing that a set or sequence of decisions once made involve costs to alter, should there be shifts in labor, technological or market contexts. Second, the differential patterns of batch vs. bulk institutional behavior (inside the firm and in associations of workers, owners and distributors) may be assessed. These preliminaries lead toward an attempt to characterize the main lines of the course staple and specialty textiles took as general prosperity (*c.* 1900) was succeeded by dimming prospects in the 1920s.

Although the options would be more evident by mid century, as early as the

[1] Alfred Chandler, *The Visible Hand* (Cambridge MA, 1977).
[2] *Statistical History of the United States* (Stamford CT, 1965, Series P92–P99), p. 412.

1810s, prospective textile entrepreneurs could orient their efforts either toward staple goods or specialty product lines. Attempts at producing cotton sheetings or woolen blankets and broadcloths aimed at a broad market for low-end plain stuffs, with such tariff statutes as were in effect pushing importers towards higher-value-to-weight goods. The scale of the potential market gave choices toward power (water, steam) and mechanized technologies a competitive purchase, and as well fostered development of national distribution agencies in Boston and New York. By contrast, specialty lines had more forbidding features in ante-bellum decades. Making fine white flannels or wool-silk shawls (Philadelphia, 1820) necessitated a focus on up-scale, chiefly urban demand which generally favored imported goods.[3] Still, production of facsimiles of Continental style goods for consumers of the "middling sort" sustained some enterprises, while the relative expense of transport for carpets opened an avenue for others. Here modest and shifting demand for any item made worker skill and technical versatility more salient than powered equipment and emphasized the value of face-to-face dealings both with workers and wholesale/retail distributors, a posture consistent with product diversity.

The Civil War experience (1861–65) indicated that distinct staple and specialty complexes operated in the American textile manufacture. The sequence of routinizing choices had led major New England cotton mill corporations to so narrow their product options that the cotton famine thoroughly deranged their activities, whereas flexibility in fiber use and institutional practice permitted Philadelphia firms to switch to all wool and wool–cotton blends whose ready sale to military and civilian buyers fueled rapid wartime accumulation.[4] Thus, an accumulation of decisions toward extremes of flexible or routinized production limited the ease with which a firm might respond to altered environments by moving away from its settled practices. Another example of this phenomenon after 1900 may be drawn from the print-cloth mills of Fall River, Massachusetts. There through a series of technological and marketing choices, plus erection of huge, heavily capitalized mills and engagement of narrow-skill workers, firms became progressively immobilized within a tight alley of the cotton staple trades, unable without decisive reorientation to cope with interregional competition in their main output lines and the result, eroding profitability. Similarly, on the flexible side, modest-scale woven carpet mills in Philadelphia, whose strengths lay in the timely meeting of diversified demand, reliance on skilled operatives and a network of supply and service contracting with spinners and dyeworks, were stumped by the consolidation of carpet demand, style shifts to plain, weft-inserted goods (axminsters) and buyers' developing market power after World War I. Only industry majors had deep enough pockets to fund the machinery substitutions and marketing/advertising drives necessary to offset the squeeze. Thus the virtues of flexibility vs. routinization were situational, each

[3] Philip Scranton, *Proprietary Capitalism* (New York, 1983), pp. 95–121.
[4] *Ibid.*, pp. 272–298.

entailing limits of a different character to the options firms could envision and operationalize.[5]

Consider five dimensions of industrial practice that help define the contours of batch/bulk manufacturing in textile (and perhaps other) sectors: raw materials, products, labor process, firm organization and management, marketing. These of course bear on technology, labor supply, finance, and location and are interrelated with one another. Indeed it is the clustering of decisions across these dimensions that gives durable shape to the industry's two divisions, the breaking of such patterns that indicates substantial difficulties, and the insufficiency of both older and modified configurations that heralds an industry-wide crisis.

The simplest to appreciate is input materials, choices concerning fiber selection and purchasing practices. Firms which confine their operations to a single fiber are well, but not utterly, positioned toward routinization. (The extreme would be a single grade of a fiber, or a single count of yarn.) Before the synthetic era, such firms could gauge reports on agricultural cycles and purchase in bulk if supply contractions were anticipated, or in lots when surpluses loom, hedging against erratic price shifts and at times reaping opportunity profits in rising markets. Simplified needs and adequate warehousing facilities here generate a *managerial* flexibility while conditioning routinized production. On the other hand, the readiness to use varied fibers, separately or in combination, the ability to alter grades with price movements and/or shun a fiber entirely if price commands or demand solicits a change facilitates productive flexibility. Purchasing in lots is standard, with the extent of orders on hand governing the scale and timing of "buys." When raw materials are reaching celestial heights, profits may be enhanced by judicious substitution of shoddy for new wool, spun silk for tram and organzine, a downgrading move that is harder to disguise in staple cloths than in style goods. Flexibility in materials demands a deeper knowledge of both fibers and fabric structures than does the routinized approach, and is far less demanding of working capital, but practitioners are more exposed to the vagaries of supply markets. They carry the risk that they will not be able to pass along rising costs in product prices, and that falling supply prices will encourage goods buyers to demand downward revisions on fabrics. A practical gain is achieved at the cost of a financial hazard, the reverse of the situation in routinized work.[6]

[5] Philip Scranton, *Figured Tapestry* (New York, 1989), pp. 323–410; John Ewing and Nancy Norton, *Broadlooms and Businessmen* (Cambridge MA, 1955).

[6] Scranton, *Figured Tapestry*, pp. 140–142, 237–239, 400–401. For an Italian parallel in the Prato district, see Michael Piore and Charles Sabel, *The Second Industrial Divide*, New York, 1984, pp. 213–216. One feature of the technological choice between ring and mule machines for cotton spinning was the capacity of mules to use a wide range of cotton varieties whereas "rings could successfully spin yarn for any given fineness from only a narrow range of cotton grades." See Gary Saxonhouse and Gavin Wright, "New Evidence on the Stubborn English Mule and the Cotton Industry, 1878–1920," *Economic History Review*, 2nd series, 37(1984), 507–519 (above quote from p. 514), with a reply by William Lazonick, "Stubborn Mules: Some Comments," *ECHR*, 40 (1987), 80–86, followed by Saxonhouse and Wright's riposte (87–94).

When yarns are considered, another element of potential flexibility is added, the range of counts used, plus the mixing of differently based yarns in creating fabrics. For example, though lace mills relied on cotton exclusively, they purchased counts from 20 to 240 hanks to the pound, the higher, finer varieties being imported from Britain. This offered an enormous flexibility in product terms. Tapestry, upholstery and carpet mills regularly used cotton, jute and wool yarns for warp, with silk, cotton, worsted, and in time, rayon, for filling, radically extending the possibilities for final goods. An extremely rationalized firm, however, would for long periods require but one or two counts of cotton or wool yarn for staple print cloths or blankets. This divergence in turn held implications for firm organization, as staple mills could envision costs savings from doing their own spinning (integration) whereas diversified enterprises relied on a *network* of independent yarn mills (and brokers) to meet their changing needs, or at most erected spinning frames to process only the most frequently used yarn types (tapered integration).[7]

With regard to products, choices about routinization vs. flexibility may be taken at four levels, which will be termed divisions, lines, types and styles of product, each a subsection of the preceding category. Divisions indicate the major sectors of the industry; woolens, carpets, knitgoods. Lines refer to the principal product segments within a division; for carpets, ingrain, wilton, tapestry, axminster, etc., in knits, seamless hosiery, full-fashioned hose, knitted outerwear, undergarments, etc. Types indicate the subordinate components of a line; within ingrain carpets, all-wool, cotton-wool blends, three-ply worsteds and such; or for seamless hose, full length stockings for women, half-hose for men, women and children, each made from various fibers or combinations. Finally, styles are item-specific, the particular range of color, pattern in knit- or weave-formation, and finishing effects which is to be found within product types; for children's seamless half-hose – mercerized white cottons, embroidered blue and whites, knit stripes or argyles, and frankly, hundreds of others under this sub-sub-sub-heading alone. For final use divisions (carpets) and lines (most knitting), these styles were in addition made in various sizes.

Despite this panoply of openings for flexibility in production it is clear that the level of styles, demand for particular goods varied dramatically. A firm might readily sell five million yards of bleached staple sheeting in a year, while another would find takers for only a hundred dozen brown-and-tan cotton argyle socks. This range informed a spectrum of product strategies in the late nineteenth and early twentieth centuries. At the rationalized extreme, entire Fall River and New Bedford (Massachusetts) factories focused year in and year out on production of several basic cotton cloths for printing, the goods southerners rapidly learned to make cheaply. Somewhat more versatile, the giant Amoskeag corporation (Manchester, New Hampshire) ran scores of patterns in woven ginghams and Patterson (New Jersey) broad silk mills generated dozens of variations on the

[7] Hiram Davis, *Vertical Integration in the Textile Industry* (Philadelphia, 1938).

blacks silks standard for women's dresses (usually achieving these effects through finishing). In Philadelphia and Providence (Rhode Island), worsted, lace, carpet, and knitting companies routinely offered buyers 50 to 200 new style designs twice annually, a practice common among Paterson ribbon mills, Passaic fine worsted plants and Connecticut silk firms as well. To achieve this diversity, flexible producers integrated sample-making into their annual schedules, accepted the challenge of anticipating fashion trends (as interpreted by clothiers and store buyers), and operated overtime in "the season" between slack periods.[8]

These product orientations were linked to different sets of management practices. Mills that "specialized" in a cluster of staple styles or types produced for and managed inventory, whether held by the factory or its distribution agency, and sought to operate at or near fully capacity around the calendar. Batch mills knit or wove their array of styles to order so as to diminish the risk of holding "dead stock" at the season's close. Whereas styles changed relentlessly and types shifted with movements in materials prices, lines were usually keyed to machinery. Ingrain looms could create an indefinite number of styles based on a range of fiber mixes (types), but they could not produce wilton or axminster carpets. Thus possession of a mixed battery of machines, and the addition of those designed for new lines of goods, characterized firms flexible at a significantly higher level than pattern or fiber. By contrast, firms hewing to a narrowly-defined style span, and routinized for its speedy manufacture, tended to follow only such technical vectors as furthered processes thus defined. Standardizing process and product yielded mid-range predictability, but as well discouraged adoption of technical advances that entailed disruption of settled balances between production segments (spin/warp/weave/finish) or settled relations between managers and either worker or distributors. Such rigidities were most evident in the case of the regionally-differentiated adoption of the battery loom, *c*. 1900, a matter to which we shall return.

At the opposite pole, operations that were diversified at the division level represented the height of flexible practice, and necessitated capital resources far beyond those available to versatile 50-loom mills. Two examples may suffice. In 1910 Philadelphia's J. & J. Dobson commanded a half-dozen mills that carried out production of carpets, woolens and worsteds for men's wear and women's dress goods, yarns for use and for sale, plushes for upholstery, and "hygienic fleeced underwear" as well.

Reported capital was $9.2 million. While the Dobsons added new departments to their original lines, across town the Bromley family deployed its accumulated capital (from success in carpets) for start-ups in lace (five firms, 1895–1910) and full-fashioned hosiery (six firms, 1916–37), moving shrewdly and early across divisional lines into growing specialties.[9]

Product flexibility at the level of styles and types entailed a close reading of

[8] Scranton, *Figured Tapestry*, pp. 74–79, 174–179.
[9] *Ibid*., pp. 33–35, 136–137, 260–265.

market potentials, the absence of risk aversion, ownership of general purpose machinery, and access to skilled labor that could readily adapt to frequent product shifts. Sheer speed in manufacturing was less critical than timely response, at least through World War I. With initial profits, such a firm could, but surely did not have to, proceed toward diversification at the lines level, or toward partial integration, without unduly stretching its material or personnel resources. However, few companies moved on to multi-divisional flexibility or sequential shifts on a scale as grand as the Bromleys and Dobsons. In addition to the above requirements, such ventures had to be based on deep capital reserves or extensive credit sources, the capacity to absorb the costs of learning to operate in new markets, a supply of workers with appropriate skills (as lace-twisting and knitting are sharply different from weaving, and from one another), as well as the ability to hire experienced supervisors and school existing partners/managers in the rigors of the division to be entered. These were daunting obstacles in the early decades of the twentieth century; understandably, the birth pangs of the cross-divisional flexible company were just as painful as those of the multi-divisional mass production enterprise, or more so. Each was pressing the logic of its approach toward wider spheres, with, given shifting contexts, quite different results.[10]

Labor process options may be approached by considering the relative significance of skill and speed in production, issues that call attention to technology, labor supply and gender, location and shop-floor authority. Firms which devote their efforts to staple goods were concerned to augment their output and cut costs, spreading the overhead "burden" as widely as possible in price-competition types (sheetings, etc.). A tight product focus encouraged technical refinements that increased loom-speed, provided more reliable and efficient power drive, and minimized down time in production.[11] In addition, routinization of production (rather than refinement of the division of labor) made it feasible to engage unskilled women and immigrants at quite modest wages, for the learning curve was quickly mastered and workers were readily

[10] Crossing divisional product lines within sectors was practiced beyond flexible textile operations. In the same era this was a frequent tactic in the metalworking trades as well. For example, Cleveland's Warner and Swazey was involved in constructing both machinery and astronomical equipment, whereas Philadelphia's James Queen and Company created scientific and educational "apparatus of all descriptions – of Glass, Wood, Iron, Brass, Alumium, Platinum, etc.," having by 1900 five area factories for instruments and microscopes, optical work, glassware, woodworking and "high vacuum" jobs (on "X-ray tubes"). See Warner and Swazey, *Some Astronomical Instruments* (Cleveland, 1909), and Deborah Warner, "James W. Queen and Co.," unpublished paper, National Museum of American History, Smithsonian Institution, 1989. In brass and bronze, it was normal for midsize and larger firms to cover the full spectrum of alloy mixes and produce as well hundreds of varieties of sheet, bar and wire goods, along with special orders. See James Morrison, "Management in the Brass and Copper Industry," *Management Engineering* 2 (January–April, 1922.)

[11] Louis Hunter, *A History of American Industrial Power: Water Power* (Charlottesville VA, 1976); William Mass, "Technological Choice and Industrial Relations," unpublished PhD dissertation, Boston College, 1985.

substitutable. Disproportionate employment of women as weavers and spinners, and men as overseers and second hands allowed gender-based authority relation to reinforce formal, rule-bound hierarchies, arbitrary systems of fines, and other disciplinary controls. The low entry barriers to employment and availability of steady work made staple mills magnets for both in-migrating rural labor and in the post-1850 North, successive waves of immigrants as well. The only durable union presence in the late nineteenth century was among mulespinners in New England, chiefly of British origin, and their oppositional stances provoked corporations to adopt improved ring-spinning machinery (post-1885), displacing quarrelsome and organized men to make way for more "reliable" women frame-tenders.[12]

By the 1890s, incremental efforts to increase speed and cut unit costs technically had about reached their limits in bulk cottons. Hence, the significance of the automatic, battery loom, introduced by the Draper machine-building firm, can be appreciated; for the new bobbin-changing device and associated stop-motion accessories in theory permitted a weaver to attend banks of 16, 32 or more looms rather than the customary six or eight, radically shrinking the per yard labor expense. As its installation spread and piece rates were pared, workers walked miles daily, patrolling the aisles between loom rows, tedious rounds punctuated only by periodic strikes against extensions of loom assignments and rate cuts, both of which symbolized the degradation of labor. Speed-up and stretch-out became the curse of the staple trades, and the stimulus to renewed labor organization efforts, in the form of both localized independent unions and the bureaucratically-managed United Textile Workers (AF of L). When near the turn of the century some New England corporations ventured beyond staples toward more flexible product mixes, positions in mill rooms making lawns or fancy worsteds were both highly sought and often vacant (highly sought, as they offered more interesting and less physically taxing work (4–6 looms) but often vacant as locally available weavers with sufficient skill to handle complex weaving were in short supply and firms declined to hike rates to levels that would draw outside talent. In dedicated bulk mills, as loom-fixers (mechanics) were charged with all aspects of machine maintenance, adjustment and repair, stretched-out weavers' technical skills either atrophied or were never developed. Thus their performance in more demanding tasks was ragged, at least initially, a problem that plagued Fall River firms.[13] In this sense, a routinized labor process achieved a certain flexibility in that workers were easily replaceable

[12] It has long been reported that militancy among mule spinners, led by British immigrants, prodded southeastern Massachusetts cotton mill corporations into concentration on ring machinery purchases after 1880. In light of the revived evaluations of these rival technologies, more technically and economically sophisticated than earlier work, the time may be ripe for investigative reassessment of this linkage. William Lazonick's work on Britain and the US has been central to these exchanges, and full bibliographic information may be gleaned from the articles cited in note 6 supra.

[13] John Cumbler, *Working Class Community in Industrial America* (Westport CT, 1978); Thomas Smith, *The Cotton Textile Industry of Fall River, Massachusetts* (New York, 1944).

and in a poor position to resist technical changes that intensified their labor. However, the trajectory of labor process shifts had a self-blocking dimension, for both skills and machinery in place were weakly positioned to adapt fluidly to profit squeezes in staple product markets. Moreover, bulk mill locales, dominated by clusters of large corporations, were unwelcome environments for mobile skilled workers and prospective "small" entrepreneurs in textiles.

By contrast, firms oriented toward style goods articulated a quite different labor process, in which skill and speed were differently construed. The alteration of products (and in dyeing, contract services) both demanded and reinforced skills of a high order. Yet speed had more to do with situational timing, quick responsiveness to seasonal rushes, than with systematic attention to throughput improvements. In order to get the goods "out the door," employers long acknowledged male workers' claims to considerable shop-floor autonomy, dealt routinely with shop committees and delegates, and consulted with weavers, spinners and knitters on tactics to overcome production snags ("kinks"). In consequence, the social relations of production were personalistic, and highly varied in content, rather than formalized in layers of corporate authority. The rough mutuality of this labor process excluded women,[14] as it was male bonding that in part conditioned the porosity of boundaries that were vigilantly guarded elsewhere. Indeed, the contested character of factory social relations, which I have elsewhere termed "fraternal paternalism,"[15] often mirrored father–son contentiousness yet coexisted with repeated tales of proprietors rendering assistance to their own factory "graduates," workers starting "on their own account," and taking pride in the latters' successes.

Piece rates and trade customs were frequent bones of contention, as they both encoded the claims and privileges of skill and hazarded profits in slack seasons. Still, properly calibrated lists and shop customs faithfully observed had their benefits for owners as well. The "bill of prices" set out rates for every class of work the mill undertook, with "extras" for special variations and "time work" pay for sample-making and attempting unprecendented styles, all arranged so that shifts from quick, simpler constructions to slow-running fancies would yield comparable earnings for comparable effort.[16] If the list was deemed "fair," earnings (and output) differentials reflected individual "efficiency" in applying skill to the tasks at hand, and within bounds were accepted as such by workers. Moreover, workers' proprietary "right" to "their" looms or knitters, frames or vats (a standard strike settlement term) together with customary expectations that weavers, etc., knew "their" equipment intimately,

[14] Gladys Palmer, *Union Tactics and Economic Change* (Philadelphia, 1932). See also, Commission on Industrial Relations, *Final Report and Testimony* (Senate Document No. 415) (Washington DC, 1916), pp. 3036–3041.

[15] Philip Scranton, "Varieties of Paternalism," *American Quarterly*, 36, 2 (1984), 235–257.

[16] Scranton, *Proprietary Capitalism*, pp. 364–365. For a reproduction of a full price-of-work list, see Scranton, *Figured Tapestry*, pp. 279–281.

quickly detected mechanical problems, and performed most maintenance (so as to keep running with minimal delays), all served to smooth production and fuel accumulation. Indeed, frequent calls for the "fixer" were judged a sign of marginal competence, and repeated production of "bad work" was a cause for dismissal, a step shop committees assented to only after examining spoiled goods.

The machine technology that informs flexible production has not been the subject of in-depth study, but in textiles both *in situ* arrays and vectors of development differed significantly from bulk production practice. Rather than the fast, simple (and relatively lightweight) looms devised for staple cottons, woolens and silks, flexible mills held mixes of slow, but versatile, 12–20 harness looms, dobby-chain and Jacquard-mounted devices whose pattern "reach" was extensive. Similarly, seamless hose and knit outerwear factories possessed a congeries of machines of different capacities and vintages, well into the 1920s, as a Federal Trade Commission survey showed.

Consistent with Nathan Rosenberg's theses, it seems that technical change was incremental and relatively continuous, rather than having breakthrough, "big bang" episodes, and was replete with creation of specific-use gadgets, only erratically patented, that modified existing machinery to enhance flexibility or quality.[17] Institutionally, it was the Crompton and Knowles loom works that produced a remarkable series of adaptable looms for carpets, silks, woolens, etc., not the bulk trade's stalwart Draper Company. And it was Crompton that established a Philadelphia branch plant, placed sample looms in mills to test their utility, and contracted to recondition and update old equipment, adding selvedge guides or stop motions, or converting it for different product lines. Demand constraints surely were central here.[18] Radically segmented product markets marked by rapid style changes made the invention and diffusion of gadgets more salient than the introduction of quantum-leap systems. In 1930 a Wharton school researcher noted that Jacquard mechanisms had undergone no significant change in the previous quarter century.[19] Meanwhile one decisive flexible technology breakthrough, the multi-section full-fashioned hosiery knitter introduced in the 1890s from Germany, experienced a 25-year lag until the demand pull of the silk hosiery craze triggered massive installations (9,000 units) in the 1920s. Another, the Raschel knitter (for outerwear) arrived in the twenties at a point when flexible firms were in deep trouble and was never widely adopted.

Though these reflections are preliminary, they encourage serious consideration of the notion that there are "multiple paths of technological change" and the related "localized learning-by-doing argument,"

[17] Nathan Rosenberg, *Inside the Black Box* (New York, 1982), pp. 62–70.
[18] J. Schmookler, *Invention and Economic Growth* (Cambridge MA, 1971).
[19] Industrial Research Department, University of Pennsylvania, "Final Report to the Manufacturers and Union Members of the Philadelphia Upholstery Industry," typescript, 1931, in Gladys Palmer Papers, PP–135, Urban Archives, Temple University, Philadelphia.

in which decisions concerning techniques influence the later learning process. Thus the decisions turn out to be far more fateful than is usually recognized because they set in motion a long-run evolutionary process linking factor prices, the choice of techniques, and the direction of technological change.[20]

In flexible production, technical change may be expected to favor skill-conserving innovations, rather than skill-displacing ones (and for a start, to have an immediate, local audience for diffusion, especially in districts of agglomerated flexible mills) and to be predominantly the incremental results of "tinkering" by skilled workers, proprietors, fixers and mill machinists. In any event, the labor process in specialty textiles was far too skill-centered and intricate for crude productivity techniques like stretch-out and speed-up or transformative inventions like the battery loom to be plausible or profitable in use.

In considering production organization and management, bulk manufacturers and flexible firms clustered their choices along broadly divergent paths. Consider issues like the extent and character of integration, finance (sources and uses of capital), site selection, incorporation, managerial expertise, and cost-finding/accounting. The hallmark of the staple textile in the North was the fully integrated factory. There a narrow product range allowed estimation of long-term needs for capital, plant and equipment. There as well the goal of steady, minimum cost production could be met readily by balancing machine sets for opening, processing, spinning, warping, weaving, and finishing so that the flows from each met the capacity of subsequent sections. The space and scale requirements for such enterprises generally entailed that their mills be purpose-built, in turn necessitating substantial fixed capital investments. These funds were raised through share sales in incorporated companies, with purchasers' assets protected through legally limited liability and major shareholders assured a policy role through their presence on boards of directors. Working and expansion capital was secured by means of short-term loans (often from individuals), advances from distributional agencies, and issuance of either additional stock, or in the post-1890 era, corporate bonds. These debts, and the expectation of steady dividend payouts, heightened the fixed costs firms had to meet and reinforced the need for full-bore production. Staple goods makers were positioned as price-takers in competitive markets, a situation that had two principal consequences. Firstly, managers, who by 1860 were mill-experienced "professionals" for hire, focussed their attention on creating techniques for cost-finding and control, so as to deal with the steady deflation of market prices.[21] Second, treasurers, the corporate CEO-equivalents and chief materials buyers, were charged with the task of shrewd fiber stock purchasing. Here the saving of half a cent per pound could across a year's time spell the difference between healthy and marginal profits. This imperative drove working capital needs and fed into

[20] Rosenberg, *Black Box*, pp. 16–17, citing and building upon the work of Ruttan, Hayami and David.

[21] Alfred Chandler and Richard Tedlow, *The Coming of Managerial Capitalism* (Homewood IL, 1985).

the calculative hierarchy of such firms. Finally, location decisions were based initially on power availability and later on transport factors. Pre-steam "big mills" sought sites with water power potential substantial enough to drive sizable integrated facilities (Lowell, Manchester, etc.) whereas post-Civil War development (Fall River, New Bedford) found coastal sites advantageous for cheap handling of raw cotton and finished goods.

By contrast product differentiation had distinctive implications. As firms could hardly anticipate the direction their product lines would take, it made little sense for them to erect fully-integrated plants. Hence the typical flexible mill was a partial-process establishment, devoted to a single stage in the overall production sequence yet imbricated in a network of related firms whose collective activity delivered goods to commercial outlets. For example, few specialty knitters spun their own yarns or handled dyeing. Instead, such firms kept their attention and their capital focused on knitting fashionable styles. Locationally, this pattern favored urban settings, within which classic external economies were available (Philadelphia, Paterson, Providence). Such industrial districts offered wide choices for contracting and as well served as magnets to draw an influx of skilled immigrant workers. Incorporation was long shunned by new-starts, for until late in the nineteenth century, it demanded special legislation approval and was costly, both forming serious obstacles to small entrepreneurs. Indeed, whether by design or as a reflex, specialist proprietors developed an anti-corporate rhetoric that stressed the honor of owners and partners, individuals whose word was backed by the exposure of their personal assets to creditors, as opposed to the insulated trickery of agents and managers who, not being "interested" in their firms, placed only their jobs or share-packets at risk.[22]

As this example suggests, specialists as managers emphasized "hands-on" guidance of production, based on a fundamental technical competence that was the mark of a "practical manufacturer." Direct relations with workers and work in process involved routine tours through the mill, in which personal knowledge of employees ("first name basis") reinforced mechanical experience. Hiring remained the province of foremen and room bosses, and unlike staple mills, was not generally transferred to personnel managers/offices in the twentieth century. Grievances unresolved on the shop-floor were taken up directly by owners, not filtered through bureaucratic layers or ignored. Skilled workers' stand-downs in the season were potentially too costly to be courted by managerial indifference. The "productionist" focus and sub-sectoral specificity of management skills, together with the proprietary nature of ownership, erected barriers against the development of generalized principles of administration. Practices in accounting, pricing, etc. were firm-specific and based more on rule-of-thumb customs than on formal investigation and dissection of quantitative indicators. Hence, turn-of-the-century advocacy of "standard costing" and financial/managerial

[22] Philip Scranton, "The Politics of Production," unpublished paper, Mellon Seminar on Technology and Culture, University of Pennsylvania, 1989.

accounting was dismissed as irrelevant to the unpredictable variations of flexible mill life. "Efficiency men" were derided; the claims to expertise that new-moded hired managers promoted threatened to devalue the craft practice of "practical" millmen and were rejected. In consequence, specialty firms had little of the financial sophistication that was common among staple goods corporations, developed across generations of experience with a shared, stable product and process base and through treasurers' contacts in associations (the Arkwright Club, National Association of Cotton Manufacturers, etc.). Even in terms of production management alone, Taylorism and time-and-motion studies made few inroads until the 1920s, and then chiefly in specialty sectors that had developed relatively standardized elements (full fashioned hosiery) and at labor's not management's initiative.[23] Not only was the "fit" between such techniques and the discontinuous nature of flexible practice quite weak, the attitudinal dispositions of manufacturers hostile to "outsiders" and their "smart ideas" remained fixed, even or especially amid a developing crisis.

Not surprisingly, the financial practices of batch textile mills were at a considerable remove from those of the bulk system. Profits were not paid out to shareholders, but were instead divided between surplus accounts and the private takings of partners. Fluidity in the boundaries of these categories enabled plowbacks into expanded capacity, machine substitution, or plant construction, as "interested" parties could be expected to supplement surplus balances with personal assets when needed. The recruitment of silent partners to add funds for major shifts long took the place of ventures into capital markets, which of course would have necessitated incorporation and dealings with powerful outsiders (securities brokers), paths rarely taken until the twentieth-century's third decades. (Before the merger era of the late 1890s, there was an extremely narrow market for industrial securities in the US, in any event.) Working capital was not

[23] Palmer, *Union Tactics*, part III; George Taylor, *Significant Post-War Changes in the Full-Fashioned Hosiery Industry* (Philadelphia, 1930). Though there is little evidence of their adoption by flexible textile firms, cost accounting systems tailored to variety production were inaugurated in the early twentieth century. Often focused on establishing "machine hour" rates to supplement labor and materials expense figures in calculating costs for individual items or short runs, they were much discussed in the machinery and machine shop trades. For the fullest development of these efforts to avoid arbitrary loading of averaged overhead charges on products which demanded sharply different combinations of power and space usage, selling expense, etc., see Alexander Church, *The Proper Distribution of Expense Burden*, New York, 1910, and *idem.*, *Production Factors*, New York, 1918, along with Clinton Scovell, "Cost-Accounting Practice, with Special Reference to the Machine Hour Rate," *Journal of Accounting*, 17 (1914), 13–27. For a re-evaluation of Church's work and the problems arising from the ultimately dominant inventory-based approach to accounting, see H. Thomas Johnson and Robert S. Kaplan, *Relevance Lost: The Rise and Fall of Managerial Accounting* (Cambridge MA, 1987). The labor and expense necessary to gathering and analyzing production cost details on multiple intermediate and final goods, along with the frequency of shop errors in or omissions of relevant data, were often cited as limitations of these techniques which were intended to inform pricing for profitability. The authors of *Relevance Lost* argue that computerization and monitoring systems make reasonable their reintroduction for the 1990s.

sought from distributors (this put the mill "in the hands" of a selling agent), but was derived from surplus accounts and personal infusions, until after 1880. Manufacturers then commenced the formation of banks in their districts that assisted in processing the steady flow of promissory notes and accommodation paper that also eased short-term, seasonal cash flows. Here was yet another expression of the drive for autonomy that fed textile entrepreneurship, for in their own banks' boardrooms millmen sat in judgment on the value of their own "paper" and that of their colleagues, rather than being subject to the disinterested standards of traditional bankers. These institutions provided as well as a different avenue for plowing back profits, here in a fashion that collectively benefited the industry, and permitted their manufacturer promoters to mobilize the savings and deposits of district workers and shopkeepers so as to oil the mechanisms of the neighborhood economy.[24]

What options for marketing their goods did textile companies consider? How were choices mapped across the range of production orientations? In the last half of the nineteenth century, two avenues were readily available: engagement of a commission house or selling direct. In either case, one's customers were a shifting mix of wholesalers, jobbers, downstream manufacturers (cutters-up in apparel), retailers, and eventually, converters, who bought staple goods for later printing, dyeing and finishing (first in cottons, and later in hosiery and rayon broad goods). Major commission houses dated to the early Lowell years, and tended to serve bulk manufacturers, taking their entire output, unloading it at the going price, advancing part of the anticipated receipts on delivery of goods, all for a commission on the final billings, plus interest on advances. Not infrequently, commission agencies held stock positions in integrated mills, binding producers and distributors closely together, giving the latter considerable influence over the former. Dependence on advances to settle wage and materials accounts pushed mills toward maximum output even when market prices were disappointing, a practice that exerted a collectively deflationary pressure.

At the other pole, selling direct involved taking individual responsibility for

[24] Banks created by industrial proprietors to service their own credit needs and provide magnets for the savings of industrial district residents were common in American cities, *c.* 1880–1930. A dozen such enterprises were started in Philadelphia alone. Their operations were sustained structurally by the state-level organization of American banking, related restrictions on branch establishment (both of which constrained competition), and availability of legitimization through meeting federal criteria for "national bank" status. Processually, they employed networks of trade contracts to manage risks in the 1890s (Scranton, *Figured Tapestry*, pp. 90–94), and were invigorated by buoyant urban industrial economies from the late nineties through World War I. During the 1920s, however, clusters of local industrial banks that were significantly involved with the fates of manufacturer directors experienced substantial losses in crisis-prone districts. Before the general banking crisis of the 1930s, such institutions were serially merged with more diversified (and generally older, larger) regional banks in order to avert failure. Philadelphia National Bank absorbed at least five of them by 1929. Many of those still free-standing collapsed during the Great Depression. See Philadelphia National Bank Collection, Accession No. 1658, Archives, Hagley Library, Greenville, DE, and J. E. Holdsworth, *Financing an Empire: The History of Banking in Pennsylvania* (Chicago, 1928), IV.

reaching purchasers, a task of extreme difficulty in national markets for staple goods. However, by the 1870s specialties flowed heavily through the New York commercial nexus (later Chicago would become a secondary center). Batch operators' early practice of carting samples of seasonal styles about the New York wholesale district devolved into more or less formal "openings" by the 1890s, week-long expositions to which buyers flocked from far points twice yearly to compare goods and prices before placing orders that would be run and shipped. To reach distant jobbers, retailers and cutters outside the market melee, a firm member or employee might take sample cases (or trunks for carpets) on the road, seeking "fill-in" orders that could round out seasonal capacity use. Though not likely less expensive than commission operations, selling direct gave the mill control over pricing (and adjustments) as well as a quick, face-to-face means by which to assess the viability of current styles. In between was the domain of agency work, a form which expanded after 1890. Firms might establish a permanent office in New York to service accounts, receive buyers privately at openings, and gather intelligence on the latest style trends. Were resources insufficient to maintain such an establishment, they could contract with an agent who handled several, non-competing lines for resident New York representation, and perhaps also arrange for agency road men, who carried one mill's upholsteries, another's damasks and yet another's laces into separate territories. Staple goods mills that chose to meet declining margins by moving gradually toward new product lines added such agencies to their marketing plans, and valued their advice on style movements. Similarly, flexible mills might sell direct at openings and engage agency men, rather than their own staff, for road work. In this context, the traditional commission house went gradually into eclipse, to be replaced by the converter of staple goods. Openings and road selling, on the other hand, diminished the importance of New York general wholesalers in specialty trades, for by 1900, batch firms could easily reach cutters small and large as well as far-flung jobbers and retailers. Each shift had important consequences.[25]

Commission houses did not take title to the goods they marketed, and thus all cost/price risks were borne by mill corporations, for periodic gluts and demand ebbs penalized full-running producers while commission houses collected transaction fees whether sales were profitable or not. Further, interest and storage/insurance charges ate away at the returns expected beyond pre-sale advances, especially when final prices faded. These expenses and uncertainties were sharpened both by the long 1890s depression and the rising volume of southern staple goods in cottons. Enter the converter. This intermediary purchased unfinished "gray goods" for cash or on short terms (30–60 days), then contracted for further processing (especially printing patterns for women's wear), selling the fabrics ultimately to cutters-up. Initially involved in marketing the output of southern mills, New York converters became steadily more

[25] Paul Cherington, *Commercial Problems of the Wool and Worsted Industries* (Chicago, 1916); "Final Report . . . Philadelphia Upholstery Industry," Palmer Papers; Scranton, *Figured Tapestry*.

prominent as they obviated the risks and dependencies commission house relations entailed. Houses linked powerfully to Yankee boards of directors kept some mills captive, to be sure, but such ties only deepened their long term difficulties. By the 1930s, these tandems had collapsed into insolvency, whereas converter activity came to dominate staple marketing.[26]

In specialty lines, bypassing the wholesaler to sell yet-more-direct to retailers, jobbers and cutters unleashed an omni-directional tug of war. Manufacturers perceived situational advantages in dealing with downstream buyers, only in part offset by the costs of reaching them and of processing a welter of smaller orders than general wholesalers had customarily placed. These expenses, it was expected, would be covered by mills realizing wholesale prices (rather than the lower mill-to-wholesaler rates) on such sales, but their new customers likewise sought price gains, anticipating that eliminating the wholesaler would reduce fabric prices to the mill rate. Wholesalers claimed their service capacities for both parties were being undervalued, and delighted in the tangles between mills and retailers/cutters even as their clientele decreased. Though it deserves more careful study in its own right, the outcome by the 1910s of this shift was an increasingly complex and rancorous marketing "system" in which general wholesalers were marginalized, and specialty jobbers came to play a vital role alongside the departmental buyers for great retail establishments and chain stores. Contention over price and trade practice swelled to a climax in the 1920–21 "inventory depression," whose outcome set the seal on transformations in market power relations that had been in process for two decades. We will return to this matter.

This thematic depiction of the divergent courses of bulk and batch, staple and specialty, routinized the flexible textile manufacturing represents a schematic conceptualization of the larger movements in this trade's industrialization. Yet insofar as it has attempted to expose structural and processual distinctions at several levels, this presentation has elided the question of narrative exposition and has focused more on polar opposites than on a sequential spectrum of patterned choices.

The second main section of this essay will thus present a narrative that sketches the historical course of the US textile industries, *c.* 1880–1930s, deals with the associational activity of business and labor and their conflicts, and to a degree fills in the gap between extremes of staple and specialty production.

For roughly two decades after the American Centennial (1876) the nation's textile industry conformed to the bi-polar segmentation detailed above. On one hand, a substantial group of firms committed itself to bulk manufacture of staple

26 For contrasts between this process and the advance of direct selling in staple food and toiletry lines, see Susan Strasser, *Satisfaction Guaranteed*, New York, 1989. For canned goods, soap and razor blades, heavy consumer advertising of brand name items proved feasible, but this was implausible for intermediate products such as cotton sheetings or print cloths. Several (chiefly southern) producers moved in the 1920s toward establishing final-use identification for sheets and towels, following the mass advertising path with fair success (Roland Marchand, *Advertising the American Dream*, Berkeley CA, 1987, pp. 123–127).

cotton and woolen goods. Initially concentrated in New England and rural towns across the mid-Atlantic states, this phalanx was augmented after 1880 by cotton mills scattered across the southern Piedmont. On the other hand, a second array of companies handled the batch production of seasonal and specialty textiles. These firms were spatially centered on either flank of New York City, especially in New Jersey and the Philadelphia area and to a lesser degree in Connecticut and Rhode Island. Whereas competition *among* firms within trade divisions of each segment was often intense, market contests *between* the two were scarcely imaginable. They were separate along other avenues of potential rivalry as well. In materials, the supply of carpet wools and knitting yarns was of little moment to thousand-bale purchasers of raw cotton and middling grade wools for print cloths and blankets. Their capital needs were structured differently, and their labor forces were far from interchangeable. Despite sharing the same industry heading, the rationality underlying accumulation was different, focused respectively on price vs. product competition, fast vs. versatile technologies, capital- vs. skill-intensive production systems, homogeneous vs. particularized demand, etc.

When markets contracted, in the recessions of the 1880s or the general slump after 1893, each of the two sorts of firms responded in ways that derived from their formats for production. Routinized bulk firms continued to run full for as long as possible before suspending, accumulating stock of their standard types, shading prices to protect market shares, a strategy related to their need to cover fixed costs, their reliance on distributors, advances, and the absence of easily implemented, flexible alternatives. For their part, batch firms had a horror of running stock in seasonal styles which might be unsalable when markets revived. Instead, when buyers ran for cover, they promptly went on short schedules, filling only such orders as dribbled in, defended price levels and set about fashioning samples of new items for the coming season. Thus in the depression of the mid-1890s, established Philadelphia firms with low fixed costs survived by scrambling after scattered orders, laying off workers and calling them back as lumps of trade were secured. In a deflationary environment, rate-cuts were negotiated at the firm or sub-sectoral level in a haphazard fashion, but short work and lowered pay did not generate labor migration. Alternate sites offered no better prospects than did riding out the slump. Firms newly-started in the run-up to the 1893 crash did fail, but the core networks rested intact when revival developed in 1897–98.[27]

Hard times brought two lasting effects to flexible trades, however. In depression, buyers consistently sought longer terms for payment and lower goods prices, pressures manufacturers resisted in better seasons. To secure orders, millmen compromised on price and terms, offsetting the costs concessions entailed by "fiddling the goods," using their skill to substitute cheaper materials (shoddy, low-grade cotton) or shaving the number of threads per inch in weave formations. These expedients, aimed at preserving some measure of

[27] Scranton, *Figured Tapestry*, pp. 136–170, 226–227.

profitability, made the labor process more difficult, as low-end materials spun badly, the resulting yarns broke frequently in the loom, etc. Piece-workers found earnings thus further diminished, and often demanded time-rates for such "filthy" jobs, which injured both their pay-packets and their pride in craft. As one proprietor later observed, the depression forced American specialists to learn to "manipulate" their goods, a dubious, but durable achievement. Second, in market relations, the practice of placing sizable orders at seasonal openings by a "sample and duplicate" pattern. Cautious buyers now purchased single "cuts" of various styles (24–50 yards), made apparel samples from each variety, then forwarded larger duplicate orders for only those that appealed to their clients. This added expense for mills (no money could be made in running single cuts) and lengthened the season, as sample orders had to go out early enough that duplicates could be run and delivered in time for clothiers to meet seasonal dates for shipment to intermediaries and retailers. Prosperity brought an end to piecemeal buying in furnishing fabrics and knits, but the sample and duplicate pattern persisted among clothiers, to the annoyance of millmen.[28]

Among staple mills, the 1890s depression intersected with three developmental vectors: (1) the rapid growth of bulk-oriented cotton corporations at Fall River and New Bedford, where capacity reached 180,000 pieces weekly (7.5 million yards) by 1893; (2) the steady deployment of southern staple mills, and their capture of the coarse goods market; and (3) the displacement of woolens by combed yarn worsteds.[29] During the slump, Yankee cotton mills were pinned between overhanging and depreciating inventories and high fixed costs, a situation aggravated by southern competition in low grade lines and aspirations toward production of finer staple cloths. Four courses for action were open for discussion. Individually, mills could undertake to increase their cost-efficiencies, in the short run by paring staff and installing rate cuts, in the longer run by adopting the new battery looms, and other technical improvements, stretching work assignments, and refining accounting practices. Second, retiring old equipment to shrink capacity, compressing financial obligations as well as adopting cost controls. Third, firms might associate to defend prices, set common labor policies, and act in concert to manage inventories. Last, enterprises could diversify either spatially or in product terms.

Each approach found favor in some quarters. Rate-slashing was general, but was soon perceived as self-lacerating in the north. Southern mills could retain their regional wage advantage with parallel reductions that had only to meet Piedmont agricultural wage levels, whereas the pre-depression state of staple mill worker earnings in New England had already aroused the ire of political reformers and regulators. Technical shifts, particularly towards the "automatic" battery loom, mandated sizable capital expenditures in difficult times, a step taken far more typically by southern firms, and one that added to their long-term

[28] *Ibid.*, pp. 237–239.

[29] Cherington, *Commercial Problems*; Melvin Copeland, *The American Cotton Manufacture* (Cambridge, MA, 1913).

prospects. Stretch-outs on older or slightly modified equipment (warp-stop motions) brought sharp resistance from workers, especially in the Massachusetts coastal centers. Hence mergers and associated actions were hot items, along with branch-planting and to a lesser degree, product diversification.

In the climate of the merger boom, *c.* 1895–1904, scarcely a season went by without news of a fresh textile promotion. Few came to fruition, and of those that did, only one (American Woolen) ever prospered. The reasons for this dead-end are not hard to find. Firms vastly inflated their valuations in merger scheming; the proposed company would be so loaded with fixed obligations that only a near-monopolistic pricing potential could have sustained them. Corporate leaders scrambled for position in the organizational plans, sharpening personal rivalries that blocked action. Merged firms (in cotton duck and cordage) soon discovered that in staple textiles scale economies beyond the single large mill were systematically elusive, while their inability to gain control of the bulk of sectoral capacity defeated price management.[30]

In association, mills could collaborate on policy initiatives without experiencing the hazards of merger. Most impressively, at Fall River corporations created a "selling committee" that coordinated production and pricing of print cloths in grades not yet widely made elsewhere, and extended their alliance into labor matters through implementation of a "sliding scale" for piece rates that rose or fell with the margin between materials costs and market prices (*c.* 1896–1904).[31] Labor's demands for substantial increases after 1900 and some mills' willingness to grant them rather than take strikes in revived markets brought the system to an end.

Spatial diversification was a strategy individual firms adopted in order to compete with southerners on their own ground. It was more widely heralded than practiced, but several major Lowell and Lawrence corporations (the Massachusetts and Pacific) erected Dixie subsidiaries, sending waves of panicky reportage through the trade journals. The apparently steadier course of markets early in the 1900s blunted this direct outflow, but two decades later, the Pacific's southern mills continued to show profits, while its original New England plants were awash in red ink. Product shifts were in general much more quietly handled. Outside Fall River, which hewed to the bulk staple track, northern corporations began in the 1890s to experiment with running specialty goods that could be made without major equipment changes or employment of highly skilled workers. Such goods (lawns, lenos, simple curtain fabrics) might be marketed to the burgeoning mail-order houses, to specialty wholesalers or jobbers, using conservative stylings that allowed for inventory reserves and quick delivery. Such goods encroached on the turf of batch manufacturers, but being targeted at the low-style end of developing mass consumption, had little direct or immediate impact on flexible firms.[32]

[30] Naomi Lamoreaux, *The Great Merger Movement in American Business* (New York, 1985); Ralph Nelson, *Merger Movements in American Industry* (Princeton NJ, 1959); Chandler, *Visible Hand.* [31] Cumbler, *Working Class Community.*

[32] Scranton, *Figured Tapestry*, pp. 174–180, 381.

More profound was the implementation of a similar strategy by the merged American Woolen Corporation. Created from the melding of over twenty-five separate factories with William Wood's vast Washington Mills (Lawrence), American included plants oriented toward staple woolens and worsteds (overcoatings, clays and serges) as well as the Fletcher mills (Providence) known for their command of fancy stylings in worsteds. Setting up its own marketing agency, with separate departments for staple and specialty trade, Wood's enterprise rapidly became the pace and price setter in the sector, a phenomenon viewed with a mix of fear and satisfaction by individual batch operators. At the semi-annual openings, American's prices set the market, and buyers could not budge the "big company" from its lists. Smaller rivals had indeed to meet its price levels, but when this was done at the level of types, styling touches could win many an order. Further, American Woolen concentrated its efforts on goods for which longer runs might be secured than were satisfactory for a "pure" batch firm, goods that became known as "semi-staples" of the sort that still parade about as subdued business suitings. This left ample room for specialists, undercut buyers' pressure for price concessions, and brought order to early 20th century woolen and worsted markets. Moreover, American challenged the sample and duplicate system by warning frequently at openings that the firm would *raise* its list prices after two weeks, for the Lawrence giant wanted to fill its books as fast as possible and get on with production. Buyers reluctant to make commitments in bulk, sticking to sample and duplicate, veered toward other mills' offerings, but had little choice but to pay prices at or near the levels American set. Though it was never huge enough to "hog" the entire market, American Woolen's ability to offer long terms, volume discounts, and rebates pulled mass market clothiers toward its lines at the same time as its price leadership firmed levels for all lines of goods. Fears about how big American would get were thus blended with satisfaction that the Lawrence-based merger's encroachment worked to sellers' general advantage.[33]

Trouble for batch specialists did, however, materialize as American Woolen solidified its place at the center of the trade. Between 1905 and 1914, the "big company" employed comprehensive cost-finding procedures Wood had developed in rebuilding the Washington during the 1890s, techniques that became the basis for an organizational and informational advantage over rivals in either seasonal or staple lines. Close costing and tactical marketing permitted it to offer prices that shocked the trade, settings that traditionally-managed firms could not meet without losses. As a number of such mills became troubled, Wood picked and chose among them, acquiring selected mills until American had 59 plants in the early 1920s. The duplication of capacity at multiple sites also played a key role in labor relations, for Wood sat out strikes in Providence by forwarding orders to comparably equipped factories and revenged himself on Lawrence's victorious strikers (1912) by channeling layoffs that derived from slack 1913 markets in their direction. Its expanding market power also enabled American to

[33] *Ibid.*, pp. 181–186.

refuse cancellations on orders booked, a perennial "evil" that afflicted all specialty divisions. American would deal chiefly with stable, reliable clients, leaving the more fly-by-night clothiers to trade with (and victimize) lesser companies. By applying management and marketing techniques drawn from staple routinization to seasonal goods, American Woolen secured for itself the most desirable segment of the fancy worsted trade, balked labor, and transformed market relations among all parties associated with wool and worsteds.[34]

In general, the incremental advance of the "mixed-output" approach to textile production, *c.* 1890–1914, filled in the empty middle ground between batch and bulk manufacturing by identifying elements within the array of batch-produced lines that could be shaped into semi-staples and woven for stock (cottons) or on seasonal orders (AW). This dynamic posed a long-term problem for flexible operations. While their versatility enabled them to venture into novel patterns, constructions and color effects, market success quickly brought counteractive forces into play. Not only did firms who hit style trends on the mark have to contend with the prompt creation of imitations by flexible rivals, encouraged by cutters who procured one firm's samples and "shopped" them to rival mills seeking lower prices. In addition, were a novelty to take hold for more than one season, the more ominous movement of mixed output major firms seeking to incorporate the product within the scope of their work could be expected.

As staple mills stumbled toward glut and decline, (a process temporarily arrested by war demand, 1914–18) and mixed output firms encroached on the middle ranges of seasonal lines, flexible firms had a number of options to evaluate. None of the leads that could be derived from the staple sectors' 1890s struggles were of much value. Mergers, promoted in carpets, worsteds, and spinning at Philadelphia, were never realized, as proprietors' attachment to autonomy was resolute. Nor were pricing committees effective, as the multiplicity of partial process firms in each sub-sector obstructed collusion. Sectoral trade associations were both local and weak, long unable to maintain solidarity even against striking workers. Relocation south made little sense, for it was within the urban complexes that skilled labor was concentrated and the nexus of institutional interchange and support had been forged (banks, suppliers, links with spinners and dyers). No technological "fix" of the sort the battery loom suggested was evident within these trades, as incrementalism was the rule in technical change. Even the new "standard" accounting practices seemed foreign and useless; how was one to determine "standard" baselines in a pre-computer era given the diversity of products, suppliers, machinery, etc., without an Herculean effort? "Efficiency men" might serve well for analyzing tonnage billets and rails, but they and most of the practices associated with the management "movement" were rejected as neither designed for nor applicable to flexible mill operations.

[34] *Ibid.*, pp. 208–209, 291; Edward Roddy, *Mills, Mansions and Mergers* (North Andover MA, 1982).

What then did appeal to flexible mills that were affected by the first phase squeeze? (It should be stressed that this did not strike important flexible divisions: Jacquard work in upholstery or draperies, hosiery and knits, lace, trimmings and narrow goods, *c.* 1900–14, but chiefly cottons, wool/worsted, and carpets, where a set of northern mergers created an analogous situation.) Firms built on product competition generally responded with product-centered initiatives. Three moves were basic and plausible, a focus on high-end specialties, stylistic novelty in all grades, or descent toward "manipulated" low-end goods. The first two could be combined, as at Passaic's Forstmann mills, but the third was a degraded spillover from depression years, a use of flexibility that made Philadelphia's Manayunk into a disreputable "shoddy and mungo district." A second approach was to shift lines, moving toward narrow goods in cottons, toward plushes and mohairs from woolens, a frequent evasive tactic in Philadelphia. Most dramatic was capital restructuring, a sector shift into new classes of products. Such efforts demanded deep pockets, such as were available to Philadelphia's Bromley clan, which reinvested carpet profits into lace between the mid-1890s and 1911, then moved massively into full-fashioned hosiery a decade later, closing first carpet mills and later consolidating the lace ones behind them. Similarly, in Rhode Island, lace-making became an outlet for sectoral capital shifts, promoted especially by a feature of the 1909 tariff act that permitted duty-free machinery imports for two years. A different structural adaptation that involved only modest expenditures yet enhanced quality and shortened lags between orders and shipments was establishment of a captive dye-house attached to a weaving or knitting mill. This partial integration in turn created problems for independent master dyers, who turned from work with woolens to attend the needs of an expanding knitting trade. Similarly, the advance of southern cotton spinning and decline of demand for woolen weaving yarns led batch spinners to adapt their machinery for running the softer yarns knitters sought. Through the war years, batch operations used their versatility to good effect; encroachment yielded innovation by practical manufacturers, not a swing toward rationalized practice and managerialism.

To this point, reference to collective behavior by capital and labor has been fleeting and politics and the state little noted at all. As we approach the heated environment of World War I and its aftermath, these matters can be neglected no longer. Association among the principals in Yankee staple corporations commenced with Boston's Arkwright Club, blossomed into the National Association of Cotton Manufacturers, and spawned the 20th century Cotton Textile Institute.[35] Initially concerned with the cotton trade, tariffs and technical practice, the staple association steered toward matters of management rationalization in the later 1800s. The CTI's emphasis on standards and research extended this trajectory into "scientific" analyses of business conditions, policy options and fiber/fabric technics. Domination of the NACM by regional interests

[35] Louis Galambos, *Cooperation and Competition* (Baltimore, 1966).

brought southern staple operators to form their own groupings, especially devoted to opposing state/federal regulation, which would have diminished their advantages vis-à-vis New Englanders. In wool, a parallel and somewhat more heterogeneous national association aped the NACM, even to promoting an ill-starred Wool Textile Institute in the 1920s; but in general its more diverse membership could agree only on spirited defenses of protectionism.[36] These three clusters, together with the Silk Association of America (founded *c.* 1880),[37] constituted the durable core of "national" trade bodies controlled by large-scale enterprises.

By contrast, associations of batch manufacturers were intensely local in character, unsettled in membership and thus often ephemeral. Attempts to unify them regionally, as in the Philadelphia Textile Manufacturers Association, were repeatedly ineffectual, due triply to proprietors' unwillingness to forward sufficient funding, to take direction from any central authority, and to the insistent conflicts of interests between established firms and fledgling rivals and between sub-sectoral clusters (e.g. wool yarn spinners and wool yarn using weavers/knitters). Social/political clubs *were* persistent (Philadelphia's Manufacturers' Club, the Hamilton at Paterson), as they provided a base for intercourse and deal-making, as well as a forum from which impassioned support for high-tariff Republicanism could be mounted. Institutions of direct service to localists did gain substantial constituencies: the aforementioned banks, Joseph Grundy's Manufacturers' Insurance Company (an offshoot of the Pennsylvania Manufacturers' Association), the Matthew Quay statewide political machine, Republican of course.[38] Unlike the regional metal manufacturers' association, which provided durable and diverse services, the PTMA was only able to hold its members by providing a comparable, immediate-use benefit (a reference bureau for non-union workers much inferior to that in metalworking), not by promulgating plans for collective research, management self-education or technical updating.[39] The diversity of batch production and its segmentation, together with the values of autonomous localism cherished by proprietors, limited the scope and reach of this class of trade associations, which in Philadelphia included eleven sectoral groups below the level of the regional PTMA. Furthermore, one goal of vital concern to twentieth century batch operators, only achievable through collective action, was ruled unlawful by the federal government. Though

[36] F. A. Taussig, *The Tariff History of the United States*, 8th edn (New York, 1931); Copeland, *Cotton Manufacture*; Arthur Cole, *The American Wool Manufacture*, vol. 2 (Cambridge MA, 1926).

[37] The SAA's minute book is held by the Pastore Library Archives, Philadelphia College of Textiles and Science.

[38] J. Roffe Wike, *The Pennsylvania Manufacturers' Association* (Philadelpha, 1960), esp. chs. 1, 2, 12.

[39] Howell Harris, "Employers' Collective Action in the Open-Shop Era: The Metal Manufacturers' Association of Philadelphia c. 1903–1933," in Steven Talliday and Jonathan Zeitlin (eds.), *The Power to Manage? Employers and Industrial Relations in Comparative-Historical Perspective* (London, 1991), pp. 117–46; Scranton, *Figured Tapestry*, pp. 347–359, 368–374.

its enforcement was erratic, enactment of the Sherman anti-trust statute (supplemented by the 1913 Clayton Act) created an environment in which collective activity by ostensible competitors was viewed as likely collusion in restraint of trade. Thus attempts by flexible firms to agree on minimum prices and more important, standard selling terms, in the post-World War I context of enhanced buyers' market power, were deemed illegal by the Federal Trade Commission in cases involving several woodworking sectors. This cast a further pall over the potential of trade associations to act institutionally in any fashion that might alter market relations in textiles as well, for associations could not lawfully govern the practices of their constituent members. Hence, even efforts to press sectorally clustered firms to adopt common accounting practices, thereby establishing a shared foundation for pricing decisions and theoretically averting accepting low-ball orders that entailed losses, stalled and were abandoned.[40]

Labor organizations displayed a spirit similar to that found among manufacturers. Insofar as long-lived unions were to be found in staple divisions, they were centered on the United Textile Workers, a New England amalgam of craft associations with national ambitions and a base among English-speaking operatives, both native and immigrant. A "bread and butter" union, the UTW shared the operating assumptions of the American Federation of Labor and developed a centralized bureaucracy that "authorized" strikes and promoted orderly collective bargaining. The UTW was thus intensely hostile to radicalism either in the advent of local militancies or the revolutionary syndicalism of the Industrial Workers of the World, which appealed to a multi-ethnic mass of textile workers largely ignored by the UTW. The IWW's inspirational fervor mobilized American Woolen's Lawrence thousands for a brilliant, but ephemeral victory in 1912, a triumph facilitated in part by Wood's and area politicians' tactical blundering. However, the IWW's organization weakness and its tendency to apply the same analysis to all conflicts yielded a crushing defeat in Paterson only a year later. There was in the "silk city" no single, hated figure like Wood, whose defeat would signal a general victory for labor, but instead an emergency confederation of several hundred shop and mill proprietors, who like local officials were determined to avoid the errors that had humbled Lawrence. This formless array baffled IWW tacticians, and the Paterson strike disintegrated from within, stilling the voice of industrial unionism in textiles for two decades.[41]

If the UTW's main strength lay in northern staple centers, it did not ignore the Mid Atlantic batch districts. However, in enrolling "independent" local craft bodies to augment its forces, UTW leaders found themselves facing a set of agendas and practices opposed to its hierarchical style. Unlike the settled monotony of bulk manufacturing, the constancy of change in batch mills

[40] A. D. Neale, *The Anti-Trust Laws of the United States* (Cambridge, 1970); Federal Trade Commission, *House Furnishing Industries*, 3 vols. (Washington DC, 1923–25); Scranton, *Figured Tapestry*, pp. 378–379, 388–389.

[41] Ann Tripp, *The IWW and the Paterson Silk Strike of 1913* (Urbana IL, 1987); Steve Golin, *The Fragile Bridge: Paterson Silk Strike, 1913* (Philadelphia, 1988).

provided a setting for endemic contests between workers and owners, contests institutionalized through firm-based shop committees and sub-sectoral clusters of shop delegates. Factory conflict surfaced in frequent stand-downs over grievances and sector-wide efforts to establish, adjust and renew bills of prices for the diverse tasks involved in making styled and seasonal goods. There was order in this seeming chaos, but an order, and even a language, separate from that of UTW business unionism. Thus conflicts between Philadelphia locals and national leaders were as much a part of the labor scene as contests in the shops. During the Wilson administration (1919–21), national officers gained a legitimacy previously withheld, dealing routinely with federal mediators (U.S. Conciliation Service) and policy makers (War Labor and Industry Boards). In return they were expected (and undertook) to ensure labor peace for the war's duration, a job that necessitated control over members' use of the strike weapon. In batch centers, such control was utterly lacking; the UTW president raged in letters to federal officials about the unruly, unpatriotic behavior of Philadelphia's textile workers, who proved as resistant to taking central direction as were their employers. In a massive 1917–18 woolen mills strike, the UTW attacked and attempted to displace local leadership, then collaborated with mediators and owners to defeat the strikers, in order to clear the ground for rebuilding orderly locals. Instead, the UTW was discredited and unions in area woolens collapsed.[42]

Other batch-based unions remained independent of the UTW, active either narrowly within the city (Jacquard loom fixers) or in regional craft federations (Wilton and Brussels Carpet Weavers, Amalgamated Lace Operatives). Strong until the post-war reaction, they associated with UTW locals in the city-wide Textile Council, whose structure mirrored that of the Manufacturers' Club and the PTMA. The Council managed a meeting house (the Kensington Labor Lyceum), sponsored social events and organized political initiatives, but was not endowed with authority to govern the labor movement in textiles, treat with employers during strikes, or act other than as a clearing house for information and a forum for debate. In sizable strikes, such as the 1903 battle for shorter hours, shop delegates elected emergency executive committees, yet settlements flowed from the bottom up, engineered by cadres of shop delegates at the firm level. Political power was thus broadly diffused, rather than concentrated in an institutional officialdom, well into the present century, a pattern that fit the specificity and diversity of labor processes and work sites in batch manufacturing, but ran against the grain of institutional development in the larger society. By the end of World War I, the patterns and practices of this quasi-Jeffersonian democracy of autonomous agents and small-scale collectivities was becoming increasingly incomprehensible to national union leaders and state representatives attuned to orderly administrative rationality, in which "superiors" managed "subordinates" rather than the other way around. Moreover, as batch sectors entered a long, gradual decline, local labor organizations positioned to maneuver

42 Scranton, *Figured Tapestry*, pp. 303–316.

through seasonal and cyclical boom and slack periods failed to devise effective means to cope with the structural shifts of the 1920s. Lacking mechanisms to manage and plan beyond the realm of situational immediacies, lacking tested ways and means for addressing long term problems in collaboration with employers (whose associations became enfeebled in any event), advocates of localist decentralization discovered with dismay the increasing marginality of practices that had served them well for two generations. In 1914, however, these shocks lay ahead, and any unease about the adequacy of current formats for industrial relations was overshadowed by the initiation of far greater conflicts across the Atlantic.[43]

The European war opened at a fortuitous moment for the American textile industry. Staple sectors, particularly in cottons, had looked anxiously for export markets since the late 1890s as a release for persistent excess production. However, tariff protection entailed pricing levels uncompetitive in world markets,[44] and Americans had seen the promise of the China trade slip away (into Japanese hands) while being unable to meet British or German prices in other spheres. Secondly, of particular moment to speciality firms, the new Democratic administration had, in fulfilling its pledge to reduce tariffs, concentrated textile cuts in high-value lines long well-defended by Republicans.[45] Third, the economy was laboring through a recession widespread enough that the federal Commission on Industrial Relations questioned public officials at its hearings about the extent of privation and unemployment in cities. Though the commencement of hostilities further deranged markets for several months, by the late fall of 1914, an

[43] In an industry as diverse as textiles, it seems that none of the three forms in which unionism was advanced (localist autonomy, nationally federated crafts, and industrial unions) could have been universally adequate, though each had its spatial and temporal configurations of success. (The industrial version, balked in the IWW case, was revived in the 1930s by the textile segment of the Congress of Industrial Organizations, and made gains through targeting a number of the largest employers. Further advances have been rare in the post-World War II decades, even given its merger with the leading apparel union, forming the Amalgamated Clothing and Textile Workers.) Beyond the question of matching union structure with sectoral dynamics, it is also evident that one cannot read off conservative or militant policy from union form alone. For example, though many Philadelphia locals regarded UTW officials as hopelessly reactionary, the head of the independent Beamers and Twisters' Organization denounced the same UTW figures as "radicals . . . a bunch of reds." (William Thomas interview, in Palmer papers, PP–135, Urban Archives, Temple University.) Moreover, socialist Emil Rieve headed the Full-Fashioned Hosiery Workers' visionary campaign to regulate brutal competition through thoroughgoing unionization, yet had to overcome a Philadelphia insurgency that put defense of local conditions ahead of achieving a national militancy that promised long term gains. (Scranton, *Figured Tapestry*, p. 493). For perspectives on "institutional" vs. "rank and file" approaches to British labor history, and vigorous exchanges, see Jonathan Zeitlin, "From Labour History to the History of Industrial Relations," *EcHR*, 2nd. series, 40(1987), 159–184, and "'Rank and Filism' in British Labour History," *International Review of Social History*, 3 (1989), 42–61; Richard Price, "'What's in a Name?': Workplace History and 'Rank and Filism,'" *ibid.*, pp. 62–77; James Cronin, "The 'Rank and File' and the Social History of the Working Class," *ibid.*, pp. 78–88; and J. Zeitlin, "A Rejoinder to Price and Cronin," *ibid.*, pp. 89–102.

[44] Saxonhouse and Wright, "Stubborn Mules," pp. 92–93.

[45] Patrick Hearden, *Independence and Empire* (DeKalb IL, 1982); Taussig, *Tariff History*.

industrial boom was under way, a surge that would last nearly six years. American agents for the warring parties parceled out huge orders for military textiles, and competition in "third world" markets faded as European firms abandoned exports. Staple mills rebounded to show their best profits in a decade, and flexible ones captured contracts for blankets or knit leggings, converting part of their facilities to staples to share in the booty. With American entry and consequent constraints on style lines, plush mills shifted to khaki cloth, carpet mills to cotton duck and tenting, as had been the case in the Civil War.[46]

Inflation accelerated, making all calculation of profits difficult and pressing workers to demand sizable wage hikes, which as ever lagged behind costs of living. Workers contending with runaway food, fuel and rent charges had little patience with manufacturers' moaning about wild bidding for raw materials, the escalating price of dyeing and so on. In Philadelphia, wartime strikes proved to be angry affairs; Bolshevism was soon everywhere feared. The war's end would surely bring deflation, depression and further struggle, if not social revolution.

Three of these four expectations were met, but they surfaced at a time and in a form that was unanticipated. Wartime shortages of goods for civilian demand and the possibility of harvesting opportunistic, inflation profits led users and distributors of textiles to seek strong inventory positions. Manufacturers' price increases were little resisted; indeed, buyers at times placed multiple orders for the same goods, taking everything that was delivered, while mills buried in the avalanche began proportional allocations to spread goods equitably among their clients and keep all their accounts active. However, in the summer of 1918, as deadlock on the Western Front ended, accumulated concern about the coming deflation dried up orders. Buyers now prepared to limit their losses when peace brought a collapse of overheated demand and thus prices. At first, it seemed they had guessed correctly, for in and after November, textile markets stalled, buyers' offering prices plummeted, retailers' inventories steadily shrank.

A combination of factors proved them to have been wrong early in 1919. The state played a double role, first by gradually releasing soldiers (averting mass unemployment) with back pay and a clothing bonus (pumping cash into the apparel trades), and second, by declining to dump its huge stocks of warehoused raw wool, commissioning instead a Philadelphia auction house to conduct an extended series of sales that cushioned materials markets by setting minimum acceptable bids. Civilian demand for style and specialty goods (especially women's wear, hosiery, furnishing fabrics and carpets) in short supply for several years added to demand pull. Distributors shorn of stock returned to block purchasing on a grand scale. Prices spiraled upward again; batch and mixed output mills ran overtime. (Staple firms were not so fortunate; the cotton duck and blanket markets were dead.) For a year, distributors demanded stock as fast as it could be flung onto trucks and trains, and just as their shelves groaned with the mass of it all, the collapse arrived. Heralded by the Federal Reserve Bank's

[46] Scranton, *Proprietary Capitalism*, pp. 272–298; and *Figured Tapestry*, pp. 284, 288–289.

move to curtail credit, and made visible to the public in John Wanamaker's stunning announcement of a storewide 20 percent reduction (allowing the great retailer to liquidate before the rush), the steepest deflation in a century caught most distributors painfully exposed. Of course, they canceled outstanding orders for goods, sending shivers through the industry, leaving all parties wondering when, where and how a bottom would be found. But more crucially, retailers first, then wholesalers and jobbers, determined to install buying practices that would guard against a repeat performance of this dismal show.[47]

The core element of the distributors' offensive was termed "hand-to-mouth" or program buying in the 1920s, but it was only the most remarked-upon of a battery of weapons brought to bear on producers. The crunch of 1920–21, aptly named the "inventory depression," hammered home a central paradox to distributors. Called upon to service customer needs by having what was wanted when it was wanted, retailers needed huge and diverse stocks, whose carrying costs were substantial and in style lines, whose perishability was dramatic. At the same time, competition among these service sector firms was becoming crushingly expensive, as advertising, amenities provision (from packing and delivery to lounges and other comforts for clients), and personnel requirements escalated. From 1905 to 1920, department stores found operating expenses rising from 20 to 30 cents of every sales dollar, yet distribution showed none of the gains in work productivity that industrial firms achieved.[48] Something had to give, and in retailing, what gave was the relative autonomy of departmental buyers, whose intuitive reading of style trends in soft goods was now deemed less important than budgetary control over their purchasing. Management analysts had before the war urged retailers that return to capital could be enhanced in an environment of flat productivity and rising operating expenses only by shrinking inventories and increasing their rates of turnover. This was the basis for frequent, small-scale, "hand-to-mouth" buying, for narrowed product lines, and intersected, from the direction of accounting simplification, with the development of gapped price-lining (goods at $10, $15 and $20) rather than a profusion of different figures. These confluent recommendations had a sudden and general appeal in 1920, and at the urging of their national trade associations, retailers rapidly adopted program buying and turnover control measures, supervised by "merchandise managers" and cost accountants. The turnover cult spread to smaller stores and specialty shops as well, but in style goods, a problem remained.[49]

If stocks were to be cut, and close track kept of them so as to file timely replenishment orders, someone upstream had to have on hand just what was most in demand. Someone had to assume the inventory burden, but neither

[47] Scranton, *Figured Tapestry*, pp. 316–339.

[48] Susan Benson, *Counter Cultures* (Urbana IL, 1986); Twentieth Century Fund, *Does Distribution Cost Too Much?* (New York, 1939), esp. figures 1–3.

[49] Benson, *Counter Cultures*; James Fri, "Changes in Marketing Since 1900," *The Manufacturer*, 11 (December 1929), 3–9; James Chittick, *Silk Manufacturing and Its Problems* (New York, 1913), pp. 285–286.

clothiers nor wholesalers and jobbers were eager for the opportunity. Their advance buying of fabrics and mill-finished textiles (hosiery, carpets) tapered off rapidly, for the constant, unpredictable stream of "hand-to-mouth" buying added costs without providing benefits. All turned to the batch manufacturers, calling upon them either to run stock in advance of demand or to manufacture goods "instanter" on call. Wholesalers who did agree to place orders for seasonal arrays now would do so only at sharply lower prices, needing to widen their margins to offset risk assumed. Millmen soon realized the ruinous implications of the new pattern. If they created stock, it might well never be called for, and was quickly devalued, having to be "thrown on the market" at season's end for whatever price job-lot "vultures" would offer. Were they to persist in the "make to order" business, constantly shifting loom setups (at $7.50 per loom, roughly), there was little money to be made from stacks of tiny orders. Moreover, both wholesalers and retailers exerted constant pressure for price reductions, even after the national economy recovered, seeking to have savings from declining raw materials costs (or wage reductions where instituted) passed along to them. Lacking a powerful institutional voice, political clout or legal standing, flexible producers were reduced to pleading, "Have a heart for the manufacturers!" in their trade journals and at endless association meetings. Little wonder that, perceiving the individual isolation of the millmen, buyers proceeded in the 1920s to add to their advantages a cluster of predatory trade terms (extended payment dates, return privileges, downward price ratchets on contracts should materials prices fall before delivery) that increasingly demoralized manufacturers accepted glumly. Ideologically, as these maneuvers were paraded by their authors as means to reduce the costs of distribution, an issue Hoover's Commerce Department, the US Chamber of Commerce and the American Marketing Association agreed was a major national challenge, industrialists' pleas were ignored as self-serving attempts to cheat consumers, and their attention was directed to the example of Henry Ford. In the 1920s, the universality of mass production practices became gospel; those who could not read the Word, or disbelieved it were cast from the garden of business discourse.[50]

Naturally enough, flexible textile manufacturers by mid-decade experienced what they nicely termed "profitless prosperity," plenty of business at prices and under terms that barely covered costs, or indeed failed to do so. As discrete closures and public bankruptcies multiplied, three contemporaneous developments helped seal their fate. First, the era of high ornament in textile stylings began to give way to design simplifications, first in Art Deco and ultimately in the vogue of monocolor carpets, etc., whose Bauhaus plainness nicely meshed with the standardization of taste during the Great Depression. Secondly, the rapid advance of chain retailing further homogenized demand and added weight to price pressures. Finally, in a partial regional restructuring that stemmed from the stagnation of southern staple cotton manufacturing, new Dixie firms were

[50] Scranton, *Figured Tapestry*, pp. 387–398.

launched to create cheaper versions of lines in the last vital flexible divisions (hosiery, upholstery and curtains). Though a handful of firms found a way out of the impasse through links to mass production textiles users (GM and Chrysler for auto seatings), the vast majority exhausted their capital, plant and machinery until their proprietors' death or insolvency forced suspension and liquidation. Those that survived until World War II briefly again enjoyed profitable years, only to fold during one or another of the post-war recessions. In Philadelphia, peak 1920s employment dropped by 1939 to half its former level, and by the early 1950s half of the residual had vanished.[51]

What of the staple trades, north and south? Symmetrical with the dominion price concerns gained in seasonal textiles, non-price competition emerged as a route out of the swamp for the most resilient staple firms. Wamsutta and Pepperell in New England, Cannon in the South opened assertive mass advertising campaigns to establish their brand names as distinctive, appealing directly to consumers and bypassing the entire intermediary mess. Retailers complained that such firms were undertaking to make their stores "merely depots" for storage of goods, but for a few firms the tactic worked.[52] For the rest, the 1920s and 30s were bleak indeed. Twenty-four New Bedford corporations shut down by 1939, about two-thirds of the industry; Fall River was even more a desert, Lowell a ghost town, and at Manchester, the century-old Amoskeag plant liquidated, vaporizing 10,000 jobs. Southern mills were squeezed as well; the reduction of overall capacity did little to alleviate staple mills' lack of control over pricing their fabrics. For this reason, the New Deal's National Recovery Administration, with its promise of state-supported codes governing prices, wages, hours and trade terms, was greeted with no little enthusiasm by cotton manufacturers. The Cotton Textile Code was the first created and approved under the 1933 statute, but weak enforcement and doubts about the constitutionality of the enabling legislation, invalidated in 1935 by the Supreme Court, diminished its effects. Southern liquidations and consolidations followed apace, with empire builders like Spencer Love (Burlington) and Royal Little (Textron) creating branch-plant networks of firms picked up at large discounts from their initial capitalizations. These clusters had break-even points far lower than scrambling competitors, were modestly diversified across segments of bulk and semi-staple lines, and booked massive orders and profits during the Second World War. Over the ensuing decade, rationalization resumed, as orders were shunted to the most technically advanced plants, while those with unionizing proclivities or antiquated layout and machinery were shuttered.[53]

Even the center did not hold. American Woolen stumbled badly in the mid-1920s. Its patriarch, William Wood, committed suicide following his son's

[51] *Ibid.*, pp. 410–499.

[52] H. B. Hough, *Wamsuta of New Bedford, 1846–1946* (New Bedford MA, 1946); Marchand, *Advertising*.

[53] On southern textile difficulties and regional restructuring, see Gavin Wright, *Old South, New South* (New York, 1986), chs. 7, 8.

death, though the firm's slipping fortunes may have contributed to his self-destruction. Two trends in raw materials prices, combined with buyer's market power, help account for its beleaguered situation. First, during the war raw wools rose to $1/lb. and despite a brief fall in the 1920–21 depression, stayed at or near this level for the rest of the decade. Meanwhile, figures for cotton, silk and rayon fiber moved steadily lower through the early 1930s. In consequence, wool-using goods became relatively more expensive than they had been at any point in sixty years. This, in a buyer-dominated market, brought a contraction in the demand for woolens and worsteds, and a determined search for substitutes. In women's wear, cottons and rayon came to the fore; rayon prices, four times those of wools in 1920 actually dropped below wool by 1931–32. American tried brand-advertising, volume discounts, and other tactics, but could not avoid the pressure to run stocks for clothiers' call, stocks which accumulated as assets on company accounts until multi-million dollar write-downs could no longer be avoided. When the extent of losses on inventory was at last acknowledged in 1927, American stock values plummeted and the "big company" began auctioning off mill after mill, bringing in depression but paltry sums into corporate coffers. With advertising ineffectual (American sold its "Rams' Head" yardage to clothiers, but shoppers bought suits and coats with store or maker labels), export opportunities negligible, and Wood's managerial genius gone, American Woolen languished.[54] With the domestic market well-insulated for decades after 1930 (imagine the swath free trade would have cut), the textile industry settled into a drab, no-growth, low- or no-profit pattern of stagnation and the management of decline. Flexible firms that once had taken the lead in product development faded into insignificance or occupied peripheral niches (lace, woolen hosiery). Technical innovation stalled; the industry's machinery builders went out of business or shifted to service other sectors.[55]

There was little weeping at and indeed much disguising of this loss of industrial vitality. Outside the narrow band of high-style women's wear, demand homogenized steadily in apparel and furnishings textiles from the 1930s through the 1960s, a phenomenon congenial to the industry's subservience to mass retailing and standardized consumption. When in the last dozen years, a trend toward sharply segmented markets emerged in apparel, American producers lacked the skill, the technology and the managerial practices necessary for an inventive response. In both mill goods and finishing clothing, the laurels went to imports,

[54] Cole, *Wool Manufacture*, vol. 2; Anthony Zappalla, "A Financial History of the American Woolen Company," unpublished MA thesis, University of Pennsylvania, 1947.

[55] For documentation of the decay in American textiles, see Brian Toyne *et al.*, *The Global Textile Industry*, (London, 1984), pp. 70–84; and US Congress, Office of Technology Assessment, *The US Textile and Apparel Industry: A Revolution in Progress* (Washington DC, 1987). Toyne established the substantial employment loss (from 1.3 million jobs in 1945 to 800,000 in the early 1980s), low profitability and thin investment in textiles (up in real terms only 27 percent, 1963–80, vs. an all manufacturing average of 300 percent). The OTA notes that the US textile machinery trade is so far behind the technology wave that 92 percent of its 1980 sales represented replacement parts (p. 74).

yielding another wave of contraction that has reduced southern textile employment by 300,000 in the 1980s and has stimulated renewed calls for tariff barriers as a last-gasp defense.[56] Having haltingly adapted the textile trades to the prevalent mass consumption norms of mid twentieth-century America, the successors of Lowell, the Bromleys, Wood and Love now confront a maze of global dynamics which promise to restore importing merchants to the centrality they enjoyed before the triumph of "domestic manufacture" in the early years of the republic.

If we are prone to trace the textile industry's "failure to thrive" to its inability to achieve technologically-driven, and organizationally-monitored throughput advances, the preceding account may serve as a cautionary tale. Firstly, there never was a singular "textile industry" in the United States. In actuality, the textile manufacture's two chief nineteenth-century divisions evolved into a baroque array of product and process constellation before a twenty-year miasma shook out all but the mass-demand survivors. Secondly, the state was heavily implicated in the construction of alternatives for action. From protection to the NRA, the Sherman Act to the state-level differentiation of factory regulation and labor law enforcement, governments set both enabling and constraining boundaries for firms and associations, and more broadly, set terms of discourse that privileged hierarchy over personalism, a severe if subtle blow to flexible practice. Thirdly, evidence of the cultural dimension of industrial transformation is evident on every side. The autonomous dispositions of batch proprietors and workers, once a source of initiative and innovation, became crippling liabilities. Undifferentiated notions of competition as simple and good blurred perception of both its darker implications and of the diversity of its components, forms and temporal and spatial manifestations. After all, in the trade's bulk production divisions, Yankee treasurers' assumptions about their superiority and the slothful inefficiency of southern workers fed into their decisions and self-justification. Similar premises undermined New England labor organizers' efforts in Philadelphia and, more critically, in the South. These dispositions were crucial to the erosion of batch and bulk sectors alike, for though the former eroded catastrophically by mid-century, the latter, despite the adoption of modern managerialism, have faltered repeatedly in an increasingly global economy. Bulk production did not triumph in American textiles so much as it stumbled onward

[56] Jeffrey Arpan *et al.*, *The U.S. Apparel Industry: International Challenge, Domestic Response* (Atlanta GA, 1982); OTA, *U.S. Textile and Apparel Industry*; James Lardner, "The Sweater Trade," *The New Yorker*, January 11, 18, 1988. Though Arpan notes import penetration in staple apparel led US firms to shift toward more "fashion sensitive" lines in the 1960s (p. 9), it was clear by the mid-seventies that fashionable imports were rising steadily (p. 80). By the late 1980s, OTA found import penetration in seasonal "private label" goods near 80 percent of a market that accounts for nearly half the apparel trade (pp. 16–17). Lardner's essays document the shortcomings of American firms offered contracts in the "short life" high fashion arena, a third of the market, and the flexibility of Hong Kong firms that marry quick turnaround, instant communications, and air transport to open the door to work on high-value lines for the US.

toward defeat in a race whose boundaries and entrants changed as the decades rolled by.

Two final points are worth noting. Given this account, the notion that "rational" economic choice is a useful analytic concept becomes problematic, for in historical contexts (rather than in decision-spaces abstracted from place and time) non-rational components of the choice environment are not separable from but enter fully into the framing of alternatives and the rules by which they are to be evaluated. These "rationalities" are themselves differentiated, yet historically patterned due to their formulation in concrete socio-cultural settings,[57] of which batch and bulk textile enterprises, districts and networks are examples. Further, it is evident that in American textiles there was a decisive conjuncture, *c.* 1918–21, during which relations, values and practices were profoundly altered, valences reversed, and the latent vulnerabilities of the industry's multiple segments became suddenly apparent. That collapse in New England was rapid and in Philadelphia more attenuated speaks again to their differing assemblies of rules and resources for "rational" action. Surely the potential for disarray, the presence of the virtual in the actual, lay but dimly recognized in the industry's previous structures and relations; but it was in an historical moment of multiply or over-determined crisis that portentous shifts materialized, loosing the cycles of decay. That subsequent business and policy debates were (and are) conducted in incommensurate languages of illusion[58] should be no great surprise, but it might well be instructive to pursue the links between textile's course and that of other industries whose prospects have more recently dimmed. Might there not be room for a fresh "textile paradigm," a successor to the discredited Whiggish practice of regarding this industry as the template for American industrial development? Such a motif would invite us to discard any notion of an immanent logic to industrial change, to seek out the situational particularities of workplace social and technical relations, market and financial practice, together with their cultural and political components, so as to assemble a contextualized and inclusive portrait of temporal and spatial developments, rather than letting a fragment of the industry stand for the whole (as was so long the case with staple cottons in textiles). It would offer not a model, but an agenda for inquiry which would engage rather than discount diversity, problematize rather than take as given institutions, language, competition and markets. It would encourage us to identify and analyze conjunctures in which latent weaknesses became palpable, strengths evaporate, and decline or restructuring gathers momentum, without presuming that a neat line divides the internal considerations (factor costs) from the external (state policy) or even that the firm is the universal unit of analysis. It would acknowledge the embeddedness of contests over power and authority in human relations without embracing any *a priori* ranking of their multi-layered incidences derived from competing general

[57] Burton Klein, *Dynamic Economics* (Cambridge MA, 1977), pp. 23–41.
[58] Anthony Wallace, *St. Clair* (New York, 1987), ch. 4.

theories. Such an approach would decenter the study of industrial history, unhinge it from its received categories and priorities. Yet though it is yet so radically open and amorphous as to mock the meaning of "paradigm," such an approach refined through exploratory practice may hold promise as a means to address the dilemmas of confronting historical intentionality without recourse to teleology or reductionism.

8 The small-holder economy in Denmark: the exception as variation

Peer Hull Kristensen and Charles F. Sabel

The exception as variation

The economic conflicts of the late nineteenth century in Western Europe and the United States look more and more like the clash of two projects of industrial modernity. For most of our own century the tumult of the period was regarded as the death rattle of the local communities of "pre-industrial" or "pre-capitalist" peasants, artisans and shopkeepers overwhelmed by competition with mass-production factories and large-scale farms that dominated world markets through vast transportation networks. Today it seems more accurate to see many of these groups of small holders as advocates of a kind of owner-operated capitalism in which small units of production cooperate in the production of specialized goods. Their production costs were not as low as those of the mass producers. But their products were often still competitive because in many markets customers would pay some premium for a specialized good that satisfied their needs more fully than the standard item.

Correspondingly, the mutualism of Proudhon,[1] the Liberalism of Gladstone and J. S. Mill,[2] the Populism of rural America,[3] and the urban Progressivism of Brandeis,[4] all are being reinterpreted as efforts to define an institutional setting that encouraged cooperation among small holders on the one hand and protected them on the other from large producers who might use their power in financial or product markets against smaller competitors. Indeed, at the end of

[1] K. Steven Vincent, *Pierre-Joseph Proudhon and the Rise of French Republican Socialism* (New York: Oxford University Press, 1984), and *Between Marxism and Anarchism: Benoit Malon and French Reformist Socialism* (Berkeley: University of California Press, 1992).

[2] Eugenio F. Biagini, *Liberty, Retrenchment and Reform: Popular Liberalism in the Age of Gladstone 1860–1880* (Cambridge: Cambridge University Press, 1992); Avner Offer, *Property and Politics 1840–1914: Landownership, Law, Ideology and Urban Development in England* (Cambridge: Cambridge University Press, 1981).

[3] Lawrence Goodwyn, *Democratic Promise: The Populist Moment in America* (New York: Oxford University Press, 1976).

[4] Philippa Strum, *Louis D. Brandeis: Justice for the People* (Cambridge MA: Harvard University Press, 1984); and *Brandeis: Beyond Progressivism* (Lawrence: University Press, 1984); and *Brandeis: Beyond Progressivism* (Lawrence: University Press of Kansas, 1993); Gerald Berk, "Neither Markets nor Administration: Brandeis and the Antitrust Reforms of 1914," in *Studies in American Political Development*, vol. 8 (New Haven: Yale University Press, 1994).

our century their programmatic emphasis on voluntary cooperation as the precondition of the economic autonomy of individuals and groups seems a prescient criticism of the limits of bureaucratic coordination and accordingly hard to dismiss as an atavistic refusal to own up to the insurmountable advantages of large-scale organization.

Yet however plausible the strategies of the small holders and the flanking programs of institutional reform may now appear, nothing can change the fact that both were, at best, inconclusive failures. It was not the small holders' strategy of flexible specialization but the principles of mass production that defined by their competitive success the image of industrial modernity even for producers who felt, because of the particularity of their circumstances, compelled to deviate from them. Similarly, legal regimes of almost unrestricted corporate sovereignty that imposed few limits on the concentration of capital won out over regimes of regulated competition that protected small-holder interests. These failures were inconclusive in that many agglomerations of flexibly specialized producers continued to survive in niche markets either as specialist producers of capital or final goods or particularly agile subcontractors to mass producers; and the project of regulating competition lived on as well in the form of prohibitions on some forms of monopoly or legalization of certain forms of explicit concertation among firms with complementary interests. But if the small-holder project appears robust, even vital, when judged by the historical winners' implicit claim that any system of industrialization but their own was utopian folly, still it led and leads a ghostly existence when viewed in the everyday light that shines from worldly successes.

Denmark was the exception. There in the decades before World War I peasant small holders built a technologically innovative cooperative movement that outcompeted estate-owners and urban financiers in virtually every segment of the dairy, egg and pork products industries. In so doing they created demand for particular kinds of capital goods that contributed to the modernization of the small-shop sector of industry as well. This movement produced no encompassing programatic document of its own, nor even extensive commentary on contemporaneous debates on the future of industrial society. Nonetheless, in areas as diverse as the law of agrarian property, the regulation of trademarks, patent-rights, or the certification of craft skill it regulated competition so as to limit the corporations' possibilities for securing long-term control of markets through the exercise of power while allowing small holders the freedom to organize their own affairs. Hence the Danish system of production inspired sympathetic contemporaries by institutional example. The compendious, pre-war handbook of the cooperative movement, Hertel's *Andelsbevægelsen i Danmark*, was translated and adapted for the British public by a Gladstonian Liberal.[5] Alice Brandeis, Justice Brandeis' wife, translated and published a tract

[5] Harald Faber, *Co-operation in Danish Agriculture* (An English Adaptation of *Andelsbevægelsen i Danmark* by H. Hertel) (London: Longman, Green and Co., 1918).

on Danish cooperative democracy.[6] Certain Danish institutions such as the "control societies" that cooperatively monitored the relation between the breeding and feeding of livestock and its economically relevant characteristics were widely imitated by dairy and cattle raisers in Northern Europe who saw the results. If there was any country where the historical alternative to mass production was successful enough to be treated as a matter of fact, Denmark was it.

This essay looks at why it was precisely in Denmark that the small-producer project came to dominate a national economy, how that small-producer system actually functioned, and how come, nearly three-quarters of a century after they were founded, the dairy and slaughterhouse cooperatives that were the core of the small-producer movement went into decline. The main focus is on the why. Our argument is that distinctive as it was, the Danish outcome resulted from the combination of various pieces of early modern agrarian and industrial society found in similar form in many other, analogous national economies of the day. Combined in a single, Danish system, these pieces of course took on through their relation to each other characteristics they did not exhibit when otherwise linked. But no single constituent piece of the Danish system was singular in the sense of having developed beyond the reach of the formative forces that shaped the like pieces of the surrounding societies; nor was their combination in the small-holder project the result of some cataclysmic break with previous Danish history or, for that matter, an abrupt deviation of Danish history from the path followed by many of its contemporaries. Put another way, Danish exceptionalism was a variation on many common themes, not an anomaly. Normally, exceptions prove rules by showing that clear-cut deviations from cardinal principles produce aberrant, deviant results. The Danish case shows that the "laws" of economic development are themselves so unruly, so elastic in what they rule in and out, that bending many of them almost imperceptibly produces an unfamiliar, apparently singular outcome.

Thus, in section II we show that Danish mass-producers followed the same strategies as their better known and more successful counterparts, and that their small-holder opponents had a strong family resemblance to the independent peasants of much of Western Europe. In this sense the struggle between them was a microcosm of the larger struggle between the visions of industrialization; and the Danish outcome was determined by many small features of institutional design, themselves the result of prior political conflict, rather than any fundamentally distinctive features of the protagonists. In section III we extend this argument to the transformation of the traditional artisans into a class of craftsmen ultimately allied in their own small-firm economy with the world of the agricultural cooperatives.

Intertwined with discussion of the "why" of Danish developments is dis-

[6] Alice G. Brandeis (trans.), *Democracy in Denmark Part II: The Folk High School* (Washington DC: National Home Library Foundation, 1936).

cussion of the "how." Here the focus is on the principles of design that the small holders applied in institutionalizing cooperation among themselves. The cooperative movement was a movement in that it grouped engaged citizens convinced that they were members of a community of fate and willing to make common political cause to defend it. But no one was more aware than the members of this movement themselves that their sense of fellowship was not sufficient to regulate complex, continuing cooperative relations in production, where even well intentioned differences in the interpretations of rights and responsibilities could easily lead to divisive disagreement. They therefore gave extraordinary attention to building institutions that monitored everyday activities not only to detect honest mistakes or cheating, but also to improve the performance of the individual cooperators and hence the competitiveness of the cooperative. Two principles, we argue, guided their efforts.

The first was democracy: decisions were by majority vote of the members, regardless of the relative size of their holdings. This guaranteed that rules would not favor larger or otherwise distinctive classes of cooperators over others. In a situation where, as we will see, the very creation of a cooperative competitor to established mass producers created incentives for the weakest members of the cooperatives to switch sides, this was indispensable. The second principle, widely applied but seldom stated as an explicit rule, was to tie acquisition of the information needed to monitor compliance with the cooperative's agreements with the acquisition of information useful for improving performance. Thus in determining, for example, the fat content of a member's milk and hence its value according to an agreed compensation schedule, the cooperative also gained knowledge of the performance of a particular milk cow. Combined with information on breeding and feeding, this could be used to improve yield and quality and reduce production costs. Thus we argue that the Danish success rested not just on a fortunate concentration of historical developments, but also on an institutional inventiveness that allowed the Danish small-producers to link maintenance of their solidarity to continuous improvement of their joint competitive condition.

In section III, finally, we look at the "how come": If the Danish system of cooperation was so robust, how come the cooperatives were riven by internal conflicts and shaken by contests for control of markets with fellow institutions in the 1960s and 1970s? The answer, we argue, has to do with the shortcomings of the system of federation that connected cooperatives in the same or related industries. Whereas the members of each cooperative were connected to each other and to their manager-agents by the principles of democracy and what might be called learning-by-monitoring, the cooperatives were connected to each other only by loose rules of confederation. So long as markets were stable or stabilized by administrative regulation, we argue, this was of little consequence. But as markets began to shift, creating uncertain possibilities for expansion but also making inaction potentially risky, the managerial elites of the cooperative began maneuvering for position. Unconstrained by any higher-level institutions, they touched off a series of self-reinforcing fights that have shaken the foundations of the cooperative movements. The turbulence they have created, however, has not worked to the benefit of either large or small units of production; and hence this time of troubles illuminates from a new perspective the importance of institutional regulation in stabilizing cooperation without casting doubt on the viability of the small-holder project under a favorable variety of circumstances.

Tietgen's trusts and the mass production strategy

From 1870 to World War I, the leading Danish industrialists tried to build a mass production, mass-consumption economy. Their deliberateness recalls the zeal of such American counterparts as J. P. Morgan. Like him, they organized industrial trusts which controlled their own raw materials, transformed them into finished products by means of continuous process technologies, and distributed the goods through allied transportation firms and sales outlets. The Danish trusts were controlled by three banks and the banks were controlled by a small group of Copenhagen financiers, estate owners, merchants and industrialists. Had all this group's plans succeeded, the country's economy would have become a confederation of giant, mass production firms. But at crucial moments the strategy of centralization was checked by opposition from small-holder peasants and small-firm craftsmen with an alternative plan of industrialization. The result was a stalemate. The trusts' origin and strategy are the themes of this part; the success of the small holders' counter project the subject of the next.

In the last third of the nineteenth century Denmark was ripe for industrialization. After forty years of exporting wheat to Great Britain, the country's farmers were rich. Their wealth and increasing specialization in agriculture made them both a source of capital[7] and an attractive market for the consumer goods which they no longer produced themselves or purchased locally.

The technological know-how on which industrialization depended was also widely available in the country and traditional guild barriers to its application were widely removed. Beginning in the late eighteenth century, the Danish crown followed the mercantilist example of France and Prussia and founded establishments to produce weapons, clothing, paper and porcelain by the most sophisticated techniques of the day. A loose coalition of royal officials, military officers, big merchants, leading artisans and university professors saw to it that Denmark applied foreign technological developments through the first half of the nineteenth century.[8] The Royal Danish Academy of Science and Letters (founded 1742) was reorganized to better serve this purpose in 1815; a polytechnical institute was founded in 1829 to create a technical elite that could

[7] Richard Willerslev, *Studier i Dansk Industrihistoire, 1850–1880* (Copenhagen: E. Harck). p. 190; Ole Hyldtoft, *Kobenhavns Industrialisiering 1890–1914* (Herning: Systime, 1984), p. 117.

[8] H. M. Kisling, *Industrien og dens pionerer i Danmark* (Copenhagen: Borgen, 1978); Camillas Nyrop, *Bidrag til den Danske industris historie* (Copenhagen: Ursins Efterf, 1873); Hyldtoft, *Kobenhavns Industrialisiering*, p. 46.

operate outside the restrictions of the old guild system, and an industrial association was formed in Copenhagen in 1838 to coordinate attacks on the guild rules which still regulated most branches of production with projects to introduce modern technology into the economy.[9] Two decades later, in 1857, the guilds were stripped of their legal powers and enterprise was officially free.

Much of the physical infrastructure of industrialization, moreover, was also in place. In the 1860s, a modernizing coalition similar to one that had boasted Danish technology mobilized British capital to construct a railroad network. As in Prussia, Russia, the Austro-Hungarian Empire, Italy and France, the network was laid out to consolidate trade within the nation while obstructing entry of foreign merchants and armies. In practice this meant diverting trade from Hamburg to North Sea ports on the west coast of Jutland.[10]

The cumulative result was a virtuous cycle of industrialization. As transportation costs dropped and trade restrictions loosened, markets expanded sufficiently to justify investment in expensive, new continuous-process technologies. From the 1840s on, the first brewers, distillers and paper manufacturers to adopt the advanced technologies increased their market share at the expense of competitors who continued to serve local markets by traditional means.[11]

These transformations culminated in and were soon reshaped by the creation of the three merchant banks which dominate Danish finance to this day: the Privatbank in 1856, the Landsmansbank in 1871, and the Handelsbank in 1873.[12] Inspired by the French *Credit Mobliér* and German *Gründerbanken*, these banks collected capital from court circles, foreign lenders and the wealthy national bourgeoisie and then used the money to finance new projects or business consolidations. As in the new German Reich and in the United States later in the century, it was banks which tried – often successfully – to introduce order into industries where a mad rush to deploy unpredictably expensive equipment or fear of such a rush produced a ruinous mixture of frenzy and paralysis.

The first and greatest of the Danish financiers was C. F. Tietgen. A cabinet maker's son, he began his career as a merchant's apprentice, completed his training in Manchester, established himself as an importer and exporter, and became – at age twenty-eight – the first director of Privatbanken in 1856. Tietgen's grand goal was to secure Denmark's economic independence – and his bank's fortune – by organizing domestic industry so efficiently that it could compete in export markets. Three complementary strategies served this end.[13]

The first was to secure control of the crucial industrial infrastructure. Through the 1860s, Denmark's international telegraph connections ran through Hamburg. After Bismarck's North German Reich seized control of the Danish

[9] Camillas Nyrop, *Industriforenngen i Danmark* (Copenhagen, 1888); Johs Kristensen, *Dansk Ingeniørforening genmen 50 år, 1892–1942* (Copenhagen: Dansk Ingeniørforening, 1942).

[10] Aage Aagesen, *Geograkiske Studier over Jerbanerne i Danmark* (Copenhagen: Det Kgl. Danske Geografiske Selskab, 1949), Kulturgeografiske skrifter 5; Willerslev, *Studier i Dansk*, p. 237.

[11] Willerslev, *Studier i Dansk*, pp. 124ff. [12] *Ibid.*, pp. 221–227.

[13] J. Lehmann, *C. F. Tietgen. Hans liv og virke* (Copenhagen: Jespersen & Pio, 1929); M. Vibæk, *C. F. Tietgen og hans samtid. Krisen 1857* (Copenhagen: Schönberg, 1943).

duchies of Schleswig and Holstein in a brief war in 1863–64, the Danish authorities were determined to secure independent access to international telegraphic traffic, and especially to Great Britain. Tietgen used his Manchester connections to organize the import of the necessary cable, and in the immediate aftermath of the war founded a company to lay it.[14] During the same period he consolidated numerous steamship companies into a monopoly – *De Forenede Danske Dampskibsselskaber* – which controlled "everything that then steamed under the Danish flag."[15] This company eventually dominated the Danish share of the crucial trade route between Great Britain and Denmark. Finally Tietgen purchased a few of the provincial private railroads which the state allowed to connect to its own trunk lines.[16] But because he had been preempted by the state, Tietgen never achieved anything like the control over domestic transportation exercised in the United States by the Morgan Bank; and this shortcoming, together with his inability to control trade in foreign bottoms hindered him and his group as we shall see, from monopolizing crucial sectors of domestic industry.

Tietgen's second strategy regarded the reorganization of particular branches of industry. The guiding idea was to buy or establish the most modern continuous-process factory in a given industry, drive down prices through low production costs and bank subsidies, and force competitors to sell out to Tietgen's firm. Following such consolidation, production capacity was to be expanded only as the growth of domestic and export markets assured that new plants could be steadily utilized. All this was to be financed by the highly profitable sale of a minority interest in the new corporation through a public offering of stock. The bank would exercise control through the shares it continued to hold. Tietgen's circle was fully aware of the cost advantages of vertically integrated, continuous flow production; but they found early on that those advantages could only be realized under stable conditions.[17]

The history of the Tuborg bottled beer factory, founded in 1873, illustrates how the circle of Privatbanken learned what to do and not to do. The original idea was to construct a complex consisting of a brewery and bottle factory, but also including fertilizer and sulfuric-acid plants, that used byproducts of the brewing process. Most of the output was intended for export.[18] But the market proved less stable than expected, and within a few years the firm had to reduce its fixed costs by closing the bottle and fertilizer plants. Nonetheless, the brewery survived; and by 1891 it had become the core of The United Breweries, which included 11 other previously independent firms. The Danish beer monopoly was complete when the last major independent, Carlsberg, joined the corporation in 1911.

Tietgen's sugar monopoly, on the other hand, started with the purchase of two refineries. The next step was to construct several factories for the production of sugar syrup, and to guarantee these a supply of sugar beets through signing

14 Lehmann, *C. F. Tietgen*, pp. 154ff; Ole Lange, *Finansmaend, Stramaend og mandariner* (Copenhagen: Gyldendal, 1978). 15 Lehmann, *C. F. Tietgen*, p. 205.

16 *Ibid.*, pp. 246–252. 17 Willerslev, *Studier i Dansk*, pp. 229ff.

18 Lehmann, *C. F. Tietgen*, p. 222.

long-term contracts with growers. During economic downturns De Danske Sukkerfarikker bought up refineries and syrup plants from the competition, incorporated the efficient ones into the firm and closed the others. That there was no over capacity in the industry was due to "D. D. S. F.'s self-interested policy of not expanding."[19] The story of the Danish United Alcohol Factories, founded in 1881, is similar.

Not all of Tietgen's efforts were, however, directed towards the reorganization of continuous-process industries. The major exception was the acquisition in 1872 of Burmeister & Wain, the leading Danish producer of machinery, ships and marine motors.[20] Here a central motive was to cut the costs of the steamship monopoly's vessels while simultaneously stabilizing the engineering firm's market for its products.[21] But mass production principles were never applied to B & W's organization as rigorously as in the Privatbank's continuous-process industries; paradoxically, in fact, the only departments of the firm which did rationalize production in this way were those which turned out machinery for the regional agricultural economies which later became the chief obstacle to the monopolists' expansion.

The third strategy was also the boldest and least successful. It aimed at nothing less than the reorganization of Danish retailing. The idea, put forward in 1893, was to establish through the sale of stock, a chain of retail outlets which would bypass the traditional network of wholesalers and retailers and distribute directly from producers in various industries to consumers across the country. This would have allowed Tietgen to cut costs to the consumer while simultaneously raising the producers' prices and expanding their markets. His system of monopolies would have been less dependent on the vagaries of the export markets as mass production was more closely tied to mass consumption in the home market. But the resistance of the provincial distributors proved insurmountable; and even broad segments of the Copenhagen commercial elite were disquieted enough at the prospect of such a thoroughgoing concentration of power that they attacked Tietgen vociferously in the capital's business clubs.[22]

Tietgen's failures did not, however, mark the outer limits of the monopoly movement. Alongside his empire there arose similar monopolies or cartels in the paper, match, coffee-roasting, jam, cement, sulfuric acid, margarine, cable-and-wire, telephone equipment and bicycle industries.[23] Often these were connected with the introduction of new technologies and were often developed in Denmark

19 L. V. Birch, *Sukkerets historie* (Copenhagen: Gad, 1909), p. 132; Lehmann, *C. F. Tietgen*.

20 Camillas Nyrop, *Bidrag til den Danske industris historie* (Copenhagen: Ursins Efterf, 1873), pp. 3–63. 21 Lehmann, *C. F. Tietgen*, p. 216.

22 Jul. Schovelin, *Tidens Hjul og Tietgen* (Copenhagen: Nordisk Forlag, 1929). This conflict had a powerful effect on Tietgen's general plans for the reorganization of Danish society. Among other things it caused him to give up his masterplan to recognize firms not by trade or line of business but by size, with large corporations set against small enterprises (*ibid.*, pp. 562–563).

23 Fred Nørgaard, "Folkenhøjskeler, Efter- og Afterskoler," in S. Dahl (ed.) *Danmarks Kultur ved år 1940, vols. I–VIII* (Copenhagen: Det Danske Selskab. Det Danske Forlag, 1942), pp. 124–126; Willerslev, *Studier i Dansk*, p. 233; Sevend Aage Hansen, *Økonomisk Vækst i Danmark, Vol. I: 1720–1914* (Copenhagen: Akademisk Forlag, 1972), p. 289; Hyldtoft, *Kobernhavns Industrialisering*, p. 380.

(F. L. Smidt's rotating oven for the continuous production of Portland cement). They were also among the first to use advertising and other mass-merchandising techniques to increase domestic sales. A well-known example is Otto Mønsted's margarine concern which, besides its Danish plants, established the world's largest margarine factory in London.[24] These firms were if anything, more aggressive in their application of modern methods than the Privatbank's: F. L. Smidt employed 44 civil engineers in the development of production equipment,[25] and just as the export market was used to justify large-scale production of consumer goods, so too was it used to absorb production of long series of the specialized machinery developed to make these goods. In 1913, for example, Denmark imported roughly as much machinery as it exported, and one third of its exports were cement machines while another third were machines for food processing.[26]

There were in addition efforts to develop mass retailing, though these were nowhere as systematic as Tietgen had wished. The new department stores were founded by merchants with key positions in traditional putting-out systems – networks of domestic procedures who used simple tools to work up raw materials supplied by the merchants – which made clothing, shoes, and household articles. The Magasin du Nord, for example, began as a dry-goods store in Aarhus in 1871, set up branches in various provincial cities, and expanded to Copenhagen in 1876, where it rapidly became the city's largest retailer. Soon the firm was producing clothing for its own account. Crome & Goldschmidt developed in the same way, while two other large Copenhagen department stores, Messen and Illum, originated in the capital.[27]

Collectively, the influence of Tietgen's circle and its successors was substantial yet circumscribed. This group dominated the new joint-stock companies in Copenhagen which aggressively pursued a strategy of expanding into export markets by obtaining economies of scale in large, capital-intensive domestic plants.[28] The three most prominent banks were represented on one third of the boards of directors of the country's 90 joint stock companies in 1905, and the third accounted for 56 percent of the total assets of the publicly held corporations. By 1914, their share of publicly traded capital had risen to 61 percent and the kronor value of their holdings had doubled. A study in 1923 indicated that this percentage was still increasing. Just how much of this growth was associated with a shift to large-scale production is hard to say. But a crude indicator is the rapid growth of employment in the largest firms. Between 1897 and 1914, for example, the number of workers in firms with more than six employees – the upper limit for *håndværk* or artisan shops – increased by 45 percent from 78,000 to 115,000, while the number of workers in firms with more than 100 workers increased 85 percent from 34,000 to 63,000.[29]

[24] Ejnar Cohn and H. N. Skade, *Margarinendustrien 1883–1933* (Copenhagen: Otto Mønsted A/S, 1933), pp. 63–66; Hansen, *Økonomisk Vækst i Danmark*, p. 294.
[25] Hyldtoft, *Kobenhavns Industrialisering*, p. 316. [26] *Ibid.*
[27] Hansen, *Økonomisk Vækst i Danmark*, p. 233. [28] *Ibid.*, pp. 232, 294. [29] *Ibid.*, p. 287.

Such indicators aside, the industrialists and engineers in the technologically dynamic firms conceived of themselves as an elite with distinct interests. For example, in 1910 Alexander Foss, a director of F. L. Smidt and a leading figure in the increasingly self-conscious and well-organized engineering profession formed an Industrial Council (*Industrirådet*). Its purpose was to represent the cause of the modern corporations in opposition both to the more traditional manufacturing firms organized in the old Industrial Association and the artisans who dominated the Common Council for Industry and Crafts.

But while the power of this new elite was hardly illusory, neither was it irresistibly expansive. Despite its industrial successes and rapid introduction of the newest technology, Denmark continued to secure its prosperity through agricultural exports. As late as 1914, agricultural exports were 10 times greater than industrial exports; agriculture provided 30 percent of the total national income compared with 10 percent for industry[30] and 38.5 percent of the economically active population was employed in agriculture.[31] Moreover, this large agricultural sector and the provincial artisan and merchant class to which it was connected proved remarkably able to evade the bankers' control. Tietgen's sugar monopoly was the exception, not the rule. The trust's efforts to organize the dairy and slaughterhouse industries on similar lines were partial successes at best. Before the large firms could take control of the processing and dictate the conditions of supply to the farmers, the latter, as we will see in a moment, collectively organized processing plants of their own.

Barred in this way from domestic expansion, the bank circles could extend their reach either through international trade or speculation. They tried both. It is indicative of the domestic stalemate that by 1914 – seventeen years after its founding – the largest Danish firm was the Østasiatiske Kompagni, a shipping and trading conglomerate which also established domestic plants to process such raw materials as soy beans which it imported.[32] The surge of speculation came later in the 1920s, as the Danish economy rode the waves of frenetic rationalization and stagnation which broke over the Western European economies in the aftermath of World War I. With little prospect of productive, long-term investment, the banks and the wealthy circles closest to them sought short-term profits in stock-market manipulations that shocked the public when they came to light in 1922–23.[33]

The Copenhagen elite was thus unable to realize its pretensions to direct and speak for the national economy. Its failures were the mirror image of the successes of the cooperative movement of peasant small holders in the provinces, to which we turn next.

30 *Ibid.*, p. 11. 31 *Ibid.*, p. 203. 32 Lange, *Finansmaend*, pp. 164ff.

33 Søren Mørch, *Det Store Bankkrak. Landmandsbankens Sammenbrud, 1922–23* (Viborg: Gyldendal, 1986).

Yeoman society and the cooperative movement

By the late nineteenth century, Danish agriculture was dominated by a large and wealthy class of small freeholders. Their cooperative banks, stores, dairies, slaughter houses, marketing agencies and purchasing bureaus were the envy of cooperative movements the world over, and the immovable object that, in contemporaries' eyes, overcame the monopolists' apparently irresistible force. This freeholder society had deep roots in the particularities of Danish feudalism, and especially in an implicit alliance between the crown and the peasantry. The monarchy protected the direct cultivators against the nobles' attempts to usurp their rights, on condition that the small holders remain on the land and refrain from expanding at the expense of their neighbors. The resulting property regime then became the matrix for a series of bold institutional experiments, partly inspired by foreign examples and constantly informed by indigenous religious beliefs, through which the Danish peasantry placed its legal prerogatives on a competitive economic foundation. The savings-bank movement of the 1860s to 1880s, the cooperative movement of the 1880s and 90s, and the cottagers' movement of the turn of the century mark the crucial steps in this transformation of a particularly fortunate servile peasantry into a proud and innovative yeoman society.

The feudal matrix

In the late middle ages, Danish feudalism, like the other feudalisms that stretched from Russia to England, was essentially a four-cornered struggle among nobles, the crown, peasants, and urban merchants and artisans over the division of the agricultural surplus. The nobles needed a king to coordinate defense of their semi-autarkic estates against foreign enemies and occasional revolts of the peasant cultivators. But the nobles wanted to limit their and their serfs' military service, and to keep the greatest possible share of the feudal dues collected in kind, cash, or labor from the cultivators. This meant reducing the crown's take to a minimum either by subverting or changing the laws governing military service and taxation – by force if necessary.

The crown needed the nobles as a military elite, as tax collectors on the estates, and as the nucleus of a central state administration. But it also feared them as competitors for the peasants' harvest and potential usurpers. Hence the monarch's periodic efforts to limit substantially the nobility's claim on the peasants while incrementally increasing the latter's royal taxes. Hence too the monarch's periodic efforts to extend the royal domains – often at the nobility's expense – and to populate them with serfs owing dues and military service directly to the monarch.

The urban merchants and artisans needed the crown and the nobles as customers but feared their predatory caprice. The towns' strategy therefore was to guarantee the crown (or the most potent local authority) an annual payment in return for protection from marauding nobles and an exemption from all other obligations. Sufficiently extended, this exemption ultimately amounted to the right to self-government in return for tax payments sufficient to finance protection from marauding nobles.

The peasants, finally, depended on all the others in wars, but were at their mercy during peacetime. If the nobles taxed them too much, they sought relief from the crown; if the crown taxed them too much, they sought relief in the nobles' protection. If the crown and nobility agreed in a division of spoils which was cumulatively unbearable for the peasants, they tried to escape to the cities, where residence conferred freedom from feudal obligations. The outcome of these conflicts and alliances depended on local circumstance. Two polar outcomes defined the range of possible long-term developments.

England marked one end of the spectrum. There the nobility so weakened the crown that conflicts between peasants and lords became essentially local affairs, resolved estate by estate, or rather region by region. From the late fourteenth century on the peasants' bargaining position improved as labor became scarce, partly because of the Black Death, and partly because the growth in demand for agricultural products created greater demand for skilled husbandry. In the end, successful peasants and modernizing estate owners secured abolition of the crucial feudal legal restrictions on the use of property. This cleared the way for a redistribution of land ownership in the countryside, whose beneficiaries were, depending on regional particulars, gentry and nobles, lease-holding yeomen, or small holders. Their gains, of course, were at the expense of a new mass of dispossessed peasants who either worked the land as day laborers or sought better conditions in the emerging industrial cities.

Prussia and Russia marked the opposite extreme. The crown and nobility cooperated much longer there in exercising a condominium over the peasantry. In the fourteenth century, they reacted to the labor shortage caused by the plague by successfully reimposing feudal restrictions on the peasantry's freedom of movement. Historians call this the second serfdom. In the seventeenth and eighteenth centuries, the autocrats created standing armies during long wars, then used these armies to assert control over a bloodied nobility and disoriented, geographically isolated merchant class. The result was a peasant class tied to the land yet so restricted in its freedom that it had neither incentive nor opportunity to invest in more productive methods.

Danish developments arrived at the midway point between these two extremes.[34] The state and nobility were strong enough to prevent the breakdown of the estate system with its characteristic assignment of tax and military obligations

[34] These interpretations are based on Skrubbeltrang's detailed chronicles. See especially his book from 1978, *Det Dansk Landbosamfund 1500–1800* (Copenhagen: Den Danske Historiske Forening, 1978), but also his earlier writings on cottagers: *Husmaend i Danmark gennem 300 ar* (Copenhagen: Det Danske Selskab, 1942); *Den Danske Husmandsbefægelse* (Copenhagen: Det Danske Selskab, 1943); *Husmænd og Inderste. Studier over Sjaellandske landboforhold i perioden 1660–1880* (Copenhagen: Munksgaard, 1940).

to small holders. This ruled out the concentration of land, aggravation of rural class differences, and migratory flows characteristic of England. But subject to this general limitation, the peasants were able to gain more and more secure control over the returns to their investments. This ruled out the creation of the estate systems typical of Prussia and parts of Russia, where a servile peasantry absolved its feudal obligations by working the lords' lands. In Denmark the peasantry had the freedom to prosper from the cultivation of its land, but on the condition that this prosperity not be used to alter the fundamental distribution of agricultural property.

From the late Middle Ages to the eighteenth century, however, Denmark's situation did not differ substantially from its closest neighbors. A fourteenth-century proclamation commanded peasants who had fled crown lands to seek more profitable property arrangements on the estates to return to their original plots. In the next century, the nobles in the east of the country (Sjælland and the islands, Bornholm excepted), sought to end their competition for scarce labor by legally tying peasants to their birth places (*stavnsbåndet*). In 1660, in the aftermath of the 30 Years War that devastated Central Europe and Scandinavia, the country followed the example of Frederick the Great Elector of Prussia and created an absolutist monarchy. Between 1730 and 1733 Christian VI gave nobles in the whole country – Jutland in the West as well as Sjælland in the East – the right to force peasants to choose between cultivating a plot on their home estate or serving in the national militia. This amounted to an extension of the *stavnsbåndet*; and the country seemed to be proceeding rapidly, if belatedly, down the path of the second serfdom.

Two closely related circumstances diverted Danish agriculture from this path. First, there was an extreme shortage of peasant labor as a consequence of the Thirty Years' War and the subsequent Great Nordic War, concluded in 1720. Second, Denmark's central position in the Baltic and North Sea trade routes, and particularly Jutland's proximity to the Dutch Republic and the free city of Hamburg allowed peasants to supplement their income by weaving woolens and other artisanal by-employments[35] – or to run off to freedom in the Dutch merchant marine.[36] The peasantry could thus make the very best of what might otherwise have been a disastrous situation. Peasant families in Jutland frequently bought unencumbered titles to land out of their supplementary earnings or from savings accumulated abroad. Peasant families in Sjælland slowly bargained up their share of the returns from their investments.

Between 1780 and 1790 these bargaining successes were ratified in new laws. The crown established neutral commissions to determine fair compensation for improvements undertaken during a peasant's tenure on a plot. The nobles' rights to demand free labor on their own lands were more precisely defined and limited.

[35] Allan Madsen, *Tekstilproducktionens Historie under overgangen fru feudalisme til kapitalisme. Bidrag til tekstil- og beklædningsindustriens historie i Danmark 1600–1850* (Copenhagen: Copenhagen University, Institute of Cultural Sociology, 1978), p. 224.

[36] Skrubbeltrang, *Det Danske Landbosamfund*, pp. 186, 187.

Their right to oblige peasants to choose between accepting tenure on a smallhold and enrolling in the militia was abolished. But at the same time the crown restricted the right of the increasing number of freeholders to purchase land from less prosperous neighbors. Royal commissions divided abandoned farms among existing holdings so as to increase the probability that families of cultivators would stay on the land without threatening the livelihood of others through expansion.[37] The small holders were thus not given clear title to the land they worked. But the estate owners were so restricted in their privileges that they had reason to sell their residual rights to the cultivators. All this was achieved through manipulation of feudal and absolutist legal forms, in the name of the anti-mercantilist liberalism that dominated so much reform at the end of the eighteenth century. But the outcome was something close to the classical republican ideal of a society of small freeholders.

Agrarian cooperation

Just as the monopolists created merchant banks and joint stock companies to realize their vision of highly centralized production, so the small holders created in the space opened by property reform local banks, schools and cooperatives to realize their ideal of a decentralized, communitarian economy. Taken together, these institutions formed a state within the state, and the small holders viewed them as such. When import of cheap American and Australian wheat ruined the Danes' lucrative grain business in the 1880s, they used their cooperatives to redirect production to profitable markets instead of demanding from the state the same tariff protection that many other European agriculturists obtained. By the 1890s, the small-holder economy was a constituent fact of Danish social and economic life. All the political parties from the Højre (Right), with many supporters among the remaining owners of large estates, to the Social Democrats with their growing base among urban workers were thus forced by self interest to pursue strategies which extended and reinforced the farmers' republic.

Construction of these autonomous institutions was stimulated by and contributed to land-reclamation projects begun in the late eighteenth century. It took institutional form with the foundation of a rural banking system.[38] In the early nineteenth century, estate owners and provincial merchants founded numerous *Sparekasser* (savings banks) for their own immediate convenience and to reduce local expenditures for poor relief by encouraging the other rural classes to save for their old age. By mid-century the flourishing grain trade made these *Sparekasser* so rich that they were lending substantial sums to farmers who were improving their land or purchasing clear title to it. Almost 7,500 new freeholds were

37 Hans Jorgen Jensen, "Landarbejderspørgsmålets udvikling i Danmark fra ca. 1870 ca. 1900," in Povl Engelstoft and Hans Jorgen Jensen (eds.), *Bidrag til arbejderklassens og arbejderspørgsmålets histories i Danmark fra 1864 til 1900* (Copenhagen: Gyldendal, 1931).

38 Henry Bruun, *Den Faglige arbejderbevægelse i Danmark indtil år 1900. 1. del til ca. 1800* (Copenhagen: Gyldendal, 1938).

established between 1850 and 1860, and just over 6,000 more from 1860 to 1873.[39] The more active *Sparekassen* became in these transactions, the more the small holders insisted on creating financial institutions under their own control. Beginning in the 1860s, they founded small savings banks in the *sogne* or parishes (the smallest unit of local government with substantial rights of self determination). These served primarily young farm hands and maids, and artisans just beginning to accumulate the capital to buy land or tools and start a family.[40] Between 1864 and 1874 the number of such banks increased from 17 to 253.[41] Larger banks were set up to serve whole regions (*Landbosparekasser*). Credit unions (*Kredit-and hypotekforeninger*) were also established during this period to jointly finance farmers' mortgages on favorable terms without being dependent on urban creditors.[42] At the same time these institutions encouraged the peasants to advance economically, they served as schools for democratic self-administration. Self-help was becoming synonymous with collective action.[43]

The propagation of this idea was the central task of another network of institutions created in these decades as well: the *højskoler* or high schools. The high-school movement was inspired by the ideas of J. F. S. Grundtvig, a philologist turned priest and parliamentary agitator. Like the Brothers Grimm, Grundtvig was part of the broad early nineteenth-century Romantic movement that tried to establish folk identities through the historical analysis of language and myth. The decisive episode in Grundtvig's development was a visit to Great Britain in the early 1830s intended for the study of Anglo-Saxon manuscripts. He was both shocked and exhilarated by what he saw of British captalism: shocked by the egoism and poverty it fostered, exhilarated by the possibilities for self-development it revealed. He returned to Denmark to preach the spiritual virtues of an economy which allowed the folk to develop its communitarian values without interference from a capitalist class or a bureacratic state.[44] For that to happen, he argued, Danes – and particularly those closest to their agrarian past – had to study both their traditions and those elements of modern science and technology required to construct a competitive economy according to traditional principles. The chief instrument of this education was to be the high school: a private, locally supported institution offering instruction in these areas to adults but also and above all the teen-age sons and daughters of small holders who had finished their formal, public education but not yet assumed control of the family farm. The first *højskole* was founded in 1844; by 1863 only seventeen had been opened, and some of these were already closed. But after the loss of Schleswig-Holstein to Bismarck's new Reich in 1864, a nationalist revival swept the country; by 1870, fifty-three new high schools had been founded[45] and the high-school movement, with its message of communitarian modernization

[39] Niels Clemensen, *Sparekassebevægelsen i Danmark 1810–1914*, p. 33. [40] *Ibid.*

[41] Willerslev, *Studier i Dansk*, p. 190.

[42] Th. Thorsteinsson, "Kredit-og Hypotekforeninger," in Dahl (ed.), *Danmarks Kultur*, pp. 168ff.

[43] Carl Iversen, "Sparekaser," in Dahl (ed.), p. 251.

[44] Dan Charlier Christenen, *N. F. S. Grundtvig. To diologer om højskolen* (Holbæk: Forlaget Odin, 1983). [45] Nørgaard, "Folkehøjskoler," p. 200.

beyond state and market, became an organizational template for the regeneration of rural Denmark.[46]

The culmination of this collective self-assertion was the formation in the 1880s and 90s of the *Andelsbevægelsen* or cooperative movement, linking rural producers and consumers cooperatives to local savings banks. This system was on the one hand an extension of the institutions just discussed, and many of its leaders were *højskole* graduates, teachers or *Sparekasse* officials. But on the other hand it was a response to the fall in grain prices which began to effect Denmark – later than many of its European competitors – in the late 1870s and with slowly increasing severity until 1895. Even before the arrival of the cheap grain, Danish farmers had begun to shift from the highly competitive cultivation of cereals to the production of meats and dairy products whose high quality partially sheltered them from market fluctuations. But the collapse of the grain price made this shift imperative, and the preceding organization of the rural economy made it feasible. Thus when the Right took up the protectionist cause of the largest estate owners in the parliamentary debates of 1885–89, the *Venstre* (Left) party never swerved from its self-confident allegiance to free trade: in the Rigsdag session of 1886, *Venstre's* speaker replied to the finance minister's demand for a tariff on grain by saying that the way of the future was not to sell grain at cheap prices, but to use cheap grain to produce expensive butter, cheese, bacon and ham.[47]

The organizations which made it possible for small holders to do this grew so rapidly, were connected so complexly, and depended on so many technological innovations that – as with all such explosive movements – it is impossible to reconstruct their collective history with accuracy. Indeed, their bewilderingly rapid diffusion is a good demonstration of the breadth and depth of the preceding waves of institutionalization and the disposition to collective action they had fostered. And these favorable conditions most probably explain why the small holders beat large estate owners, provincial merchants, and the Copenhagen mass producers in the race to switch markets. In any case, the vitality, technological creativity, and democratic character of the movement are easily documented.

The first cooperative dairy is generally considered to be the Hjedding Mejeri in Ølgod in West Jutland, a center of the small freeholders and hence the cooperative movement.[48] The farmers evidently saw themselves as pursuing prudent economic strategy, not undertaking an utopian experiment.[49] By 1890, there was 699 cooperative dairies.[50]

[46] *Skoler* and "enlightenment" were considered to be aims in themselves in this period. Hence the creation of free-schools for children and evening schools for adults, along with the founding of technical or craft institutions serving the needs of particular economic groups.

[47] Hans Jørgen Jensen, *Dansk Jordpolitik 1757–1919* (Anden Del, Copenhagen: Gyldendal, 1945), pp. 387–388.

[48] A. Axelsen Drejer, *Den Danske Andelsbevægelse* (Copenhagen: Martins Forlag, 1929), pp. 130ff.

[49] Ole Hyldtoft, *Det Industrielle Danmark, 1890–1914* (Herning: Systime, 1981), pp. 173ff.

[50] *Ibid.*, p. 185.

Technology, democracy and learning

The success of these dairies depended on a series of technological innovations, many of which were apparently inspired by the desire to make small holdings competitive in their new markets. The most prominent example is the Magdlekilde centrifuge, developed by L. C. Nielsen in 1879 to separate cream from whole milk. Earlier experience had shown that vibration of whole milk during transport from scattered small dairy farms to the common dairy resulted in a reduction in the amount of recoverable cream. This made collective butter production prohibitively expensive as compared to the costs of production on a single large estate, where transport distances to the main dairy were substantially shorter.[51] Nielsen's centrifuge solved this problem by increasing the amount of cream which could be separated out at the central dairy. A device for determining the fat content of milk assured all producers that the contributions to the dairy's intake were fairly reckoned.[52] Without such inventions it seems likely that the shift to dairy production would sooner or later have led to a concentration of production on large, geographically compact estates.

The success of the cooperative dairies depended more fundamentally still on their democratic constitution. Shares of the annual profit were calculated according to each farm's contribution to the total output. But each member farm had one vote in the cooperative's supervisory council, regardless of size. In all decisions regarding the cooperative's future, the principle was "to count heads, not heifers." The members guaranteed each others' debts, and the cooperatives were constituted for a term, renewable only by the members' vote.[53] However much democracy was rooted in the ideas of Grundtvig and the *højskole* movement out of which the cooperatives grew, it also made good business sense for the new cooperatives. Particularly in their earliest and most vulnerable years the new institutions were dependent on the loyalty of their economically weakest members. It usually took several years for a new cooperative to establish a reputation for reliability and high quality; during this teething period revenues were likely to be low and, given the need for continuing investments in improvements, distributions to members were to be lower still. Established, large-scale competitors frequently tried to take advantage of the resulting strains by offering above-market prices for products pledged to the cooperative. If they found enough takers, the new venture could be crippled. But absent such inducements, members pinched by the start-up costs could be tempted to improve their situation at the expense of the fellows by cheating: the use of cheap seed or provision of old eggs could cut production costs without being easily detectable. The result, however, would be precisely to

[51] Drejer, *Den Danske Andelsbevægelse*, p. 126.
[52] Hyldtoft, *Det Industrielle Danmark*, p. 176.
[53] Drejer, *Den Danske Andelsbevægelse*, pp. 145–146.

jeopardize the reputation for quality and reliability that the cooperative was trying to establish.[54]

The obvious solution to both problems was for the members to bind themselves at the outset to act in the interest of the cooperative. In the case of new ventures this meant the members' pledge to deliver the relevant farm products exclusively to their cooperative for a term of years; for both new ventures and going concerns the members had to agree to respect whatever rules the cooperative as a whole regarded as necessary to safeguard production. Given that the weakest members of a group could hardly be expected to join in a risky venture if the more powerful ones could make rules to shift the risks disproportionately onto them, majority rule by simple or qualified majority, without regard to the distribution of productive assets, was a precondition for constituting the cooperatives. That way the weak could profit from the dependence of the strong on the success of the joint enterprise.

Rules, even democratically decided rules, were by themselves plainly insufficient to the task of regulating cooperative production. Contracts between cooperatives and member farmers detailing in several pages what kind of seed was to be used for a contracted crop, what could be planted on the plot reserved for that crop before and after the contracted planting, what could be planted adjacent to that plot at any time, and referring the grower to still more detailed guidelines regarding the treatment of seeds, seedlings and maturing plants illustrate the general problem: The member was in effect an independent subcontractor responsible for producing, without supervision, a complex product, some of whose defects might not be detectable by inspection upon delivery. Anyone ingenious enough to master the intricacies of farming could find ways to cut corners; and if some tried, then others would be drawn by prudence or greed to follow suit, and the cooperative would be at risk.

Monitoring of the rules was therefore necessary as well; and as the centrality of the device for determining the fat content of cream to the development of the cooperative dairies showed, the Danes were aware of this from the beginning of the *Andelsbevægelsen*. But the cooperatives went farther, turning a prudent necessity into a productive virtue. It was necessary to check the quality of products to insure compliance with the rules and protect the common reputation. But close examination of livestock, crops or dairy products also allowed for comparisons to lead to identification of superior varieties; combined with information about inputs and care, the same examination could shed light on the effectiveness of various methods of cultivation and livestock raising. The cooperatives had an overriding motive to deter cheating; and every member had an interest in benefiting from the knowledge acquired in connection with monitoring. By systematically collecting the supplementary information that made the results of straightforward monitoring of great economic utility, the cooperatives simultaneously reduced the probability of successful cheating, and

[54] Faber, *Co-operation in Danish Agriculture*, pp. 72ff.

reduced the temptation to try by giving members an above-board way to increase their yields and hence their income. The proliferation of information-gathering systems, often in connection with precisely organized competitions for best crop, cow or pig, testified to the efficacy of this solution.[55]

The cooperatives and the state

The dairy cooperatives quickly became the central nodes of an extraordinarily complex network of complementary activities. Sometimes the complimentarities had technical origins. For example, the extraction of cream at the dairies left the farmers with large quantities of skim milk, which made excellent swine fodder. Pig production increased, and it was an obvious step to found cooperative slaughter houses. Ten of these were established between 1880 and 1890, sixteen more by 1900, and another fifteen by 1914, for a total of forty-one. In that year the number of cooperative dairies had reached 1,168.[56] Other complementarities were the straightforward result of business logic. Thus the more the cooperatives produced, the more fertilizer, fodder and machinery they purchased. New cooperatives were established to buy these wholesale and distribute them to member farms. The more the cooperatives produced, the more they earned and invested. Because many of the *Sparekasser* had by the end of the century become affiliated with the three Copenhagen banks, new cooperative banks were established to pool resources and reduce the cost of credit. The more sophisticated and higher quality their cooperatives' production became, the greater their need for consultants to advise members on the newest techniques and control the quality of final products. And the more complex the network of institutions became, the more important it was to establish higher-level instances coordinating them locally, regionally and nationally.

It was this system which gave substance to Grundtvig's folk romanticism. It succeeded almost without appealing to the state in halting the monopolists from

[55] *Ibid.* It has been forcefully argued that producers' cooperatives of the type considered here are viable economic institutions while other forms of employee ownership are not because the producers' interests are identical: as dairy farmers have a common interest in securing the highest price for processed products made with raw materials they furnish, they plausibly share an interest in jointly owning the processing facilities as well. Where inputs vary and the connection between inputs and output is less clear cut, the argument continues, ownership of the conversion facility by the input suppliers can lead to conflicts of interest, with inefficient results (see Henry Hansmann, "Ownership of the Firm," *Journal of Law, Economics and Organisation* (Fall 1988)). But when, as in the example above, monitoring simultaneously produces efficiency improvements, this argument need not hold. As the parties jointly discover new solutions to common problems, their understanding of what is jointly possible changes, and – in the appropriate institutional setting – their several interests become aligned. The operation of the modern Japanese firm is arguably an example of such transformative coordination of diverse producers' interests. For this case, and an extended discussion of learning by monitoring generally, see Charles F. Sabel, "Learning by Monitoring: The Institutions of Economic Development," in Neil Smelser and Richard Swedberg (eds.), *Handbook of Economic Sociology* (Princeton: Princeton University Press, 1994).

[56] Hansen, *Økonomisk Vaesk i Danmark, Vol. I*, p. 218.

centralizing production through the market. Tietgen's circle, for example, tried to create a slaughter-house monopoly in the 1890s, and in 1914 there were in fact eleven more corporate dairies than cooperative ones. But by the 1920s, the cooperatives accounted for 70 percent of the industry's production, not least because of their superior research, quality-control and expert services.[57] The cooperatives established their own coal-import firms and cement factories to protect themselves against trusts in these industries. They established consumer cooperatives partly as a defense against the efforts of the mass merchandisers to gain control over consumption in the provinces, and partly to re-establish within the cooperative movement itself that direct connection between customer and producer which was characteristic of local economies before the rise of mass production.[58]

But if cooperative self-help was the rule, it was hardly an inviolable principle. The small holders marshaled national parties and administrative agencies to their defense when local action was insufficient. When Otto Mønstred mixed butter in his margarine and dyed the blend butter yellow in 1887–88, the cooperatives declared a butter war. The danger was that Mønstred's "artificial butter" would undermine the cooperatives' reputation for unadultered products abroad, as well as reduce their domestic market. Their solution was a law which defined butter and margarine precisely, required that "artificial butter" be marketed as margarine, that margarine be dyed yellowish but not butterish, and that its butter-fat content be declared on the package. The law also granted the minister of internal affairs to determine whether and under what conditions margarine could be exported.[59]

The small-holders were, finally, all the more disposed to seek help from the state in fundamental questions regarding the definition of property rights and the distribution of land. The central and long-standing issue here was the status of the *husmænd*, the landowners whose plots were too small to secure their families' subsistence. This class did not benefit directly from the late eighteenth-century reforms, and its changing, labile allegiances – now to the rural society, where its labor was indispensable during periods of peak activity, now to the cities where it might be incorporated into a factory proletariat – threatened to change the balance of forces in Danish society in the late nineteenth century.

The problem, however, was much more acute in the east of the country than the west. In Jutland, the *husmænd* question seems to have largely resolved itself. There, we saw, the peasants had by-employments as weavers, fishermen, masons, teachers, millers or tailors. Just as the large farmers used income to buy clear title to their land; so the landless were able to buy plots or increase their size. As early as the last quarter of the eighteenth century many Jyder (inhabitants of Jutland) changed their census designation from farm hand or

[57] Hyldtoft, *Det Industrielle Danmark*, p. 192; Drejer, *Den Danske Andelsbefægelse*.
[58] L. A. Godsk, "Brugsforeningerne," in Drejer (ed.), pp. 57ff.
[59] Ejnar Cohn and H. N. Skade, *Margarinendustrien 1883–1933* (Copenhagen: Otto Mønsted A/S, 1933).

artisan to *husmænd* with land.[60] Through the nineteenth century the *husmænd* continued to expand their holdings by buying pieces of larger farms or, more typically, acquiring plots in fields that were opened for cultivation through massive drainage projects. The upshot was that the rural population in Jutland increased rapidly and stayed put. Whereas the population in the country as a whole increased by 96 percent between 1801 and 1901, the population in two of Jutland's major jurisdictions, Ringkøbing *amt* and Ribe *amt* increased by 263 and 233 percent respectively in the same period. At the height of the agricultural crisis in the late nineteenth century, the rate of emigration from these districts was half the average for rural jurisdictions in Denmark as a whole; and all of these developments were closely related to the sharp rise in the number of small holders – a 63 percent increase between 1850 and 1870 in Ringkøbing, for example.[61]

In Sjælland to the west, in contrast, the *husmænd* problem became more serious and more openly political in the course of the nineteenth century. In the decades before 1870, the small holders prospered from the corn trade while the *husmænd* continued to scrape by. Class differences became more pronounced as the wealthy farmers absented themselves more and more from the society of the small artisans, farm hands, and maids with whom they had traditionally passed much of their working day and leisure time. As the agrarian crisis of the 1880s drew the landless to the cities, the prosperous farmers and the estate owners feared labor shortages at planting and harvesting time.

The solution to the problem, and the keystone in the arch of institutions that rose protectively above the farmers' republic was the *Husmandslov* of 1899, by which the state agreed to subsidize substantially the purchase of small farms and the equipment necessary to operate them. What was remarkable about the law was that it was supported by all parliamentary parties. For the estate-owners in the Right and the prosperous small holders in the Left, the law was onerous because it limited their own possibilities of expansion. But this disadvantage was outweighted by the possibility it created to keep labor on the land while blocking the spread of Social Democracy from city workers to the farm laborers. For Danish Social Democracy meanwhile, support for the law was in fact a way to win the allegiance of the *husmænd* by furthering their ambitions to full inclusion in rural society. In the light of contemporaneous developments in related societies, the outcome of these political calculations was hardly foreordained. For example, the German Social Democrats (except in the Southwest, where conditions resembled those in Denmark) assumed that small-holder agriculture was doomed to give way to the concentration of production on large estates. Hence they were determined to organize the farm workers as a rural proletariat, rather than encourage their transformation into small holders. What better

[60] Skrubbeltrang, *Det Danske Landbosamfund*, p. 409; Verner Bruhn, "Myter og mønstre i historien om Vestjyllands udviklingshistorie," in Flemming Just (ed), *Vestjyllands Udviklingshistories ca1750–1914* (Esberg: Sydjydsk Universitetsforlag, 1984), p. 15.

[61] Ben Iversen, "Bonden i Vestjylland i 1800-tallet," in Just (ed.), pp. 53ff.

measure of the self-evident centrality of the farmers' republic to Danish society than the concurrence of political parties that their conflicting interests were best served by strengthening it?

Property in skill: the artisans and craftsmen

Alongside the agrarian republic there was another estate of small holders, the artisans and craftsmen. Their property was the knowledge of tools, materials, and techniques which made them independent of any one market or employer. By the outbreak of the First World War, they too had built institutions – particularly a network of technical schools – which allowed them to defend their place in Danish society by constantly renewing it.

The experience of the artisans and craftsmen paralleled the experience of the independent peasantry. The Danish state by accident and design helped create the agrarian property order which made the struggle for independence depend on founding collective institutions. The state's attacks on the guilds and its encouragement of new centers of technical training similarly made the defense of individual skill a common enterprise. Compromises among factions within the artisans' own estate, moreover, were as essential to their success as to the agriculturalists. The *husmænd* slowly distinguished themselves institutionally from and then allied on more advantageous terms with the larger free holders; the journeymen (*svende*) – artisans who owned nothing but their skills and tools – formed craft unions in the late nineteenth century to defend their interests against the independent master craftsmen (*mestre*) or factory owners for whom they worked, then joined these latter in expanding and overseeing the schools which guaranteed the regeneration of skill. Moreover, just as the farmers' self-organization united the choices of the large corporations, so too did the artisans': the more skilled the workers became, the greater the firms' incentives to choose work arrangements and make products which depended on those skills. The crafts' political defense of their skills thus helped create an economic demand for them.

But beyond these parallels, there were crucial direct connections between the formation of the two groups. The slow consolidation of the agrarian republic first undermined the artisans' position by opening centers of production beyond their traditional jurisdictions. Yet the farmers' progress eventually contributed to the reconstitution of the crafts by creating large new markets for their skills in the countryside. As the urban and rural crafts fused, the skilled workers with their national institutions and traditions blurred in fact the apparently ineradicable distinction between the provincial agrarian republic and the Copenhagen world of monopolies and mass production. In town and country, in many large firms and most small ones, industrial work in early twentieth-century Denmark was still craft work; and self-employed or not, the craftsmen felt themselves part of an estate as autonomous in its way as the rural cooperatives.

Roughly a century earlier, Danish artisans were divided into three principal groups. The largest was composed of traditional craftsmen – shoemakers, bakers, smiths, and so on – who plied their trades in local, urban markets that were elaborately defined and protected by guild regulations. Outside of Copenhagen, these craftsmen were established in a network of royally recognized trading cities or *købstæder*. The second, much smaller group was the artisan elite which supplied the court and maintained extensive contacts with foreign merchants and craftsmen. This elite group saw itself as the natural leader and representative of the first. Yet in virtue of its greater expertise, experience, and ambition, it was much less constrained by the rules governing products and production methods than those for whom it claimed to speak. The third and most amorphous group were the rural artisans. Depending on their economic and geographic circumstances, they produced for different combinations of local and long-distance markets. Many, we saw, turned to craft work only in the off seasons to supplement their agricultural income. But in any case they escaped control of the urban guilds, who therefore feared and reviled them as unfair competitors. It was the relation of the artisan elite to the court circles and the state on the one hand and the emergent centers of provincial production on the other which led to the extraordinary consolidation of skilled workers in Denmark.

The artisan elite was itself a hybrid. One part consisted of artisan-artists who supplied the nobility's demands for luxury and art: Cellini was their spiritual master. Another part consisted of artisan-merchants. These were wealthy tradesmen who had built up or inherited firms much larger than those normally permitted in their respected crafts. Their successes depended on entrepreneurial capacities to build up local production networks and trading relations, but also on technical prowess and royal patronage, which conferred exemption from the guild rules regarding both the size of establishment and the use of technology. For example, in the late eighteenth century, members of this part of the group were granted the privilege to establish royal manufacturies producing porcelain, armaments, chronometers, and the other technologically demanding goods of the day. Finally, the group included clerics, state officials and scientifically versed aristocrats, all of whom wanted to further the nation's economic interests by enlightening – that was the omnibus term – the mass of artisans through a program of combining moral uplift with technical instruction. In Denmark this elite was powerfully influenced by the example of nearby Hamburg (from where it drew many advisors through the early nineteenth century); but there were analogous hybrids in cities as different from Copenhagen as Paris.

At one pole within this elite were those who saw the guilds as the antithesis of Enlightenment. The most extreme advocates of this position wanted to abolish the guilds and let the competitive struggle determine the artisans' use of the new technological and economic possibilities. At the opposite pole were those who held that private and state-supported instruction would lead to reform of the guilds from within. In that case they eventually could be integrated into the Third, bourgeois, Estate, which in the aftermath of the French Revolution was inevitably seen as a crucial political force. Under the pressure of foreign competition and out of concern for domestic order and posterity, this Copenhagen elite combined elements of both positions into a series of decisions between 1800 and 1880 which molded the Danish artisanry and the skilled workers that it in turn formed.

The substance of these decisions recalls the reform of agriculture. In the crafts, too, *de facto* property rights were redistributed so that small holders could mutually protect their livelihoods, but on condition that each made fullest possible use of the economic opportunities. On the one hand, the state moved – slowly – to abolish restrictions on competition. This program culminated in the *Næringslov*, which became effective in 1862 and granted Danes the right to pursue any occupation they chose, subject only to such regulation as might be established by public authorities. On the other hand, the state encouraged the creation of vocational schools, gradually assumed partial financial responsibility for their extension and maintenance, and – above all – granted the artisans' representatives the right to participate in the administrative supervision and ongoing reform of this vocational education. By granting the artisans rights of substantial self-determination within a state-supported training system, the public authorities in effect allowed the guilds to reintroduce key elements of their former control over labor markets – for example, completion of a well defined apprenticeship as the precondition for many attractive careers. But the artisans' rights were now exercised within the control of (more) competitive markets where their control over their own skills was no longer directly connected to control over the establishment and organization of production units. Several episodes illustrate the evolution of this compromise.

Activities at the Royal court in 1800, for example, anticipated much of the subsequent debate. In that year, N. H. Massman, pastor at the German church in Kristianshavn, established a Sunday school to teach artisans arithmetic, reading and writing and eventually technical drawing and ship construction. Massman used church facilities to house the school and financed it from private contributions, the proceeds of public concerts, and a grant from a foundation for the advancement of Danish industry established by the testament of a state counsellor in 1793.[62] Almost immediately his efforts were imitated in provincial *Købstæder*. Also in 1800, however, the king issued an ordinance which forbade the guilds to do anything which "bore the stamp of the old days' wild and rough customs," commanded apprentices to study at the Royal Academy of Arts (which had previously been charged with providing them a technical education) and abolished the journeymen's obligation to complete an itinerant period of *Wanderjahre* before establishing a shop of their own. As compensation for this change in the rules, the artisans were to have the right to stand for election to the thirty-two-man patriciate which was responsible for much of Copenhagen's government. Indeed, henceforth, half

[62] Nyrop, *Bidrag til Dansk Håndvaeker-undervisnings histories*, pp. 57–58.

of this council was to be composed of "artists, factory owners, and artisans."[63]

A second burst of reform came in the late 1820s, when Copenhagen was still suffering from the destruction of the Danish fleet by Great Britain during the Continental Blockade (1806 to 1814), the rural economy was stagnating, and official impatience with the artisans' privileges was at a peak. A commission originally established to improve the guilds' training system became so fascinated instead with French and German models of combining science and industry in a radically new way that it proposed creation of a wholly independent institution: *Den polytekniske Læranstalt*. Its goal was to educate a new technical elite which would ultimately substitute the old master craftsmen as an entrepreneurial class.[64] The polytechnic did not break the power of the guilds; but throughout the rest of the century the civil engineers it trained competed – often successfully – for influence over the emerging system of technical education with representatives of the craft.

The next reform episode was the creation of the *Industrieforening* (Industrial Association in 1838) *Håndværkerforeningen – København* (The Copenhagen Artisans' Association) in 1840, and – as direct extensions of the latter – the creation of *Det Tekniske Selskab* (the Technological Association) and the *Håndværker-Læreanstalt* (Artisans' Education Institution) three years later. The *Industrieforening* was clearly inspired by *laissez-faire* ideas; its goal became the promotion of technically modern crafts free of guild restrictions – what came to be called a *kunstindustrielle* (applied arts) program. The other institutions had in contrast more ambiguous aims. All were founded by J. L. Kramp. A cabinet-maker's son, Kramp was forced to enter his father's trade. He then apparently decided to realize his own ambitions to enter high bourgeois society by helping the most dynamic artisans to prosperity and social respectability. His Teaching Institution was aimed at providing instruction in as many crafts as possible. It soon became a model for and consultant to similar institutions in the provincial *købstæder* and the organizational point of reference for the artisans determined to retain control over the modernization of their crafts. Kramp himself was vehemently opposed to the *NæringsLov*. But his opposition did not grow from attachment to traditional privileges. Rather he wanted state supervision of *laissez-faire* in Denmark to be coordinated by men of his circle, not by polytechnicians.[65]

By the late 1870s, Kramp's position was winning out. General prosperity had protected the artisans from the potentially disruptive effects of the *NæringsLov*. But when times turned bad for the artisans in the 1870s, they demanded state

[63] *Ibid.*, pp. 55–56.

[64] Svend Winther Kristensen, *Kampen om mælken: En erhvervshistorie om Mejeriselskabet Danmark* (Landbohistorisk Selskab Odense, 1942), pp. 1, 16–17; Bruun, *Den Faglige arbejderbevægelse i Danmark indtil år 1900*, p. 8; Peer Hull Kristensen and Jan Annerstedt, *Innovative Kapabiliteter i de Skandinaviske lande* (ISP, Roskilde University Center, 1980), pp. 180ff.

[65] Nyrop, *Bidrag til Dansk Håndvaeker-undervisningshistories*, p. 172.

support for the technical schools; and the only institution with the knowledge of provincial conditions required to administer the public funds approved for this purpose in 1876–77 was Kramp's *Tekniske Selskab*. Three years later the artisans formed a peak association – *Fællesrepræsentation for dansk Industri og Håndværk* – which lobbied successfully for increased subsidies. From then on, despite changes in its legal status and the inclusion of other groups, *Det Tekniske Selskab* and its successor organizations became quasi-public bodies which slowly reasserted the artisans' control over skilled labor in Denmark by defining and limiting the distribution of certificates of technical accomplishment.[66] The culmination of this line of institutional development was the creation in 1907 of the *Teknologisk Institut*. This agency, founded at the urging of the *Fællesrepræsentation*, acted as a technical consultancy to artisans and small firms, trained and accredited master craftsmen, and distributed investment subsidies.[67] Thus as urban and rural crafts reacted to the pressures of mechanization in the decades before World War I, the artisans used these institutions as a means of forward defense, and were inevitably transformed by their participation in them.

The situation of the various crafts was extremely diverse, and several did not survive the transition to mechanized production. Shoemakers and tinsmiths, for example, lost their markets to cheap, mass-produced goods. In the garment, textile, knitwear, glove, and cigar industries, the introduction of cheap machinery and the availability of low-wage labor resulted in the creation of putting-out systems similar to those in the sweated trades in London, New York or Hamburg: A merchant organized a network of small masters, who in turn organized a network of home workers, and the more powerful struggled to pass the costs and risks of production to the weaker.[68] The metalworking, machine-building, printing, lumber, and furniture trades, on the other hand, flourished under the new conditions.

The Copenhagen furniture industry was exemplary, both in that it exhibited in the clearest form traits characteristic of the others, and in that through Kramp's Tekniske Selskab it presented itself as a model for the artisanry as a whole. The core of the industry was a group of furniture stores established by the leading master craftsmen. Besides furniture, these stores sold interior decorations ranging from curtains and carpets to iron fittings. They were supplied partly by goods from the masters' own workshop; but the largest share of their stock came from small subcontractors in this industry or from the shops of other masters and *their* subcontractors in other industries. This vast network of small producers renewed itself technologically by introducing better and better special-purpose machines. Beginning in the last decades of the century, these were driven by

66 Erik Rasmussen, *Velfærdsstaten pa vej 1913–1939, Politikkens Danmarks Historie*, vol. 13 (Copenhagen: Politikens Forlag, 1971), pp. 74, 81ff.

67 Aa. S. Schou (ed.), *Fagskolen for Håndvaerkere og mindre Industridrivende* (Copenhagen: Teknologisk Institut, 1906–31).

68 Willerslev, *Studier i Dansk Industrihistorie*, pp. 97ff; Hyldtoft, *Kobenhavns Industrialisering*.

fractional horsepower gasoline and electric motors. In 1897 only four firms employed more than fifty workers.[69] Individual shops specialized in the production of, for example, chairs, frames, beds, or baby carriages; but all benefited from the services of cooperatively owned lumber mills and wood wholesalers. The trade's attention to quality and style made it a forerunner and instigator of the early twentieth-century *Skønvirke* (Pursuit of Beauty) movement. *Skønvirke*, like similar movements in Austria, Germany, France and Great Britain, aimed to foster cooperation among artists, architects, and artisans so as to raise the aesthetic value of commercial products. It was in large measure due to the movement's efforts that the Danish silverware and graphics industries – all organized on the furniture model – established their reputations for excellent quality and design.[70]

The new craft unions played a crucial role in stabilizing these small-firm production systems. The same competitive pressures which led master craftsmen to lobby the state through the *Fællesrepræsentation* in the late 1870s and to coordinate their wage and training policies by founding branch organizations (the forerunners of the employers' associations of the 1890s) also led the journeymen to transform local artisans' associations into trade unions that represented their interests as against the masters'. For many journeymen, in fact, the unions were a substitute guild system whose role was to protect them against employers who had abandoned traditional principles of equity.[71] There were numerous conflicts between the two groups through the 1880s; but by the end of the decade these conflicts were resolved through the definition of two closely related and overriding common interests.

First there was their joint interest in regulating the conditions of wage determination in each industry. If any shop started to undercut the others by paying lower wages or using inferior materials for the production of the same goods, the others could be forced to follow suit. The result, familiar to contemporaries, would be shoddy products and destruction of the tissue of cooperation which assured the industry's long-term capacity to produce innovative and high-quality goods. An obvious way to avoid this possibility was for the union to police the industry, thus assuring that all firms paid standard wages and maintained acceptable working conditions. Again the furniture makers were exemplary: In 1885 the journeymen succeeded in obtaining a 20 percent wage increase in return for an agreement to refuse work in shops paying substandard wages.[72]

The second joint interest regarded regulation of entry to the labor market as a whole. The skilled workers were as interested as the independent master crafts-

[69] Hyldtoft, *Kobenhavns Industrialisering*, p. 179.

[70] Vilhelm Slomann, "Kunsthåndværk," in Dahl (ed.), *Danmarks Kultur*, pp. 83–91.

[71] Bruun, *Den Faglige arbejderbevægelse i Danmark indtil år 1900*, pp. 280–287; Georg Nørregård, *Arbejdforhold indenfor Dansk Håndvaerk og Industri, 1857–1899* (Copenhagen: Gyldendal, 1943/77), pp. 173–175.

[72] J. Jensen and C. M. Olsen, *Oversigt over Fagforeningsbevægelsen i Danmark. I tiden fra 1871 til 1900* (Copenhagen: De Samvirkende Fagforbund i Danmark, 1901), p. 38.

men in both guaranteeing and controlling the supply of journeymen. Their unions therefore joined the *Fællesrepræsentation* and the *Industriforening* in securing passage of a new Apprenticeship Law in 1889. This legislation stopped short of reintroducing journeyman certification as a precondition for skilled work (*that* became law in the 1920s). But it did oblige masters to send their apprentices to technical schools, thereby strengthening the position of the latter as the arbiters of skill and establishing a *de facto* definition of minimal craft qualifications in each trade.[73]

In the following decade, however, this cooperation was endangered by a series of labor conflicts provoked by large Copenhagen firms determined to control wage costs and the organization of trade unions. Employers' associations in the small-firm industries participated only grudgingly in the lockouts of those years. The workers response was to federate the single craft unions into a peak association – *Den Sammenvirkende Fagforeninger* – (DsF, later renamed the *Landesorganisation* or LO). This federation pooled its members resources and acted politically through an alliance with the Social Democratic Party to meet the employers' attack. In 1899, after a major lockout, the employers' association agreed to recognize the DsF as its negotiating partner. In return the latter accepted the employers' fundamental right to organize production.[74] Yet this agreement did not change the pattern of cooperative labor relations within particular industries and shops.

Denmark thus entered the twentieth century with a two-tiered system of industrial relations uniquely clear in its articulation. At the national level a militant trade union confederation closely allied to a Social Democratic Party was already elaborating the negotiating rules which would become typical of the post-World War II Keynesian welfare states. At the local level craft unions and employers bargained collectively according to principles of equity and notions of economic organization which both traced back to their common past in the guilds. Although the prevailing international Social Democratic orthodoxy had it that workers were easiest to organize in the large factories deemed to be production units of the future, Denmark's experience showed otherwise. The country was widely regarded as the "home of small industry."[75] But because its system of small-firm production depended on union cooperation, it also had an extraordinarily high rate of unionization. In 1910, 52 percent of Danish workers were trade union members, whereas the organization rate was 32 percent in Germany, and between 20 and 24 percent in Great Britain and the other Scandinavian countries.[76]

[73] Rasmussen (ed.), *Bidrag til Dansk Håndvaerkerundervisnings Historie*; Henrik Fode *et al.*, *Håndværkets Kulturhistorie*, vol. 4 (Copenhagen: Håndværkerrådets Forlag, 1984), p. 70.

[74] Jensen and Olsen, *Oversigt ovr Fagforeningsbevægelsen i Danmark*, p. 156.

[75] Bjarne Hastrup, *Håndværkets Økonomiske Historie, 1879–1979* (Copenhagen: Håndvaerkerrådets Forlag, 1979), p. 214; Hansen, *Okonomisk Vaekst i Danmark*, p. 287.

[76] Hansen, *Okonomisk Vaekst i Danmark*, p. 253.

The converging interests of small shop and small farm

By the end of the century, finally, the growth of rural industry was clearly dovetailing with urban changes, although it had earlier appeared that the rural crafts were the enemies of the urban guilds. Here it is useful to distinguish two periods, the first covering the wheat-export boom from roughly 1835–80, and the second running from the early 80s through the construction of the cooperative system down to the First World War. During the first period, rural and urban crafts were in conflict; thereafter they began to cooperate to the point of fusing into a single institutional and production structure.

The wheat-export boom led to increasing specialization of rural households. The more the small holders relied on their growing agricultural income, the less it paid to spend time supplying themselves with clothing and other consumption goods and the more opportune they found it to buy those goods in the market. And the less autarkic the farm household became, the greater the opportunities for part-time artisans to dedicate themselves more completely to craftwork. At first they apportioned their time among related trades – in construction, for example. Then, as markets expanded still further they concentrated on one in particular – for example, masonry or carpentry. Artisans who found themselves in competition with mass produced farm equipment or household articles could sell and service the mass producers' wares.

With a relatively secure subsistence income from these sources, and the proceeds of residual farming activity, the rural artisans were in an excellent position to compete with the urban trades in the production of furniture or garments. For the rural producers additional production meant an almost proportionate increase in profits, whereas the urban producers had to use a high fraction of the revenue from each order to cover the fixed costs of staying in business. Tension between the two groups probably reached a peak during the crafts' downturn of the 1870s. Competition for scarce work was naturally fierce, and the rural producers took full advantage of their more diversified sources of income and the lower wage levels in the countryside (itself a consequence of the workers' own multiple sources of earnings). To make matters worse, the migration of farm girls to the cities provided the urban sweat shops with just the kind of low-cost yet skilled and reliable labor which made competition from the traditional crafts hopeless.[77]

The reorientation of agriculture and the explosive growth of the cooperative movement from the late 1880s reversed the situation. These created a huge demand for skilled workers in the countryside to build and maintain the new cooperative dairies and slaughterhouses, to construct the increasingly specialized farm equipment which the strategy demanded, and to supply the consumption needs of the prosperous small holders. The flow of migration reversed accordingly. Instead of unskilled farm girls going to the cities, it was now skilled urban

[77] Willerslev, *Studier i Dansk*; Hastrup, *Håndværkets Økonomiske Historie*; Hyldtoft, *Kobenhavns Industrialisering*.

workers who moved to rural areas. The technical schools, which had already begun to reach the *købsteder* in the 1840s and 50s, now spread rapidly to the smaller provincial towns and the tiny *stationsbyer* which grew up along the feed lines of the railroad network. By 1900 there were 130 technical schools in the provinces.[78]

The cumulative result of these developments was a division of labor between country and city which depended on a common supply of jointly defined skilled labor. Metalworking firms in Copenhagen, for example, specialized in the production of equipment used in or required by the small shops throughout the country. These products included universal machines for forming metal and wood, the fractional horsepower electric motors to drive the universal machines, and the electricity generation equipment to drive the motors. Often routine production of standard machines would be combined with skill-intensive production of capital equipment in a single firm. For example, T. B. Thirge, who had worked with Edison in the United States, developed an extremly robust electric motor especially for rural use, as well as generation equipment adapted to the needs of rural communes; the latter bought both at the same time and sold the motors to local producers on credit.[79] Even Burmeister & Wain, once as symbolic in Denmark of large-scale industry and the unskilled industrial proletariat as Fiat in Italy or the Ford River Rouge plant in the United States, also combined mass and craft production: on the one hand it produced the Magdlekilde centrifuge for use in the Danish cooperative dairies and for export; on the other hand it produced ships and marine diesel engines to customer specification.[80] Metalworking firms in the countryside produced specialized agricultural and food processing equipment using the capital goods they purchased from Copenhagen.

In comparative perspective, artisans and skilled workers in Denmark had achieved a unique combination of workplace and institutional autonomy. German workers, for example, had in fact a similar role on the shop floor. But their unions were organized according to industrial principles – all workers in a single branch were grouped in the same organization, regardless of their level of skill. German craft unions which cooperated with small masters on the Danish model had been crushed by powerful centralized unions in their drive to create the institutional forms appropriate to the presumptively inevitable mass producers of the future. The fate of the Solingen cutlery trades was exemplary. And these disputes of course helped separate the skilled workers from the artisans, despite their many continuing common experiences. In France a system of small-firm production similar to Denmark's continued in areas such as Lyons and St. Etienne. But here too the new workers movement broke decisively in the late nineteenth century with guild traditions that had lived on despite state attempts to suppress them during and after the Revolution of 1789. The new French syndicalist organizations with their vague revolutionary demands were

[78] Nørregård, *Arbejdsforhold indenfor Dansk Håndværk og Industri*, p. 292.
[79] Tage Kaarsted and P. Boje, *Thomas B. Thirge primus* (1983), p. 24.
[80] Hyldtoft, *Kobenhavns Industrialisering.*

wary of cooperation with either the state or the new industrialists. In the United States, the craft unions affiliated with the American Federation of Labor resembled their Danish counterparts in the building trades or metalworking. But as mass production gained ground in the United States, these unions covered an ever smaller and less representative fraction of the work force; and the same victory procession of mass production undermined more and more of the independent masters who formed in Denmark such a crucial pole of the world of skilled work. Thus in industry as in agriculture, an unforeseen combination of old principles and new institutions allowed the creation of a class of small property owners convinced of the need to defend individual autonomy through collective political action.

An inconclusive victory

But some fifty years later, in the late 1950s, the cooperative core of this small-holder economy was coming visibly undone. First cooperative dairies, then the cooperative slaughterhouses began to combine into larger and larger units abandoning in the process many of their original constitutional features and becoming in fact and law corporations. The corporations in turn fought with one another and the remaining cooperatives for control of their respective markets. The drive for economies of scale was always evoked to explain and justify the transformations and conflicts. It appeared that the very organizational principles against which the small holders had rallied in defiance of which they had ingeniously developed the cooperative system, had finally proved their invincible efficiency. In the end, it appeared, they made of Danish exceptionalism an exception that does prove the rule of the standard logic of industrial development.

There is, however, an alternative explanation for the disarray in the co-operative sector that focusses not on questions of productive efficiency. Rather it concerns on the one hand the character of the confederal organizations linking the cooperatives in one branch or sector to one another, and, on the other, the relation of these confederations to various state-protected markets, created at the urging of the cooperatives themselves to help stabilize their sectors. From this perspective the key to understanding the reorganization of the cooperative sector is not the logic of mass production but the way in which a national regulatory system well suited to one set of circumstances can become so detached from developments that it creates just the kinds of opportunities for disruptively self-regarding behavior that it was designed to limit.

By way of conclusion, then, we trace general features of this regulatory regime as it arose in the 1930s and show how these, in combination with the particularities of the cooperatives' connections to each other, created the occasion for disputes that were presented as conflicts over efficiency principles but whose outcome cannot be explained by reference to them. Our aim is to show how the historical limits of the overarching institutions of the small-holder economy came to shake the foundations of that economy without casting doubt on the principles upon which those foundations themselves rested. The focus is on the dairy sector where reorganization – revealingly, as we will see – begins earlier and leads to more clear-cut results that in the slaughterhouse industry.

The uneasy post-World War I truce between the provincial small holders' republic and the Copenhagen world of mass production was transformed in desperation into institutionalized cohabitation during the Great Depression of the 1930s. As international markets collapsed, the large capitalist countries formed semi-autarkic economic zones with their colonies, protectorates and partners. Denmark too was forced into bilateral agreements that fixed the relation between its exports to and imports from particular countries. These agreements required the state to centralize control over the nation's trade in a single agency, and set prices in markets for key tradables; and the political precondition for this expansion of state authority was a comprehensive domestic bargaining system which assured respect for the vital interests of each group in Danish society in dealings with foreigners.

The framework for this cohabitation was the *Kanslergadeforlig* (Chancellor-street compromise) elaborated, eighteen hours of bargaining at the apartment of the Social Democratic Chancellor, Stauning, from January 29 to 30, 1933. Procedurally the *Kanslergadeforlig* established the state's obligation to encourage private interest groups such as farmers' associations to settle their differences, and then provide a forum for combining the local compromises into a comprehensive settlement. Substantively, the compromise represented an alliance of farmers and workers against large industry. The Social Democrats agreed to a devaluation and a program of agrarian reform that would raise the workers' cost of living; the *Radikale-Venstre* and the *Venstre* (which participated indirectly in the government by abstaining in parliament on votes regarding such key proposals as this one) agreed to help the Social Democrats defend against wage cuts by legislative means.

The new laws effected first were on the political organization of the economy, then the pattern of production. The *Kanslergadeforlig*, devaluation, exchange controls and a commercial treaty signed a few months later with Great Britain created a system in which most business in Denmark's internationally exposed economy required state authorization. To get such authorizations, firms and farms turned to the many organizations which, we saw, had represented their interests in politics, or they created new ones for this purpose. The economy became *gennemorganiseret* – organized through and through – as productive units became more tightly linked to associations and the associations more tightly linked to one another and the state. A system of cooperation was established whose operation would shape its goals even as those goals were being influenced by changing political and economic conditions in the country as a whole.[81]

[81] Peter Helby, *Erhvervsorganisationerne i 30ernes undenrigsøkonomiske regulering* (Speciale, Roskilde University Center, 1981).

The consequences of this emergent system were dramatically evident in agriculture. The quota on Danish imports contained in the commercial treaty with Great Britain meant that the Danes had to limit, for example, domestic production of pigs to avoid driving down the price of their own exports through overproduction. The government and the producers' associations agreed to ration production rights on the basis of the previous year's sales, with pig keepers receiving a number of "swine cards" redeemable at a fixed price upon delivery of the corresponding number of pigs. Cards could be sold to assure that low-cost producers actually kept the pigs. But because pig keeping was traditionally of greater importance to *husmænd* than to the larger freeholders, the distribution of swine cards was skewed in their favor. In return, the *husmænd* agreed to a "grain import fee" – a tariff by another name – which buttressed the domestic price of the larger farmers' cereals. The "fees" went to a fund in aid of small farmers unable to cover the costs of fodder. The system mocked the formal rights of property holders by jumbling subsidies and expropriations. Yet at the same time it guaranteed the foundations of the property order as a whole by creating insurance funds and review procedures which assured each group that its costs and revenues would be kept in tolerable balance.[82]

In the dairy industry the regulatory innovations of these years created a distinction between "city" or "consumption" cooperative dairies and "rural" or "production" cooperatives that became in time an invidious distinction between rich and poor. By the late 1930s competition for outlets among urban dairies led to such inefficient overexpansion of distribution systems that consumers were paying high prices for products that paid little in return to the owners of the cows or the dairies to which they delivered. The urban dairies founded the *Danske Konsummælkmeijeriers Fællesrepræsentation* in 1939 to seek redress. A year later they found it in the *Mælkelov* (Milk Law) and subsequent administrative decrees. These aimed to remedy the situation by allowing urban dairies to form price-fixing cartels on condition that the cartel agreed to reduce the number of member producers. The export-oriented rural dairies were left to their own devices on the reasonable assumption that the combination of stability provided by the trade agreement with Great Britain and opportunity created by the Second World War would secure their prosperity even in the absence of additional domestic stabilization.

This system worked acceptably well through the mid-1950s. In the immediate post-war years the continuing access to the British market together with the possibility of supplying former belligerents whose own agro-industrial facilities had been severely damaged kept the rural dairies prosperous. But beginning in 1955 Britain opened its market for dairy products to all producers and the Danish cooperatives had to compete with other suppliers just as many other traditional Danish export markets were limited by import restrictions imposed by counties rebuilding domestic industries. To make matters worse, the domestic price of butter had been linked in the early 1950s to the export price

[82] Rasmussen, *Velfærdsstaten pa vej 1913–1939*, pp. 428ff.

declines led immediately to reduction in revenues from domestic sales as well.[83]

By 1958 the cooperatives in Southern Jutland, the area with the largest concentration of milk production, organized countermeasures. Their instrument was a planning commission (the *Planlægningsvdvalget* of 1958) formed under the auspices of the cooperative's traditional confederal organization, the *Danske Mejeriforeningers Fællesorganisation* (DMF). Although this planning commission was formally dissolved in 1964, subsequent developments followed the trajectory it defined. The enlarged cooperatives that resulted from its work in turn successfully pressed for relaxation of the rules sheltering urban markets from rural competition and for changes in the cooperative association's accounting practices that facilitated fusion of assets. Administrative extension of markets and encouragement of concentration led to the formation of larger units with the resources necessary for further expansion.

The process culminated in 1970 with formation, on the initiative of the Southern Jutland firms at the center of the whole concentration movement, of *Mejeiriselskabet Danmark* (MD), which today controls half the national milk production. MD so dominates the *Fællesorganisation* that the DMF is widely regarded as a wholly owned subsidiary of its nominal member. The urban dairies, which contributed to the conflict by making gestures in the direction of expansion when the rural producers were facing great difficulties, are today marginal actors in a landscape dominated by MD as the foremost advocate of large production units and economies of scale.

To this point, of course, the story of the dairy cooperatives reorganization can be read as a cruelly ironic demonstration of the futility of the small holders' attempts to find an historical alternative to large-factory mass production. The fact that it is the rural cooperatives themselves who realize the advantages of mass production and apply it to disentrench small producers surviving in sheltered markets testifies from this point of view to the power of reason and self-interest to overcome utopian illusion and political manipulation.

Three reasons weigh heavily against this interpretation. The first is that the large producers are not more efficient than the smaller ones. Measured net of all costs but depreciation and interest – both of which reflect a firm's capital structure rather than its actual operating efficiency – earnings per kilogram of sweet milk were virtually the same for all size-classes of dairies except the smallest (less than 9.9 million kilograms of milk processed per year) through the 1980s,[84] and this despite evidence that the largest dairies are in fact concentrating an ever larger share of their total production in ever larger production facilities.[85] If fact has triumphed over fantasy, where are the facts?

[83] Svend Winther Kristensen, *Kampen om mælken: En erhvervshistorie om Mejeriselskabet Danmark* (Landbohistorisk Selskab, Odense, 1986), pp. 11ff.

[84] Villy Søgaard, *Spildt Mælk?* (Esbjerg: Sydjysk Universitetsforlag, 1990), p. 159, table 14.3.

[85] *Ibid.*, p. 160, table 14.4. Earnings net of interest and depreciation show only marginal advantages to size; and a regression with this net result as the dependent variable and the logarithm of the annual production volume as the dependent variable "explains" only 8 percent of the total variation in net earnings as the result of size of plant. The result is significant at the 95 percent confidence level (*ibid.*, p. 161, table 14.3 and n.4).

Second, and by way of confirmation of the view that the prospect of vast economies of scale had not been realized, even observers who regard the concentration movement as a natural phenomenon note that MD will likely have to decentralize control over operations that it has been busy centralizing. To make the large plants run more efficiently and to make them flexible enough to respond to shifting markets it is apparently necessary to operate them more like the independent firms they recently were.[86] Third and finally, although the slaughterhouse sector production units were traditionally larger than dairies because of economies of scale and indivisibilities, the concentration movement set in later and led in many cases to disastrous results. So the small-holder sector that presumably knew the most about the possibilities of achieving economies of scale was especially timid in pursuing them – and, in the event, with good reason.

Economies of scale do not appear where they are sought; those who know best where to look for them are hesitant to join the search; a principal practical lesson of concentration is the rediscovery of the virtues of decentralization – none of this vindicates a triumph-of-hard-facts view. But the hard fact remains that for want of a better suited system of regulating their conflicts and sharing their burdens even the heirs of the most extensive experiment in small-scale production in our century abandoned, at least for a time, belief in the possibility of their project.

There is reason to think that they will revisit these matters, and better reason to think that the small holders' legacy lives on in other parts of the Danish economy.[87] But if the historical alternative to mass production elsewhere suffered an inconclusive defeat, in Denmark, for the time being, the most that can be said is that it – like every other form of successful economic organization – won only an inconclusive victory.

[86] Kristensen, *Kampen om maelken*, p. 235.

[87] Peer Hull Kristensen, "Industrial Districts in West Jutland, Denmark," in Frank Pyke and Werner Sengenberger (eds.), *Industrial Districts and Local Economic Regeneration* (Geneve: International Institute for Labour Studies).

Part III

The resurgence of flexible production

9 In search of flexibility: the Bologna metalworking industry, 1900–1992

Vittorio Capecchi

This essay deals with a particular case of the flexibly specialized industrialization characteristic of the province of Bologna. The history of the engineering industry from the beginning of this century allows us to understand better the initial premises, take-off, stabilization and future potential of this development pattern. Of special interest within this history is the formation of an urban industrial subsystem, similar to Marshall's definition of an "industrial district," in the automatic packaging machinery sector, the most successful example of flexible specialization in Bologna. Both direct influences (on the engineering industry) and indirect influences (on the overall development of industrialization) will be taken into account, and a distinction will be drawn between endogenous factors (actions produced by agents present in Bologna) and exogenous factors (actions caused, for example, by the national government or the international market).

This history can be divided into three periods: (a) 1900–1945, when the province of Bologna was predominantly agricultural and its economy precarious, even if a number of factors can be identified which explain both the later development of the engineering industry and its already visible characteristics, including those of the earliest packaging firms; (b) 1946–1980, when Bologna industrialized rapidly, with the expansion of the engineering industry and the formation of the automatic packaging machinery subsystem; (c) 1981–1992, when the effects of international competition and the transition from mechanical to electronic technology became clearly felt, the packaging subsystem entered a phase of relative stabilization, other engineering firms began a transition which could lead to the establishment of new subsystems, and previous lines of industrial development were closing down.[1]

At the end of the essay, a typology of current changes in the Bolognese engineering industry is advanced, in order to facilitate international comparisons and to permit more general reflections, bearing in mind, as Hirst and Zeitlin

[1] On the socio-economic history of Bologna, the following studies have been particular useful: L. Arbizzani, *Sguardi sull'ultimo secolo, Bologna e la sua provincia 1859–1961* (Bologna: Galileo, 1961); Various Authors, *Cultura popolare in Emilia Romagna*, 6 vols. (Milan: Silvana Editore, 1983); W. Tega (ed.), *Lo studio e la città: Bologna 1888–1988* (Bologna: Nuova Alfa editoriale, 1987).

write, that "the complexity of the past helps us to recognize that there are a variety of options in the present."[2]

1900–1945: the beginnings of the engineering industry in Bologna and the establishment of the first packaging firms

If we take as a point of reference the statistics published by the Ministry of Agriculture, Trade and Industry,[3] it may be observed that in 1897, there were only 22,000 industrial workers, male and female, in the province of Bologna, just 3,000 of whom were employed in the metal and engineering industry. So at the beginning of the century, the province of Bologna was characterized by an almost exclusively agricultural production, as can be further seen in the fact that half of the workers referred to in the statistics were connected with hemp preparation, straw processing and straw hat manufacture.

Census figures from 1921 and 1931 tell the same story. This delay in industrialization was accompanied by a high level of unemployment and illiteracy. In 1905, Andrea Costa (Socialist Member of Parliament) claimed that 60,000 out of a total of 114,000 farm laborers in the provinces of Bologna, Ferrara, Ravenna and Forli were unemployed and that 5,100,000 additional working days would be necessary in order to ensure a minimum living wage at current rates. In 1902, the average illiteracy rate in the province of Bologna, calculated for inhabitants over six years old, stood at 38% (34% for males and 42% for females); the comparative figure for Emilia was 46%, much higher than the figures for Piedmont and Lombardy (18% and 22%), though much lower than those for the south, where the illiteracy rate was between 65% and 80%. Between 1900 and 1945, illiteracy decreased and industrialization spread, but these were relatively slow changes which, together with the figures for unemployment and emigration, might well mean that the history of social and economic development of Bologna should be read in terms of endogenous and exogenous "delays" and "obstacles" to industrialization. However, a whole series of positive factors, both endogenous and exogenous, were already exercising direct or indirect influence on the engineering industry and help to explain the quantitive and qualitative development of Bolognese industry in the following period, from 1946 to 1980.

Developing research into the positive factors involved in a certain kind of industrialization or industrial sector demands an answer to a precise question. The first half of the twentieth century in Italy was a period of great technological, political and social change (the transition from water to steam power and electric energy, two world wars, the growth of the workers' movements and left-wing political parties, fascism and the liberation struggle, etc.) but, even if we take into account local variations, these changes were common to the whole country.

[2] P. Hirst and J. Zeitlin, "Flexible Specialization versus Post-Fordism: Theory, Evidence and Policy Implications," *Economy and Society*, 201 (1991), 25.

[3] Ministero di agricoltura, industria e commercio, *Notizie sulle condizioni industriali della provincia di Bologna* (Rome: Tipografia nazionale di G. Bertero, 1899).

What, then, were the specific elements within this changing structural framework, which distinguished the province of Bologna from other regions, leading to the development there of flexibly specialized industrialization?

Factors explaining the characteristics of the first packaging firms and subsequent development of the Bolognese engineering industry

If we begin with the endogenous factors which had an overall influence on industrialization in the province of Bologna during this period, it is not difficult to identify the elements which had the most positive impact on its specific characteristics. First of all, it is important to keep in mind the history of socialist administration which characterized the countryside and villages around Bologna at the end of the nineteenth century and for the first fifteen years of the twentieth century, constituting a rather special case of "rural socialism" effective at the level of both enterprise and unions (mutual assistance societies, cooperatives, active union participation of agricultural workers, etc.). For example, in the town of Imola, Andrea Costa (who became the first socialist member of the Italian Parliament in 1882) and Luigi Sassi (who became its first socialist mayor in 1889) encouraged the formation of cooperatives not only in the building industry and amongst agricultural workers but also in the engineering, ceramics and printing trades, directly creating a basis for local industry. This increase of socialist power and the parallel spread of the cooperative movement extended to other towns in the province of Bologna, so that at the beginning of the century we have the image of a "red province" which was slowly laying siege to the provincial capital.

A further important factor was the presence of entrepreneurial experience in the countryside and villages around Bologna. For example, the 1931 Census clearly shows that 60 percent of agricultural workers in the province of Bologna had some kind of entrepreneurial experience: 40 percent belonged to families working in small sharecropping enterprises, 16 percent to landowning families and 6 percent to families of tenant farmers. This entrepreneurial experience was not only oriented towards the market for agricultural produce (cereal, wine, vegetable and animal products) but also towards the industrial market; the processing of hemp and straw in the countryside around Bologna led to the wide diffusion of weaving (especially of hemp) as part of the normal activities (usually carried out by women) of the peasant family.[4]

[4] The analysis of the Bolognese countryside also draws on A. Broccoli, *Chiamavano pane il pane* (Bologna: Edagricole, 1979) and Broccoli, *Il bello della festa. Storia di nonne nel mondo contadino* (Bologna: Clueb, 1988), which provide a very vivid picture of this area at the beginning of the century. A conference held in Bologna entitled "Between Research, Politics and Economics: the Agrarian Society from its Origins to the Giolittian Era" has been invaluable for an understanding of agricultural politics during the first half of the twentieth century. The conference proceedings underline the contradictory role played by the Agrarian Society in Bologna at the beginning of the century. On the one hand, it assisted landowners technically; on the other, it wanted to "educate" but not "teach" the peasants, fearing that more learning might lead to conflict with the landowners.

Analyses of industrialization in Emilia, such as the study by Adele Pesce,[5] have pointed out that one of the specific features of the Bolognese area was the different relationship between men and women visible in the double hierarchy of power (female and male) within the sharecropping family, as well as in the important public positions held by women. Thus, for example, Argentina Altobelli, after being nominated provincial director of the *Federterra*, the union of agricultural laborers and sharecroppers, became national secretary and led their struggles in Italy from 1904 to 1925. Pesce's research poses a more complex question regarding the relationship between women and flexible specialization. She points out that one of the central characteristics of flexible production, that is its consideration for the needs and demands of the individual client, can more easily be attributed to female experience (women who look after the family take into account the needs and demands of the individual members) so that the growth of flexible industrialization might occur in an area where women are more powerful within the family, even though there was limited access for girls to schools teaching specialized skills, and hence a female social character might prevail.

In addition to these positive factors affecting rural areas and the towns around Bologna, others can be identified in the case of the city itself. The *Camera del Lavoro* (Chamber of Labor) was created in 1883 and by 1914 Bologna had a socialist mayor who became well-known for encouraging the establishment of cooperatives, public housing, mutual assistance funds, a people's university, etc., which helped to combat illiteracy and create a favorable climate for small enterprise.[6] Furthermore, Bologna had a long industrial tradition. As Carlo Poni demonstrates,[7] it was known as the "city of silk" during the second half of the sixteenth century, when a good part of the population of 60,000 worked in the industry. Poni describes the complexity of this industrial district and its crises at the end of the seventeenth century, when it succumbed to the greater flexibility of the Lyonese manufacturers, but this experience of the decline of a flourishing industry is important for later developments, in that it drew attention to the lack of industrialization during the late seventeenth and eighteenth centuries for which Bolognese industrialists and intellectuals attempted to compensate in various ways.

In order to understand subsequent industrial development, it is also important to keep in mind the history of urban intervention in controlling the water supplies of the Bolognese plain and the industrial tradition linked to agriculture: not only

[5] A. Pesce, "Un 'altra Emilia Romagna," in A. Pesce (ed.), *Primo rapporto della Regione Emilia Romagna per la Commissione per la realizzazione delle parità fra uomo e donna* (Milan: Franco Angeli, 1990).

[6] For an overview of this subject, see Arbizzani, *Sguardi sull'ultimo secolo*, and Various Authors, *Il sindacato nel bolognese. Le camere del lavoro di Bologna dal 1983 al 1960* (Bologna: Ediesse, 1988).

[7] C. Poni, "Sviluppo, declino e morte dell'antico distretto industriale urbano (secoli XVI–XIX)," in W. Tega (ed.), *Storia illustrata di Bologna*, 7 vols. (Milan: Aiep Editore, 1990), II, pp. 321–380.

was there a textile industry using hemp or straw as its raw materials but also food processing and distilling industries. This industrial tradition intersected with artisanal activities (such as the manufacture of leather, shoes, iron, etc.), as well as with other activities connected with the university (such as printing and publishing) or with the cooperative movement (such as the building industry).

The presence of a university[8] played a major role in the development of Bolognese industry. This nine-century-old institution is an essential point of reference for the socio-economic development of Bologna during this period because of the presence of an important tradition of experimental physics and of an active department of engineering. Professional associations (the Agrarian Society, founded in 1807; the Medical and Surgical Society, 1823; the College of Engineers, 1880, etc.) were also important, as they directly influenced the development of industry in the area and interacted with the University itself. For example, the Agrarian Society of the province of Bologna played an important role in founding the College of Agriculture, which was provisionally approved in 1901 and confirmed in 1911.

In addition to these factors influencing overall industrial development, we should add another, very important one that specifically influenced the flexibly specialized engineering industry: the presence of the Aldini-Valeriani Institute in Bologna,[9] which played a vital part in the formation of the packaging machinery subsystem. Its foundation was connected with the names of two intellectuals, one of whom was born in Bologna and the other in Imola during the second half of the eighteenth century: Giovanni Aldini, an experimental physicist, and Luigi Valeriani, an economist. Both felt the effects of the crisis in the silk industry and thought it essential to extend technological knowledge in other areas, particularly engineering. At the beginning of the eighteenth century, Aldini went to London and Paris to study the major innovations of the time, and on the basis of the resulting information, he had built a series of teaching models of machines and instruments, realizing what one might define as a conscious transfer of technological knowledge. This collection of models was left to the city of Bologna, together with the estates of the two intellectuals, to lay the foundations for the first technical school for artisans. Later, the Aldini Cabinet for physics and chemistry was enlarged with teaching materials and kept up-to-date between 1863 and 1877 under the directorship of Sebastiano Zavaglia, a doctor, self-taught engineer and inventor. It was Zavaglia who continued Giovanni Aldini's initial project of disseminating knowledge of mechanical physics through lessons and demonstrations, building links with a European network of leading instrument makers and the most up-to-date institutes of technical

8 For further information on the University of Bologna, see the essays in Tega (ed.), *Lo studio e la città*.

9 The story of the Aldini-Valeriani Institute has been analyzed in detail in the book by the Commune of Bologna, *Macchine scuola industria* (Bologna: Il Mulino, 1980), and in Roberto Curti, "La formazione di una nuova cultura meccanica," in W. Tega (ed.), *Lo studio e la città*, and Curti, "Insegnare la macchina," *Scuola officina*, 2 (1992).

instruction, not to mention pioneering projects for the production of agricultural machinery, the application of steam power, the improvement of hydraulic engines, and the provision of telegraph, testing and consulting services for Bolognese industry.

The Institute of Arts and Crafts, founded in 1878, was a three-year post-elementary school with two sections, one mechanical (more highly developed) and the other physical-chemical. Evening and Sunday schools for engineers, electricians, chemists and bricklayers were also founded. The idea behind all this was not simply to pass on knowledge of how to operate the machines, but also to provide the technical background (drawing, geometry, engineering, technology) needed for their design. Two important aspects of this school, which trained future entrepreneurs for the engineering industry, should be kept in mind. The first is that this school was born of a three-fold agreement between the city of Bologna, the University and representatives of local industry and crafts, who discussed its "philosophy" at length. The second aspect is that, by training people both to operate and to design machines, traditional working-class schools were superseded and a completely new entrepreneurial career path was opened up to the sons[10] of workers and artisans. In this case, one may certainly speak of an endogenous factor, in that all the main actors were from Bologna, but there was also a transfer of knowledge from other areas which does, from this point of view, constitute an important exogenous factor.

Exogenous factors can also be identified if we consider state intervention. The Italian state followed an educational policy from the time of the Casati Law, passed just after unification, until the Gentile reform put into effect during the Fascist period, which delegated the development of elementary and technical education to individual towns. Thus a significant imbalance was created between the regions of Central and Northern Italy and those of the South. The poorer southern towns made little investment in elementary and technical education and even within regions like Emilia Romagna, Tuscany, and the Marche, there were wide variations, depending on whether or not the town in question created a certain type of school. So the absence of a national policy for technical education, which left the individual town responsible for choosing which schools to create,

[10] The history of female careers and technical schools for women can be found in A. Pesce, "Trajectoire des femmes dans la famille ouvrière," *Annales des Vaucresson*, 26 (1987), 80–97, and Pesce, "Un 'altra Emilia Romagna." In order to gain some idea of how different the situation for women was, it should be recalled that in Bologna at the end of the nineteenth century, Argentina Altobelli promoted the foundation of a School of Trade and Crafts for Girls which later took the name of Elisabetta Sirani. But it was structured in a completely different way from the Aldini-Valeriani. Resources were much more limited and above all, there was a different relationship between abstract knowledge and manual skill. In this school, as Pesce writes, the acquisition of knowledge by females was not culturally legitimated and hence was not accompanied by the transmission of abstract knowledge. It was a school, therefore (drastically reorganized during the Fascist period) which strongly reflected traditional labor roles and power relationships between men and women, whose final aim was really to direct young women towards employee status. It was not a school which had the explicit aim of acknowledging and favoring autonomy, creativity and enterprise.

constitutes a very important exogenous factor. But there were also two far more active and direct examples of state intervention: (a) the presence of the state in the running of two important engineering firms in Bologna as well as in the *Manifattura Tabacchi* (State Tobacco Manufacture), which ordered machines from Sasib, an important local enterprise; (b) the impact of orders to the engineering industry during the two world wars for military production. The Bolognese engineering industry undoubtedly profited from these orders and was strengthened by the experience both quantitatively and technologically.

The beginnings of the engineering industry in Bologna and the production of the first packaging machines

At the beginning of the twentieth century[11] there were in Bologna only three private metalworking factories employing more than 100 workers: Calzoni (200 employees), the Italian-Swiss Society (165 employees) and Gaetano Barbieri (120 employees), as well as two larger state concerns: the Pyrotechnic Laboratory (more than 100 employees, male and female) and the Italian Company for Southern Railways (376 employees). Thus we have a very small nucleus of enterprises which expanded during the first half of the century for two reasons: firstly, the Aldini-Valeriani Institute facilitated the diffusion of a "mechanical culture" capable of organizing and designing flexible, customized machines; secondly, before each of the two world wars, the Italian state drew these Bolognese firms into military contracts, reinforcing their structure without altering their versatility and productive flexibility.

The engineering industry in the province of Bologna between 1900 and 1946 can be divided into three main sectors. The first consisted of extremely diverse specialized products: from apparatus for radio and film (Ducati) to the manufacture of surgical and orthopaedic products (Rizzoli). The key point is that the Bolognese factories in the first part of the century displayed great versatility (for example the Calzoni factory manufactured goods ranging from hydraulic systems to aircraft) and it was only after the Second World War that they began to specialize in particular product lines, while still retaining their flexibility in responding to clients' needs. The second sector was linked to the Bolognese passion for the motorcycle. Bologna was producing GD, MM (Mazzetti and Morini) and CM (Cavedagna Mario) racing cycles, etc., and Morini and Cavedagna themselves were not only engineers who owned their own firm but

[11] A summary of the history of industrialization in Bologna (not only of the metal and engineering trades) may be found in V. Capecchi, "L'industrializzazione a Bologna nel novecento," in W. Tega (ed.), *Storia illustrata di Bologna*, IV, pp. 341–360, and V, pp. 161–180. An important analysis of industrialization in Emilia from 1920–1940 can be found in P. P. D'Attore, "Una dimensione periferica. Piccola industria classe operaia e mercato del lavoro in Emilia Romagna 1920–1940," *Annali Feltrinelli 1979–1980* (Milan: Feltrinelli, 1980), pp. 685–763. Other significant contributions appear in P. P. D'Attore and V. Zamagni (eds.), *Distretti, imprese, classe operaia: l'industrializzazione dell'Emilia Romagna* (Milan: Franco Angeli, 1992).

also skilled racers. A third sector was that of agricultural and industrial machinery, consisting entirely of special equipment designed to customer specifications: agricultural machinery (Calzoni, De Morsier), machine tools (Barbieri, Minganti), canning and food packaging machinery (Acma), cigarette-making machinery (Sasib), and machinery for making tortellini and other types of pasta, tiles and matches, bottling machines and metal box manufacture for food products like mortadella, etc.[12]

Only two of these factories were producing packaging machinery during this period.[13] The first, which was to play a very important role in the future of the packaging industry, was Acma (Automatic Machine Construction Company), founded in 1924 by a Bolognese businessman, Arturo Gazzoni. Gazzoni had successfully launched a great variety of mass consumer products at the beginning of the century with a huge publicity campaign (his most important book was entitled *Sell, Sell, Sell*) including "De Giovanni's Nerve Tonic," "Sun King Cough Pastilles" and "Idrolitina," a powder for carbonating drinking water. These products were hand-packed on traditional Taylorist production lines, and Gazzoni had the idea of building machinery to automate the packaging process. He persuaded his partner, Gaetano Barbieri, to invest capital in this enterprise, and Acma was founded. Technical planning was entrusted to a young designer, Bruto Carpigiani, who had studied in Swiss technical schools and attended electrical engineering courses in Bologna (probably at the Aldini-Valeriani Institute). Carpigiani took other young men into his office, often from the same Institute, though none of them was a qualified engineer, and he managed to overcome foreign competition, even introducing some notable mechanical inventions, such as the "Z wheel" which permitted a rotating movement to be subdivided into a large number of phases. The establishment of Acma was essential for the later development of the packaging machine subsystem in Bologna, for as we will see, it was this "mother" company which spawned the technicians who would create its most important constituent firms.

The story of the second company, Sasib, is somewhat different. It was not, in fact, established for the production of packaging machinery. The name Sasib (Societa Anonima Scipione Innocenti di Bologna) was registered in 1933, by its eponymous ex-worker founder, and specialized in electromechanical signaling

[12] The history of the Bolognese industry for canning foods products such as mortadella is described by Antonio Selvatici, "Alle origini del distretto meccanico bolognese. L'industria della mortadella alla fine dal XIX secolo," *Scuola officina*, 2 (1992), 20–23, who has placed its origins at the end of the nineteenth century, earlier than the beginning of the packaging machine industry.

[13] A detailed history of the early packing industry was written by Aurelio Alaimo in Alaimo and V. Capecchi, "Lindustria delle macchine automatiche a Bologna: un Casodi specializzazione flessibile," in D'Attore and Zamagni (eds.), *Distretti imprese, classe operaia*. The part dealing with the 1946–81 period written by Capecchi has been reproduced here. The history of the later period from 1981 to today was originally published in V. Capecchi, "La storia dell'industria del packaging bolognese," *Scuola officina*, 2 (1992), 10–19. The excerpts from interviews were published in Alaimo and Capecchi, *ibid.*, and in a special number of *Scuola officina*, 1 (1992), on "Packaging Valley," edited by Alaimo, Bergonzi and Capecchi.

equipment for the state railways on the basis of connections with the Swedish company, Ericsson. Packaging machine production was introduced in 1937 when the State Tobacco Manufacture ordered ten cigarette packing machines patented by an American firm, the American Machine Foundry. These machines were delivered in 1940 and other orders followed, thus beginning a production which would continue after the war.

1946–1980: the expansion of the Bolognese engineering industry and the formation of the industrial subsystem of the packaging industry

The early post-war years were dramatic for the province of Bologna, which was emerging from the world conflict with houses and factories destroyed, problems of unemployment and poverty, and widespread political clashes. At the first local government elections in 1946, the people of the city brought to power the left-wing parties – the Communists (PCI) and Socialists (PSI) – the former by a wide majority, the latter with a substantial presence. This result, which was largely paralleled throughout Emilia Romagna, contrasted starkly with the rest of Italy. After the 1948 general election, in fact, a conservative government came to power controlled by the Christian Democrats (DC), rigidly Catholic, pro-American and backed by big landed and industrial capital, even if the party also had deep popular roots.

In this contradictory situation, when Bologna and the other cities of Emilia-Romagna began to develop their "diversity," the local population and those in power opted for an alternative development path to that of the national government. This was a project of social and political reform with a wide popular basis and supported by a vast intellectual elite, closely connected with the growth of a particular type of industrialization, based on flexible specialization. Its strength lay in the transition from industrial wage labor to small-scale entrepreneurship, as part of an extraordinary process of social mobility made possible by the favored type of industrialization: small firms with family management, reviving the artisanal traditions of the town and the entrepreneurial traditions of the countryside.

The process of industrialization advanced very rapidly during the next thirty years. The percentage of industrial workers in the economically active population rose from 31% (1951) to 43% (1971), before falling back slightly to 41% (1981), with a compensatory increase, after 1970, in industrial services. There was a parallel increase in the percentage of the industrial workforce employed in the metal and engineering sectors in the province of Bologna: from 30% in 1951, to 41% in 1971 and 47% in 1981. The total number of employees in these sectors increased from 22,000 in 1951 to 80,000 in 1981 and the number of firms increased from 3,000 to 6,800. The local political situation during this period was extremely stable, with the two main parties, the PCI and PSI in control of local government, while at national level the PCI increased its power, reaching a high point of 34% of the votes in the administrative elections of 1975 (to 35% for the DC). There was a parallel increase in the number of union members after the successful workers' struggles of 1968–69.

Factors in the development of the Bolognese engineering industry

During this period, the foundations were laid of that nexus between economic, social, cultural and political development for which Bologna gained an international reputation as an extraordinary laboratory of experimentation and innovation. After the war, the economic policy of the Christian Democrats favored the great industrial complexes, especially in Northern Italy, using Marshall Plan funds. By the end of 1951, for example, firms in Piedmont and Lombardy had received 25 percent and 28 percent respectively of total dollar loans for the purchase of equipment, whereas those in Emilia had received only 0.7 percent. No assistance was envisaged for artisanal enterprises and cooperatives.[14] Furthermore, the national government withdrew orders from the Bolognese factories which had been engaged in military production. Ducati, which had employed 7,000 workers during the war, had just over 1,000 in 1973: the number of workers at Sasib decreased from 1,100 to 500, at Minganti from 1,500 to 400 and so on. There was also a high level of political repression in the factories. As Luigi Arbizzani points out, between 1947 and 1957, 15,000 workers were prosecuted for "labor struggles," struggles for democratic rights and strikes.[15] 8,000 of these were convicted, for there had been many violent clashes between workers and police resulting in injuries, and in some cases, fatalities.

The Christian Democratic national government was identified by most of the Bolognese workers, who had fought against fascism and voted Communist, as an opponent whose economic policy was not helping local industrial development policy and which was not supporting workers' interests in the factories. The climate of repression which was part of the sharp exacerbation of class conflict facilitated the emergence of a political polarization between "friends" and "enemies." In Bologna an opposition developed between "us" and "them": with "them" as the Christian Democratic government and the huge Taylorist-Fordist factories of the north and "us" as the left-wing coalition, its electoral supporters, the vast majority of the trade unions, the artisans belonging to the Communist–Socialist CNA (National Confederation of Artisans), the cooperatives of the Federcoop, as well as a complex interplay of cultural forces embodied in different organizations and institutions (from party sections to the intellectuals involved in the periodical *Emilia* founded in 1949). This "us," unified by the Communist and Socialist values expressed by their political

[14] See D. Melossi, "Lotta di classe nel piano del lavoro," in Various Authors, *Ristrutturazione capitalistica e piano del lavoro* (Rome: Editrice sindacale italiana, 1977), pp. 80–97, on post-war government politics.

[15] L. Arbizzani, *La costituzione negata nelle fabbriche* (Grafishe: Galeati Imola, 1991).

leaders, may be considered the main point of reference for the endogenous factors which most strongly influenced Bolognese industrialization as a whole.

In general terms, it was during the 1950s and 60s that this influence became evident, both as a network of solidarity linking the different actors (the policies of local government were based on informal relations, whereby workers and small business enterprises came into close contact with the local administration) and as a process of legitimation of the transition from wage labor to small-scale entrepreneurship and self-employment. The fact that this legitimation came from a Communist party, created a very different situation from that in countries like Spain and France, where its counterparts approved only of wage labor. In the 1950s, this legitimation was followed by a series of policies which influenced industrialization directly or indirectly. The so-called "Emilian model" applied in the Bologna area led to "qualitative" town-planning which not only prevented building speculation (safeguarding the historical city center and the green areas in the surrounding hills) but also activated an intensive policy of social-service provision of which helped a high percentage of women to enter the labor market, thus indirectly supporting the families setting up small enterprises.[16] Other policies influenced industrialization directly, such as financial support for the city's technical schools and for university research, holding the price of gas below the national average during the 1950s,[17] the construction of industrial estates in the 1960s and 70s[18] and, especially in the 1970s, the establishment of industrial service centers, coordination and incentives for the expansion of professional training for both waged and self-employed work.[19]

The role of the trade unions in this period was very interesting. On one hand,

[16] See S. Brusco and M. Pezzini, "Small-scale Enterprise in the Ideology of the Italian Left," in F. Pyke *et al.* (eds.), *Industrial Districts and Inter-firm Cooperation in Italy* (Geneva: International Institute for Labour Studies, 1990), on the relationship between the Communist Party and the politics of development of small enterprises. On this subject, C. Trigilia, *Grandi partiti e piccole imprese* (Bologna: Il Mulino, 1986), has drawn comparisons between the role of the PCI in regions such as Emilia Romagna and the DC in the Veneto region. One of the elements helping the enterprise to emerge is the formation of a local community with a network of solidarity between the main actors. On the wider relationship between the process of social mobility and the development of flexible specialization in industry, see the chapters by V. Capecchi in M. Barbagli *et al.* (eds.), *La mobilità sociale in Emilia Romagna* (Bologna: Il Mulino, 1987); the continuity in the relationship between countryside and industrialization typical of Emilia Romagna and the province of Bologna is described in G. Fuà and C. Zacchia (eds.), *Industrializzasione senza fratture* (Bologna: Il Mulino, 1983).

[17] For information on gas, see Antonio Campigotto and Roberto Curti (eds.), *Il solequinon tramonta. L'officina del Gas a Bologna (1846–1960) (Bologna: Ed. Grafis, 1991)*.

[18] See the essay by G. Negretti, "Le prime esperienze degli anni '60, i villaggi artigiani," *Parametro*, 133 (1985), 14–35, for an analysis of the creation of industrial estates. This may be compared with the description by A. Rinaldi, "La sinistra e l'industria diffusa: il ruolo delle istituzioni locall," in D'Attore and Zamagni (eds.), *Distretti imprese, classe operaia*, pp. 125–153, of similar initiatives in Modena and Reggio Emilia.

[19] See V. Capecchi, "Formation professionelle et petite entreprise; le développement industriel à specialization flexible en Emilie Romagne," *Formation-emploi*, 19 (1987), 3–18, and S. Brusco, "Small Firms and the Provision of Real Services," in Pyke *et al.* (eds.), *Industrial Districts*, pp. 142–159.

they were tough, politically minded unions that had to struggle against repression in the workplace in the early years. On the other hand, however, they were also part and parcel of the "Emilian model," and as such were sensitive to the needs of the skilled worker seeking to become a small enterpreneur. The Bologna unions, (especially the Communist–Socialist CGIL or General Confederation of Italian Labor, which had the vast majority of members) were therefore both conflictually and contractually orientated, trying not only to safeguard working conditions (wages, health care, etc.) but also to ensure that the best investments were being made by firms to benefit overall growth.[20]

So the endogenous factors identified for the period 1900–45 intersected with these later ones. But new exogenous factors which emerged after the Second World War were also important. We have already noted how the state strongly affected Bolognese industry during the two wars through military orders. The interruption of these orders and the consequent decline of skilled engineering work may be seen as an obstacle to local industrial development, and was indeed experienced as such by the redundant workers. These redundancies did, however, have the positive effect of releasing skilled manpower which was able and willing to set up independent firms. So an exogenous factor which was in itself negative proved later to be a positive factor.

Another factor was probably the most important exogenous influence on the development of the Bolognese engineering industry, especially its packaging sector. After the war a series of industries (food, pharmaceuticals, cosmetics, soap, tobacco, chemicals, etc.) felt the need to introduce automatic equipment and mechanize the phases which until then had been carried out by hand on traditional Taylorist lines. It was not just a question of lowering labor costs. As Georg Borgstrom has written, "Packaging is a key factor in the retailing of foods. The introduction of retail packaging has revolutionized food selling by (1) facilitating handling, (2) relieving the food trade of cumbersome weighing, (3) permitting self-service, (4) providing facilities for on-the-spot advertising and (5) providing protection against food contamination."[21]

This very specific demand from the international market meant that it was in the packaging industry, above all, that machinery production developed in the province of Bologna. In order to clarify how the Bolognese engineering industry developed, a distinction must be drawn between two main tendencies: (a) the emergence of individual firms which might or might not have been able to generate small groups of firms of the same type and (b) the formation of true

[20] For an analysis of the trade unions' role in the process of flexible industrialization in Emilia, see V. Capecchi, "The Role of Unions in Innovative Small and Medium-Sized Enterprises: A Case Study of Flexible Specialization in Emilia Romagna," in S. Tokunaga *et al.* (eds.), *New Impacts on Industrial Relations: Internationalization and Changing Production Strategies* (Munich: Iudicum, 1993).

[21] Reported in G. Borgstrom, "Food Processing and Packaging," in M. Kranzberg and C. W. Pursell (eds.), *Technology in Western Civilization* (New York: Oxford University Press, 1987), II, pp. 386–402.

"industrial subsystems." The definition of an "urban industrial subsystem"[22] is very similar to the term "district" used by Marshall to identify a flexible production system organized by a group of small and medium-sized firms, of which one part produces for the market and the other is part of a subcontracting network, without there being, however, any rigid division between the two. Such production is characterized by its overall complementarity, so that internal competition between the firms in a district is lower than competition from outside. The term "urban industrial subsystem" was introduced because what Marshall meant by the "district" was an exclusive connection between a certain type of production and a certain place (for example in the sixteenth century one may speak of a silk district in Bologna because Bologna was the "city of silk"; today we can speak of Prato as a textile district because the Prato area is completely dominated by the textile industry, etc.). In Bologna, however, engineering is not the sole industry and it does not consist of a single type of production. On the other hand, it is true that the Bolognese packaging industry developed as an industrial subsystem with all the main characteristics of a district, except for the fact that it was not the only type of activity in the area. Beyond the packaging industry, some local engineering sectors consist of a series of separate enterprises which may or may not generate a limited number of firms of the same type, while others organize themselves in smaller subsystems.

In Bologna, among the separate enterprises which may or may not give rise to subgroups of firms of the same type, there was a Società Scientifica Radio Brevetti Ducati founded in 1926, which closed after the war; the redundant technicians then formed Ducati Electronics (1954) and Arcotronics (1962) for the production of condensers. Castellini, founded in 1933 to manufacture dental apparatus, which became successful at a European level during this period, provided the technicians who went on to form companies with the same type of production: Marposs, founded in 1952 by the engineer Mario Possati, began the production of measuring apparatus for grinding machines and

[22] The debate on the industrial district in Italy refers to G. Becattini, "Dal settore industriale al distretto industriale: alcune considerazioni sull'unità di indagine dell'economia industriale," *Rivista di economia e politica industriale*, 1 (1979), 7–21, and Becattini, *Modelli locali di sviluppo* (Bologna: Il Mulino, 1989). For the history of the districts in Emilia Romagna, see: S. Brusco, *Piccole imprese e distretti industriali* (Turin: Rosenberg and Sellier, 1989); V. Capecchi, "Petite entreprise et économie locale, la flexibilité productive," in M. Maruani *et al.* (eds.), *La flexibilité en Italie* (Paris: Syros, 1989), pp. 271–285, and Capecchi, "A History of Flexible Specialization and Industrial Districts in Emilia Romagna," in Pyke *et al.* (eds.), *Industrial Districts*; and D'Attore and Zamagni (eds.), *Distretti, imprese, classe operaia*. A discussion of other Italian regions can be found in M. Regini and C. Sabel (eds.), *Strategie di riaggiustamento industriale* (Bologna: Il Mulino, 1989). For information on the current international debates, see generally Pyke *et al.* (eds.), *Industrial Districts*, and within that volume J. Zeitlin, "Industrial Districts and Local Economic Regeneration: Overview and Comment," pp. 179–294; M. Storper and B. Harrison, "Flexibility, Hierarchy and Regional Development: The Changing Structure of Industrial Production and their Forms of Governance in the 1990s," *Research Policy*, 20 (1991), 407–422; M. Storper, "To limits to Globalization, Technology Districts and International Trade", *Economic Geography* 68 (1992), 60–93.

became one of the most important manufacturers of electronic products; Datalogic, founded in 1972 by Romano Volta, an engineer whose studies represent a connection with the packaging industry (even though he qualified at the Aldini-Valeriani Institute and his first product was electric/electronic palletizer controls for beer crates).

The manufacture of small motorcycles which developed between 1945 and 1975 with about ten factories producing different types of machine (M. Ducati, Moto Morini, Malanca, Minarelli, Malaguti, etc.) and the network of subcontractors specialized in the manufacture of engines, frames, handlebars, etc. may however be defined as a small "subsystem." In this case internal competition was less severe than competition from the rest of the national market or from abroad.

Formation of the packaging industry subsystem

Before describing how this particular subsystem developed, it is important to specify the characteristics of the machines produced. According to the stage of the production cycle in which they are utilized, these machines can be classified as: (a) dosing machines, which can be found at the beginning of the cycle, immediately after the processing of the product, and which insert it into a container: for example, machines which insert tea into filters, medical products into capsules or liquids into a bottle; (b) wrapping machines, which may operate with a number of different stages: for example, in the case of the tobacco industry, cigarette production is followed by packaging, cellophaning and insertion into cartons; (c) packing machines, which carry out the final operation of cartoning the packaged products so that they may be transported and stored in large quantities. The definition adopted by the Central Institute of Statistics – packaging machines – identifies these three main phases, each of which corresponds to a large number of different techniques: for example, closure systems with metal wire, flexible film wrapping, boxing with card, cardboard or tin strips, etc.

As well as different products, production stages and techniques, there are also a very wide range of specific demands from the firms: single or double wrapping for sweets and toffees, soft or rigid packing for cigarettes and so on. Thus an extremely differentiated demand for single machines and complete production lines developed, and competition between suppliers was based not only on costs but, above all, on ability to meet clients' needs. During this period, certain characteristics of the machines became particularly important (and we will see how these demands were modified in more recent times): speed of execution of the various production stages, effectiveness of quality control for product preservation and dosing, compactness (less space occupied for an equal number of operations or the capacity to automate a larger number of phases with the same machine).

But this wide productive differentiation was handled by an engineering technology which gave designers maximum versatility. The main technological distinction during this period was between alternating-motion machines (in which the product passes through a series of stations where the various operations occur) and continuous-motion machines (in which the product remains motionless and the machines perform the various operations on the product itself). The latter is faster but the machines take up more space, are more rigid and require considerable maintenance. It should be noted, however, that both systems existed in the same factory and were created by the same designers. There was no qualitative difference between the two, just a different use of the same engineering technology.

So there existed a series of basic rules which were gradually applied and which, once learned, permitted designers to leave one firm to create another in which machines for different products were manufactured. At this point it is important to show how the formation of a subsystem of hundreds of firms emerged from just two between 1924 and 1945. The predominant strategy of new-firm formation enterprise has been "mitosis": technical designers leave a "mother" firm, establish other enterprises and so on. The "mother" firm *par excellence* was Acma, which produced ten technicians who went on to form other enterprises.

Among the most important was Ariosto Seragnoli, who founded GD, one of the most innovative firms in the sector. Seragnoli had only attended courses in mechanical drawing and worked at Acma as an electrician. There, he had designed a relay cell and asked the director of the technical department for promotion in order to develop his innovation. This request was turned down and Ariosto accepted the offer of his cousin, Enzo Seragnoli, to move to GD. GD had been founded in 1923 by Ghirardi and Daolio (hence its name) and specialized in bicycles and motorcycles, as well as manufacturing parts for truck engines. In 1939 Enzo Seragnoli became director and in addition to traditional products, the firm grew on the basis of military contracts (machine-gun parts and engines for the army and air force). At the end of the war the problem of reconversion arose and Enzo Seragnoli asked his cousin Ariosto to join GD and launch the production of automatic machinery. After some teething troubles, Ariosto Seragnoli's creative ability rapidly made its mark. The first machine produced by GD was an automatic hydraulic packer for chocolate bars, which was exhibited at the Milan Trades Fair in 1946, and in the 1950s, when hydraulic machines had been definitively replaced by completely mechanical ones, GD gained a firm hold in the international market. The machines which put the seal on its success were the 2,500 and the 5,000: machines for wrapping sweets (from soft toffees to double wrapped candies) or for inserting sweets into a stick. When these machines came out, they were able to wrap 350 sweets a minute (500 a minute today) whereas Acma was making machines which wrapped only 160 a minute. At the beginning of the 60s, GD began producing cellophaning and packaging machines for cigarettes, thus entering into competition with Sasib. This business strategy of producing machines in direct competition with the "mother" firm or other firms in the sector was, however, an exception. The most common strategy was to design machines aimed at other types of product, thus widening the offerings of the industrial subsystem.

Let us consider the other firms which emerged from Acma. Giovanni Preci, a fitter, set up his own firm which specialized in producing machines for making *croccanti* (nut toffee) and only later packaging machines. Natalino Corazza, an ex-student of the Aldini-Valeriani, had also been a fitter at Acma from 1935 to 1948 and he set up his own business step by step: first with repair work for automatic machines, partly sustained by jobs for Acma, and then, in 1955, by starting his own enterprise, specializing in the production of machines for packaging stock cubes, butter and soft cheese. The main competition for the stock cube machine was German and these firms made machinery both for dosing and wrapping the cubes. In 1956 Natalino Corazzo managed to overcome his rivals by creating a monobloc machine (for both dosing and packaging) and the machine was so designed as to be completely sealed, thus losing no oil. We will see that Corazza's wife, Maria, played a fundamental role in the history of the firm, and she took part in its management from the very beginning. When her husband died in 1975, she took over from him.

Other important technicians to leave Acma were two designers both of whom, curiously, were qualified surveyors: Martelli and Billi. At the beginning of the 1950s, Antonio Martelli founded Cam (Martelli Automatic Constructions) and specialized in the production of cartoning machines, aimed especially at the pharmaceutical sector. His activity inspired other designers to follow in his footsteps. One of them was Dante Mantovani: a qualified mechanic who had received his diploma in 1939, he never launched an independent firm but played (and still plays) an important role in the design of automatic machinery for the Martelli group. Agostino Billi set up his workshop in about 1954 (the exact date is uncertain because in many cases the beginnings of an enterprise were informal) and almost immediately obtained support from Carle and Montanari, a firm established in Milan at the beginning of the century, which was specialized in the construction of machines for sweet and chocolate manufacture and aimed to offer its clients a complete production line. So the Bologna branch of Carle and Montanari was created and Billi became its technical manager. In this case too, the new firm later attracted technicians workers from Acma.

One case which is interesting because of its atypicality is that of Wrapmatic, born in 1960 when a group of technicians found an investor. He was Vittorio Gentili, the owner of Panigal, which at the time was also sustaining the activity of another small firm in the sector, Zucchini. Two of the designers were ex-employees of Acma, Sergio Stanzani and Paolo Cassoli. Wrapmatic had trouble getting a foothold in the automatic machine market, perhaps because it lacked an original design idea at the outset, and it finally specialized in the production of wrapping machines for large reams of paper and toilet tissue. Cassoli also began his own business in this field in 1964 in collaboration with a young apprentice (Giorgio Bonafe) who was simultaneously attending evening classes at the Aldini-Valeriani. Cassoli's story is typical of many careers which led from Acma to an intermediate employment experience and then to the creation of an autonomous firm. Another example is that of Andrea Romagnoli, who qualified as a mechanic at the Aldini-Valeriani in 1947, worked at Acma as a designer from 1948 to 1953, went to GD and then only after an important period in the technical office directed by Ariosto Seragnoli, founded Ima (Automatic Machine Industry) in 1961, with his brother-in-law Renato Taino, backed by the capital of Giuseppe and Luigi Vacchi. Ima was one of the most successful firms in the subsystem, launching an Easter-egg machine in 1961, the C.20, a machine for filling tea-bags in 1967, and then in the 1970s machines for the pharmaceutical sector, such as the C.60 in 1976, a blister machine with integrated covering for solid oral products which conquered an important part of the international market. The succession of working experience was even more varied in the case of Dino Lullini, another Aldini-Valeriani student: Acma, Carle and Montanari, Zanasi, followed by the creation of Farmomac in 1961. This firm entered a new market, the cosmetics industry, with a machine for filling perfume bottles. Lullini had already designed machines for the pharmaceutical market as an employee of Zanasi (the capsule-filling machine, which inserts into capsules the exact dose of medicine) together with Aurelio Ricci, his partner at Farmomac who then left in 1969 to create Famar, a manufacturer of blister machines.

We have now reached the second generation of enterprises created by technicians leaving the firms formed by ex-Acma employees, and this second generation entered the market during the 1960s and the early 1970s. There are many examples, and here too the general tendency was to find new markets and further specializations. For example, two technicians at GD, Tartarini and Rabbi, left in 1962 to found Fima, which began with packaging machines for a special type of chocolate (the liqueur-filled "boero"), before shifting production towards machines for bundling yarn and thread. Gino Rapparini left Cam in 1963 to set up Ica, which specializes in machines for dosing granular products. An ex-employee of Corazza, Flavio Zambelli, left to create his own eponymous firm in 1969, which produces machines for packing cheese with cellophane and labeling, and Renato Grande, again from Corazza, found his own market niche in 1970 manufacturing bagging machines for the building industry (liquid cement or sand) and for agriculture (fertilizers).

The next wave of start-ups during the 1970s and 80s confirms these trends, with firms established in areas of production which are fairly marginal compared with those formed during the preceding period. One might recall, for example, Anidriti and Minelli, who originally worked for Fima (which had in the meantime become New Fima) and who founded AM in 1977 to produce machines for packaging school rulers (a case of the third generation) or the story of Franco Barilli who left Ima in 1980 to found Mab, which produces machines for cartons and boxes ordered by Ima itself.

There are few exceptions to this pattern of entrepreneurial development. One is the case of Zamboni, a firm established at the turn of the century for the production of machines for making pasta. Like many others, this firm was involved in war production and after the Second World War began to manufacture machines for packaging pasta, completing its own production cycle with designers from outside the sector. Another exception was Zanasi, founded in 1949 by the brothers Luciano and Amilcare Zanasi in partnership with Fallaci (all three trained at the Aldini-Valeriani). The three partners had had previous experience as manual workers in various factories and Zanasi's first enterprise was the construction of moving parts for piston engines. In 1954 there was a change of production thanks to a commercial intuition of the two Zanasi brothers. This is how Luciano Zanasi describes the change:

In 1954 I was in charge of the administrative, financial and commercial side of the firm and I began to realize that competition was so strong that either you had to make disproportionately large investments or there was no profit margin. I said to my brother, "Look Amilcare, we need to find another activity with our own product." And what was our strong point? That below where we had lived as children, there lived a certain Matteuzzi who was technical director of Alfa Pharmaceuticals, which was then just beginning and whom we knew well.

"I went to him and said, 'We need our own product. What do you recommend?' I'll always remember. He had a handful of gelatine capsules on his desk which were completely unknown at the time. He took them and said, 'Look, Luciano if you can make the machine to fill this product, you've made your fortune.'" In this case, the initiative came from someone outside the sector. The designer of the Zanasi machine for the pharmaceutical sector was, as mentioned before, Dino Lulli, previously with Acma and Carle and Montanari.

These entrepreneurial histories allow us to define more precisely the dynamics of this flexibly specialized industrial subsystem. The first point to make is that technical competence has been of fundamental importance for the subsystem's development. Most of the founders of firms had previously been fitters or designers, whose mastery of production techniques enabled them to become proprietors. This was a particular type of competence: in almost all the firms, the technical management was not entrusted to an engineer but to a designer (very often the founder himself) who had taken courses at the Aldini-Valeriani, but only in some cases had he actually taken or received a specialist diploma. Competence therefore derived from theoretical knowledge acquired at technical school (you must learn to draw in order to design). The particular type of mechanical production designed to customer specifications also presupposed a type of skilled worker who knew how to draw: not only did he know how to operate the machine, but also how to design it, or at least how to make an active contribution to its design. So in these decades, too, we find the technological characteristics typical of Acma's pre-war experience. The reflections of one of the few engineers active in the sector who occupied a managerial or organizational position explain the situation well. As Gianluigi Gamberini, director of Sasib from 1963 to 1984, and later director of Wrapmatic, remarks:

The reason why there are very few engineering designers in the automatic machine sector is linked to the fact that its design is mainly mechanical, requires little theory and a lot of experience. Automatic machines work with very little energy, as little as 3 h.p. Now the Faculty of Engineering teaches you to design machines which require a lot of energy because an engineering faculty has construction science in mind above all. But if you have to design an automatic machine, all those calculations are useless and it's only when you start to talk about vibrations or electronics that an engineering training becomes important. Today there are many engineers working on electronic applications for automatic machinery. In electronics, the transition period from theoretical knowledge to practical application is very brief, whereas in order to be able to design an automatic machine you need at least two or three years' practical experience, and it's only when you have ten years' consolidated practical experience that you become a good designer.

So we can understand how, from a technical point of view, it was possible for technicians with no higher education to design machines which nevertheless beat international competition. The most striking example is that of Ariosto Seragnoli of GD, who had only taken courses in mechanical drawing, while his chief collaborator, Riccardo Mattei, was self-taught, with a single brief experience in the factory. The testimony of Andrea Romagnoli provides a vivid image of the technical office at GD in the 1950s:

How was it organized? There was Ariosto, who had a very strong character and was technical manager; he directed the projects and had a great ability to get the most out of his employees, although he wasn't an expert draughtsman. The whole technical office at GD consisted of young men . . . six or seven altogether, 25 or 26 years old, with a lot of determination to make up for, in one way or another, their lack of technical knowledge. None of these colleagues had their diploma, they had done the basics, all at the Aldini I think. When I went there I had finished at the Aldini-Valeriani Institute, I had the highest qualifications. But I was the only one in the group with a diploma.

These young men, however, could test their materials at the Aldini-Valeriani, and, if need be, they could also turn for help to the engineers at the university. So the contribution of the engineers – if there was one – was relevant not so much at the design stage as for particularly technical problems which required specific expertise.

To understand how it was possible to set up a competitive enterprise, we remember that in such a differentiated market it was not difficult to find – or invent – one's own specific niche and obtain the first orders applying if necessary for loans (at least in recent years). And in recent decades, as before the war, a sound knowledge of the market was one of the strong points of the service fitters. This is summed up well in the testimony of one of the "minor" entrepreneurs, Franco Barilli, founder of Mab in 1980, who again draws attention to the traditional contrast between fitters and designers:

In the 1970s, I wasn't a draughtsman at Ima but I had connections with the technical offices where I worked after hours. Being a fitter has taught me a lot. Learning to draw depends both on practical experience and on having seen machines in other factories, including those of the competitors. Draughtsmen who have spent all their time in the technical office may be able to design better but they've seen fewer machines in production and they don't know the problems of wear and tear that certain materials may have, they don't realize the tricks you need to know to design a machine. A fitter has more practical ability. Perhaps he's not so good at doing calculations but he has greater ability to design a new machine. The beginning of many firms in this sector comes from doing an experiment, just as a joke. It's still a sector where you can start from zero if you take a few risks. Someone who buys a special automatic machine knows that he's buying an object that doesn't exist, that it must be made to measure and so he facilitates the financing of the project by paying in advance. If things are well-planned and delivery dates respected, external suppliers can be paid a little later, for a while you keep the bank happy and in the end you manage to close the circle with other people's money. This is a little game that all the firms in Bologna play, big and small.

One of the most important features of the internal rules of a flexibly specialized industrial subsystem is the balance between relationships of conflict and collaboration: a series of relationships which must be taken into consideration within each firm, between each firm producing for the final market and its subcontracting network, and between the different firms themselves. As far as internal relationships in the same firm are concerned, the political conflict in Bolognese factories at the beginning of the 1950s was very harsh, culminating in 1953 with 166 redundancies for political reasons, over 1,000 suspensions and thousands of penalties of various kinds. A militant unionist at Sasib[23] wrote at the time:

Sasib's best workers are leaving the firm: this is a consequence of the politics of repression and low wages adopted by the management. The employment of apprentices took place in a rather discriminatory manner by means of information of a political nature and blackmail.

There are numerous other examples and testimonies of this kind in smaller firms in the automatic machine sector which were founded by entrepreneurs of working-class origins. But parallel to these stories of conflict and tension in the factories, we have other stories of collaboration between skilled workers and managers of the new enterprises, frequently confirmed by oral testimonies. This is how Gaetano Bortolotti, for example, describes work at GD during the early 1950s:

Ariosto (Seragnoli) was a man who valued his best workers, who did his duty but who also had some very good qualities. But however skilled they were, they were still only workers, and it was nothing to him to give them 5 lire an hour or an

[23] G. Brini, *Sasib (Amf) Story* (Bologna: Edizioni Galileo, 1969).

increase of 10 lire. Once Bortolotti called me – I was a member of the internal commission. He pointed his finger at me and said: "You know why I left Acma. It was because Carpigiani refused to make me a white-collar worker. Listen: I want to reward my best workers like this. They stay where they are, as turners, fitters, grinders, slotters, and I'll promote them to the third technical level A." So he promoted them . . . fired them and took them on again as white-collar workers. Whoever deserved promotion, in his opinion, became a third-level employee, as a draughtsmen in the technical office, not as an accountant in administration, no, a draughtsman. The first time there were 11: fired and automatically taken on again . . . and then others. He started in the 50s. There were quite a few especially in the assembly shop. In a few years, nearly all the senior assemblers had been promoted. Turners, too, that's what I did . . . When I became a white-collar worker, there had already been fifty or so promotions.

Seragnoli's attitude was not so exceptional. Unlike factories which in those years were guided by Taylorist and Fordist principles, (like Fiat, partly, or Weber in Bologna), the flexibly specialized firms depended on a high level of collaboration between technical management and a wide range of skilled workmen. Confirmation is provided by a 1975 study I coordinated for the FLM (Federation of Metal and Engineering Workers). In the flexibly specialized packaging sector in Bologna, the percentage of workers at levels IV and V (skilled levels) was 58 percent; in the sector producing car parts this percentage dropped to 22 percent. The higher proportion of skilled labor in the flexibly-specialized firm leads to collaboration with designers of working-class origin and there are often friendly bonds between them and a nucleus of workers and draughtsmen. Thus when examining class conflict in Bolognese factories in this sector, it is important to consider not only the most visibly conflictual aspects, but also levels of profound collaboration which grow up inside them and contribute to the firms' success.

This search for collaboration is even more visible if one analyzes the relationship between each firm and its network of subcontractors. This is a very important aspect of the functioning of the entire subsystem. In fact, firms producing for the market, both big and small, always decentralize a part of their production (often, in the smaller firms, only assembly is carried out) and this is possible because of the large number of engineering firms that are able to do specialized mechanical work, with whom the purchasing firm has a relationship of strong collaboration. Sometimes a capable employee is actually encouraged by the mother firm to set up on his own, backed by loans in the form of payment for the first machines and a guarantee of orders over a long period. In the early fifties, alongside the traumatic exodus of workers who had been fired from the major firms (who constitute the nucleus of this sector of specialized engineering production) there was also a negotiated exodus which led to the diffusion of a mass of small metal and engineering enterprises which strengthened the industrial system as a whole.

There are also relationships of competition/collaboration, tending more towards collaboration, in firms producing for the final market and in this sense we can speak of the Bolognese packaging sector as an industrial subsystem similar to the Marshallian district. In some cases the above-mentioned stories show levels of competition between the firms in the sector. As most of these firms were founded by technical workers, it was inevitable that they sometimes designed machines of the same family as the firms they came from, although they elaborated more sophisticated solutions (the most significant example of this kind is that of Ariosto Seragnoli's GD and Acma). But in most cases (both in the second and third generation of firms), this was not true. It was the nature of automatic machines themselves which made this choice possible, as Gino Rapparini, founder of Ica, explains:

My experience at Cam Martelli was a bolt of lightning because I realized that the production of automatic machines was my sector, those were my machines, in the sense that they did things for which you needed no fixed rules. The important thing was to have ideas and a little initiative.

So what resulted was a choice of designs in related fields, or at least in production areas where it was possible to use acquired experience. Hence the automatic packaging subsystem turned to the production of machines for the food processing and pharmaceutical industries, cosmetics and chemicals, covering case by case different phases of the production cycle from initial packing to the so-called "end of line." We will see that later, too, the overall level of competitiveness between different groups which developed in the Bologna area was much lower than that with foreign rivals.

One last consideration regards the almost exclusively male characteristics of the sector. The emphasis on the technical designing flair of the founder of a firm has excluded women, who were already discouraged from going to technical schools and therefore without the necessary basic knowledge. But in some cases, women have nonetheless played an important role in administration and finance. The testimony of Maria Corazza is significant. At the time she was director of sales and administration in the firm founded by her husband, Natalino Corazza, who headed design and machine production.

My husband would say, "I think if we do it like this, we'll manage" and I always encouraged him . . . He wanted to buy three machines for making soft cheese, which had been damaged in some factory or other. The salesman who had these second-hand machines wanted to be paid in cash, and he wanted a lot, too. My husband talked and talked but didn't get anywhere because he wanted a million and a half in cash which we just didn't have. So I went to the grocer's, who's still there, and while we were talking he said, "Signora Corazza, I sell mortadella, not bolts like you do." "Don't mention them," I said, "it's not a good moment." And he said, "Are you short of money?" It was as if he'd read my thoughts. Anyway, to cut a long story short, he had just come into a lot of money, his mother or father had died and the children divided the booty – and he had this money. Well in just an hour I'd found the money my husband needed to buy his machines.

This episode is typical of the climate of social solidarity characteristic of the "Emilian model" at the time. It should be pointed out that Maria Corazza's career as a woman (as previously stated, she later took over the firm after her husband's death) was carried on by her daughter, Valeria, who is now vice-president, going against the stereotype which tends to regard managerial responsibilities in the engineering industry as exclusively male.

During the period 1946–80, this complex ensemble of characteristics and contributory factors led to the construction of a very substantial subsystem of enterprises. In the 1981 census, there were 283 firms in this sector in Emilia Romagna, with over 9,200 employees, and a survey carried out in 1984 estimates that 68 percent of all workers in the Italian packaging industry in Italy were employed in Emilia Romagna, while 70 percent of the latter worked in Bologna. Hence Bologna became well-known in the financial press as "packaging valley."[24]

1981–1992: stabilization and change in the engineering and packaging industries

This period was marked by great changes in the organization of work, international competition with the emergence of the "Japanese model," and by the increasing use of new electronic and information technologies. Many of these changes began in the 1970s but it was in the 80s that their effects were felt more clearly, as a phase of relative stabilization followed the previous phase of rapid industrial development. Other important changes also occurred during this period. The increasing success of trade unions and left-wing parties at the national level during the 1970s was reversed during the 1980s. Bologna still had the highest level of union density of any Italian province, but membership was falling even in the engineering industry.[25] So, too, were the membership and votes of the PCI, which was rechristened the PDS (Democratic Party of the Left), while a group of hard-line militants split off to create a new party called Rifondazione Comunista, underlining the conflict with the Socialists.

Bologna had thus moved a long way from that "us" which had characterized the 1950s and 60s, and local government was less able to take medium-term measures in the face of the new conflicts and problems that were emerging. Contradictions which had been present earlier now emerged in a more complex and widespread way: between women and men, between social classes and between generations. Women rightly demanded greater power in political parties, unions and public bodies, etc., while the emancipation project of the

24 See, for example, the *Sole-24 ore* dossier of May 25, 1992 on Bologna's Packaging Valley.

25 See A. Accornero *et al.*, "Les formes de la syndacalisation en Italie," in G. Bibes and R. Mouriaux (eds.), *Les syndacats européens à l'épreuve* (Paris: Presse de la Fondation Nationale des Sciences politiques, 1990), pp. 94–124, on the decline in trade union membership. For an analysis of trade unions in relation to the history of flexible specialization in Emilia, see Capecchi, "The Role of Unions."

1950s and 60s proved insufficient to keep pace either with structural changes in the lives of a new generation of women (whose academic careers were now longer than men's) which increased their expectations and strengthened their identity and goals, or the emergence of a new culture of sexual difference which radically challenged male notions of equality, postulating the construction of a society tailored to the needs of both sexes.

Factors of stabilization and change in the Bolognese engineering industry

Among the factors affecting the city's industrial development, beyond the exogenous economic and technological trends mentioned earlier, new contradictions may be identified resulting from the success of the preceding period. In the province of Bologna, greater economic prosperity resulted in one of the lowest birth-rates in Italy, so that today there is a drastic reduction in the size of the 15–29-year-old cohort, leading to a shortage of skilled workers, especially male, for the engineering and building industries. The economic boom has attracted immigrants for the first time from Morocco, Tunisia, Egypt, Senegal, Pakistan, etc. These immigrants (about 15,000–20,000 people) have found work only in less-skilled occupations and local authorities have not yet been able to help them to qualify for the skilled manual jobs in high demand.

Furthermore, the industrial success of earlier years made many Bolognese firms feel so secure that they are investing very little in professional training, research and development. This reduced tendency to innovation has interacted with two other factors (one endogenous and the other more exogenous) which have emerged as crucial in the most recent debates on the future of Bolognese and Emilian enterprises. First, there are fewer links between industry on the one hand and technical schools and the university on the other in electronics than had been the case for mechanical technologies. Despite the fact that the University of Bologna has a Department of Electronics and Information Technology which is at the forefront of international research, product innovations in fuzzy logic, for example, have been commercialized through US firms since no local partners were available. Secondly, firms have had difficulties in finding skilled workers with post-high-school diploma and post-graduate specializations due to the delays in reforming the educational system in Italy (the secondary school still has not been reformed and still operates under the old Fascist regulations, and short post-diploma university courses were only approved in 1992). Another significant problem is that of generational succession in firms centered on the owner-designer-inventor, often of working-class origin, rather than professional managers. Generally, these proprietors would like to pass on the firm to their children (especially males), but have not always been able to find the managerial capacities for future development within the family. These generational problems have led to the sale of Bolognese firms to national or multinational groups, signaling the absence within the region of an adequate financial structure for merchant banking and venture capital operations. Finally, industrial prosperity and a high standard of living here led to greater environmental pollution, due to increases in traffic, use of chemicals in agriculture and inability to control industrial waste products.

Faced with these difficulties, local and regional government have formulated plans, appointed commissions, and attempted solutions. Three structures have favored the process of stabilization and prevented a period of recession from following the rapid development phase: (a) a series of centers for real services to the flexible firms, tailored to the specific needs of the various areas and sectors; (b) an efficient organization of professional training both before and after the diploma based on collaboration among the various actors (unions, small business associations, local authorities, etc.) which compensates at least in part for the failure to reform the state educational system; (c) a generally positive role of the unions in facilitating the process of industrial development.[26] These measures have made it possible to introduce new information and electronic technologies into provincial industry, and have led to a fairly well-coordinated industrial policy and lack of employment problems (facilitated too by population decline). All of this has had a certain influence, but the overall lack of product innovation which is felt today by all concerned (local government, unions and business associations alike) should not be underestimated.

In the light of these circumstances, how can we characterize the overall situation of the engineering industry during the 1980s and 90s, leaving aside the packaging industry? If we consider the separate firms which have generated only small competing groups in the area of Bologna, only those which have managed to become "network-enterprises" have flourished and developed. This term *network-enterprise*[27] refers to the articulation of a single legal entity into a group of small and medium-sized firms (generally not more than ten or twelve) which are divided into relatively autonomous factories and units with a planning and design center. It is not, therefore, a rigidly hierarchical, vertical structure, but a more horizontal one, which creates structures abroad in order to consolidate its own markets. This term *network-enterprise* should be distinguished from the term *macro-enterprise network*, which refers to large corporations and multinational groups which are obviously of very different size to the Bolognese *network-enterprise* even if they follow, from this point of view, the same logic.

Among the firms discussed earlier, there are two main local examples: Marposs and Datalogic, which had succeeded in manufacturing high-quality electronic goods, exporting to Japan and the United States, and have consolidated themselves as *network-enterprises*. Marposs has managed to expand its production to include a wide range of electronic measuring instruments for machine tools (grinding and turning machines, lathes, machining centers,

26 See Capecchi, "Formation professionelle," and Brusco, "Small Firms," on the role of real services to industry and on professional training in Emilia Romagna and the province of Bologna; see Capecchi, "The Role of Unions."

27 The definitions of *network-enterprise* and *macroenterprise-network* are analyzed in the context of the Italian debate in the studies of G. Sapelli, "Gruppi di impresa e trasformazione della sovranità popolare," *Stato e mercato*, 3 (1987), 345–373, and G. Dioguardi, *L'impresa nell'era del computer* (Milan: Edizioni sole-24 ore, 1986).

transfer lines) and various kinds of controls (from numerical to optoelectronic). The changes during this period may give some idea of the transformations not only at the level of production but also at that of organizational structure. For example, in 1953, the Marposs in-process measurer for external grinding was developed using electronic technology with thermonic valves, in one basic version and very few variations. In 1968 electronic technology with semiconductors was used, which provided for 5,000 different versions which could be customized according to the buyer's needs. In 1988, the use of miniaturized integrated circuits and the automation of the process meant that the client could chose from 110,000 different models with the same delivery time (four weeks). So we are dealing with an extremely flexible product which is no longer customized. The client in fact simply fills in a questionnaire about the type of product required and the answers indicate the 110,000 possible variations. The range of possible products is very wide but completely predictable and procedures are all automated. So the organization of work for the manufacture of these products is very different from that in the automatic machinery subsystem which are still customized. The ability to anticipate the full range of marketable products makes it possible to automate procedures which, in the case of automatic machine production, still remain partly non-standardized and require a much more intense and informal involvement on the part of the workforce. With these automatic procedures, Marposs has entered the international market and presently exports 85 percent of its products, more than half of which go to the technologically strongest markets: the United States, Japan and Germany. The connection with these areas has led to the construction of research, sales and assistance centers in Schweinfurt, Boston and Tokyo, with a strong sales network and other factories in Italy in addition to the one in Bologna.

This success in technologically competitive areas has also been achieved by Datalogic, which presently has two production lines: optical/electronic controls for automatic movement machines and optical bar-code readers. Today, Datalogic is a network-enterprise with several factories in Italy and three research and production centers in Japan, Germany and the United States. One of the Bolognese factories is among the few in Italy manufacturing industrial lenses, only 60 percent of which go to Datalogic itself, allowing us to speak of productive diversification.

The success of Marposs and Datalogic show the development potential of a Bolognese electronic-instrument industry, but also its limitations, as only a few small firms (Boconsult and Copeca) have developed around these two successful enterprises. The reasons for this lack of growth are interesting, for, according to some of the entrepreneurs interviewed, they are linked to the presence of a strong packaging industry subsystem. These firms have generated a demand for electronic components which has been met by small Bolognese firms, often owned by university professors and researchers. These small enterprises have often ended up as part of the packaging subsystem and not as points of reference for a new subsystem. Thus, so far, these new entrepreneurs have not been able to overcome a relationship of close local dependence (a stable relationship of specialized subcontracting) and so develop successful products of their own on the national and international market, independent of their connections with the automatic machine industry in Bologna. So, even though conditions are favorable (a university department specialized in electronics and information technology, two important enterprises for electronic production, etc.) it is difficult to foresee whether a new subsystem will be created or whether this subordination to the strongest subsystem will persist.

On the other hand, one should also note the decline of the small subsystem involved in motorcycle manufacture. In this case, flexible production in Bologna has run into direct competition from the large Japanese macro-enterprise networks (Honda, Yamaka, Suzuki, etc.) which are able to produce both racing cycles and standard machines with variations (flexible catalogue production, the ability to manufacture a wide but predetermined menu of possible options). In the face of this competition, all the motorcycle factories of the earlier period (M. Ducati, Moto Morini, Malanca, Minarelli, Malaguti) have undergone radical transformations. The long professional experience of M. Ducati in racing cycle manufacture was taken over by an outsider (Castiglioni Giovanni of Varese) who launched the Cagiva model (taken from the first two letters of his name, surname and city of origin) and today Cagiva is a successful network-enterprise with several factories (not only in Bologna but also in Rimini and Varese), and collaborative agreements with Japanese firms (who have brought up other of the group's plants) to establish its presence in their domestic market. The Cagiva motorcycles are extremely competitive (winners of the Paris–Dakkar, and racing successfully in the large engine category) so that, in Bologna, the Cagiva factories are continuing a long tradition of specialization and success. But on the whole, the Bolognese motorcycle subsystem has entered a period of crisis. Malanca has closed down completely, Franco Morini and Minarelli have stopped production of finished machines and make engines for Suzuki and Yamaka, so that only very few marginal producers, apart from Cagiva, have survived, such as Malaguti with its scooters.

At the same time, however, the motorcycle subcontracting network has not been dismantled. Indeed, the pre-existing firms have been saved from closure by reconversion into a subcontracting network that is available not only for local enterprises but also for foreign firms (even if their factories are based in Italy). An interesting point reported by the trade union emerged during a negotiation: in exchange for permitting Cagiva to penetrate their domestic market at a certain moment, Japanese motorcycle manufacturers demanded that the firm also use their domestic subcontracting network. This demand indicates an important trend in this sector too: the internationalization of subcontracting.

Changes in the packaging industry

From 1981 to 1992, the number of firms and employees in the automatic packaging machine subsystem has tended to stabilize, even though this had not meant a loss of competiveness in international markets. A study by Biggiero reports that in the period 1978–85, Italian production of these machines, concentrated almost exclusively in the Bologna area, increased its share of the international market from 12% to 20%, whereas Germany, its strongest competitor, dropped from 38% to 34%, the United States from 13% to 10% and Switzerland from 9% to 7%.[28] Bologna's packaging valley has thus established itself well on the international market and data published by Ucima (the Association of Automatic Machine Constructors) confirm that national turnover (to which the Bolognese firms contribute decisively) rose from 1,000 billion lire in 1983 to 2,500 in 1991, a figure which remained stable in 1992 in spite of the general crisis.

This effective resistance to international competition (especially from Germany) has been possible thanks to important changes which have taken place in the subsystem's structure. The main change which has enabled international competition to be successfully overcome is the concentration of several firms in groups, forming network-enterprises as previously defined. One can use this term especially in relation to the GD, Ima, Marchesini and Martelli groups, and it is important to understand their structure in order to determine the relationship between such network-enterprises and the other individual firms of the subsystem. Take, for example, the Ima group, which today has about 1,000 employees, excluding its subcontractors. This group's starting point was the Ima factory (established in 1962) which specialized mainly in tea packaging machines and blister machines for the pharmaceutical industry. It later widened its range of machines for the pharmaceutical industry taking over Zanasi, in 1985, and Farmopac in 1986. Its cycle of machines was then completed when Win Pack was purchased in 1988 and Cestindue, which produces vertical packing machines, was created in 1990. Then in 1991 a merger took place with Sassa, producers of packing materials, to form Sassapack, aimed at developing new types of packaging that take greater account of consumers' ecological concerns. All the acquired firms were in Bologna, which is where the new firms also have their offices. In the Ima group a new sales division, Soteco, was also created, which is relatively autonomous and has sales and maintenance branches in six countries (the United States, France, Great Britain, Germany, Austria and the ex-USSR). The logic which led to the formation of this network-enterprise is clear. In the face of international competition, it is essential to be able to offer the client a complete cycle of machines from dosage to final packaging. There must also be reliable maintenance and sales structures abroad and research in complementary directions (an eye for the type of packaging and not just for the machine that makes it) becomes important.

Given the importance of network-enterprises, can one still speak of a subsystem in the packaging industry resembling the Marshallian district? The

[28] L. Biggiero, *Il commercio internazionale di macchine automatiche per dosatura, confezionamento e imballaggio* (Bologna: Materiali Ervet, 1987).

answer is yes. In the province of Bologna, we still have fifty or sixty other small to medium-sized firms around the network-enterprises which continue to sell machines on the national and international market, and the products of the network-enterprises and the single firms are still relatively complementary. There is a project, for example, to establish a permanent Trade Fair in Bologna for all the machines produced and the network firms, which have a better structure of maintenance and sales abroad, often market the products of the single firms (for example Soteco, of the Ima group, also sells the products of five other Bologna firms).

It is also true, however, that a subsystem of individual firms with a high degree of internal mobility (as in the first half of the post-war period) is very different from a subsystem with network-enterprises like the existing one, with rigid internal hierarchies which are difficult to overcome. From the study by Biggiero and Forni,[29] it appears that in 1984, the packaging industry in Emilia decentralized 50–60% of its mechanical operation, 60–70% of its electronic work, 8–10% of its commercial and marketing functions, 5–7% of assembly and testing and 3–5% of design. On average, these percentages have probably not changed much since then. What is changing is the stabilization process which also involves the subcontractors. At the moment, some of the firms that do mechanical or electronic subcontracting in this subsystem tend to work permanently with particular groups and firms with access to the final market. In addition, there are a limited number of suppliers who in turn have had to take over sales and administrative functions, guaranteeing delivery times and production quality for other subcontracting firms. Thus subcontracting becomes more specialized, again with the definition of stable internal hierarchies and thus an overall decline in the fluidity of the subsystem.

These changes have not only affected the relationships between manufacturing firms, and between these firms and their subcontracting network. The relationship between the different functional areas is also changing. The particular history of this subsystem led to dominance of design (closely connected to production) compared with sales and marketing; so all that had to be done to beat competitors, it was supposed, was to make machines that were faster and more efficient. Today, the more innovative firms (such as Ima) have realized that requirements are different and those responsible for sales and marketing have gained decisional power over those responsible for design and production (although the process is still in its infancy, given the strong influence exercised by the founder-designers). As Daniele Vacchi[30] director of quality, image and communications at Ima, writes:

Previously, the main orientation (in the packaging industry) was to the product. For example, if technicians believed that for a given machine it was possible to move from

[29] C. Biggero and F. Forni, *L'industria delle macchine automatische per la dosatura confezionamento e imballagio* (Bologna: Ervet, 1987).

[30] D. Vacchi, "Quale marketing per l'industria del packaging," *Scuola officina*, 2 (1992), 28–29.

one speed to a higher speed, marketing would determine whether or not there was a certain place on the market for this innovation, but wouldn't ask too many questions about the different requirements for that machine that would emerge from the market over the medium–long term. Today [. . .] marketing is much more market-oriented and there is an attempt to try and understand and meet the buyer's expectations at all levels (needs, dissatisfactions and even their perceptions of the offerings of other firms), to establish an increasingly close and stable interactive relationship with the client . . .

These changes in entrepreneurial mentality are not, however, easy to put into practice. The viewpoint of the owner-founder still tends to prevail and continuity of the enterprise is not always positive. The sale of some of these firms to German competitors gives some indication of the problems inherent in this process of generational succession in the packaging industry. In the last few years, Wrapmatic and Famar have been sold to Korber and Macofar to Romaco. If entrepreneurial mentality changes, then it becomes possible to grasp the importance of marketing and the different demands of clients. Thus Daniele Vacchi writes:

Going in this direction, a general change may be observed in the type of packaging machine which is demanded: in the 60s and 70s the demand was for faster and faster machines; in the 80s, for machines that were more reliable and electronically controlled; in the 90s the demand is for machines that are very simple to operate, that are very flexible and which can be programmed to pack products of different sizes, as many firms have more products in small series.

This demand for increasingly flexible machines, so that the same machine can package products of different sizes, interacts with a technological change that has taken place in the 1990s and which may completely revolutionize the design of automatic packaging. This change is a more widespread application of electronics. This was introduced in the early 80s when engineers began to enter the packaging industry, and until recently, electronics has been used to improve control apparatus and make certain phases, for example the dosing of pharmaceutical products, more and more precise: the cost of the electronic components of a machine may be as high as 30–40 percent. Today, in many industrial sectors such as the food and cosmetics industry, competition is such that the firms continually offer the consumer a wider product range in small–medium series which vary continually.

The Bolognese packaging industry is responding to these market demands with the application of electronics not only for "service" but also for "handling" (this type of machine is already being produced by Ima and Corazza). Mechanical handling, which constituted the center of the automatic packaging machine, has been replaced by electronic handling which allows the movements to be varied, thus lending the machine greater flexibility. The design revolution involved is comparable to that which took place in the transition from the traditional machine tool to the numerically-controlled machine tool. It is a completely different way of conceiving the machine and the more innovative firms realize this; in 1992, four of them (GD, Ima, ICA and Cassoli) drew up a plan, christened PUMA (Unified Project for Automatic Machines) at the University of Bologna to deal with the question of new electronics. The subsystem is in a period of relative stabilization (no new firms are being created) but not of crisis or stasis, because it is, on the whole, managing to face complex changes like those of mechanical and electronic flexibility.

In order to clarify further the characteristics of the packaging machine subsystem in Bologna we can examine the results of a study I carried out in 1991 on Japanese firms operating in this sector (especially Omori in Koshingaya and Shibuya Kogyo in Kanazawa). First of all, research into the Japanese packaging industry has confirmed many of the ideas developed by David Friedman[31] on flexible specialization in Japanese industry as a whole. Furthermore, the Japanese machine tool industry studied by Friedman has many similarities with the packaging industry, for in both cases the prevalent structure is that of the small or medium specialist enterprise.

The comparison between the packaging industry in Bologna and that in Japan leads to two observations. The first is that, whereas there are significant differences in the organization of work in factories like Toyota and car manufacturers like Fiat, the differences between Ima in Bologna and Shibuya Kogya in Kanazawa are much smaller. In both cases, we have the typical story of small enterprises, that of the owner-designer, and in both cases the working methods are strongly characterized by maximum collaboration between skilled workers and technicians, most of the manufacturing process being decentralized. In Kanazawa, as in other parts of Japan, the professional dedication of the skilled labor force far exceeds the limits of normal working hours, but this also happens in Bologna if it is a question of meeting contractual delivery dates. The attachment of the skilled workers and technicians to their firm is much more than a question of formal working hours.

The second observation concerns why, at present, the Japanese packaging industry is scarcely competitive internationally, most of its production being directed at the home market and many products being imported from Bologna or from Germany. The answer of Japanese entrepreneurs to this question points in two directions. First of all, there is general agreement that the customization of the machine and its continual technical improvement mean that those firms with a network of clients who stimulate changes in design and production are at a considerable advantage. It is very difficult for a competitor who is new to the market to "copy" an experience of continual customization and technological improvement based on a relationship between client and producer that has lasted many years. Another answer is that Bolognese machines are still dominated by mechanical flexibility, a technology in which Japanese industry is thought to be

[31] D. Friedman, *The Misunderstood Miracle: Industrial Development and Political Change in Japan* (Ithaca, NY: Cornell University Press, 1988).

weaker, though its competence in electronics is greater. We have seen that there is currently a shift from mechanical to electronic flexibility in the Bolognese production of packaging machines. One might ask, then, whether this shift, made necessary by the client's demands, might not in the long run prove advantageous for the Japanese packaging industry.

Towards a typology of industrial development

This final section proposes a typology of Bolognese industrial development based on a series of criteria and indicators which should highlight the changes that have occurred, as in other recent classificatory schemes such as that advanced by Storper and Harrison.[32] This typology could be used in two ways: as a synthesis of research on industrial development in a particular area (the province of Bologna) and as a set of criteria for a comparative analysis of flexible industrialization in different countries and periods. In order to facilitate comparison, this proposal is divided into two parts.

General changes in the Bolognese engineering industry

The starting point for the typology is the distinction made by Piore and Sabel[33] and by Sabel and Zeitlin[34] between mass production and flexible specialization; we have already shown that Bolognese industrial development, which took off between 1946 and 1980, was largely characterized by a flexible specialization engineering sector. Considering the three phases into which this development process has been subdivided, the key axes proposed for classifying its content are temporal and spatial: the temporal dimension of continuity/discontinuity (with a further distinction between more recent industrial history and the more distant origins of industrialization) and the spatial dimension of positive linkages/obstacles with the further distinction between endogenous and exogenous factors.

As far as the temporal distinction between continuity and discontinuity is concerned, it may be applied to the Bolognese engineering industry especially as regards the previous history of industrialization (in this particular case, that of the silk industry). If we take, for example, the story of the engineering industry of Kanazawa in Japan, for example, as analyzed by Masaykuki Sasaki,[35] it can be seen that in this area too, there was a tradition of silk manufacture which industrialized in the Meiji era, together with the production of textile machinery, thus beginning an interrelationship of textiles and engineering which continued after the war

[32] Storper and Harrison, "Flexibility, Hierarchy."

[33] M. J. Piore and C. Sabel, *The Second Industrial Divide* (New York: Basic Books, 1984).

[34] C. Sabel and J. Zeitlin, "Historical Alternatives to Mass Production: Politics, Markets and Technology in Nineteenth Century Industrialization," *Past and Present*, 108 (1985), 133–176.

[35] M. Sasaki, "Endogenous Development of Urban Economy in Kanazawa," unpublished paper, Kanazawa University, 1989, pp. 1–13.

when production moved from rayon to synthetic fibers and Kanazawa became a world leader in the production of polyester fibers as well as machines for their manufacture. The story of the Bolognese engineering industry is somewhat different, for it emerged as an alternative, an interruption in the decline of the silk industry which occurred at a later historical period than that in Kanazawa. This analysis of industrial development in historical terms may also be applied to older industrial areas. Both Bologna and Kanazawa are "peripheral" areas of industrialization in comparison with older, more important areas like Milan, Turin or Tokyo. But unlike engineering production in these areas, however, there is no continuity with the past in Bologna or Kanazawa, with the emergence of new metalworking products and a sharp separation, still evident in the two areas today, from the large industrial groups of Italian and Japanese engineering.

Further specification of the industrial development under examination derives from a spatial analysis based on a study of the *positive linkages/obstacles* in the three phases considered.[36] The term *linkages* derives from Hirschmann's work,[37] which shows very clearly how difficult it is to distinguish between favorable and unfavorable factors which have influenced industrial development. In the history of Bolognese engineering, too, it is difficult to separate positive linkages from obstacles in the initial phases of development. Indeed, there is a constant temptation to consider the obstacles encountered in this period as favoring development, much in the same way that one ends up thinking of a difficult childhood, paternal conflict, lack of affection, etc. as positive factors in the biography of a successful businessman. Obviously, it is much easier and more convincing to identify obstacles in phases of stasis.

The main *positive linkages/obstacles* above-mentioned are illustrated in table 9.1 and the positive linkages have been divided into endogenous and exogenous ones. This more complex type of analysis differs from studies which attempt to classify industrial development in a simplified manner with an exclusive use of endogenous or exogenous factors. These are always present, of course, in considerations of *positive linkages/obstacles*, in the form of state intervention and the national and international context of certain industrial processes in a certain place. This interplay of *positive linkages/obstacles*, in turn identifiable mainly as endogenous/exogenous, may however, at least in part, be disentangled and the analysis proposed in table 9.1 is an attempt in this direction, offering the possibility of comparisons with other cases.

[36] For an attempted analysis of the relationship between obstacles/opportunity in the industrialization of Emilia Romagna, see V. Capecchi, "The Informal Economy and the Development of Flexible Specialization in Emilia Romagna," in A. Portes *et al.* (eds.), *The Informal Economy, Studies in Advanced and Less Developed Countries* (Baltimore: The Johns Hopkins University Press, 1989), pp. 189–215; and Capecchi "Changing Patterns of Small Entrepreneurship," in B. Strumpel (ed.), *Industrial Societies after Stagnation of the 1970s* (Berlin: Walter de Gruyter, 1989), pp. 245–265.

[37] A. O. Hirschmann, "A Generalized Linkage Approach to Development with Special Reference to Staples," *Economic Development and Cultural Change*, 25 (1977), and Hirschmann, *Rival Views of Market Society and Other Recent Essays* (New York: E. Sifton Books, Viking, 1986).

Table 9.1 *General changes in the Bolognese engineering industry*

Type of industrialization	Flexible specialization
Continuity/discontinuity with regard to local industrial history	Discontinuity regarding the former silk industry
Continuity/discontinuity regarding older industrial areas	Discontinuity regarding engineering industry in Milan and Turin
Positive linkages/obstacles 1900–1945 phase (preliminary)	*Positive endogenous linkages*: socialism in city and countryside; relationships between men–women; university and technical schools; previous industrial experience; proto-industry in the countryside *Positive exogenous linkages:* State in Bolognese engineering firms and state orders for war production *Obstacles* State and local policies against the spread of education
Positive linkages/ obstacles Phase 1946–1980 (development)	*Positive endogenous linkages* Collaboration between actors in the "Emilian model" for transition from employment to widespread independent enterprise *Positive exogenous linkages* International market demand for flexible engineering products *Obstacles* Lack of state reform and state policy more favorable to Northern regions
Positive linkages/ obstacles Phase 1980–1992 stabilization and stasis	*Positive endogenous linkages* Collaboration between actors for services to industry and professional training; reinforcement of main subsystem with network-enterprises *Positive exogenous linkages* Demand for international market for flexible products *Obstacles* More international competition; difficult generational transition in firm ownership, lack of credit–finance structures; limited relationship with the University and development of new technologies; little investment in research and development and lack of medium–long term strategies; decline of "Emilian Model"; lack of state reform

Table 9.2 *Changes in the Bolognese engineering industry: specific aspects*

Relationship between enterprise/ and territory	Formation of an area specialized in packaging machines
Relationship between firms with access to final markets	In packaging machine subsystem transition from single firm to network-enterprise
Relationship with sub-contractors	In packaging machine subsystem greater stability of sub-contractors who are entrusted with administrative roles
Choice of products	In packaging machine subsystem, product development "from above" and "from below"; product selection according to market demands prevails
Type of product	Special production with customized flexibility and special production with catalogue flexibility
Type of worker	From skilled worker with prospects of setting up own enterprise to skilled worker with more difficult career mobility
Work relationships	"Spirit of collaboration" persists in packaging machine subsystem. In other specialized firms, tendency to more formal relations
Relations among local actors	Presence of several actors and permanence of collaboration; decline of the "Emilian model" makes it difficult to elaborate medium–long term innovative strategies

Specific changes in the Bolognese engineering industry

An analysis of specific aspects of a particular type of industrial development ought to facilitate a comparison both with other types of flexible industrialization (which today have changed since the early phase) and other types of mass production (this too being greatly modified as far as labor organization and production are concerned). A first distinction concerns the relationship between enterprise and territory, in the sense that it is possible to have a specialized, or only partly or non-specialized area for a certain type of production. It is clear that in the Bolognese engineering industry, the area of specialization is that of automatic packaging machine manufacture, even though there are other types of flexibly specialized engineering production, just as Kanazawa specializes in machines for synthetic fibers, and so on.

This specialization may take the form of an industrial district or urban industrial subsystem, following the definitions discussed above based on the relationships between competition within this group of enterprises and competition on national and international markets. Two subsystems have been identified in the Bolognese engineering industry, the large packaging sector and the much smaller motorcycle sector, and these have had different fates. The former has maintained itself through the creation of network-enterprises, the latter been radically restructured, leaving only a single network-enterprise and a group of subcontractors.

An analysis of the Bolognese engineering industry illustrates the transition from subsystems based on separate firms (though of different sizes) to subsystems with several network-enterprises. One might ask whether this transition can be generalized for other districts or subsystems and whether the presence of a network-enterprise is the only means by which a subsystem can successfully compete on international markets (since it would be possible in principle to establish interfirm consortia, integrated sales centers, etc.) In the most important subsystem of the Bolognese engineering industry the formation of several enterprise groups has led to a loss of internal fluidity, as too has the crystalization of hierarchical relations among the subcontractors themselves.

An additional classification of subsystems derives from the manner in which different products are chosen. Hirschmann distinguishes two processes of selection; one "from above," whereby the industrialization process extends from the product to the machines which produce it, etc. The other runs "from below," when one line of products stimulates the manufacture of others and so on. On the basis of this distinction, the formation of the packaging industry subsystem was in its initial stage of the former type (the food industry required equipment to mechanize its processes) and then developed into the latter type because the presence of an initial line of machines stimulated the production of similar machines for different products and so on.

One can add a further classification which takes cycles of technological innovation into account. The products of the Bolognese packaging industry were selected on the basis of market demand (the need of the food industry, for example, to mechanize its various phases) and so it is possible to speak of a production dependent on market demand. But even departing from these requirements, there has still been a continuous accumulation of technological knowledge culminating today in the transition from mechanical to electronic flexibility. In this case we are dealing with what Michael Storper calls a "technology district" with "product-based technological learning by which is meant the ability continuously to reinvent, differentiate, improve and reconfigure products through a dynamic development of key, specialized production skills and equipment."[38] Thus the Bolognese packaging industry is one of those districts which, according to Storper, are able to withstand "globalization."

The situation changes when there are vital technological innovations in the

[38] Storper, "Limits to Globalization."

development phase and it is only possible to understand them in certain research centers and specialized universities. Technologies such as fuzzy logic, for example, are only available today in a limited number of institutions and the enterprises closest to these technologies can design products which, in this case, are not chosen on the basis of market demand but of the potential development of this new technological knowledge. The stasis in the Bolognese engineering industry, apart from packaging, is a consequence of its distance (with a few important exceptions) from sources of technological innovation and of the difficulty of developing new products (with available technology) on the basis of differentiated needs which may come from different groups of people, particular groups of enterprises, etc. The packaging industry subsystem has avoided this stasis (although it is necessary to talk in terms of stabilization rather than the expansion which characterized the earlier period) because it is a subsystem of "product-based technological learning" forced to innovate continually.

It is also important to define the types of products, entrepreneurs and workers necessary for their manufacture. First, we can distinguish between "specialized" and "mass products" simply on the basis of world sales figures; world production of cars or motorcycles has a much higher overall turnover than a specialized product like packaging machines. But, it is also necessary to formulate more precise definitions, since, as far as Bolognese engineering production is concerned, several types of production may be identified: *prototypes* (racing motorcycles), *specialized production with customized flexibility* (packaging machines which in turn can be divided into customized machines for individual products or customized machines for a range of differently shaped products), *flexibly specialized catalogue production* (like the Marposs products discussed earlier), *flexible mass production* (motorcycles for the consumer market). Again, this production can be subdivided according to the varying importance within it of electronic and mechanical engineering (as in the case of the Marposs products and the latest packaging machines, where electronics has become predominant). The important point is that the two types of Bolognese flexible specialization (customized and catalogue) involve significant differences in work organization. Not only does customized flexibility influence work organization, but the close relationship which must be established with the client gives rise to an accumulation of knowledge which is more difficult for a new competitor to acquire.

Specialized production with customized flexibility in the packaging industry influenced the first generation of entrepreneurs who mainly began as skilled workers and later acquired experience in technical offices and in assembling machinery abroad. The transition for this generation to a managerial type of organization with a more equal balance of power between design, production and sales has not been easy, nor has it been easy for the new generation of entrepreneurs to establish themselves, coming as they do from university and post-university training which has introduced innovations into electronic production not previously conceived for the packaging industry. Production with customized flexibility involves the presence of skilled workmen whose active engagement, very often informal, is necessary to achieve the aims of the firm. This explains why a "spirit of collaboration" is required within the individual firms, and between the firms and their subcontraction. The Bologna unions can therefore speak of "co-determination"[39] because this is the direction that production is following. In the case of flexible catalogue production, forms of automation are possible which create areas of "blocked mobility" within the firm. Work is more formalized, a "spirit of collaboration" less desirable and less usual. If the heart of flexible specialization remains customized flexibility and the heart of mass production can at the most become catalogue flexibility, the differences between the two organizational modes remain very large. It remains to be seen whether or not flexible specialization will itself become catalogue flexibility.

A final element of the proposed typology regards the relationship between the various actors shaping the industrialization process (state, regional and local government, unions, business associations, University). In the period from 1946 to 1980, there was an intense collaboration between local actors in the province of Bologna, so much so that what has been defined as the "Emilian model" emerged in the 1950s and 60s. In the most recent phase, collaboration between local actors still exists, and they are still able to intervene, which is more than can be said for most Italian provinces. But there is much more political uncertainty, a loss of the "us" mentality and it is much more difficult for local protagonists to propose middle to long term strategies. So social innovation will influence the development of technology and enterprise in the years to come.

[39] The term *co-determination* is used by the Bolognese trade unions and taken from the actions undertaken by Scandinavian unions. This term indicates the attempt of the unions to negotiate over the effects which technological changes may have on work organization before the changes actually occur. See Capecchi, "The Role of Unions."

10 Local industry and actors' strategies: from combs to plastics in Oyonnax

Jean Saglio
Translated by Cyprian Blamires

The industrial development of the area round Oyonnax, in the French Jura close to Switzerland, represents another chapter in a very long story. The current vigorous development of the plastics moulding industry there owes a great deal to the renewal of an age-old tradition of small businesses specializing in working first with horn and later with celluloid. Throughout its history local industry has managed to keep on adapting to constant changes in markets, production techniques and the populations that provide its labor force. But the processes of adaptation, often fraught, have frequently involved profound changes both in the local way of managing industry and in local society.

The presence on the spot of many of the actors involved in these changes enables us not just to reconstruct the economic history of this area, but also to attempt to understand how the actors themselves construct their perception of their situation, and what meaning they attribute to industrial development and to their experiences of conflict and cooperation. The study of an example like this enables us to establish the main lines of an economic sociology adapted to French situations of this very particular kind.

Industrial development and urban growth in Oyonnax[1]

Typical explanations of the implantation and development of an industrial city will refer either to specific features of its site – coal deposits, sources of hydraulic energy or water with particular chemical qualities, or some other resource above or below ground – or to the strategic location of the site in a position that makes it the key to an important through route. These are the kinds of arguments normally used to explain how a particular industry was established in a particular place.

In the case of Oyonnax it is difficult to identify any such first causes, for it is a locality singularly lacking in natural or commercial assets. There neither is nor

[1] Data on the history of Oyonnax before the First World War are taken from the works in the bibliography by S. Barcellini, B. Dominjon Bombard, N. Favre Brun, R. Lebeau, Y. Lequin, H. Puel and J. Saglio. The unpublished memoirs referred to may be consulted in the town's library. For the period after the First World War I have also utilized oral sources in the form of interviews on the spot.

ever has been any exploitation of minerals, and the soil supports little other than boxwood and in modern times fir plantations. Where the water situation is concerned the local river system with its limestone geological structures has little to offer in the way of energy supply. Finally, though the site is indeed on a through route, it is hardly conceivable that the road between the French towns of Nantua and Saint Claude could have had crucial significance in terms of the explanation of industrial development. Not even on the micro-regional level does Oyonnax seem ever to have been a trading center, site for a fair or deal-making venue prior to 1960.

To account for the development of their town the locals tend to refer to factors having to do with the local identity, but even a sketchy demographic analysis makes this explanation look very doubtful, at least on a straightforward level. The population of the urban area has never been made up solely of individuals who could boast of a long, local family tradition. On the contrary the industrial dynamism of the town has attracted newcomers for years. The resulting demographic mixture is a reflection of the problems involved in the construction of a common local identity. What does it mean to be an Oyonnaxian? Is it a matter of belonging to one of the few families with their roots in the locality? The evidence is that such a reply is totally unsatisfactory. To answer it we need rather to analyze how a local sociability was constructed and in successive periods transformed into an economic resource that local companies could mobilize.

Two centuries of industrial history At the beginning of the nineteenth century, even though its agriculture gave poor yields, this austere mountainous district of Haut Bugey not far from Geneva was nonetheless relatively populous. However, the industrial growth that was to take its rise in the first half of the nineteenth century and carry on developing right up to modern times was focussed exclusively on the village of Oyonnax, and while the overall population of the area remained unchanged or indeed even decreased, Oyonnax itself grew into an industrial town with a population of about 25,000 in the early eighties. Broadly speaking therefore we can speak of the Oyonnax Valley as growing demographically and industrially in the heart of a region that was in general on the decline.

There was no regularity about this movement. A study of the growth of the population of Oyonnax from the beginning of the nineteenth century up to our time reveals three separate periods of relatively rapid growth and two different phases of stagnation, if not decline. What brought the growth about? Was it just the usual phenomenon in France of depopulation of rural areas in favor of small urban centers – in which case the movement would have difficulty in maintaining itself today simply because of the exhaustion of local demographic resources? Or was it industrialization that made this town a more powerful center of attraction? Questions like this suggest the need for a closer study of the new populations: who were the new Oyonnaxians of these growth periods?

There is no question about the expansion of the town of Oyonnax over the period of two centuries since the French Revolution. Since the early twentieth century, at least, this urban growth seems to have involved the attraction of migrants from outside the area – a development encouraged by the fact that in the last thirty years at least there have been relatively substantial demographic outflows from the town. A place with a shifting population, Oyonnax does not then look like the archetypal small French town developing slowly on the basis of the exploitation of local agricultural resources and local trading activity.

The beginnings of local industry

Oyonnax has a very long history. Like the neighboring village of Dortan it began with a monastery grain depository, the property of the monks of the town of Saint Claude thirty kilometers away, and became a proper village around the year AD 1000.

Agriculturally, the situation remained unpromising until the forest economy really began to develop with the rise in the price of wood at the beginning of the nineteenth century. At the Revolution the agricultural censuses show this area to have been one of the poorest in the Ain *département*. The wealth of the canton's farmers was assessed at a quarter of that of their Bresse counterparts. Agriculture, including animal husbandry, was not even viable enough to guarantee survival for a population whose density at the period was relatively high. Assessing the state of local agricultural activity, René Lebeau paints a very dismal picture: "In cereals, in oleaginous crops, in textile crops – always the same poverty. Haut Bugey farming was inadequate all down the line, and there was a dependence on neighboring districts, especially for cereals and most of all for corn – the principal import."

Paradoxically, it is these very shortcomings in the local agriculture that provide us with a key to understanding the development of local industry and the strength and permanence of local community structures. For the poverty of the soil meant that just in order to survive the inhabitants of this part of Haut Bugey were obliged even before the French Revolution to find supplementary ways of making money: "industries of necessity" to use Lebeau's terminology. In Haut Bugey in general the commonest of these "industries of necessity" was the temporary winter emigration to card hemp in neighboring districts. But the Oyonnax area – the town and its immediate surroundings – was an exception, for it was a breeding-ground for small local artisanal-type industries.

The beginnings of the "industry of necessity" Work on boxwood combs had a long history in the area, for it had already been encouraged by the monks of Saint Claude with the support of at least some of the ecclesiastical authorities. In the seventeenth century there were workshops for wood turning in a little cluster of villages – Bellignat, Dortan, Oyonnax and Arbent. These activities were integrated into daily life. In the summer each man was a farmer, in the winter a comb-maker and in the spring a carrier. Though production was organized and integrated within the family cell there was always a certain level of cooperation, and at the time of the Revolution the comb-makers formed a group to create the first "union" of comb-makers to lobby the public authorities.

At this period the village was still just like the other local communities. A 1797 census gives the figure of 800 inhabitants. But certain demographic features were already suggestive of the shape of the future industrial town. There were workshops for wood turners in the village and its neighborhood before the Revolution, and in 1801 there were twelve comb-makers and twenty-two workers engaged in the industry.

From the beginning of the nineteenth century the village began to grow and diversify. While the neighboring villages specialized in woodworking – as they still do even today – the Oyonnaxians were already working with other materials at the end of the eighteenth century. This may perhaps have been the result of a relative paucity of boxwood. Whether or not that was so, by 1808 work on *clampons* or bovine horns had developed there and the statistics tell us that in that year the town produced 300 kg of items made from boxwood and 400 kg of items made from horn.

Up to the middle of the century growth was continuous. In 1820 there was a population of 1,158 representing a far from negligible 45 percent increase over twenty-three years, or 1.6 percent per annum. Textile industries developed in the area after 1830 and remained in existence up to the end of the century. In 1831 35 horn flatteners, 44 manufacturers and 130 workers were active in the comb industry. At that time "turning and comb-making," notes Lebeau, employed practically the entire population: "600–700 out of 1200 inhabitants." The population of the town grew to more than 3,000 in 1850, an increase of 159 percent or an annual growth of the order of 3 percent. In all probability, growth was not absolutely regular over this period, for crises were already occurring.

This demographic explosion was not solely a matter of the birth rate – as we have seen, Oyonnax had a long tradition of immigration, though the extent of such immigration in the nineteenth century is difficult to assess with any precision. It is probable that over this period the contribution to the workforce by the local rural population was significant – witness in particular at the end of this first period the virtual absence of names not native to the locality. Local tradition claims that certain families came down from the nearby Retord plateau to work as wheelwrights but fairly quickly became mechanics.

It is worth pausing here to note that the Oyonnaxians had already mastered two specific technologies: that of working with hardwood – especially turning techniques and the cutting and polishing of boxwood objects – and technologies for the working of heated plastic materials. There are two ways of working on *clampons*: a process of sawing and turning makes it possible to work on the solid tip of the horn with which various small objects – buttons in particular – can be made. It is equally possible to work on the wider hollowed-out portion of the horn to make hollow cylindrical objects such as napkin rings. But the main way the hollowed-out part was used was in sheets created by the work of the flatteners. They sawed the hollowed-out part of the horn following a generating line of the cone and then flattened the piece at high temperature to give it the form of a sheet from which combs could be cut. Horn was thus a natural thermoformable plastic, the first to be worked in the area.[2]

Our demographic analysis has shown the first industrial development of the town occurring in the years 1800–50. The new commercial structures that emerged probably corresponded to two requirements arising from the development of the traditional artisanal industry: on the one hand there was a need to develop sales networks – something that was quite beyond the traditional colportage system; on the other hand there was a need for increased supplies of raw materials – especially of horn, for the amount produced locally by the scanty pool of livestock was quite insufficient.

If we accept the argument of Yves Lequin, the impetus came mainly from the local firms:

> Between 1820 and 1830 they gave new outlets and new vigour to a routine tradition which was disappearing elsewhere. Warehouses in the towns sent out combs, buttons and boxwood objects made in all the surrounding villages and the expansion of this trade stifled small-scale watchmaking and the cutting of precious stones in the Nantua area. In 1848 combs were what Oyonnax was famous for and 1450 workers, for the most part men (915) were working on them in the canton, while in the canton of Nantua 200 people were working on the manufacture of knick-nacks. If we add in the 300 woodcutters and sawers, the total numbers involved amounted to nearly 2,000 persons.

Diversification and new outlets But it does not seem that this factor internal to the comb industry is sufficient on its own to account for this first phase of industrial development. The thesis put forward by René Lebeau (which admittedly covered a larger area than just the canton of Oyonnax) flatly contradicts what appears at first sight to be the case. According to him "it was not a case of traditional old industries reemerging, but of new ones, for the most part imported, becoming established, and often in areas where the workforce had previously been composed entirely of rural artisans." Textiles in particular really took off in the Nantua–Oyonnax region from the 1830s. It is true that there was a silk throwing works in Nantua before the Revolution, but by 1820 it was on the decline and indeed virtually defunct. It was in 1828 that Piguet "came back from Groissiat to his home village, set up several looms there, then planted mulberry trees and established a silkworm rearing-house on his land." Subsequently there was considerable development of the textile industry throughout the area, at least

[2] Historical technological descriptions of the modes of fabrication in Oyonnax, including those employed in the relatively distant past, were obtained in the course of discussions with specialists teaching at the Lycée Arbez Carme.

up to the early 1860s. "In 1863," says Lebeau, "there were 4,000 shuttles on the go between Oyonnax and Nantua."

The traditional industries partially adapted to this new industrial configuration. According to an 1860 text the turners of the village of Arbent "had developed a monopoly of the supply of rollers, weft spools and bobbins to Lyons and Saint Etienne." In 1835 the existence of a cotton spinning works at Dortan was recorded. But what actually developed in this sector was a pattern of home working rather than a real manufacturing industry.

This first industrial advance was not without its crises and difficulties. The manufacture of wrought luxury combs was particularly sensitive to the vagaries of fashion and to the repercussions of general economic conditions. In the face of such problems the response of local industry was to rely on the collectivity and the public authorities: in 1839 at a period of crisis the municipal council decided to build a church "to provide work for destitute workers." The same scenario recurred in 1848, when the municipality decided to take out a loan of 30,000 francs to provide assistance for the unemployed. Thus we see the development of a distribution of roles between entrepreneurs and the community: the latter allowed the former to develop their industry in times of growth, but intervened in bad times to attenuate the difficulties of the workers. It may be noted at the same time that from this period onward economic development was based on a plethora of small local businesses and not on the growth of a concentrated industry either of local or external origin. The Oyonnaxians did business with each other.

Growing pains and first renewal: 1850–1936

This initial phase of growth in the first half of the nineteenth century was followed by a period of stagnation – almost of recession – in the 1860s. There was a decline or at least a slowdown of growth that was perceptible in the demography: in the early 1870s the town had less than 3,000 inhabitants, i.e. slightly fewer than twenty years earlier. But this growth took off again at an ever faster rate for a period of just over fifty years, with the First World War merely prompting a slowdown rather than stagnation or recession, and the population peaked in 1926 at 11,500, which represented a growth of 28 percent or 2.4 percent annually over the period 1870–1926.

The working population This population growth resulted from an influx of labor from well beyond the neighboring localities and well beyond the local rural workforce. In the 1901 census, 41% of the population was born in the commune, 52.4% in the canton, 76.6% in the *département*, 5.8% in the Jura *département*, 13.1% in the rest of France and 4.5% abroad. At this date then more than one Oyonnaxian in six was born outside the Ain and Jura *départements*. Already the town's attraction was being felt beyond France's borders: the proportion of foreigners in Oyonnax grew regularly between 1896 (2.5%) and 1911 (7.1%). There was however a time lag between urban growth and the growth of the population of foreign origin: the global population grew pretty substantially after 1870, but it was only towards the turn of the century that foreign labor began to be attracted. During this period the immigrants were for the most part Italians, and Christiane Barcellini estimates the proportion of Italians in the foreign population of Oyonnax at between 77% and 83% from 1911 and 1938.

In the 1906 census only 2,300 persons, or 30% of the overall population, were of local rural origin. The average age of the population was low and its provenance was fairly varied: "One is struck," notes Lebeau, "by the large number of locals hailing from the Massif Central: Creuse, Corrèze, Indre, Nièvre, Puy de Dôme, Loire, Saône et Loire." And he adds this remark: "They were not usually comb-makers but merchants, or else they carried on small trades (shoemaking for example)." So while the town was certainly attracting a lot of immigrants from elsewhere in France or from abroad, they were not directly employed in the main industry.

The renewed advances after 1870 continued to depend basically on the comb industry: there were 550 workers in 1872 and 3,000 in 1900: 41 manufacturers in 1873, 200 in 1890 and 300 in 1900. In 1901 53% of the 6,140 inhabitants officially declared an occupation (73% of the men and 29% of the women). For the active male population the breakdown is given in table 10.1.

The industrial workforce comprised more than 71% of the men, 61.6% of whom were workers. We have however excluded artisans and traders – some of whom may have worked for the industry too – and merchants and commercial agents, whose activities were closely linked to it.

From the beginning of the century the industrial vocation of the town was quite clear. The emergence of the celluloid industry that was introduced to Oyonnax at this time merely accentuated the movement.

Industry and its environment A new problem appeared in the mid nineteenth century – that of energy. Local hydraulic capacities could not satisfy the appetite for energy generated by industrial development, and one casualty of the energy shortage was the textile industry. The comb-makers had two essential energy requirements: they needed to be able to heat the sheets of the flattening presses and they also needed mechanical energy to operate the presses. The solution they adopted towards the end of the period was a collective one. It was the "Industrial Motor Society," more usually known as "the Steam" which provided the artisans with the two energy sources in a special building where they could rent "cabins." This installation went into service in 1865.

The same principle of a collective solution to the problems of the production environment was embodied in the establishment at the same period of a friendly society for insurance against horn workers' occupational illnesses: it was the "flatteners' bank." Working conditions in this sector were in fact particularly

Table 10.1 *Male employment in the 1901 census (percentage)*

Comb industry workers	52.3
Building workers	6.3
Turners	1.1
New working class	1.9
Manufacturers	9.5
Merchants and commercial agents	2.4
Artisans and traders	15.5
Servants	2.8
Farmers	2.8
Civil servants and administrators	3.1
Liberal professions	0.4

unhealthy. Before being treated the horns had to remain in water for quite a long period. This foul water was an excellent place for the culture of every kind of microbe and the flatteners suffered from many ailments, apparently including tetanus.

There was a crisis in the period 1866–73 which probably had a two-fold origin. On the one hand, there were the vagaries of fashion – a factor frequently mentioned by different authors: by introducing a fashion for hairnets the Empress Eugénie made life much harder for the comb industry. On the other hand, the problems the textile industry was experiencing did not allow it to make up for the decline in the comb industry in the area. This led to a crisis situation locally and between 1866 and 1872 the population fell by around 10 percent.

Cooperation and conflict After this preliminary rapid survey I shall attempt to describe a relationship apparently very specific to Oyonnax between the phenomenon of competition and cooperation and the particular political features of the town.

In a way cooperation was a local tradition. It is true that the local peasant farmers owned the land they cultivated, but the area that was managed collectively was nonetheless quite substantial: in 1850 Lebeau notes that it still comprised 54 percent of the land surface (though of course these were not necessarily the best soils!). To an extent the beginnings of industry were marked by this tradition. Problems relating to the production environment (whether pre- or post-production) were often handled collectively, either in a cooperative manner (flatteners' bank, manufacturers' association, Industrial Motor Society) or by delegation to the public authorities (crisis management). This meant that there was local regulation of competition between manufacturers by a strict delimitation of the areas of competition, and that there was a kind of brake on moves by individual companies to gain a monopoly.

This mode of managing industry and its collective problems was reflected in the local development of a republican opposition with a cooperativist political ideology. Oyonnax showed left wing tendencies from an early date. In 1848 it had a republican majority and it opposed the Empire in 1852. At the plebiscite of May 8, 1870 there were 563 "no" votes and 206 "yes" votes, which made Oyonnax one of only two towns in the Ain *département* where the "no" votes carried the day. Under the Moral Order the opposition continued to attract majority support and conflicts with central power bear testimony to this. Thus Louis Dupuy, a republican notary and anticlerical who was mayor from 1868 to 1881 was removed from office by the government. Vigorous support from the population enabled him to regain his position in the full glow of publicity. At the turn of the century a Socialist Youth branch was founded at Oyonnax, and after 1914 the Socialists enjoyed an electoral majority locally.

The political militants were not however drawn from the traditional working class, like their counterparts in the great industrial centers. Whether republican or socialist the town council remained dominated by manufacturers and members of the liberal professions. In the 1874 Assembly the 20 persons elected included 8 comb-makers, 2 workers, 1 tinsmith, 3 merchants, 2 notaries, 2 doctors, 1 usher and only one farmer. In the 1904 Assembly the 23 persons elected included 14 manufacturers, 3 merchants, 2 landowners and 4 "miscellaneous." Between 1881 and 1919 the succession of mayors who followed the notary Louis Dupuy were all manufacturers.

The cooperative movement exerted a strong influence on these local elective office-holders. Thus the manufacturer Louis Bolle who was mayor from 1904 to 1919 actively participated in the creation of the "Oyonnaxienne" cooperative in which militant socialists asserted the operating principle of "one man one vote" against the capitalist principle.

The second phase of urban and industrial growth In the years between the end of the 1870 War and the mid-1920s the town enjoyed significant demographic and economic growth and underwent a whole series of transformations. From a demographic point of view we have seen that the population almost quadrupled over this period, rising from just over 3,000 to 11,500 in the 1926 census. This growth took place thanks to a big dose of immigration, both from the immediate locality and also from the other areas in France and from nearby foreign countries – Switzerland and especially Italy.

This demographic growth was of course linked to economic growth as reflected in the turnover of Oyonnax industry, which was estimated to be worth 3.5 million francs in 1878 and grew rapidly to 10 million in 1897, 12 million in 1902, 20 million in 1912 and 160 million in 1924.

This indicates a relatively unstable rate of annual growth: it was of the order of 8 percent in the period between 1878 and 1897, fell to 1 percent between 1897 and 1902, pulled back to 3 percent up to 1912 and climbed to 4.5 percent between 1912 and 1924. These figures are probably not very precise, but they do indicate that there was very strong growth in the first three decades, then a phase

Table 10.2 *Growth of industrial production 1878–1924*[3]

Dates	Total turnover for town	Average turnover per company
1878	100	29
1897	414	44
1902	437	43
1912	583	55
1924	976	67

of relative stagnation at the turn of the century, followed by a fresh takeoff before the First World War which regained its impetus after the war and up to the mid-twenties.

This period of growth was accompanied by a marked trend towards the creation of new companies; 41 were established in 1873, 120 in 1880, 200 in 1890, 300 in 1900, 310 in 1913 and 400–450 in 1925. Again we see a concentration of growth in the early years of the century. A survey of turnover (average turnover in thousands of francs at 1900 values) by company shows the same concentration (table 10.2 last column). The companies were not very large however – in 1900 only two of them employed more than fifty persons.

In this period technological advance was producing major upheavals in local industry. First of all there was the virtual disappearance of the textile industry, which left the totality of industries associated with comb-making in a dominant position in the area.

Then there was the transformation of energy sources. The Industrial Motor Society had opened its doors in 1865. From 1889 the town was one of the first to benefit from a provision of electricity thanks to the installation of a line from the Charmines Falls on the Oignin. Soon a second plant at the Mortier Falls on the Ain supplemented the first. After some initial hesitancy, the use of this source of energy developed rapidly and by 1900 90 percent of Oyonnax machine tools ran on electricity.

Then there was also the development of communication systems – the railway from La Cluse to Saint Claude was opened in 1881 and in 1906 the telephone made its appearance with a line from Oyonnax to Saint Claude.

The most significant innovation at this period was perhaps however in the area of raw materials. In 1869 the Hyatt brothers had discovered celluloid in the USA. This is a cellulose nitrate obtained by treating cotton in a bath of nitric and sulfuric acid and stabilized in camphor. This was the first modern "synthetic plastic" that was thermoplastic – i.e. that could be moulded at high temperatures. By the use of the appropriate paints or dyes it could be given all sorts of finishes and made to look like horn, ivory, etc., but it had one very serious defect –

[3] In francs at their 1900 value. Taken from the index of wholesale prices in the *Annuaire statistique de la France* for 1954, retrospective French part, cf. pp. 47–88. Base 100 in 1878.

made of nitro-cellulose, it was a highly inflammable and indeed even explosive product. Its use did not really develop till after 1887, the date when industrial chemists managed to produce substantial quantities of it at prices that made it much more profitable than horn. Serge Barcellini tells the story of how it was introduced:

A few Oyonnax comb-makers exhibited their products collectively at the 1878 Paris Exhibition. Lucien Verdet, assistant to the Mayor and comb-makers, was delegated by the town council to study the exhibition and see what local industry could learn from it. On his return he gave his fellow citizens an account of the first attempts at working with celluloid. A few manufacturers then produced some coral-type bracelets in two, three or four rows and in the form of snakes or bands for various Paris stores. The use of celluloid still did not become general because its cost was too high at 40 fr per kilogramme and because the workers were suspicious. Though the small manufacturers took to celluloid fairly readily, the workers apparently opposed it on the pretext that this unknown material ruined the tools. It was only around 1886 and 1887 when the new manufacturing processes had brought the price of celluloid down to 12 fr that its use spread.

For ten years or so Oyonnax manufacturers bought their celluloid from four factories: two French ones (Saint Sauveur in Calvados and Monville in Seine Inférieure) and two German ones (Mannheim). Curiously, according to Barcellini "at the 1900 Universal Exposition almost all the Oyonnax products on show were made of horn," and yet in the year 1899 646 tonnes of raw celluloid had been unloaded at Oyonnax railfreight station.

The Oyonnaxians' response to the new (rather fragile) development was to pull together. In December 1899 a subscription was opened to build a factory that could make the celluloid locally, and more than 600,000 francs was raised in three hours. A cooperative finance society, the *Oyonnaxienne*, was created. Initial output from the factory in 1902 was 250 kg per day, and this rose to 1,000 kg in 1906, 1,500 kg in 1910 and 2,500 kg in 1913. At the same time two other factories had been set up locally, but under a more classic juridical form: one by an established firm, the Convert company, which opened a factory that was producing 500 kg per day by the beginning of the war, and the other by Bellignite (same level of production). Finally the Lyonnaise Celluloid Company was turning out 830 kg per day by that date.

The story of the establishment of the *Oyonnaxienne* is a testimony to the continued hold of cooperative practices in local industry at the beginning of the present century. But at the same time other types of industrial policy bear witness to the emergence of more "classic" strategies among the local capitalists. Contrary to what had happened with the 1878 Exhibition, when the Oyonnaxians' contribution had been inspired and coordinated from the town hall, local manufacturers exhibited at the 1900 Exhibition on their own initiative. Likewise the strategy of the Convert Company differed markedly from the local model of cooperation–competition. Founded in 1830 as a trading company, it bought up a small comb-making business and by 1897 had become one of the two biggest local factories – one of the two that employed more than 50 persons. Its strength was further boosted by the establishment of a celluloid factory and then by the employment of qualified chemists – something extremely unusual in Oyonnax.

The survival of the artisanate and mechanization However alongside this first movement towards concentration of labor in factories with the emergence of a local industrial capitalism there was also a corresponding movement towards the development of the artisanate. We have seen that the latter had been at the heart of the initial industrial development of the town in the first half of the nineteenth century: in 1831 it was estimated that two thirds of those involved in making combs were independent workers. The energy supply problem had led to a certain concentration – geographically at least – of production, which now began to take place in the "factory cabins" available in "The Steam," with some of the workers located there working for themselves and some having the status of employees. But the need to group together in one place had been very badly received by the workers who, in 1870 – five years after the opening of the installation – had gone on strike to protest in particular against "being forced to work on top of each other in the factory cabins" and against the modalities of remuneration (payment in merchandise). The arrival of electricity made it possible to come back to traditional forms of artisan working and in 1896 it was noted that in the previous six years more than 300 workers had left the big establishments (though this does not necessarily mean that their juridical status as workers had changed). In 1901 a quarter of the comb workers were working for themselves.

There were thus two contrasting movements at this time: on the one hand an undeniable concentration of companies (which however remained only medium-sized or smaller) and on the other the persistence of an artisanal tradition. However the trend towards specialization was widespread as workers able to perform the totality of operations involved in the manufacture of a comb on their own were replaced by workers each of whom was specialized in one of the twelve stages of the manufacturing process. This meant, as Serge Barcellini observed, that after 1900, the *de facto* monopoly previously enjoyed by workers of local provenance began to disappear and immigrants began to become comb-makers too.

It was largely mechanization that made this movement possible. The structure of local industry encouraged a lively competition between the different local manufacturers. It is true that "artistic creation" enabled some to have a secure income, but this could only be ephemeral, since copying was common. The chief preoccupation of the worker working for himself was therefore to produce more and more quickly, and it was from the artisans themselves that the movement towards mechanization came. Its beginnings date from the 1870s and the first machines were for cutting or fluting. They were made on the spot by mechanics who were manufacturers as well and the Colettas, the Barnoux and a few others both manufactured machines for themselves and marketed them. With the advent of electricity these relatively light machines could be installed in large numbers of separate workshops.

To begin with it was simply a matter of mechanizing the tools traditionally used by the local workers and substituting mechanical for human energy. But fairly soon these do-it-yourself mechanics also sought to transform modes of operation and in particular to utilize the heat ductility of thermoplastics to try and inject them. Thus at the start of the century an Oyonnaxian took out a patent for an injection screw. But such innovations could not be implemented in industrial manufacturing for the heated horn was not plastic enough to be injected while celluloid has an unfortunate propensity to transform into an explosive when heated under pressure.

This mechanization was structured as much by the movement towards the concentration of capital as by the survival of local artisan traditions. Thanks to cooperation the "pieçards" – a local name given to piece-workers who had an intermediate status between that of the paid worker and that of the artisan working for himself – were able to survive as long as they equipped themselves with machines. Taking account of prices and the simplicity – indeed rusticity – of the tools concerned, such a move was quite feasible and many went down that path. The only domain where substantial investment was required – the manufacture of raw materials – saw the concurrent development both of a cooperative form and of a private form of company, with a clear preeminence accorded locally to the former over the latter on the level of production.

Industrial development, division of labor, specialization, mechanization – so many factors that profoundly transformed Oyonnax industry's labor requirement. The artisans and the small industrialists asked the town council to establish a school. This opened its doors in 1879 as an elementary school providing a general education. The spread of mechanization and the relative economic difficulties of the turn of the century were an incentive to extend it. A mechanics' workshop opened in 1904 and the school thus became a "Practical School." In response to a lack of skilled labor for the different operations involved in working with celluloid a second technical section was added around 1910 – a plastics section that trained fluters, cutters and smoothers.

The war and the post-war years The First World War and the immediate post-war years wrought a series of transformations in local industry but at least to begin with these did not block the economic and demographic growth of the town.

Changes in raw materials first of all.[4] The problems of working with celluloid led the chemists to experiment with other synthetic plastic materials. They were able to obtain casein by treating milk; treated with water it produced a thermo-

[4] These technological points were established in the course of discussions with speakers from the Lycée Arbez Carme.

plastic that had the disadvantage of being putrescible, but by treating it with formalin they avoided this problem – and this is how Galalith was born.

Military needs, in particular the need for aeroplane windshields, encouraged the search for a material less inflammable than celluloid. The Germans succeeded by treating cellulose not with nitric acid but with acetic acid, which produced the very much less inflammable cellulose acetate. After the war this produce was manufactured on a grand scale at the Le Péage factory in Roussillon and marketed under the name of Rhodoïd. It was a non-injectable thermoplastic that the Oyonnaxians quickly began to work with even before the problems with its manufacture were all resolved. To begin with they were only able to make tri-acetate, a relatively unstable product that did not age well, and it was only during the Second World War that techniques were developed for making the much more satisfactory bi-acetate.

It was at this period too that phenol formalin resins appeared. These, unlike the other synthetic products produced previously, are thermoset and not thermoplastic: under the action of heat and pressure they harden by polymerization and reticulation in three dimensions. The best known of these resins was Bakelite, the trade mark of a product made by a factory in Douai.

The manufacturing constraints for these new materials were different – particularly the constraint of safety considerations. The processing of celluloid, a highly inflammable material, had necessitated the construction of highly compartmentalized little workshops in single-storey buildings with north-facing windows to prevent the outbreak and especially the spread of fires. These new materials made possible the construction of bigger and more concentrated workshops.

The specific features of thermoset materials did not pose any insoluble new problems. It had long been known how to use hot presses to flatten horn or celluloid; what was needed now was not the old flat presses but hollow moulds, and the technique was quickly assimilated. From the beginning of the twenties such presses were installed at Verchère and Collet Janco.

But the main problems of the twenties and thirties were product markets. The fashion for short hair was a big blow for the comb industry and diversification of products was therefore once again the order of the day. Faithful to their fancy goods tradition, the Oyonnaxians got involved in the production of yoyos and innumerable other sorts of knick-nacks, but the need to diversify also incited them to take an interest in the markets of their neighbors in the Southern Jura, and around 1923–24 the production of plastic glasses was undertaken. To begin with they bought the hinges and side-pieces from Morez, restricting themselves to making the lenses and assembling the glasses. Fairly soon, however, they were involving themselves in the manufacture of the side-pieces. As for the hinges, they did not manage to develop the manufacture of these little metal assemblages in the locality.

The second difficulty in the area of markets came from the rise of protectionism. For many years local industry had been strongly orientated towards export markets, with half of all exports passing through England. The export quotas established by certain countries and the rise of new producers meant that after 1927 the external markets collapsed, and between 1929 and 1932 their value fell from 175 million to 30 million francs. Diversification in the direction of cheap mass production products was not enough to prevent Oyonnax industry from falling on hard times.

Conflict management However the internal social system was not really affected by this crisis and the transition from worker to artisan via "self-employment" continued to be the preferred path to promotion. There continued to be tariff negotiations between representatives of those giving the orders and the employers on the one hand and representatives of the local trades union on the other, with recourse to mayoral arbitration where necessary.

The mayor was René Nicod and he was a key personality in local political life. Born in 1881 in St. Claude he joined the Socialist Youth in 1898. In 1901 he came to Oyonnax to work for the workers' printing press there and very soon became the leader of the local socialists. He was behind the strike movement that shook local industry and especially "the Oyonnaxienne" in 1913 and he was even arrested on that occasion. During the First World War he took an active part in the organization of resistance to the "Union Sacrée" (Sacred Union) and in this way made the acquaintance of militant unionists like Monnatte and Merrheim as well as dedicated internationalists like Martov and Trotsky. In 1919 he was elected Mayor of Oyonnax, a post he continued to hold till he died. In the same year he headed the poll at the legislative elections in the Gex-Nantua constituency.

This victory was all the more remarkable in that it occurred in the national context of the victory of the National Bloc and the constitution of the Blue Horizon Chamber. In 1920 at the Tours Congress he followed the majority of his friends in opposing the Union Sacrée and took part in the foundation of the Communist Party. He lost his deputy's seat in 1924 but won it back in 1936. In 1938 he rejected the "spirit of Munich" and in 1939 denounced the Nazi–Soviet Pact, thereby incurring the enmity of his fellow party members. In 1940 he refused to vote Pétain full powers and was arrested and interned for the duration of the war. He did not return to Oyonnax till 1945. In spite of the hostility of other communists who regarded him as "Nicod the traitor" his local standing remained high and he was regularly re-elected mayor till his death in 1950.[5]

Local solidarities survived the crisis. The town hall had a key role in organizing help for the most deprived. The mayor supported the union, which he provided with premises near to the town hall. This union was one of the oldest in the Ain *département*; it was the comb workers' union set up at the end of the last century and successor to the Flatteners' Bank. Nicod himself had been one of its most active militants, especially at the time of the 1913 strikes. Throughout the

[5] On René Nicod, see H. Puel and J. Saglio, *Oyonnax* (GLYSI, 1979).

interwar period its practices continued to be marked by "reformist" and anarcho-syndicalist tendencies. It helped to find jobs for workers. It played the role of a labor exchange. It negotiated "piece prices" with employers' representatives. Two old union militants, close to the Communist Party, still remembered this period: "People came to the Bourse du Travail (labour exchange) when they had a problem. They paid their subscription and could get information about what jobs were available. Before the Second World War it was mainly the piece workers' union and the Bourse du Travail tradition was very strong."

The premises of the Bourse du Travail were located behind the Town Hall in the stables of an old hotel. In 1934, 1935 and 1936 by agreement with Mayor René Nicod a big hall was made available to the union members which they did up themselves. Two offices were opened there, one for the Plastics Union and the other for the Local Union, plus a hall that could hold 150 to 200 persons for meetings. All this work was done by militants in the construction industry.

In this tradition the union brought together local workers with very different statuses. The same work could in fact be done at home or in factories and it was paid either by the piece (in which case the worker was a self-employed or employed "pieçard") or by the hour. The home workers was not necessarily independent and might be paid as an assistant, while the factory worker might (less often) be working for himself under supervision or else by the piece for the factory boss. To complicate the picture still further it was not unusual for the same person to operate under different guises, for example as a paid factory worker for a part of the day and working for himself – perhaps even employing others (sometimes on the quiet) for another part.

In a situation like this the union served both to press wage demands and to give information about possible employment or about existing subcontracting markets and was organized on a branch basis. Members were none too committed: "pieçards came to the union to get their stamps every two or three months" remembered the old militants. One of the essential functions of the union lay in keeping up pressure on bosses and the public authorities by organizing "agitation." On the basis of broad economic solidarity and political militancy it succeeded in federating a world whose interests could sometimes seem rather disparate: "in spite of this the union formed a whole. [. . .] The pieçards and artisans went back a long way. They could command good wages. You see the work involved a lot of different operations. They earned a good living. [. . .]" The employers did not remain inert in the face of the crisis of the thirties. They strove to transform their production and change their markets once again. Forsaking their traditional comb and celluloid markets, the Oyonnaxians tried to tailor their manufacturing to the vagaries of fashion and in particular involved themselves in the manufacture and marketing of items like yoyos. They also fought to boost the productivity and the competitiveness of their companies. Many tried to inject the available plastic materials, especially celluloid, thereby risking explosions or fires.

The plastics boom of 1936–75

Demographic evolution The population of Oyonnax fell in the thirties. In fact the 1931 census showed that the population had been falling since 1926, and the decline continued up to the 1936 census (10,166). The population was more or less the same in 1946 (10,156): it was after this date that urban growth took off again. In the thirty years after the Liberation the town's population multiplied by a factor of 2.3. It reached 23,180 in 1975, since when it has remained relatively stable (22,739 in 1982 and 23,721 in 1990). If we examine growth rates in the periods between censuses it is clear that differences are relatively sharp: as an annual rate and for the town alone demographic growth was 1.8% from 1946 to 1954, 2.3% from 1954 to 1962, 4.9% from 1962 to 1968, and then fell back again to 1.9% from 1968 to 1975. Since that date there has been virtually no growth – in fact a slight fall up to 1982 with minimal growth since that year. But for this last period the figure is not strictly comparable to the extent that urban growth then spread to neighboring communes: the urban area had 28,107 inhabitants in 1982 and 30,500 in 1990.

This brief summary of the global population figures does not however give the whole picture. Even in the seventies, despite the growth of the overall population there was both an inflow and an outflow and on each census date (1968 and 1975) about 30 percent of the current population was new, with foreigners representing only about a third of the new influx. The outflow of population was not directly addressed by the census, but we can arrive at some estimate of it if we assume that the natural growth is to be attributed integrally to the stable resident population, which leads to the conclusion that 28 percent of the persons present in 1968 had left Oyonnax by 1975. The same calculation for the earlier period between censuses ('62–'68) gives a rather lower figure, but still of the order of 20 percent. In other words, even in a period of strong demographic growth inward and outward population movements were both relatively substantial.

Among the new residents there was a particular category of some importance: foreigners. We have seen that during a period of relatively substantial immigration – the sixties and seventies – they represented about a third of the total additions to the Oyonnax population from outside. An examination of the data, not in terms of flows but in terms of presence, throws some further light on the situation. The percentage of foreigners in the population was at a minimum in 1954 (2.5 percent). The decrease this represents does not signify that a large number of foreigners from the pre-war period had left, for there were a substantial number of naturalizations in the immediate post-war period. The percentage of foreigners then grew rapidly in the sixties (14.2 percent in 1968) and seventies (20.4 percent in 1975 and 27.1 percent in 1982), this last figure being calculated for the whole urban area and not just for the commune.

The same time lag as that noted for the preceding period may be observed in this last period of demographic growth. While the percentage of foreigners reached a low with the 1954 census when the town's population takeoff was under way, it increased greatly in the seventies when growth had already leveled off.

In ethnic composition this second wave of foreign immigration was not like the first, for the diversity of nationalities of origin was much broader. Out of the 5,155 foreigners in the 1975 census covering the urban area only 14 percent were Italians and fewer than 1 percent came from other EC countries, while 32 percent were Portuguese, 22 percent Spanish and 15 percent Algerians.

However, these variations in the demographic composition of the town did not affect its industrial character. Oyonnax was still a town of industrial workers. In the 1975 census the proportion of the population in work was high at 50.5 percent (it was 41.1 percent over France as a whole and 42.3 percent for the Rhône-Alpes region at the same period). Compared to the national average, the percentages of those in work were high for both sexes: 60.1 percent (compared with 53 percent nationally) for men and 41.9 percent (compared with 30.3 percent nationally) for women. The age structure of the population may have had something to do with this, and in addition there is the fact that Oyonnax was a town with a high proportion of immigrants from within France and outside it; this had the effect of boosting the categories of the population in which the percentage of those in work would be high.

What did those in work do according to the 1975 census? Workers, skilled and unskilled, made up just under a half: 48.6 percent. Their average level of skills tended to rise from one census to the next but remained lower than the comparable national average. The middle echelons of industry, with 4.3 percent employed, were few in number, even though their percentage was rising. Artisans (4.6 percent) were rapidly decreasing in number as were those employed in industry. Office workers, representing 18.5 percent of those in employment, were increasing rapidly and significantly in number.

The industrial consensus at the Liberation As we have seen the Oyonnaxians had adopted mechanization and the new raw materials available in the thirties but they still had not managed to inject plastics. This they finally achieved after one of their number learnt in the course of a trip to the 1936 Munich Exhibition that the Germans had finally made an injectable plastic material. The innovation was adopted very rapidly by local businesses, although the Second World War and the Occupation put a brake on the economic revival that had developed from this new technique. At the same time a local banker continued to allow artisans to work for themselves and financed – though not without heart-searching – the purchase of new machines.

With the Allied Landings in Normandy local members of the Resistance occupied the Town Hall and on June 7, 1944 created a provisional committee with the task of administering the town. This committee then set about relaunching industrial activity and on June 20 an agreement was signed with representatives of the unions and the employers. This agreement provided for a living wage, payment a month in advance and the establishment of a fund to enable payment of wages, including for the piece-workers. A later agreement stipulated that only those who had taken part in the Resistance or who had been deported would be allowed to set up in business on their own account.

To a considerable extent the existence of such an agreement is explained by local circumstances and the specific features of the local system of industrial relations. The town, which had been the theater for some spectacular Resistance exploits during the Second World War, was liberated by local Resistance groups who were well acquainted with the way it functioned economically and socially when they took over. The local employers had often come from the working class via the system of self-employment, and had remained fairly close to the Resistance groups, sufficiently at least to be able to meet round the negotiating table straight after the Liberation without agreement being prevented by political hostilities or bitterness. So although there was class consciousness there remained a strong feeling of local solidarity, as a militant of the period observed:

> At the Liberation, there was a local agreement between employers and employees. It was possible here to the extent that there had been very little collaboration in the neighbourhood. Of course there had been an industrial black market. But the basic idea of this agreement was that "we could understand each other," which fully corresponded to the Oyonnaxians' state of mind [. . .] It was our small-town spirit that enabled us to establish a minimum wage before such a thing ever became official.

This is confirmed by two CGT militants:

> By the terms of an agreement with the Employers' Association [those working for themselves] had to organize to have materials. Within the CGT there was a section of home workers which had managed to obtain raw materials and orders for products. Frankly there weren't too many raw materials about at the Liberation. The Employers' Association had provided facilities for prisoners of war and deportees – they could set up as artisans.

The former Resistance fighters managed to integrate quite well: one of the prerogatives they allowed themselves was a monopoly on traditional self-employment, in preference to taking over such jobs as there were. It is a seeming paradox that these Resistance fighters, for the most part regarded as on the Left or even the extreme Left, took great care to keep open the option of becoming small employers in a system that was totally capitalist. More than twenty years later, some of them who had succeeded were still labeled "left wing": it was in their companies that militants fired from more classic companies found work and some of them can still be found on the list of candidates for the town council submitted by a Communist leader in 1977.

The consensus on the importance of an industrial relaunch was not restricted to militants and Resistance members. The local banker was involved in his own way and took it upon himself to guarantee the purchase of a consignment of raw materials without referring to his superiors in Bourg en Bresse. The operation was a success and contributed to the takeoff of local industry.[6]

[6] Several discussions drew attention to this event which has been confirmed by the present officers of the Bank.

In the early fifties, prospects grew brighter for the whole of local industry. The demand for plastic products grew even faster than the overall growth of French industry. With all their basic know-how for this type of work local companies found themselves in a good position to meet the burgeoning demand. The appearance of new types of plastic did nothing to hinder this advance since the Oyonnaxians had no fear of experimenting to make the desired products in the required qualities.

Social conflict within the union world It was at the time of this notable economic growth that the social bases for traditional regulation were put under pressure. The communist militants had been very shaken just before the Second World War by the position adopted by their leader, Mayor René Nicod. The latter, faithful to the internationalism of his militant origins, had denounced the Nazi–Soviet Pact and had therefore been excluded from the Communist Party. For those who remained in membership of the party at the Liberation, he became "Nicod the traitor."

This kind of hostility could be found within the CGT. The orthodox Communists, and especially the young militants from other towns who had come with the mission of bringing this over-corporatist union into line, condemned traditional wage negotiation practices, which mixed up dependent workers with self-employed or even small subcontracting employers. In the context of the Cold War and the class struggle, such practices were denounced by them as smacking of class collaboration. In 1982 a CGT leader still remembered his astonishment on his arrival in Oyonnax in the late Forties:

> I came from Bellegarde and I was completely gobsmacked by these practices. I just could not understand it. So when an employer's father and brother came to the union to sort out an arrangement about unpaid holidays and overtime, I did not know what to do. It was the same thing in the firms where there were disputes. The boss had been close to the workers and their wives. It was often the boss's car that took the wife to have her baby in Nantua. It was only Couvert's that had a real boss's dynasty and a lot of strikes. But what surprised me most of all was this mixture of the anarcho-syndicalist tradition with bouts of industrial action that were never followed up. [. . .] There was also the flower festival tradition. That was a time of class collaboration. It was for the elderly, we made carts and paper flowers; everyone worked together, district by district. When we condemned it people told us "you don't come from round here." [. . .] I remember kicking colleagues on the ankles in negotiation meetings. The fact is that everyone around the table knew each other.

Internecine conflict within the political forces of the Left had begun at the town hall right after the Liberation. At the 1945 local elections there were two lists on the Left, one Communist by affiliation and the other on the Socialist ticket led by René Nicod. The right-wing list got so few votes that in the 1947 elections there were only the two left-wing lists. Each time it was the list led by Nicod that won the day.

This political conflict moved to the union level too. The battle within the union was fierce. The great national issues that were preoccupying the union movement at this time with the split between the CGT and the CGT-FO did not really have any local repercussions. Internal debate was focussed on discussions about agreements with the Employers' Association to ratify or improve on national agreements. Up to 1950 the traditional practice continued according to which the union took part in the negotiation and signing of wage agreements for the whole of local industry. It was also very much concerned with following through the application of these agreements. Thus in Oyonnax the Conseil de Prud'hommes had an important role, especially in dealing with disputes about overtime. The two militants, part of a union new wave whose strategy was closely modeled on that of the Communist Party, remember this period of conflict very well:

the moulders were employers, but they had the option of being self-employed. They could pay for a press very quickly and moreover they did a lot of overtime. The main claim was for an hourly rate. [. . .] But at the time there was more respect for human dignity than today. Hard work, good workmanship and conscientious application were respected. The bosses pushed a bit over hours. There was an enormous battle at the Conseil de Prud'hommes over overtime payments at 25% or 50%. [. . .] Every Monday at the Prud'hommes there were these endless meetings.

The turning point within the CGT was in the years 1949/50. The conflict set the young communists who had often come from elsewhere against the old local militants who were closer to René Nicod. The younger men tried to replace the old Bourse du Travail type of operation by a system that was more focussed on the union's individual company branches. They preferred to have lists of claims put in for each business, especially in the biggest of the companies, rather than negotiate at the level of the Employers' Federation. They secured support from employees of other companies by agitating in the town and appealing to financial solidarity.

Gradually these militants from outside took control of the union organization and managed to impose their new policy which placed the emphasis on struggle within companies over hourly wage problems. Union membership dropped substantially on the way – according to one of the former militants it dropped from 800 members in 1945 to 214 in 1949. But these young activists were there mainly to correct union policy rather than to increase membership, and in fact their agitation did pay off in the end. They organized a consultation with the key category of press workers, the moulders, who were paid 100 francs per hour at the time. The claim they put in called for a 50 percent increase. A strike was declared in one of the biggest companies: Injecta. The moulders won an initial increase of 20 centimes. They went back to work but their example was contagious and other companies were hit by the movement. Again the Injecta moulders went on strike and this time attained their objective of 150 francs. The union leaders thus managed to unleash a wave of company disputes resulting in substantial wage increases for the moulders. On the other hand they did not succeed in having these increases extended to cover other categories of employees or women, and they very quickly encountered great difficulties in maintaining an active presence in the companies.

The employers response This was because the employers, faced with this victory and the new circumstances it created, were not slow to react. Profiting from the economic euphoria of the plastics boom, the bosses went off and started taking on substantial numbers of new staff, chiefly immigrant workers with little interest in the unions, while developing various strategies to eliminate traditional union militants by isolating them in their work, helping them to set up on their own, or simply just sacking them. Sometimes, as with the local manufacturer of injecting machines, the employer met the union threat by setting up two different companies: one to manage the personnel, and the other to manage economic and industrial activity. As the second had very few employees and those it had were carefully screened, it became impossible for the union to obtain reliable information on the company's state of health or its future plans.

The union branches set up by the young militants were hard put to it to resist this economic euphoria, which led to a lengthening of the working day by overtime; nor were they well placed to withstand pressure from the employers. The most stubborn lasted for four or five years. In the meantime the union movement made hardly any impact on the new immigrant employees. Union strategy was all the more difficult for local workers to swallow in that the employers' riposte, with its offer of overtime or self-employment on a subcontracting basis, fitted in with the old tradition of piece-work under which no one really knew the difference between a worker and an employer. Buyers from elsewhere who came to Oyonnax to meet suppliers were often very surprised to see injection presses located in cellars and employees busy with glueing work on kitchen tables above.

Certain businesses set up by former employees grew very rapidly indeed – Pontaroli for example. The oldest companies like Lacroix had big presses. Here the unions put up more of a fight and when they sensed a grievance they did not give up. The heart of the union's task was delegation: "But," said the union official, "the employers got round us very quickly because their strong suit was the knowledge that they knew how to pull the wool over the workers' eyes. I knew an employer like that who was in with the Red Cross and who went on holiday with other workers' sons." Such conflicts not only resulted in a certain fragmentation of the union but affected social relations in the town profoundly. There was no returning to the earlier situation in which everyone – including new immigrants – soon got to know one another. The cleavage between employers and employees grew sharper. This development the militant unionists remember very clearly:

> Suddenly we noticed that Oyonnax was no longer a big village. In the main street we saw the employers go down one side and the employees go down the other.
>
> [. . .]This immigrant population was not very interested in union matters. They

came here to save. They went along with the locals' system of self-employment. The way the employers approached the immigrants they took on was to say; "work when you like." This fitted in with the old Oyonnax mentality where to be accused of being an "idler" was a terrible insult. I remember the whole period in the union when we used to fight over safety issues in regard to people losing hands. The employers came to defend themselves in the cinema and we had a dialogue between the CGT and the employers under the aegis of the occupational medicine people . . . we were employed by Curtet on Viennese presses that were known to be very dangerous. It was in 1952–3. When I protested to the boss the fellow next door said "you're a coward" and yet he himself had lost a hand in my press. It was a real massacre. And at that time there was no safety system specified for that press.

Soon enough the union, with its political base already reduced as a result of René Nicod's supporters no longer wanting any part in overall local regulation through piece price or wage negotiations on the local level, found itself marginalized. It was resurrected – though only partially – in 1968 on the basis of a program of support for immigrant workers inspired by the militancy and dedication of two nuns working in a factory – which is singularly ironic when one remembers the anarchist and anti-clerical origins of the local union tradition.

On the political level the tradition of the Left stayed alive for longer. At the death of René Nicod the Left was still divided, but the non-communists managed to hold on to the town hall up to the late fifties. A notable conservative, boss of a local cardboard manufacturing company linked to the plastics and glasses industry, ruled the town hall for three successive mandates up to the mid-seventies. The political Left had still not disappeared but the militants of the time recall that in the early sixties it was easier to get together a demonstration to oppose the Algerian War than to fight for wage claims.

Local regulations The relative marginalization of union actors did not signify the end of all mechanisms for the collective local regulation of industrial activity. To begin with, all the actors involved continued to mobilize on the collective issues of urban growth. This was an area for collective administration and the town hall whether Right or Left continued to play the predominant role. Up to the mid-seventies and the outbreak of the international crisis, the chief objective was to provide cheap accommodation for unskilled labor, for the most part immigrant. This policy initiated by the socialist successors of René Nicod in the fifties was continued and broadened over three successive Center Right mayoral mandates up to 1977.

In the meantime local businessmen for the most part continued to implement cooperative practices, both in the area of technology and in that of trade, and this greatly limited competition in the area. Not only did information circulate easily through the network of local businesses but the practice of mutual aid on the production level continued: this involved the organization of composite subcontracting operations with periodic changes of leadership. Effectively everyone had his particular niche and the others left him to it or indeed even helped him to keep his foothold on condition that a firm supported in this way remained competitive in external markets.

Finally the fact that (at least up to the late seventies) the unions no longer participated so directly in wage regulation as in the interwar era did not mean that there was no regulation of employment at all at the local level. Admittedly such regulation was for the most part informal and customary; it bore little resemblance to the kind of formal collective agreements found in other branches of French industry. Customs defined with precision the relations between ethnic origin and occupation, they made it possible to know the usual levels of remuneration (which the local actors called "market prices") and they gave an indication of the pathways and careers that each could pursue.

From fancy goods to industrial parts From the mid-seventies, the local Oyonnax industrial system underwent a fresh transformation. The glasses industry, linked to the plastics moulding industry since the twenties, went through a grave crisis from which it never recovered and in the eighties the number of companies making glasses could be counted on the fingers of one hand. In the face of the rise of contact lens producers in countries with low labor costs Oyonnax solidarity fragmented and there was an abandonment of the collective tactics that had been its strength; instead businesses surrendered to the seductive charms of market *coups* and engaged in ferocious competition. At the same time the plastics moulding industry itself underwent a transformation; most companies abandoned the mass fabrication of fancy goods and toys and looked for promising gaps in the market for technical plastics and automobile subcontracting – i.e. "industrial parts" requiring greater technical competence and a greater rigor both in the manufacture itself and in relations with customers and commercial partners.

Conflict among employers and national negotiations Such a transformation can be read as an adaptation to the changes that were affecting the external environment of Oyonnax. But at the same time this movement wrought a considerable change in the internal relations of the actors with each other. A rapid study of the history of this transformation will make it possible to gain a better understanding of how this change occurred and of the positions adopted by the various actors.

In the early seventies Oyonnax production accounted for about one-third of the total output of the French plastics moulding industry. The other two-thirds was produced in various parts of the country with a concentration in the Paris region and a few less substantial concentrations in other localities. What particularly distinguished Oyonnax industry from the rest of French industry were its working methods. The moulders were immigrant men and the machines were old and amateurish. Elsewhere the machines were more modern and the press workers tended to be women. Oyonnax companies responded to competition by doing overtime while their rivals elsewhere stuck to a forty-hour week.

The classic line in the classified ads that Oyonnax employers of the period put in the papers to attract labor was "hours to suit."

With the sudden rise in the prices of raw materials after the 1973 oil crisis and the increasing power of certain countries where the cost of labor was low, national industry chiefs felt very threatened. They sensed a possible escape via a conversion to products that had high added value, that were demanding in terms of technical competence, and that were linked to hi-tech industries. Not only was Oyonnax very weak in all these sectors but the strategy involved companies having to invest to equip themselves with new technically sophisticated automated equipment, and also having to employ qualified technicians. The bosses of non-Oyonnaxian companies were ready to launch into a strategy like this but they believed that to do so they first of all needed to rid themselves of competition from the Oyonnaxians, who could at any moment take markets back from them by working longer hours and making do with obsolete equipment.

The strategy adopted to carry out the conversation was based on a national alliance with the unions. The latter were rather poorly represented in a sector which they had been condemning for a long time on the score of its poor working conditions, low wages, doubtful employment practices and predilection for moonlighting. Under the leadership of a small group of Parisian industrial leaders the non-Oyonnax bosses went to the unions and suggested opening fresh negotiations at the national level with the resulting agreements to be made immediately applicable throughout the industry. (By this very French procedure an agreement reached between an employers' organization and a representative union – i.e. essentially a union linked to one of the five big national interprofessional confederations – became mandatory by a ministerial decree for the totality of the companies in the sector concerned.) At the commencement of negotiations the Parisian employers proposed an agreement involving a reduction of working hours without any lowering of wages, on terms very favorable to the employees.

The Oyonnax employers were caught in a trap. If the agreement went through they could no longer offer overtime to their employees and would not therefore be in a position to profit from the differential advantage that resulted from the employment of immigrant workers. Moreover their economic position was endangered because the agreement laid down the maintenance of wage levels as they had been before the oil crisis. To counter this initiative they tried to sign a local agreement that might allow them to plead exemption from the terms of the national agreement. The local unions were too weak to defend the immigrant workers while their national leadership saw all the indirect interprofessional advantages to be gained by signing an agreement that was advantageous on the national level. All the more so in that there was something else on the agenda of the Parisian bosses; they were committed to opening negotiations on job classifications, (thereby indirectly regulating wages and careers) and negotiations on professional training on a basis that was very attractive to the national unions.

The counterattack attempted by the Oyonnax employers did not succeed. The agreement was signed and quickly extended and then negotiations got under way. But the failure was only relative. For instead of turning back on themselves and managing their own decline, the Oyonnax employers, with the support of the new left-wing town council, embarked on the conversion of their production machinery, investing massively in sophisticated new automatic injection presses, and brought in training bodies to create a corps of technicians and plastics engineers; this enabled them to get a foothold in markets for technical and industrial parts. Their intention was that Oyonnax should become a "European plastics center."

The new deal in Oyonnax At the 1977 local elections the candidates on the Left regained control of the town hall under a Communist leader and became the driving-force behind an industrial policy for change. The main item on the agenda of the previous town hall regime had been housing and the construction of council blocks for workers and especially for immigrants. The Left abandoned this policy and used the economic crisis as an argument in favor of giving local industry the means to cope with the requirement for industrial conversion. Urban policy was focussed on encouraging an inflow of technicians and executives to assist in the conversion of local industry for the technical plastics market; large-scale projects for council estates gave way to the construction of individual houses on separate plots: a timorous cultural policy based for the most part on subsidies to existing associations was abandoned by the new team in favor of an audacious project for a prestigious cultural center able to satisfy the cultural needs of populations of executives and technicians. Moreover, as in the early part of the century the left-wing town council intervened actively in support of a professional training policy, in particular by making the local technical college that had emerged from the Professional College into an important center for the training of technical staff; in fact they set up what was the first center in France for the training of plastics engineers.

At the same time work in workshops using injection presses was ceasing to be the classic way to self-employment: thus in the seventies immigrants coming from Southern Europe and the Maghreb made up the great majority of moulding workers but they constituted only 8 percent of the new names on the craft register between 1975 and 1981. It was the skilled French workers and technicians born in Oyonnax who became self-employed moulders. They made extensive use of the most sophisticated production technologies. Thus a local company founded in 1974 by several brothers and established on the family farm numbered ten staff in 1984 including the founding brothers. It had a park of 14 fully-automated machine tools, in addition to classic machines. The whole production of moulds was controlled by computer from design to fabrication. Their customers were for the most part local, but the company also exported a good part of what it made, selecting top of the range customers, the most demanding technically, in Germany, the USA and Japan. The numerically controlled machines were in use for 95 percent of the time – i.e. they were operating twenty-four hours a day and 365 days a year. With the support of the machine manufacturers the employers had taken on massive bank debts in order to acquire this sophisticated machinery: 30 percent of their proceeds went straight to repay these debts.

The mainstream plastics moulding companies have also evolved. Several of them have become part of more powerful industrial groupings on a national or European scale. One of the most important has opened several production units close to the assembly plants of its automobile manufacturing clients. It has kept its registered office and its administration and research/testing center in Oyonnax in order (according to the director, who is in fact the son of the founder) to "take advantage of the local technical and industrial milieu."

The local bank has also adapted to the new situation. In the fifties and sixties it had abandoned its traditional strategy of personal support for local employers and the establishment of the big banks progressively deprived it of its local leadership. In the sixties, under pressure from an Oyonnax man who had become president of the regional bank, it went back to its old practice of systematically penetrating the networks of local employers. At the time of the collapse of glasses making it supported a company in difficulties "in order," said the current president, "not to have a banker's attitude [. . .] not to sully our reputation." On this occasion it agreed to treat all the small subcontractors involved in the glasses manufacturing crisis as employees and so allowed them to "ride the storm." Thus it gradually regained a foothold in the local industrial milieu and today prides itself on being regarded as one of the local century-old companies and not just as any old bank. In their discussions with local clients bank representatives pride themselves on paying more attention to the human and technical qualities of the company bosses and to their respect for local ways, rather than exclusively to their financial situation. The bank has invested heavily in the local "European Plastics Center" technical and economic project in which it has sunk a million francs, while offering securities to the Public Authorities to the tune of another 15 million. Today a commitment like this enables it to keep ahead of the other banks, who would have to do the same if they wanted to penetrate this network of information and contacts.

The actors' discourses

The different elements that at any given moment make up the background to the functioning of local industry and contribute to its dynamic equilibrium do not appear such in the eyes of the system's protagonists. Rather, each person acts according to his own logic, which gives him his place in the industrial system and enables him to participate in its collective regulation and historical evolution without enabling him to get the measure of it. The regulation of the whole is thus the creation of webs of different rationalities and not the construction of one specific rationality. The totality of these rules and positions form what we have proposed calling a "localized industrial system."[7] By this term we mean to signify that the interplay of the actors is based on an evolving collection of rules which bear not only on professional relations but also on economic and technological relations between rival firms, defining a common identity with a local base and an ensemble of collective interests, establishing the modalities for the management of local conflicts and giving the nucleus of forms of collective mobilization in confrontations with the outside world.

Within the whole the rationality proper to each actor is not deducible simply from an analysis of his position. The economic actors and especially the entrepreneurs do not align their modes of apprehension and reasoning according to the classic canons of economic thought but develop other modes of analysis and other systems of reference to guide their action. One of the mandatory references in any of the discourse of local actors is the specific character of the Oyonnaxian. This is the *Deus ex machina* to which the protagonists appeal to explain local features such as the continuing activity of innovators, the fondness for mechanical solutions, the capacity for rapid adaptation to the fluctuations of different international markets and the development of complex subcontracting arrangements or narrow specializations.

What part do such representations play in the ongoing economic behaviors of the actors? Are they just marginal considerations having little importance in the understanding of strategies and tactics? It is noteworthy that actors refer explicitly to them and integrate them into the construction of their reasoning. Thus rather than reducing these different elements to mere errors or distortions explicable in terms of a lack of historical or economic knowledge, we have tried to find in them traces of the local actors' concrete modes of reasoning. Their way of apprehending and analyzing may be very different from what is usually understood by economic discourse, but its mode of functioning needs to be understood and should not be interpreted simply as symptomatic of their lack of academic competence.

To explain their behavior and their decisions local actors frequently refer to local history – or at least to a particular account of local history – and see their practices as reflecting the continuity of that history. I have therefore selected from the empirical material available a collection of accounts that reflect this procedure. Here I shall mainly utilize the transcript of a discussion in the presence of ten observers having no involvement in Oyonnax industry. The main interlocutor (whom I shall call François) was an Oyonnaxian small employer with whom I established contact in the course of earlier research interviews. He was a member of the left-wing town council elected in the late seventies. He was accompanied by one of his friends, Jacques, a former local journalist and a good connoisseur of industry and local life over many years. This material, except in its most autobiographical parts, is quite coherent with

[7] M. F. Ravewre and J. Saglio, "Les systèmes industriels localisés: éléments pour une analyse sociologique des ensembles de P. M. E. industriels," *Sociologie du travail* (1984), 157–175.

what I was able to discover in more systematic discussions with a sample of local employers. I have focussed on this particular discussion simply because it is the most complete and one of the most intriguing.

Myths of origin

The union point of view will be represented by the history of the town as it was written by a novelist close to the Communist Party in the fifties.[8] This contains a denunciation of working conditions and the mirage of the "self-employed" solitary worker. In the discussion François and Jacques often refer to this book as a classic element in Oyonnax culture, especially when they want to emphasize certain discrepancies between their account and his.

The innovator In both accounts the industrial history of the town really began in 1936 with the arrival of injection presses. Both accounts accord an important place to the innovator who heard of the new technique and brought it to Oyonnax to be implemented. The first difference bears on the identification of the principal actor. In both cases the innovator is not to begin with a capitalist with financial assets, but a worker from elsewhere who has had to work hard to establish himself. In Vailland he is from the Auvergne. For the purposes of the novel and the thesis he is defending Vailland changes the name of Oyonnax to Bionnas and gallicizes the name of the main innovator: but it is not difficult to recognize in the features of his Jules Morel a good number of those of Hannibal Zanco, and the Oyonnaxians are perfectly well aware of that.

Jules Morel senior had come to Bionnas from the Auvergne thirty years before as a jobbing builder. With two associates he had built the workshop that was to be the original part of his factory for an artisan from whom he later bought it. (p. 82)

According to the Oyonnaxians the social distance traversed by the innovator was even greater: he was an Italian who turned up in the town one day as a child with nothing more than an address:

Jacques: Zanco turned up with a sign hanging from his neck "My name is Hannibal Zanco, I am from . . . Italy, I am going to stay with Monsieur M . . .,[9] a painter, in the rue de la Victoire in Oyonnax" . . . In those days train conductors looked after kids, like they do nowadays in planes – that's just what they used to do in trains. So that's how old Zanco came here. When he got to Oyonnax he got out at the station and asked "Where does Monsieur . . . live?" in Italian and went to his house, so he had somewhere to eat and sleep and all that, and what did he do then, old Zanco? [. . .] to begin with he was a plasterer, but he didn't feel he was making enough money [. . .] he took cement in iron buckets up to peoples' roofs with ladders, there were no cranes, there was nothing like that at all.

[8] Roger Vailland, *325,000 francs*, ed. Corrén Bucket Chastel (1955; Paris: Livres de Poche, 1975).

[9] The two names were given but on the recording they are indecipherable.

So the social distance to be covered for Zanco to become the top Oyonnax employer is even bigger in the local account than in Vailland's. The second important difference relates to the reasons for his actions and for his success. In Vailland the young *auvergnat* quickly becomes an autonomous economic actor who works with fellow-craftsmen. But in the other account the little Italian immigrant depends on help and sympathy from those around him – first the train conductors and then chance meetings in the town – to reach his destination, and someone else puts him up and feeds him.

The two accounts seem to diverge most in their presentation of the way the individual can be the bearer of the technical innovation which will transform the local economy. For Roger Vailland, the knowledge of the new procedure is an individual attribute due to a family accident: Jules Morel is the one who knows the secret and he exploits it to his own advantage.

In 1936 he had invested the whole of his savings in the purchase of an injection press, the first in Bionnas and probably in France. The artisans had laughed at this crazy mason: for them it was still only work done by hand that counted. "Plastic is special." How did he get the idea of bringing machines to Bionnas? He had married a girl from the Rhineland he had met in 1920; the Germans were the first to make injection presses; Jules Morel saw the presses in operation in a factory in Düsseldorf while he was on holiday there in 1935 – the first holiday he had with his in-laws; he knew at once that he had found the weapon he needed to conquer Bionnas [. . .] With the profits from six months' work on the machines which he operated on his own in a shed Jules Morel was able to buy a second [. . .] he brought in 6 auvergnats and set them to work in relays on the two presses which he kept going 24 hours a day. He looked after the sales side himself. (pp. 82–83)

But the account given by the Oyonnaxians lays the stress on two entirely different elements, the publicity given to technical knowledge and the gambling aspect, the risk involved in industrial ventures. To begin with, it was not Zanco who learnt the new technique, but Leon Verchère, an Oyonnaxian who had been to an industrial exhibition in Germany in 1936. And the moment when the two innovators got together was in public (in a café) in the presence of witnesses who could appreciate the importance of the occasion:

François: When they set up the injection with Verchère it was at old Jeannie's cafe, the Côtelette Salée . . .
Jacques: It was there or at The Forge – those were the two cafés; The Forge and the Côtelette Salé. [NB. These were the names of Oyonnax cafés of the period]
François: Léon Verchère was looking for 100,000 francs, it was 1936.
Jacques: A press cost 75 thousand francs. The first Eckert cost 75 thousand francs.
François: So when he played cards [. . .] – they were for ever playing card games of some desciption . . . I saw [. . .] lose his factory in one night and then win it back [. . .] . . . At that time Zanco was a painter-plasterer, he had a house on the rue Anatole France. To get the 10,000 francs so he could work with Verchère he sold the house . . . and then two years after that he bought it back. Later he sold it again, when he had to put more money into the company . . . he sold it four times I think . . .

So the foundation of the company did involve new technical knowledge. But that knowledge was not a secret. At the start it was not just shared between the two principal protagonists, it was also known to the public and in a sense put up for auction. In the novel what motivates the innovator, what causes him to act, is the thought of power over the town. In the Oyonnaxians' version what is at stake has much more to do with sociability: it is the possibility of carrying on gambling, and indeed of raising the stakes, but also of losing on a much grander scale.

Knowing how to be partners But it was still necessary to convince the potential associate of the value of a partnership. While in Vailland's version Jules Morel is immediately convinced of the value of the innovation, i.e. of its underlying economic potential, and has no need to convince any associate, in the other account Verchère really just makes a joke, and it is Zanco who has to persuade him to get involved. Here everyone is convinced of the economic value of innovation. For them the problem with getting started is not economic, it has more to do with questions about honour, *savoir vivre*, and keeping one's word. The two protagonists do have some notion of economic calculation, but what they have to do is become partners and what binds them is the gambling they do together and a promise one makes to the others.

Jacques: Let me tell you the real story; what you are saying is true but the exact way it was put was "I have a press, I would like another one, but I don't have the cash. I need seventy-five to one hundred thousand francs" and he said that to Hannibal [Zanco] even though he was very well aware that Hannibal was absolutely flat broke because he was in plastering and it was '36 and there was no work being done, he wasn't getting paid, people hadn't paid him for the work he had done. He had materials all over the place, old Zanco wasn't working any more, there was nothing, and when Verchère said that to him he said it as a joke, he told Hannibal: "You Italian, I want some money from you," because they used to call him "You Italian" like that. "I suppose you wouldn't have seventy-five thousand francs for me?" And Hannibal said "You know damn well zat I can't give you ze money, where do you suppose I could get it from?" Then apparently he went off all round the town asking for all the money he was owed all over the shop. Then there was the business of his selling the house. And he came back a few days later with the seventy-five thousand francs in his pocket, because you used to put money in your pocket, no-one had cheques. And he comes to Verchère who is asleep, because Verchère is a fantastic sleeper, he goes to bed in the early hours every day, he is a terrible drinker, a fine guy who loves rugby and some say women too. He wakes up and sees Hannibal downstairs. And he says to him "Oh, it's you."
What was Verchère's first name?
François: Léon.
Jacques: . . . then Hannibal says to him "Oh you're there are you, Léon?" and Verchère says "Yes, what do you want?," and he answers "Well, I've got ze seventy-five thousand francs for you." So Verchère says to him "You idiot, it was only a joke" – "Ah but we shook on it didn't we?" because they had shaken hands, you know when you shake hands . . . "We shook on it, I've got the money for you." Then

Verchère says to him "Pff. . ." and then he argues and then after a while he says "Well why not . . . why not. You'll do as well as anyone else. You have the seventy-five thousand, so let's go and buy another machine." And that's how Injecta Plastics got started.

For a long time this company was a kind of flagship for the local plastics moulding industry. It was the first to install injection presses, for a while it was one of the biggest in the town, it was where the whole wave of moulders' strikes started in the early fifties. For several decades it was managed more by Zanco the immigrant than by local boy Verchère. Our interlocutor underlined this by giving another definition of the particular character of the Oyonnaxians, based more on human experience than on family origins.

Jacques: Léon Verchère, a local man, old Zanco from Italy . . . Two blokes who had had a very hard life, and they were mates who used to go on diabolical benders together.

More than economic calculation and/or the lure of gain it is a gambling or wagering instinct that brings these two men together and sets them off on a common enterprise. But to get started you have to be partners, in other words you have to be able to put together necessary but contradictory qualities – witness the association of the "hard drinker" and the "hard worker." Industry is about hard work of course, but it's also about cunning and living on your wits . . .

Not all innovations are as radical as this and in any case we need to put the radicalness of Verchère's contribution into perspective. The innovation he brought back from his trip to Germany was not the injection press but the raw material "trollytul," and injectable polystyrene that could be used instead of cellulose nitrate (celluloid). But it is certainly true that this was the signal for the other Oyonnaxians to buy injection presses from local manufacturers in order to work with these new materials, which the French (Rhône-Poulenc) started to make as well, and the innovation spread to the whole industry.

The universe of competition The two accounts again diverge in their explanation of the spread of innovation and its consequences. François includes himself as one of the later beneficiaries of the spread of innovation. At this point his account becomes more personal. But Roger Vailland's entrepreneur operates in a finite world with a limited market. Once his secret is out he loses the economic advantage of his situation:

After 1940 Jules Morel stopped expanding Plastoform, even though the war had increased demand for plastic objects. Numerous competitors had appeared in Bionnas and other parts of France. The use of the injection press was becoming the rule. Ruined artisans were becoming piece-workers doing the finishing or were taken on for press work. Profit margins were growing slimmer and plastics moulding was becoming just another industry. The fact that he was the first was of no advantage to Jules Morel – in fact quite the reverse, for it meant that his presses were already out of date. (p. 84. Plastoform was the name used by the novelist for Injecta Plastics.)

Thirty-five years on there is not much doubt about the inaccuracy of Vailland's version. But he was just one among many local observers who prophesied the rapid decline of Oyonnax industry after analyzing its markets, physical resources and technological problems. Some observers were already predicting the end of the bomb industry as early as the mid nineteenth century, while between the wars people were talking about the death of celluloid, and more recently a lot of observers have predicted that automation and the diffusion of plastics moulding techniques would deal fatal blows to Oyonnax manufacturing. The novelist's scenario is close to the kind of traditional economic analyses put forward by commentators who fail to take into account the particular rules of the local industrial setup.

This is precisely the way the local union militants I questioned continue to think. The employer is seen as a classic entrepreneur whose strategies are easy to understand because they arise from one single type of behavior irrespective of time or place. In this account self-employment, which raised so many problems for Oyonnax militants, can only be a mirage that leads artisans to their ruin. In the face of the adversary the struggle has to be collective and uncompromising. Recast in this way the myth not only makes it possible to identify the opponent and his weaknesses but also reinforces group cohesion while justifying earlier exclusions as rejections of unacceptable compromise. Finally it throws light on the current choices of the local union which cannot see any salvation for Oyonnax industry except in the rapid adoption of new technologies and specialization in the most highly technical areas.

Collective myth and individual history The employers' account also corresponds to the way the actors understand and live their own practices. It is not rational economic calculation that inspires the actors to launch into business. On this point personal history merges with myth. The wager, the promise and the gamble are key elements in the story told by François to explain how he himself got involved in the new technology.

François: I began with injection in 1957 when B. began to bring out his semi-automatic presses.
Q. Wait a moment, was it because B. was making the presses or because you had a market?
François: Oh not that . . . nothing ventured, nothing gained.

How had the idea of getting involved in injection come to him? First of all he observed that his "mates," often classmates, but also deportees and resistance fighters like himself, were involved in it. And then there was a local press manufacturer. The latter did not belong directly to the same network.

François: I think he was a Centrale engineer and he married an Oyonnax girl whose parents had a comb company. When he came to Oyonnax he began making combs and then he tried to improve what Mr C. had just done, injection [. . .]

So here is another element from outside, someone who does not belong directly to the network of friends born in Oyonnax where they had gone through school and then the war together. His outsider's status is notable and even today he is one of those François uses the word "Mr." for. Relations with him are more distant.

François: One did not get familiar with B. because Mr. B. was an educated man. He was a college man, he was different.

There is then a certain social distance, but this does not necessarily mean that the person in question is someone you cannot socialize with or spend time with: in actual fact he was one of the small group of young people with whom François went skiing on Sundays. On such occasions François got to know the man better and the idea came to him to acquire an injection press. This was how an entrepreneur could get started in the injection market. The machine he got from his friend. François had no money to pay for it, but this wasn't a problem:

François: We came to an agreement.

Once he had a machine, he could set it to work to produce the maximum return.

Q: You employed someone to work on it?
François: Yes of course [. . .] at least one and then we did the second twelve ourselves.

The problem was he did not know what to make.

François: When we got the first machine in I didn't have a mould for it.

Again the entrepreneur appealed to his network of colleagues. But he was perfectly aware that the risk and the danger were his. He was going to have to show what he was made of before being accepted as a home worker, then subcontractor, and eventually maybe launching out on his own with his own product range, his own machines and his own moulds.

François: I went round my mates. There was one called [. . .] he said to "look, I've got a mould for tumblers here" and I used that. It was . . . [laughter] the problem was that the tumblers . . . it's true it was a mould [. . .] It had been messed up by the mould-makers . . . Cooling was tricky. If you gave it too much water pressure it came out . . . In fact it was a joke . . .

François was aware of the risks of going to his network of colleagues. Though he was able to get a mould from them and launch a production line, it was a poisoned chalice, an initiation test as much as a helping hand: he had to go through the test and show his colleagues he was capable of coping with defective tooling. Once this first phase of initiation was over he could then launch out as a home worker who would make objects with the materials and the mould provided for him:

François: You sell working days, that's all.

But to earn his living by home working he really had to have his wits about him.

François: For example S. would give you a home workers' mould. This mould turned out one a minute when they used it, but the trick with home working was to make it turn out 90 an hour. It wasn't too difficult – you just had to make a few small modifications to the machinery – which you undid if the mould was taken back because otherwise things would escalate; if you sent it back to them and then they found out it could do 90 an hour then the next minute they would be wanting you to do 100.

In a game like this the successful home worker demonstrated technical *savoir-faire* and mastery of the whole overall operation. Then he could move up a rung as François did, become a subcontractor and finally, one rung higher, find a range of products to make and market independently. But the basic social *savoir-faire* of the home worker was hardly any different from the *savoir-faire* of the piece-worker, the traditional Oyonnax *pieçard*. At this period self-employment did not cause a social rupture either in the networks of sociability or in individual practices, and a person could pass smoothly from one status level to the next.

Getting round the banker

The rules of internal operation produced by the totality of the actors may prove quite important and effective in ensuring the economic and social regulation of this totality. Nonetheless, it is not a self-contained universe and does not function or intend to function autarkically. There is always an external world which can be mobilized or exploited for profit.

But cunning is required to get the best response from the person external to the milieu with the needful resources: indeed in certain cases it may prove necessary to play a double game, and in the case we shall look at, we shall witness the operation of a double system of values. In Vailland's story Jules Morel can buy his machine without any problems. In the Zanco myth Zanco has to sell his house and collect problem debts to get the necessary sum together. François did not have this option: it is true that his father was himself boss of a small firm, but it was as a result of family disputes that François set up on his own, so he did not have much hope of help from that quarter. He had to go to the bankers for help when he wanted to buy a press. There were two distinct categories of bankers – or at least the local network explains their behavior by referring to two distinct models.

François: I'll tell you one thing: the Oyonnax bankers, apart from [. . .], Monsieur J., were totally opposed to the injection machines, because they thought it would be the death of Oyonnax. They wouldn't lend money to buy them.

The Communist Roger Vailland was not therefore the only one to think that the system of small local businesses would not survive technological innovation. The bankers thought the same way but certainly not because of any ideological affinity with Vailland.

To get the money to buy one and later several François, like his colleagues, had to resort to subterfuge – or rather negotiation. He did not employ exactly the same tactic with everyone he approached, and the key to success did not lie in the weight of his documentation or the persuasiveness of his argument. The whole business was more like a game, and winning the game meant getting the banker to lend to you. The tactics employed partly depended on whether or not the banker was a local man:

François: So we had them one way or the other, because ostensibly we were buying a cutting machine whereas it was actually an injection press. You could always get by [. . .] The bankers were against it. They had no faith in it.

Of course there is dishonesty here, but it is not serious as long as everyone is happy at the end of the day. The banker who gives the finance for the purchase of a classic machine is repaid thanks to the purchase of a modern machine of which he would not approve. But everyone remains pragmatic, and all that matters is that the banker gets his money back . . . years later François resorted to a similar strategem. The emergence of new and more efficient machines meant that he had to retool:

Q: But who lent you the money to buy those [new] presses?
François: Well that's what bankers are there for.
Q: They didn't refuse?

This time François played on the financial argument. By now the bankers were convinced of the worth of the new machines but they required to see a "healthy balance sheet."

François: Well we have to avoid giving them catastrophic accounts. Where loans are concerned to buy materials that is a great help. When you've got a healthy balance sheet . . .

So François was perfectly well aware of the change in the bankers' attitude. Before, they wanted you to buy classic machines (for cutting) rather than injection presses whereas now they wanted to see a "healthy balance sheet." Once again he resorted to a bit of trickery. In the first case he had claimed to be buying cutting machines, in the second his balance sheet look healthy . . . give or take a few liberties with the rules of strict accounting orthodoxy, just like the liberties he took before with the nomenclature of production machines:

Q: And were your accounts always accurate?
François: I only had one that was seriously illegal, in '74.

Though one could generally handle "the bankers" like this, the same did not go for "Mr. J." from the regional bank, whose chief was a Jurassian. "Mr. J." did not look at things the same way as the other bankers, for he was more familiar with local professional practices and adapted his strategy to the specific features of the terrain. So local workers who wanted to go self-employed had to adapt too. If their heroes were, like Léon Verchère, "hard drinkers," they were also, like Hannibal Zanco, "hard workers," and it was this quality of "seriousness" that found favor with Monsieur J. and was the way to get finance. Thus the ambivalence of the local "hard-working/hard-drinking" culture reflected a dual experience of socialization in peer groups and acceptance by the financier. The encounter with the bankers was a crucial moment in the life of the Oyonnax small businessman.

Jacques: . . . You have to understand what J. was like because J. had been an extraordinary fellow for Oyonnax people who were serious. He would have endless bargaining sessions with them and he would interrogate them: your grandfather, your father, your mother, what did they do? What has your wife done? Are you serious? Do you drink? Do you go to the cafés? [. . .]
François: It was like the confessional.
Jacques: That's just what it was like, [. . .] And in the end he would say: "Do you have a bit of money?" – "I don't have a penny Monsieur." And when he had finished . . . getting all his information, he would say "You don't go to the café – do you play boules a lot?" – "Oh no, I don't go there much" because when people got addicted to boules they stopped working. "Well then you have a wife, you've just got married, you have no money, you are serious, you want to succeed, what do you want to do in life?" – "I'll lend you money, I know you'll do well" . . . He was the banker for the Oyonnax small man who was just starting up.

So this banker was different from all the others. He was not like those small local employers who made up the network of plastics moulders, and yet he was one of them, for he knew the local ways and was capable of exploiting them to his professional advantage. But at the same time history shows that this interface between the local network and the world external to it was a locus of power and efficiency: for when the crisis of the thirties came along, Oyonnax industry was saved not just by its own inner resources but by the strange alliance of two local non-industrial actors.

François: At the height of the unemployment, it was '32 . . . if it had not been for Mayor René Nicod and then old J. at the Bank . . . he was like a private banker you know, it was as if the money was his own [. . .] without old J. who was on the far right Oyonnax would have gone down the tubes.
Q: And they got on, those two?
François: Of course, they had the whole village in their pocket.
Jacques: Each played his own part.
Q. He was on the extreme right?
François: No.
Jacques: He was just an employer of his time who had a certain way of looking at things.

At the interface of the outside world – the world of banking, finance and money-making – with the world of the Oyonnax plastics moulders, where it was more a matter of getting by, struggling and taking risks, the local banker adopted ways of working and modes of reasoning that made it possible to adapt the rules of the outside world to local habits and *savoir-faire*. This local branch manager enjoyed considerable autonomy in practice with respect to his superiors in the hierarchy. According to present-day bank executives, old J. only sent his superiors accounts once a year and those accounts were short on detail. But he gave no cause for complaint, for although his practices were not in conformity with the strict canons of banking orthodoxy, they nonetheless provided a strong basis for the development of local institutions.

Q: So he lent money but did not put any capital into businesses?
Jacques: No, never [. . .] and he took a good few knocks when loans were not repaid because there were some pretty dire situations. When they failed to pay for a consignment of combs in India or somewhere you would go to old J. and say to him: "Look Monsieur J. I've worked hard and I've done my best and now I've got nothing –" "Don't fret old fellow, I will give you money." Yes – he would give money to help someone start again because he knew the fellow was serious. "Pay it back when you can" [. . .]

This banker stuck to his own business and there was no question of him taking shares or getting involved directly in local firms. He was alive to differences in personalities and to the variety of difficulties and situations faced by the locals. At this point the account is more suggestive of classical myth than of historiography: it does not claim to be describing typical bankers' behavior, but to be interpreting a kind of behavior that is archetypal even though it is exceptional. The banker sticks to his role of financier and his precise knowledge of the local industrial milieu allows him to support projects his colleagues would never entertain. In the example above, he commits himself and so accounts his partner as well to an exchange that is social rather than strictly economic. No one knows exactly the precise conditions for the reimbursement of the loan. It was because he agreed to work outside the traditional rules of exchange and to commit himself to a more demanding system of exchange that the banker could be considered one of the two "saviors" of local industry.

Is the story of the banker as told by the retired small businessman in our discussion of any use to help us understand the current behavior of Oyonnax employers and bankers? At first sight that might seem unlikely. Old J. retired from banking in the forties to be replaced by a son who did not make the grade. By the early sixties the bank no longer had much importance locally, but then it managed to pick itself up again and today it has a clear position of leadership. It managed to acquire this position partly by playing on the continuity of its role as an institution with a very long-standing tradition of service to the locality. Today's bank executives are perfectly well aware how much their success comes down to the respect they have shown for local customs and rules and to their readiness when necessary to get involved in loss-making investments and cover abortive ventures. For them financial analysis is a secondary factor in the decision-making process that comes after an assessment of the individuals concerned and their social position. They explicitly present themselves as the successors to old J., even to the point of according only limited respect to strict financial orthodoxy. So for an Oyonnax employer today to refer to the story of old J. can be a good way of finding a banker with an attentive ear.

History and memory

Stories like this tell us much more about the rules and the modes of reasoning employed by the actors involved than they do about the real history of the town. Chronological exactitude is not their main quality, but something more important – they present the world as complete and its rules as immutable. They frequently ignore the conflicts that upset the established order of things and social relations. To reconstitute the history of union conflicts I consulted well-placed informants who were actors in this particular story as a means of supplementing the very sparse documentation. But the best informants were all effectively out of the fray when they talked to me, one because he had left the town, the second because he had retired, and the third because she was a woman with a low-grade position in the Town Hall. The discussions I was able to have with others – especially with union men who were actually protagonists in the story – were paradoxically vaguer and even downright inaccurate on the events of the period. Thus some of the extracts from statements that I have used in the foregoing account are fragments of wholes in which chronology is twisted and certain events ignored. Extraordinarily, analysis of the conflicts of the late sixties was only possible with the help of complementary studies of negotiations on the national level, despite the fact that the period concerned is so recent.

A first explanation of this might be found in the difficulty faced by a member of any organization or institution in trying to explain the earlier positions or strategic choices of his organization while still hanging on to the idea of a coherence between current choices and positions and previous ones. Thus in explaining the union's relative inertia since the war, union militants still living in Oyonnax play down the influence of the battles of the early fifties and of the union U-turn of the period. Instead, they stress the technical features of production:

> What makes local industry so special is working plastics on your own, which gives everyone the idea of being self-employed, something that would never occur to a miner or someone working for Renault. Here in Oyonnax, people think of it automatically [. . .] At the time when things were on the up it was logical and the employers pushed for it. A lot of unions were destroyed like that. First there was the overtime business: guys could do a whole load of overtime. The unions fought against this but not all of them. Only a minority. The others fell into the trap. And overtime brought along with it the idea of being self-employed. A guy would be working a lot and then say to himself: if I had a press I would work just as many hours but I would be working at home.

So the first explanatory factor invoked by these union militants has to do with production techniques; and they claim that because of this same factor it is more difficult to go self-employed today because modern machines are so much more expensive.

The second explanation they give relates to the particular mentality of the local employers, which made them encourage this move towards self-employment.

Bosses very often encouraged it. Some of them even sold presses to guys who wanted to go self-employed. In fact they would rent out the press while the mould still belonged to them. The guy did piece-work. After a certain time during which the employer deducted part of the price to pay for the press, the guy had his own press. That's what they call leasing nowadays. It was all profit for the employer – he was reselling a press he had already paid for [. . .] That's what they did at Billion's [the company that manufactured the presses]. Billion let guys take a milling machine, say, and they worked at home as moulders. Most of the people who were union militants ten years ago at Billion's are working on their own now.

Not content with getting rid of union militants by encouraging them to go self-employed the employers, again according to their union adversaries, used every means to prevent them from coming back into the companies:

The Oyonnax employer is a tough nut. The militants could only find work with sympathizers or else they had to put up with conditions they should not have tolerated. In the small firms they worked in you can negotiate but it's really a bit like class collaboration. For instance in 1968 I went to see the boss, there was no problem, he paid us over the eight strike days.
[another union man, employed in the public services] In 1968 my father-in-law said to his workers: "What are you waiting for? Get out on strike!"
The result was the fellows turned against the strikers. It was an agonising problem [. . .] At G.'s they did things differently, they had elections, guys who had complained were made delegates and eight months later they had all become foremen.

Thus, instead of reflecting on the history of union strategies today's militants prefer to locate the reasons for their present difficulties in particular production techniques and employers' strategies. So past history does not really interest them very much. Questioned about the history of the union in the early fifties, one of the main CGT militants of today acknowledged this lack of interest:

I don't know anything about social relations in Oyonnax just after the War. I wasn't there, I was in Indochina.

They are well aware that the origin of the problems encountered by the union lie in its political position but – and this is their third explanation – they attribute this peculiarity not to the political choice of the union but to the Oyonnaxians' political mentality, i.e. to the specific nature of the local situation.

Here it's the exact opposite of how things are in Bellegarde where the union movement was always more powerful than the political movement. Here, during the Algerian War, you could bring people out on the streets for Algerian independence, but you couldn't get them to do it over wage claims. We had occasional movements within individual companies but they never lasted.

Company chiefs also tend to forget this episode in local industrial life. Only a few among the oldest remember that there were a few disturbances at the time, and even those few play down their significance:

François: Oh! it lasted three hours in our company.
Q: But it did go on for a while at Injecta?
François: Ah yes! It was another story at Injecta. It was completely different [. . .] No question but at that time there were tensions between teams. You see there were two daily shifts of twelve hours, including Saturday and Sunday, seven times twelve, 84 hours, because everything was sold in dozens. It was the fault of the British.

Thus most of the current actors, including some of those who took part in the conflicts of that time, tend to minimize their significance or even to forget them entirely. And this is not the only example of collective amnesia I noticed in the course of this local study. Most of the local actors attributed to one of the mayors, an employer who was elected for three successive mandates at the head of a conservative town council, the responsibility for having launched the huge workers' housing projects in the town in the late fifties and early sixties. The only person who asserted that the decision had been taken by the previous council with a socialist majority was the town hall employee – and when we checked in the town hall records we found she was right.

How are we to understand such mechanisms for selective collective amnesia? In both cases the interpretation favored by the actors even when inaccurate or vague proved to be more in conformity with their present strategic options. Militant unionists "forget" the context of the 1952 strike, which allows them to argue that earlier union strategies involved "class collaboration," just like today's individualist strategies of those who agree to go self-employed. The local political actors "forget" that it was the socialist town councillors who launched the housing construction programmes, since to do so fits in better with the idea that all they cared about was satisfying the needs of the local employers. In both cases the construction of the collective memory is tailored to the interpretation of current strategies rather than to historical veracity.

Given this way of looking at things, the veracity of a statement – i.e. its capacity to account effectively for past history – is less important than the tactical and strategic significance attributed to it by the actor. Whatever the scientific pertinence of the features attributed by the actors to the stock collective identity, it seems important to get a better understanding of the resources that this type of consideration enables them to mobilize. In reality what we are talking about is not simply forgetting but the reconstruction of history for the purpose of producing a new way of understanding present situations. History as told by the actor must therefore be understood and analyzed more as the structure of a mythical account than as a source of verifiable information.

Such collective myths, with their distortions and their forgetting, can at the most only serve as signposts for anyone who wants to compose an objective history of local industry. But if we aim to understand the motives behind the actors' actions and the formal and informal rules that bind them to each other in cooperation as in conflict, then such accounts become more powerfully explicative than are straightforward descriptions of industrial structures, production techniques and market positions.

BIBLIOGRAPHY

Barcellini, C., *Le logement des travailleurs immigrés: le cas d'Oyonnax*, Mémoire pour le DES de Géographie, Université Lyon 2, October 1976.

Barcellini, S., *Oyonnax, la cité du peigne. Histoire d'un boom économique 1880–1914*, Mémoire pour le DES d'histoire, Université Lyon 2, 1974.

Bodin, R. P., *P.M.E. et lois Auroux*, Rapport pour le SES du Ministère du Travail (Erisht, Grenoble, 1984).

Dominjon Bombard, B., *Essai monographique sur Oyonnax et l'industrie du peigne* (1935).

Favre, Brun N., *Oyonnax, étude de géographie humaine*, Mémoire de DES, Université de Dijon (1967).

Ganne, B., "Conflits du travail et changement urbain: transformation d'un rapport local," *Sociologie du travail* (1983).

Lebeau, R., *La vie rurale dans les montagnes du Jura méridional. Etude de géographie humaine*, Institut des Études Rhodaniennes (Lyons, 1955).

Lequin, Y., *Les ouvriers de la région lyonnaise (1848–1914)*, ed. PUL, 2 vols. (Lyons, 1977).

Puel, H. and Sagio, J., *Oyonnax*, Rapport pour l'ATP Croissance Urbaine, Economie et Humanisme (GLYSI, 1979).

Raveyre, M. F. and Saglio, J., "Les systèmes industriels localisés: élément pour une analyse sociologique des ensembles de P.M.E. industriels," *Sociologie du travail*, (1984), 157–175.

Saglio, J., "Systèmes locaux de relations industrielles et innovations productives," *Cahiers STS*, 3 (1984).

Segrestin, D., "Les communautés pertinentes de l'action collective: canevas pour l'étude des fondements sociaux des conflits du travail en France," *Revue française de sociologie*, 2 (1980).

Vailland, R., *325,000 francs*, ed. Corréa Buchet Chastel (1955; 1975).

11 Producing producers: shippers, shipyards and the cooperative infrastructure of the Norwegian maritime complex since 1850

Håkon With Andersen

Introduction

Shipping has long been considered one of the most internationally competitive sectors of the modern world economy. The triumph of large-scale integrated transport systems symbolized by the super tanker is widely believed to have transformed the small shipping businesses of the nineteenth century into managerial bureaucracies based on massive investments and worldwide control networks. Formerly national systems became internationalized as the large oil companies turned their transport operations over to flag-of-convenience carriers. More recently, even the major maritime powers have established so-called international registers which provide reduced taxation and exemption from other national regulations for shipowners operating under this regime.

This standard account of the development of modern transport systems creates a paradox. Since the advent of modern shipping in the mid nineteenth century, certain small countries have managed to build up and control a disproportionate share of the world's merchant fleet. This essay concentrates on perhaps the most successful of these smaller nations, Norway. Since the late 1860s, the Norwegian fleet has on average ranked fourth-largest in the world. Given that over this period Norway had between two and four million inhabitants and was not regarded as a rich country until quite recently, this astonishing fact seems to contradict everything we know about the success of managerial hierarchies and large-scale manufacturing. The Norwegian achievement appears all the more paradoxical since it has not been based on the attraction of foreign carriers to a flag of convenience regime. Indeed, some Norwegian owners have instead fled the strict domestic shipping regime to register under the flag of other nations.

To show that this paradox is not a paradox at all we have to venture into the political economy of shipping and shipbuilding, of insurance, classification, brokerage, consultancy and research: all international activities which defy national borders today. At the same time we have to examine regions and localities that can help to explain the choices of the historical actors themselves: why they insisted on a classical liberal policy towards shipping and shipbuilding, and how they managed to form the future at the same time as they themselves were formed by the past. Rather than pursuing modernity for modernity's own sake, these actors had their own reasons for eagerly developing and adapting new technologies to suit their particular needs.

How, then, has it been possible for a small nation like Norway to maintain a central role in the international shipping system with rather limited indigenous resources? Why have not large-scale technologies, mass production and managerial hierarchies long ago either bypassed the Norwegian owners and shipbuilders as a way-station on the road towards "rational large-scale units" or integrated them into the transnational web of giant companies? It is this question which sets the basic problem for this chapter: to show that flexibility, networking and the ability to change course rapidly have continually enabled small shipping and shipbuilding firms to offset the countervailing advantages of size and wealth.

This essay does not seek to recount the whole story of Norwegian shipping and shipbuilding. Instead, we will use snapshots of four different periods with very different characteristics to show how the political economy of shipping and shipbuilding has developed since the mid nineteenth century. Our account of each period will focus on different groups of regional actors pursuing different strategies, linking different parts of what we may call the maritime complex to one another in different ways. The passage between these different periods highlights the underlying flexibility and resourcefulness of the Norwegian maritime complex in reorganizing its activities at a regional or national level to meet the changing demands of international competition.

We will begin our account with the position in the late 1980s and early 1990s to see if it is possible to understand this period in its own terms. Concluding in the negative, we will look backwards first to the period of sail technology during the second half of the nineteenth century, and then to the British production ideal which predominated during the first half of the twentieth century. In the final section we will analyze the attempt to reap the fruits of mass production through the adoption of an alternative American ideal both in shipbuilding and in shipping. With this background, it will become possible to understand current developments in shipping and shipbuilding from which we started.

Before we embark on this historical journey, however, it will be necessary to give a short account of the particularities of the maritime sector with regard to institutions, markets and the links between shipbuilding and shipping. If we look at shipping in a vacuum, with firms relating only to their markets, the history of the Norwegian maritime industries becomes incomprehensible. Hence we will argue for the need to focus on the political economy of the maritime sector rather than on individual firms confronting a harsh and unfriendly market.[1]

[1] This article is based to a large extent on my own earlier empirical research on different parts of the Norwegian shipping and shipbuilding system. On shipbuilding, see H. W. Andersen, *Fra det britiske til det amerikanske produksjonsideal – Forandringen av teknologi og arbeid ved Aker mek. Verksted og i norsk skipsbygging 1935–1970* (Trondheim: Tapir, 1989). A broader interpretation is given in Andersen, "Norsk skipsbygging gjennom 100 år," in E. Lange (ed.), *Teknologi i virksomhet. Verkstedindustri i Norge etter 1840* (Oslo: Ad Notam, 1989), pp. 75–127. On

The maritime complex

In this section we will set out some prerequisites for the understanding of international shipping and shipbuilding, together with the links between owners, brokers, builders, insurers, consultants, classification societies and research organizations. At the end we will develop a distinction between *frontline* and *supporting* groups, that is between firms directly exposed to market fluctuations, and firms and institutions whose collaborations makes it easier to survive those fluctuations.

Let us start with the technology proper, the ship. As a capital good, it has the strange property of being easily transportable all over the world, of being in fact transport itself. This means that the market for ships as capital goods must theoretically be very large and highly competitive as long as it is not regulated or restricted by national authorities or the like. Shipping is also international as long as it often involves transport between different countries: again a sign of an international and large-scale market. We would thus expect both shipping and shipbuilding to be highly competitive activities, oriented towards international markets, with the whole world as their arena – so long as they are not restricted by the regulations of national states.

If we examine the link between shipbuilding and shipping we find another interesting relationship. The shipbuilding industry delivers the marginal increase to the world fleet each year. Given stable ship technology and a liberal shipping regime this means that the shipbuilding industry is extremely exposed to the fluctuations of the shipping market. If there is a deficit of transport capacity shipbuilding prospers. But as soon as the first ships in the world fleet are laid up, prices for new ships fall dramatically. This means that international shipbuilding is perhaps even more exposed to the fluctuations of the shipping market than shipping itself.

These trends reinforce themselves in relation to the shipbuilding industry. And, with ships as a highly transportable capital good, there are few barriers to slow down the diffusion of the problems. Even if we loosen some of the assumptions (stable technology, free trade), the picture remains much the same. In principle this has always been the experience of shipbuilding and shipping. So what in one analytical tradition may be regarded as conditions favoring large-scale and specialized transport, may in another more dynamic perspective become the conditions favoring much more flexible units capable of surviving market fluctuations.

The important point is thus to include the dynamic forces of the market fluctuations in analyzing trends towards mass production or towards flexibility. It is not only a question of increased markets, production volumes and "rational" technology, but rather the combination of the technology and supply-side

shipping, insurance and classification, see H. W. Andersen and J. P. Collett, *Anchor and Balance – Det norske Veritas 1864–1989* (Oslo: Cappelen, 1989).

institutions on the one hand and the market and demand-side on the other. It is in this complex interplay that we find the forces shaping both international shipbuilding and shipping. These forces are simply the combined actions of members of the whole shipping community, evaluating and choosing strategies and tactics based on their knowledge, experience and intuitions, all products of the past confronting the future.

We have to turn to history to find sufficiently long time periods to capture all the different elements in this interplay. To make the analysis manageable we will confine our focus to a single nation. We do not claim that the relations between shipping, shipbuilding and supporting institutions in Norway are typical of international developments. Rather we contend that insight into the development of the Norwegian maritime complex can open up new ways of understanding contrasting patterns of shipping and shipbuilding development in other nations and regions.

Shipbuilding and shipping have traditionally been labeled manufacturing and service industries respectively. There are good reasons for this distinction, particularly as regards production technology. Shipbuilding has often been mixed with repair work and production of other goods. Shipping, on the other hand, is systematic transport. So even if they depend on each other in the sense that the latter is the basic market for the former, there are good reasons for separating them. What they have in common, however, is the dependence on the international market and a reliance on fixed capital investment. The reliance on investment, paradoxically, is probably stronger for shipping than for shipbuilding, at least in terms of capital intensity for a large part of the period. This characteristic combined with heavy exposure to the international markets justifies my application of the concept *frontline* firms to large parts of both the shipbuilding and shipping industries. The idea is to underline the weight of fixed capital which limits the scope of operations once invested.

The other components of the maritime complex along with the shipbuilding and shipping companies are the brokers, insurance clubs and companies, consultants, classification societies, and research and development establishments. All of these activities belong to the service sector as *supporting* institutions to the frontline firms. They are also, of course, exposed to markets, but in different ways. But most of them have not invested heavily in production equipment and many of them operate in several different markets. These two features combined give them other opportunities and other ways of managing through both upswings and downturns of the shipping markets. For these supporting institutions, too, the Norwegian experience is rather central. Oslo is today one of the most important cities for international shipbrokerage. The classification company, Det norske Veritas, is the fourth-largest classification company in the world, handling around one-eighth of the world fleet. In insurance Oslo is also an international center with some large ship-insurance businesses.

The basic argument of this essay is that the interplay between frontline firms and supporting institutions accounts both for the vitality and survival of the rather small Norwegian shipping firms and for the large Norwegian share of the world fleet. The changing relations between different parts of these groups of firms form the core of the analysis and of the outcome. Our exegetic strategy, however, will be to analyze particular periods and then draw out the conclusions at the end.

The north-west coast yards

This section examines the recent experience of shipping and shipbuilding in Norway. Three main characteristics will be emphasized: the salience of product rather than process innovation, the role of the supporting institutions, and the forms of inter-firm collaboration which compensate for the lack of in-house administrative capacity. We will largely focus on shipbuilding and the north-west coast yards since their revival has been the most striking and unique development within the maritime complex in the 1980s and early 1990s.

During the late 1970s, when international shipbuilding experienced an unprecedentedly severe depression, a cluster of Norwegian yards flourished. At the same time as several of the larger Norwegian shipping companies went bankrupt or struggled with severe problems, new firms were established and developed rapidly. It is this double picture of stagnation and expansion in the 1970s and 1980s that forms the theme of this section. We will focus here on perhaps the most surprising fact in contemporary shipping: the rise of a new shipbuilding region and new groups of expanding yards. We will return to the stagnating sector of the maritime complex in a later section.

Most of these new yards started as builders and repair shops for fishing boats in the early 1950s and were regarded as the technological laggards in Norwegian shipbuilding circles during the succeeding period. How was it possible for them not only to survive, but also to prosper and expand in a period when the large mass production yards all over Europe were closed down?

Let us start by examining the state of shipping and shipbuilding at the end of the 1980s. International shipping was being revitalized at a rapid pace, even if new organizational forms had been introduced. Most North European countries established so-called "international registers" to avoid national legislation and to take advantage of cheap labor from the Third World. The impact of this development on the Norwegian merchant fleet has apparently been dramatic, as its share of world registrations fell from 8 percent in 1976 to 2 percent in 1987 before rising to 10 percent at the start of 1990.[2] Part of the explanation is, of course, a withdrawal from the Norwegian national register to foreign registers and then back to the International Norwegian register (NIS). The Norwegian owners, however, have been eagerly buying ships for the last few years, mostly second-hand, but in 1989 several contracts were signed for new construction.

[2] Statistics from *Det norske Veritas*, annual statistical tables.

The buying and selling of second-hand ships has been one of the most profitable business activities in these years, turning the shipowners into merchants and financiers rather than managers of shipping. In the public debate discussion centered on whether a new type of shipowner had entered the industry since standards were deteriorating and turning a quick profit seemed to have become more important than sailing a fleet.

At the same time shipbuilding was also being revitalized along the Norwegian coast. At the beginning of 1990, Norwegian yards held contracts for more than one billion dollars (seven billion NOK). However, these contracts were not flowing to the older, traditional shipbuilding industry in the large cities; on the contrary, they were concentrated to a large extent in a cluster of yards on the north-west coast of Norway, in the counties of Møre and Romsdal. Most of these yards had started as producers of fishing ships and coastal vessels. They were typically organized in Vestlandske Fartygbyggjarlag (the West Norway Boat Builders' Association) and not in the more prestigious Skibsbyggerienes Landsforening (Norwegian Shipbuilders' Association).[3] Today, very few of the former large shipbuilders in the larger cities survive: they are either doing offshore work for the oil industry or have simply been closed down and, in some cases, turned into postmodern shopping malls.

The position was much the same for shipping, although there was no comparable pattern of geographical localization. There has been a large influx of new owners and a diffusion of ownership, mainly through a system of limited partnerships. The distinctive feature of Norwegian shipping is nonetheless the large number of small companies each operating relatively few ships. The more or less transient activity of buying and selling old ships has not changed this picture; on the contrary it has made it much clearer in the sense that it has become possible to establish new firms and to take advantage of the market opportunities whatever they may be.

The most spectacular part of this picture is the development and expansion of the small yards on the north-west coast of Norway. The region has a very long tradition of building vessels for the fisheries in the North Sea and the North Atlantic and for whaling and sealing in the Arctic. At the beginning of the 1960s, small yards predominated, each with their own distinctive designs which appealed to particular owners. The region also had the highest concentration of owners of such vessels and the linkages to the different yards were close and intimate. At the same time, however, competition between the yards was fierce and different designs flourished. As each yard was rather small, it lacked any substantial marketing and sales capacity, which was in any case unnecessary since different owners preferred different designs and as a consequence different yards. The capital intensity of each yard was low and their access to skilled

[3] "Perspektivanalyse for skipsbuggeindustrien," *Norges offentlige utredninger (NOU)*, 58 (1973) (part I), and 51 (1974) (parts II, III and IV). See also "Innstilling om skipsbyggeindustriens problemer fra et utvalg oppnevnt av Industridepartementet 6 desember 1963" (Industridepartementet, Oslo: 1965).

workers rather good as the region had somewhere between [illegible]00 small yards, boatbuilders, producers of ship equipment and engine[illegible]kshops.[4]

At that time the region was not considered part of the Norw[illegible]building industry proper. Only two yards were listed in a survey [illegible]hirty-two Norwegian yards capable of producing ships above 100 grt [illegible]d by the *Bransjerådet for skipsbyggeindustrien* (a public committee fo[illegible] promote cooperation and expansion within the shipbuilding industry).[5] A[illegible]er committee which examined shipbuilding problems for the Ministry of Industry in 1963 made the same distinction. The thirty-five large yards employed some 19,000 people, while 300 boatbuilders employed 4,000 people, half of which were located on the west coast, mostly in the north-west region.[6] The figures are probably too low, but they do give an indication of the relative size and standing of the two groups in Norway during the early 1960s.

The north-west coast yards continued what was considered to be a rather backward and primitive existence on the periphery through the 1960s. But this existence was not as backward as it looked. The modernization of the fishing fleet posed new challenges for the industry, while the close cooperation with the owners gave rise to a kind of user-stimulated innovation among the boat builders.[7] These trends had two major effects on the north-west coast yards. Firstly, they sheltered them from competition on the world market. Secondly, they created a kind of local development bloc, to use Erik Dahmén's phrase.[8] In other words they generated agglomeration effects in the local communities along a coastline of only some 100 km. User–producer linkages focussed on product development, increasing the operating efficiency and economy of local boat designs, and adapting them to the changing demands of fishery practice.[9]

This development bloc was not at all linked to the traditional merchant fleet and the fluctuations in the world market. On the contrary, at a time when other larger yards tried their best to meet the lower prices for merchant vessels with rationalization and Fordist strategies, the smaller yards linked to the fishing sector were highly innovative in products, but not production. As late as the mid-1970s a report observed:

> Building methods and the work process are traditional and little industrialized . . . The degree of automation and mechanization is characterized . . . as very low . . .

[4] *Ibid.*

[5] Bransjerådet for skipsbyggeindustrien, *Norsk skipsbyggeindustri* (Oslo: 1962), p. 13. The yards were Bolsønes Verft in Molde and A. M. Liaaen in Åesund.

[6] "Innstilling om skipsbyggeindustriens problemer," pp. 20–21.

[7] O. Løken, "Teknologi og lønnsomhetsutviklinga i norsk fiske 1950–1980 med hovedvekt på havfiske," Hovedoppgave in History, University of Trondheim, 1984.

[8] Erik Dahmén, *Svensk industriell företaksverksamhet. Kausalanalys av den industriella utvecklingen 1919–1939* (Stockholm, 1950).

[9] B-Å. Lundvall, "Innovation as an Interactive Process: from User–producer Interaction to the National System of Innovation," in G. Dosi *et al.* (eds), *Technical Change and Economic Theory* (London and New York: Pinter, 1988), pp. 349–369.

None of the firms are interested in closing any department due to fear of reduced flexibility.[10]

As we can sense, this observation is written from the perspective of the large-scale companies. On the other hand, it also demands an explanation since we have presented the region as an innovative user–producer bloc, and particularly since it has become the most vital component of the Norwegian shipbuilding industry, integrated into the shipping tradition where it formerly was regarded as peripheral and outdated. The development of this region is similar in many respects to that of other industrial districts discussed in this book, focussing on market strategies and forms of action that take advantage of the opportunities created by the clustering of firms. Such opportunities were not apparent to contemporary analysts and commentators, seduced by the apparent triumphs of large-scale integrated companies. The point they missed, but which was well understood by the north-west coast yards themselves, was the possibility of pursuing product innovation within a flexible and labor-intensive production process.

Up to the end of the 1950s, the herring fisheries were the most important on the north-west coast. But after a record-breaking season in 1956 which brought in 12.3 million hectoliters, the herring disappeared and only 3.7 million hectoliters were caught in 1958. This demanded a rapid shift of the fishing fleet towards ocean fisheries, factory ships and vessels with on-board refrigeration plants. There was also a steady pressure for changes in equipment, for instance different types of trawlers and seine-fishing ships.[11] These trends led to an even closer cooperation between the owners and yards in developing new types of ships and equipment.

The new types of ships also included smaller vessels for the coastal traffic, such as ferries and cargo-vessels for the North Sea. The local development bloc was strengthened as several yards and engineering workshops turned to the production of all sorts of ship equipment. A/S Hydraulikk at Brattvåg pioneered deck machinery and equipment for anchor handling, A/S Stålprodukter in Molde developed hydraulic cranes which eventually enjoyed success on international export markets, Tenfjord mek. Verksted developed integrated steering machinery for smaller and medium-sized vessels which they also exported. A. M. Liaaen in Ålesund made a pathbreaking innovation in pitched propellers and Ullstein developed a system of passive stabilization.[12]

Most of these innovations were developed locally, based on long experience with the rough climate in the North Atlantic and close links with owners and users in the region. Formal research played only a small part. Two important

[10] E. Saugestad, "En struktur og perspektivvurdering av skipsindustrien på Sunnmøre," *Perspektiver for samfunnsutviklingen på Sunnmøre*, 8b. STI, Møre, 1977, pp. 129–130.

[11] P. Larsen, *Møre og Romsdal* (Oslo: Gyldensdal, 1977), pp. 269–271.

[12] See *ibid.* pp. 267–275, and P. O. Smogeli, *Skpsbyggingsindustrien i Møre og Romsdal 1970–1980* (Oslo, 1983), pp. 148–158; M. Grimstad, *Ulstein Mek, Verksted, 50 år* (Ålesund, 1967).

exceptions, however, were the pitched propellers and passive stabilization. These were developed in cooperation with SINTEF, SFI and the Ship Model Tank, large research institutes integrated with the Technical University of Norway in Trondheim. In a sense, this cooperation marked a kind of take-off from the local knowledge base. But the initiatives and objectives derived from just this base and demonstrated a pragmatic attitude towards development problems. Commissioned research has since then become more common among the yards.

These research establishments have their own history as supply-side institutions for the yards. All were established in the late 1940s or early 1950s except the model tank which was somewhat older. SINTEF was intended to serve all sorts of industries, while SFI and the model tank were intended for the larger deep-sea shipping companies. But they were all hunting for contracts and contacts with the industry, and were not always successful in obtaining customers among private companies. Their main *raison d'être* as semi-public research establishments was the widespread acknowledgment of the so-called problem of small-scale Norwegian enterprises, which lacked the in-house research facilities of the larger multinational companies. While they did not always succeed in performing this compensatory role, their relationship with the north-west coast yards proved very fruitful once established and the cultural barriers between researchers and manufacturers had been overcome.[13]

The main problem for the north-west yards during the 1960s was that once they moved beyond the circle of local shipowners, they ran into problems with marketing, sales and credit. This problem was solved by the transformation of the Vestlanske fartygbyggjarlag (VFBL, The Association of West-Coast Boatbuilders) into an agency for marketing and financing new ships. These associational functions became important when the yards started to export vessels to other fishing nations and developing countries. They also stimulated the yards to cooperate among themselves by sharing out large orders and splitting up the production of a single ship between several yards. Thus what had earlier been regarded as fierce competitors now became collaborators in various ways. The typical pattern was one of collaboration among independent yards, but some companies, particularly Ullstein mek. Verksted, started to buy up other firms to keep up with the expansion of demand and to increase their flexibility. Thus Ullstein acquired Molde verft in 1964 and established a new yard to meet demand from the Icelandic fishing companies for thrust propellers on fishing trips. In 1970–71, they also bought three other firms, among them the Hatlo yard.[14]

In December 1969 the large Ekofisk field was discovered in the North Sea, opening up prospects for a booming oil industry. Even if the yards at the north-

[13] S. Kvaal, *Drømmen om det moderne Norge. Automasjon og virtkelighet i etterkrigstiden* (Trondheim: Hovedoppgave in History, 1991); K. Jensen, *Forskning og ny teknologi; fra mulighet til forutsetning* (Oslo: Hovedoppgave in History, 1989); P. Ostby, *Tilfellet Comtec* (Trondheim, STS Report no. 8, 1989).

[14] The yard in question was A. M. Liaaen, but it was later acquired by the Ullstein Group.

west coast were considered too small for the big orders for semi-submersibles and later for fixed installations, they became involved in several important ways. Particularly significant was the production of supply ships for the whole offshore business. Since the second half of the 1960s the Ullstein yard had experimented with this kind of ship, based on imported designs, but they did not prove well adapted to the extreme conditions of the North Sea. So, together with the semi-public Maritime Research Institute (formerly SFI) at the Technical University in Trondheim, they developed new designs. The experience of building deep-sea fishing ships was crucial in this process. Not only the hull but also the deck machinery and the propellers had to be redesigned to meet the tough requirements of the North Sea, while the previously-developed passive stabilization system also proved important. All these elements were carefully orchestrated to design a modern supply-ship which could maneuver close to the offshore structures even in heavy seas. Another research institute in Trondheim (the SINTEF Department of Automation) collaborated on the propeller system and other yards and engineering shops in the region specialized on deck equipment. The result was such a success that it became impossible for the Ullstein group to meet the demand alone. Three different strategies were adopted over time, of which two were based on the use of local resources. One strategy emphasized the subcontracting of complete ships to other independent yards, another on building sections or components in nearby yards and engineering shops, as well as collaboration with external firms on deck equipment. To some extent, as we have seen, the Ullstein group also expanded by acquisitions. In the end the group started to sell licenses to yards abroad, but such agreements typically involved some sort of mixed package from the north-west coast industry, including components such as outfitting, propeller systems, stabilizers, etc.[15] Later the yards who got into the supply-ship business perfected and diversified their product range as demand emerged for more specialized vessels (firefighting, safety, transport, diving, etc.). By then, however, they had secured an internationally dominant position in the market for supply ships.

Why were the north-west coast yards so successful in seizing this important market? A number of reasons may be adduced. First of all, the timing: up to 1973–75, all the large shipbuilders were busy building large ships. Smaller vessels under 10,000 dwt were not considered very important or profitable for the larger plants even though they demanded sophisticated product technology and hence remained rather labor-intensive in the building process. As may be guessed the amount of labor required was far from proportional to the volume of the ships as the fetishism of measurements in dead weight tons might indicate.[16]

[15] Smogeli, *Skipsbyggingsindustrien*, pp. 148–153.

[16] The problem is the measurement of volume. It is possible to develop another more realistic approach based on the surface area of the ship or on tons of steel. But both a small and a larger ship need much of the same equipment and facilities even if the quantity of steel may be different. Hence a smaller ship may take as long to build as a larger one. See Andersen, *Fra det britiske til det amerikanske produksjonsideal*, pp. 403–408 and 597–605.

Second, the possibility of cooperation between various yards and engineering shops made it feasible to build these ships without redesigning specialized plants and making heavy investments in new production equipment.

A third reason for success of the north-west coast yards was their access to an array of knowledgeable supporting institutions. We have already mentioned two research institutes at the Technical University in nearby Trondheim. One of these, SFI, had been doing contract research on ship and ship production since 1948, and alongside the commercial firm STS (Shipping Research Services) became one of the most important consultants on production.[17] Another, SINTEF, which played a key part in propeller control, had already by the late 1950s become an international leader in automation and cybernetics. During the 1970s, a number of independent consulting engineering firms (which always have been important in the Norwegian industry) also flourished on the north-west coast. The yards and workshops consciously preferred to use such external consultancy services, a solution which helped to ease the transfer of general knowledge between the firms.[18] Significantly, however, these consultants all focussed on ship design rather than on production equipment or rationalization. These two facts reflected the orientation of local industrialists towards the avoidance of unnecessary administration and formal bureaucracy, attitudes which were considered a major problem in the analyses undertaken in the mid-1970s by STI (a state institution set up to help the industry rationalize and foster development and growth).[19] A final supporting institution was Det norske Veritas, the aggressive research-based classification society, which served simultaneously as a consultant, a research center and an independent certifier of quality. As one of the major international classification societies, Veritas collaborated with the yards and provided the necessary technical guarantees for the yards and for potential buyers. This meant that owners would not encounter problems with insurance or acceptance of the ships by local authorities, transporters or oil companies.

On the market side, Norwegian owners went into the supply-ship business and contracted for the first vessels. The yards thus enjoyed a rather large home market, even a local one as fishing-boat owners switched over to supply ships. Equally significant was the large shipbrokerage environment in Oslo which could open the way for international export of ships or licensed production and advise on redesigning construction for other environments. The importance of the brokers in providing the link between the yards and the international market should not be underestimated. One brokerage company, R. S. Platou, had by 1985 negotiated 108 contracts for supply ships alone.[20]

Finance and credit also require some explanation. Perhaps the most important step was the merger of several local commercial banks into a single company,

[17] Saugestad, "En struktur og perspektivvurdering av Skipsindustrien på Sunnmøre," p. 153.

[18] *Ibid.*, p. 144. [19] *Ibid.*, pp. 152–153.

[20] T. Baalsrud and H. T. Gram, *R. S. Platou A. S. 1936–1986, Fifty Years of Shipbroking* (Høvik: 1985), p. 147.

Sunnmørsbanken, in 1975.[21] In addition, the larger shipping banks in Norway all had subsidiaries in the region. Towards the end of the 1980s one of these, Den norske Bank, after a merger, became a dominant international bank in shipping for a few years.[22] But the pressing problem for the north-west yards was less that of financing production than of financing the product. The competitive situation was such that owners could demand credit for up to 8–10 years for as much as 80 percent of the price and the yard had to provide it as part of the contract. Once more the network was put into action. The smaller yards could not do this alone, but with the assistance of brokers, the banks, a public guarantee arrangement and a public loan institution for ship mortgages, it became possible to offer these credits. Public support in the form of paying a part of the contract was only used for the ships supplied under the development aid program.

An important consequence of the supply-ship experience was the overcoming of the old fierce competition between the north-west coast yards which had existed when they were producing different types of fishing vessels. The abundance of orders provided an easy start which evolved into a system of sharing contracts and competing for new ones at the same time. On the marketing side, too, the yards developed new forms of collaboration. This was particularly important in relation to the growing offshore market which several of the yards were interested in entering, where the dominance of the large oil companies appeared to demand a common initiative. The result was that six of the largest companies set up a common marketing and sales organization as a separate limited company in 1974.[23] Although this sales organization never achieved its objectives in the offshore market – in part because of the downturn in demand after 1976, it nonetheless carved out an important role in another field: negotiation with the national authorities for various kinds of support arrangement. What was intended to be a joint marketing tool was instead transformed into an instrument of political bargaining.

Just after the new marketing company was formed, the situation became critical for the whole Norwegian shipbuilding industry. One strategy pursued by the government was to create a credit guarantee institution, the Garantiinstituttet for skip og borerigger.[24] This was not aimed originally at the north-west coast yards but was intended to save some of the larger shipbuilding firms, particularly the Aker group. Another strategy was the attempt to combine development aid with support for the Norwegian shipbuilding industry by giving or selling at a very low price new ships to Third World countries. But most of these contracts were for

[21] F. Sejersted (ed.), *En Storbank i blandingsøkonomien – Den Norske Creditbank 1957–1982* (Oslo: Gyldendal, 1982); E. Høidal, *Frå A/S Søkkelven og omegns privatbank til Sunnmørsbanken A/S Sykkylven, 1907–1987* (Sykkylven, 1987).

[22] Sejersted, *En Storbank i blandingsøkonomien.*

[23] Smogeli, *Skipsbyggingsindustrien*, p. 163.

[24] See, for example, *Norges offentlige utredninger*, NOU (1978): "Skipsbyggingsindustrien" and NOU 1978: 13 "Skipsfartsnæringen." See also annual *Stortingsmelding*, for example no. 22/1980–1981, "Om virksomheten i Norsk garantiinstitutt for skip og borefartøyer A/S in 1979."

smaller ships, such as fishing vessels, ferries and the like. Between 1976 and 1981, seventy-six such contracts were granted by the Norwegian authorities, of which thirty-six wound up in the north-west region, thanks mostly to the new marketing institution. In addition came state support for rebuilding the Norwegian fishing fleet itself, of which 13 out of 20 contracts went to the same region in 1978–79.[25]

One of the most successful of the north-west yards was the Kleven Group, which started out as a family firm, doing repair work for the fishing fleet. From 1970, however, the four brothers decided to specialize on new construction. They also sought to carve out a particular niche for themselves, settling on special-purpose vessels between 3,000 and 10,000 dwt, which proved particularly advantageous since, as with the supply ships, none of the larger firms could then produce small series effectively in this range. In this case, too, several important innovations were introduced. New methods of mechanical handling for bulk goods, particularly automatic unloading, formed the basis for several new vessels around 10,000 dwt.[26] Another success was a highly automatic and specialized cargo ship, constructed for transporting paper from Finland to the continent. Through extensive computerization it became possible to reduce the crew with four men on these ships. The protytype was a result of a collaborative research project with the buyer, the Maritime Directorate, Det norske Veritas and the Maritime Research Institute in Trondheim (formerly SFI). Other new products have included LPG/NH3 carriers, hotel ships, exploration vessels for geophysical research, vessels for fish farming and so on.[27]

The Kleven group started with repair, but switched in 1970 to new construction. This shift was atypical of the region as a whole. Most north-west coast yards consider it important not to become dependent on new construction alone, particularly if they are not very large and have not invested in expensive plant. New construction is also regarded as demanding more engineering competence as opposed to labor skills. The Kleven group does not fit any of these characteristics. They neither built up a larger engineering department nor invested heavily in fixed capital equipment. Their plan was to subcontract all these types of work including that of several major shipbuilding trades such as painters, electricians, shipwrights, carpenters and other finishing activities.[28] In design and construction they worked very closely with an engineering consultancy firm in Bergen, in research with the supporting institutions discussed earlier, in sales with brokers, in the finishing trades with other firms in the region. Even with regard to capital equipment, they were very reluctant to invest too early or too much. Instead, Kleven built a specialized engineering workshop with computerized equipment for section-building, whose services could of course be sold to other firms in the region.[29] So instead of heavy investment in production equipment the group remained flexible and opted for high technology and product innovation. This

[25] Smogeli, *Skipsbyggingsindustrien*, pp. 167–169. [26] *Jernindustri*, 10 (1983), 264–268.
[27] *Kleven – bedriftsavis for Kleven-gruppa* (Summer and Christmas 1986); "Kleven Mek Verksted A/S- Fullt belagt for ca. 2 år," in *Jernindustri*, 10 (1983); *Kleven – Design and Shipbuilding* (Ullsteinvik, 1987). [28] *Jernindiustri*, 10 (1983). [29] *Ibid.*

strategy proved so successful that they became the most profitable shipbuilders in Norway in 1990. But it was quite a different strategy from that of the successful Norwegian yards of the 1950s and 1960s, as we shall see.

The Ullstein group, the leading producer of supply ships, followed another strategy and became a more "complete" group of firms. However, this group too had its non-traditional features. The core of the firm consisted of four major yards concentrating on new construction, each smaller than the typical yards of the 1960s. However, the group also diversified into ship-equipment with six different firms by 1986, one in Scotland and one planned in Canada.[30] Among the products were integrated gearbox/controllable pitch propeller systems, a variety of thrusters, bulk handling systems, stern rollers, passive stabilizing systems, hydraulic towing pins and hydraulic anchor handling systems. But the group's activities were not confined to its own firms. Even in its advertising material Ullstein emphasized the combination of its products and those of nearby yards outside the group. It also practiced extensive subcontracting and hired out its own capacity to other firms. The whole area functioned as a network for the group.[31] In contrast to Kleven, Ullstein has invested more heavily in computer-controlled machinery, mainly for the production of ship equipment. With regard to shipbuilding, however, the two groups are more similar, combining flexible low-cost production with highly sophisticated products.

These two groups had become the largest firms in the region by 1990. They had pursued different strategies and developed different segments of the market, but their commonalities were perhaps even more striking. Both emphasized cooperation among several smaller plants rather than building one large plant. Both continued to make heavy use of supporting institutions such as brokers, classification societies, consultants, financial agencies and so on. In a way they were more like a family of smaller yards than a large-scale integrated plant. The difference between these two groups and the individual yards and workshops was not so great in terms of actual shipbuilding methods. It was more or less a question of the framework for cooperation: between firms with the same owner or between firms with different owners.

How do these examples square with broader statistical evidence on the development of shipbuilding in this region? Since the major shift in the district occurred during the 1970s, it may be helpful to concentrate on this period. This is also the period during which the pre-1974 shipping boom gave way first to recession and then to severe crisis for the shipbuilding industry as a whole. Thus the 1970s dramatically changed the conditions for both shipping and shipbuilding. Most of the larger European yards experienced grave difficulties and were eventually closed down towards the end of the decade.

Probably the best figures on developments in this region during the 1970s stem from an investigation by a geographer, Per Ove Smogeli.[32] He has

30 *Ullsteinposten* (winter 1986/1987).

31 Various sales brochures and other material from the firm.

32 Smogeli, *Skipsbyggingsindustrien*.

corrected the data from the national statistical accounts to fit what we have here considered the shipbuilding industry of the north-west coast, including the ship-equipment sector and engineering workshops operating as subcontractors. Smogeli based his investigation on seventy-six plants, of which sixty-nine were organized as independent companies in 1970 and only two had more than one plant. Over the following decade, nine new single-plant companies entered the industry, while seven additional companies had acquired more than one plant. Thus Smogeli's population had grown to seventy-one companies in 1980, with plants compared to seventy-six plants ten years earlier. Nine companies were now multi-plant groups, of which one, Ullstein, had thirteen subsidiaries and integrated facilities, including several foreign sales organizations. Of all the other firms only three had bought subsidiaries outside the county, and of these two were in the neighboring counties. These statistics document the operation of strong agglomeration effects and underline the industry's attachment to the region.

The distribution of the firm population across different industries remained fairly stable during the 1970s. A little less than half of all local firms were shipyards, and somewhat less than a third were engineering workshops, while two-sevenths were producers of equipment and miscellaneous components. A more interesting finding is the change in market position of the firms.[33] In 1970 74 percent of the plants were producing for the final market, while in 1980 only 45 percent were doing the same. The rest were divided in two parts: the first serving as internal suppliers within larger companies, the second working as independent subcontractors for other yards. But the number of firms had increased substantially since 1970 both in absolute and relative terms. Shipyards themselves accounted for most of the decline in the proportion of plants producing for final markets, while that of plants in other sectors remained almost constant. Among the subcontractors most of the work was in ship equipment, hull sections, complete hulls and miscellaneous components. These results indicate an increased interdependence between plants and firms, combined with some concentration of ownership. This interdependence may, however, take many forms, from subcontracting between legally separate firms to specialization within a larger company. But the trend towards local agglomeration and the unwillingness to expand outside the county indicates strongly the advantages of this kind of network set-up. The benefits of remaining within the region, despite its remoteness from the main national urban centers, must have been so great that they effectively discouraged external expansion.

If we change focus from firm numbers to employment, the pattern of expansion becomes clearer. Between 1970 and 1980, employment in the firms studied by Smogeli increased by 42 percent, from 4,698 to 6,674 persons. Of these only 313 were employed outside the county. Although the number of independent single-plant companies remained stable over the decade, employ-

[33] *Ibid.*, p. 104.

ment in [illegible] fell from 3,672 to 2,621, while that in multi-plant companies increase[illegible] from 1,026 to 4,053, accounting for the bulk of local expansi[illegible]roportion of employees in multi-plant groups rose significantly during [illegible], this did not reflect the development of internal managerial hierarch[illegible] the proportion of salaried staff, office workers and supervisors remaine[illegible]ly low by industry standards: 11 percent of total employment in local shi[illegible]ds and 13 percent in engineering workshops, compared to 30–40 percent in more Fordist yards elsewhere.[34] Such findings confirm our earlier conclusion that yards in this region tended to avoid building up in-house bureaucracies for production, marketing, finance and research.[35] A final significant observation is that the size of plants increased only slightly during the 1970s. While plants with 20 to 100 workers accounted for one-third of all employees in 1970, they comprised close to half in 1980. In 1970, one-fifth of the total were employed by firms with more than 100 workers, none of which exceeded 300 employees, while in 1980 9 plants had more than 200 employees. Yet the individual plants and even the multi-plant firms remain rather small, while their larger number once more underlines the interdependent structure of the regional industry.

If we seek to place the development of the north-west coast yards in a longer time perspective than that of the 1970s, we must turn to the official statistics with all their difficulties of classification and re-classification. But some trends nonetheless stand out clearly. First of all, employment grew at a far more rapid rate in the region than in the rest of the country from the 1960s through the 1980s. Thus the region's share of national employment in all types of shipbuilding increased from around 5% in 1960 to 11.5% in 1971, reaching 16.5% in 1980 and jumping to 32% in 1987.[36] There was also a significant shift towards large ships (above 100 grt) as the local yards employed 37% of the national shipbuilding workforce but only 15% of that in boatbuilding in 1987. The production of ship equipment has also increased dramatically, as regional firms' share of the national workforce rose from 37% in 1980 to 54% in 1987. One reason for this relative expansion, of course, has been the dramatic fall in total shipbuilding employment from 30,100 in 1971 to 27,400 in 1980 and down to 8,900 in 1987. So while the rest of the industry closed down in the 1980s, the yards and workshops in the region consolidated their level of employment, and even increased substantially once more after 1987.[37]

Where does this paradoxical story take us? The most "backward" and peripheral yards of the 1960s survived the crisis in a way that was considered backward and problematic. They did not follow the "modern" strategy so

[34] Booz-Allen and Hamilton International B.V., *British Shipbuilding 1972* (London: HMSO, 1973); Andersen, *Fra det britiske til det amerikanske produksjonsideal*.

[35] Smogeli, *Skipsbyggingsindustrien*, p. 133. [36] *Ibid.*, p. 74.

[37] From *Norges offentlige statestikk, NOS*, various years. From the "Industristatestikk": number of items: vessels 3,841, ships 38,411, boats 38,412, parts 3,841,437. From "Distrikstall for Møre og Romsdal," years: 1971, 1980 to 1987.

common in the rest of the industry, both in Norway and internationally: large-scale investments in specialized production facilities. On the contrary, they were much more aware of changing markets while they invested in product development and instead evolved new forms of collaboration between plants for sharing production tasks. Such collaboration could occur either between legally independent firms or within a larger group, but the basic principle appeared to be the same in both cases: small plants working together in something like a federation.

In explaining this outcome we have stressed the contribution of the firms and public or supporting institutions not on the frontline of the market. These supporting institutions were of course operating in markets, but they were not the same markets, and offered other options than those typically available to shipowners and shipbuilders. These institutions and firms provided to a large extent the services usually associated with in-house facilities and managerial hierarchies in the large-scale, heavy-investment type of industries. The main point here is that such an external infrastructure already existed, which the shipowners and shipbuilders could and did make use of in many ways.

The next two steps of our analysis will seek to account for the availability of this infrastructure of supporting institutions. How were they established and why did they not only survive but also expand? And how could the Norwegian maritime complex hold to a steady performance course even as the fortunes of its component parts – shippers, shipbuilders and supporting institutions – rose and fell sharply in different periods? To answer these questions requires a longer time-horizon: one which extends backwards to the mid nineteenth century foundations of the modern Norwegian deep-sea fleet.

Shipping and shipbuilding in the age of sail

In this section we shall explore the construction of the Norwegian maritime complex during the second half of the nineteenth century. The paradoxes of rural areas and villages with easy access to international transport created something close to a cultural form which integrated shipbuilding and shipping, craftsmen, shareowners and sailors in the same families, sometimes even in the same person. This mix of local resources and international market opportunities fostered the emergence of a variety of supporting institutions: share-ownership to raise capital, mutual clubs to provide insurance, a classification society to certify vessels' quality standards, and eventually to serve as the technological knowledge base for the industry. Brokerage, on the other hand, developed as a consequence not of collaboration, but as a logical answer to the demands of the particular trades entered by the Norwegian owners.

Ship and boatbuilding, shipping, fishing and sailing have for centuries been major occupations along the Norwegian coast. But starting from a low level in the 1840s, Norway built up the third-largest merchant fleet in the world by the mid-1870s. This activity was concentrated along the south (or more precisely south-eastern) coast, with Arendal as its main center, just as Ålesund today is the center for the north-west coast yards. Both shipbuilding communities thus took the form of regional clusters despite their separation in time and space. But while the modern yards could rely on a good infrastructure of supporting institutions in shipbuilding and shipping, the south coast communities had to establish it themselves or rely on foreign networks.

To illustrate the size of the south coast industry, we know that on a single day in February 1875 some seventy-five ships for deep-sea transport were being built at some fifty small yards along a coastline not longer than 150 kilometers.[38] The region comprised a mix of small cities, towns and villages or small-holder/fishing communities scattered along the coast. Its topography was poorly suited for large-scale agriculture, leading together with a rapidly increasing population to an extended search for alternative employment opportunities. The resulting developments were similar in certain respects to those in continental Europe familiar from the proto-industrialization debate.[39] But there were also important differences: first in terms of the specific by-employments themselves, and second in terms of the longstanding links between the region and the international economy. It has been argued, in fact, that the concept of proto-industrialization is not relevant to Norwegian farmers and peasants because of these direct links to export-orientated industries, mainly timber, fisheries and shipping.[40] The typical farmer and small holder is often depicted as a figure who combined traditional agriculture with fishing, forestry or shipping, i.e. both subsistence and commercial activities. Although the coastal districts were poor in terms of capital, they were thus rich in terms of the variety of employment opportunities open to them.

On the south coast this multitude of trades and occupations developed into a cultural form combining the different types of work at different times of the year, divided along both gender and generational lines. Women were often responsible for the majority of the agricultural work while the men were sailing or fishing. In the winter fishing and forestry predominated among the men, as did various crafts, including those related to shipbuilding. For younger sons shipping was important throughout the spring, summer and early autumn, providing opportunities for work as seamen. This exposure to foreign nations must have shaped

[38] *Det norske Veritas 1864–1914* (Cristiania, 1914), table 6.

[39] F. Mendels, "Industrialization and Population Pressure in 18th Century Flanders," *Journal of Economic History*, 31 (1971), and "Proto-industrialization: The First Phase of the Industrialization Process," *Journal of Economic History*, 32 (1972); P. Kriedte, H. Medick and J. Schlumbohm: *Industrialisierung vor der Industrialisierung* (Göttingen, 1977). For a Scandinavian parallel, see M. Isucson and L. Magnusson, *Proto-industrialisation in Scandinavia* (Leamington Spa: Berg, 1987).

[40] The journal *Scandinavian Economic History Review* devoted a special issue to the debate over proto-industrialization: no. 1, 1982. For a Norwegian perspective, see the article by S. Tveite *et al.*

the local culture of the coastline, increasing its openness, including to economic change.[41]

Shipbuilding was a labor-intensive business. The investment needed to open a yard was rather low, particularly as it was not uncommon for skilled workers to own their tools. Building a ship was like building a house, nothing came ready-made, everything had to be done at the same location. Drawings and plans were very rough, existing mainly in the heads of the leading shipbuilders and skilled workers. Thus flexibility was provided not by cooperation between firms, but rather by the movement of workers between the different building sites or yards. Hence, working in different yards combined with other activities became the normal way of life for the craftsmen and laborers. But there was little systematic training, apart from the Naval Technical School in Horten which supplied the Naval Yard (and others) with more formally educated craftsmen.[42] This diversity of work experience gave rise to flexible skill-formation and training patterns within the shipbuilding communities, which would become another rich source of future opportunities.

The foundation of the deep-sea shipping business was rather different than its continuation, emphasizing early learning processes and their adaptation to other fields. From the beginning, shipping and shipowning were seized upon by local merchants and the few merchant houses in the smaller cities as a means of combining their import and export businesses. It was the repeal of the English Navigation Acts in 1849–50 that triggered the expansion of Norwegian merchant shipping. In the succeeding twenty-five years, shipping and shipbuilding became dominant activities along the south coast. As these activities expanded, other groups were drawn into the profitable business. A key problem was that of financing the expansion, particularly when the credit system was not very well-developed.[43] One solution which became very popular was share-ownership. Several entrepreneurs raised capital by issuing shares in a ship project. Shares could be paid for in money or in kind: timber, work, equipment or the shipbuilding itself. This method of raising capital had some important advantages. It was possible to mobilize resources within the local communities which were otherwise unobtainable. It linked shipbuilding very closely to shipping as merely a first step in the project. But this rather "democratic" form of ownership also had distinctive drawbacks compared with limited liability companies. All the

[41] Among recent works in this tradition, see the publication of the so-called Skagerak-project: H. Try, A. Gustavsson and P. Ludvigsen (eds.), "Kattegat-Skagerrak regionens kulturutvikling under 1800-tallet," *Meddelelser fra Kattegat-Skagerrak prosjektet* (Aalborg/Kristiansand, 1982–1988); S. Molaug, *Vår gamle kystkultur* (Oslo: Dreier, 1985).

[42] R. Baggetun, *Horten ingeniørhøgskole – Horten tekniske skole* (Horten: Ingeniørhøgskolen, 1980).

[43] These statements about the nineteenth-century fleet draw heavily on J. S. Worm-Müller, *Den norske sjøfartens historie*, vols. 1–3 (Cristiania/Oslo: Steenske forl., 1923–1951). See also Andersen and Collett, *Anchor and Balance*, chs. 1–2. For a discussion of credit, see F. Hodne, *Norges økonomiske historie 1815–1970* (Oslo: Cappelen, 1981).

shareholders were responsible both for losses and for profits. The most important problems were that all the participants could veto important decisions and that shares were limited to a single ship. When the ship was sold or lost the business was dissolved and the remaining proceeds divided among the shareholders. A new project had thus to start from scratch. There were no means of reinvesting the capital or expanding into the ownership of several ships. The absence of such mechanisms would become crucial during the transition to steam and steel ships around the turn of the century.

The reason it was possible for the Norwegian shippers to succeed in tough international competition was mainly due to their low transport prices and the development of new carrying trades as world trade as a whole grew rapidly. The ownership system left much of the responsibility for business decisions to the captains, who were often major shareholders themselves, and the resulting flexibility and opportunism of the Norwegian fleet proved a source of competitive advantage. As the larger nations established shipping lines and conferences which oligopolized trade, however, the need for flexibility and creativity became even greater. To monopolize a route and coordinate traffic along it, shipping lines had to develop managerial hierarchies like those pioneered by the railroad companies. But such lines never became an important part of Norwegian shipping, since as ships did not need rails tramp shipping was a viable alternative. In a sense the Norwegian fleet was marginalized, becoming a kind of appendix to the major lines and trades, taking on any kind of good wherever it might be found and wherever it was to be shipped. Their marginal position pressed the owners, the captains and later the shipping companies and the brokers to become highly innovative in searching out new trades and transport opportunities. Superior information on the emergence of new trades became their defense against the larger shipping lines. But technology never became a part of this knowledge base, which remained a typical market information network.[44]

The shipbuilding industry formed part of the Norwegian maritime complex's adaptation to these constraints, as did the recruitment of sailors from the coastal communities. The absence of concentrated capital proved functional in adapting to the initial conditions of international competition, but would later become dysfunctional for the reproduction of local industry when technological change demanded larger financial resources.

Decentralization facilitated collective solutions to other problems besides finance and production. In shipping insurance is an important element as are the certificates of seaworthiness that ships had to carry. The reasons are obvious: before hiring a carrier, the shipper generally wanted to know that his goods had a reasonable chance of arriving safely, and usually to insure them. This meant that the shipper wanted some guarantee of the ship's condition, as did the company which insured the cargo. The owners, too, wanted insurance to protect their

[44] See Worm-Müller, *Den norske sjøfartens historie*, and also H. W. Andersen, "Laggards as leaders" in T. Bruland (ed.), *Technology Transfer and Scandinavian Industrialisation* (Oxford and New York: Berg, 1991).

interests in the ship itself, if not in the cargo. Some of the most interesting collective solutions were developed in this field, solutions which became the foundation for a kind of meso-level actor: the supporting institutions as I have termed them earlier. These institutions could be regarded as a kind of coordinator and mediator for the whole industry. Their development can be briefly sketched in three chronological steps starting with that of the ship insurance organizations.

Before the 1850s Norwegian shipowners had to obtain insurance through Hamburg merchants or London underwriters on purely commercial and in their all-too-expensive terms.[45] Given the marginal, cost-cutting position of the Norwegian fleet, this became a major obstacle to expansion. Thus the idea of mutual insurance clubs emerged both from economic experts and from the industry itself.[46] Part of the reason for the high insurance rates in Hamburg and London was the incidence of insurance fraud. But a local mutual club could use its social network and moral pressure to hold the losses due to fraud down to a minimum. Mutual clubs could also reduce the normal rates and instead collect more money whenever needed. These advantages gave rise to a set of local insurance clubs based on personal knowledge and social control. The clubs also arranged with inspectors to survey new ship construction to maintain standards at an acceptable level.[47]

These local clubs soon recognized the desirability of becoming large enough to absorb shocks from sudden losses among its members. But such expansion was incompatible with the structure of local social networks, and forced the clubs to formalize inspection and control routines. In addition, the clubs could only arrange for ship insurance, not for the certificates and papers necessary for shippers to insure their cargo in foreign harbours far from Norway. For this service the Norwegian owners had to rely on another expensive foreign commercial company, the Belgian and later French Bureau Veritas. The idea thus emerged of creating a Norwegian body to perform this function as well, in the shape of a domestic classification society. Such a society, the local clubs believed, would reduce the costs both of certification and classification, while at the same time relieving them of the burdens of inspection and development of formalized rules. This solution would also permit the mutual insurance clubs to expand geographically beyond their local districts.

Det norske Veritas was established by the major mutual insurance clubs in 1864 to provide cheap classification and certification to the shipowners. It was to be organized as a collective society with the clubs as members. But so long as the membership of the mutual clubs themselves consisted mainly of shipowners, the latter came to control both insurance and classification. During the twentieth century, the role of the clubs diminished with the development of other forms of

[45] W. Klavenes and K. Lorange, "Sjøforsikringens historie," in Worm-Müller, *Den norske sjøfartens historie*, III.2.

[46] Andersen and Collett, *Anchor and Balance*, pp. 18–22. See also J. Aall, *Nutid og Fortid* (Arendal, 1836).

[47] Andersen and Collett, *Anchor and Balance*, pp. 24–26.

commercial insurance, while the classification society on the other hand expanded sharply, particularly after 1950. Det norske Veritas is today the fourth-largest classification company in the world (after the British, American and Japanese) in terms of classified tonnage.

Veritas came to be the main norm-setter, norm-interpreter and control institution for ship technology in Norway and later in many other countries. However, gaining acceptance in Norway was not the same as being recognized in London, New York or Sydney. Thus the third step in its development was the establishment of the classification society's authority in the major foreign ports, a project which demanded collaboration with both shippers and foreign freight insurance companies, as well as the creation of a central office staff in Kristiania (Oslo) and a network of representatives abroad.[48] This collective institution, Det norske Veritas, grew rapidly in size and authority on questions of ship technology. But its virtues as an accelerator of technological change first became visible in the second half of the twentieth century. Until then Veritas worked more as a kind of police body, guaranteeing quality according to their own standards, nicely calibrated to those of the British Lloyd's or the French Bureau Veritas. It also became a political spokesman for the whole shipping business, at least until the establishment of the Association of Ship Owners (Norsk Rederforbund) in 1907.

Another important group which developed in age of sail were the shipbrokers. The law regulating shipbrokerage dates from 1818 and set strong limitations on their work. Brokers had to be licenced and could not run a business as a merchant, shipowner or money changer, alone or in partnership. Chartering ships in foreign waters was not mentioned, but was very difficult to do, at least in the first half of the nineteenth century.[49] During the 1860s, foreign brokers showed an increasing interest in the rapidly-growing Norwegian fleet and began to send representatives to the major harbours: around 1870, for example, London and Hamburg brokers started to use Bergen businessmen as agents. In 1869 the old law was abolished and the different shipping cities could decide when brokerage should be open to all comers.[50]

The core of the brokerage business was from the outset the administration of freight, connecting shippers who wanted a good delivered from one harbor to another with shipowners ready to take on the cargo, and then negotiating the price. This activity later expanded into the negotiation of construction contracts between yards and owners, as well as the buying and selling of secondhand ships. A third branch of activity became the negotiation of all types of insurance, thus mediating between owners and insurers. The brokerage profession came in this way to establish an international network of utmost importance for all relevant groups in shipping: owners, builders, shippers and insurers. It became the spider

[48] *Ibid.*, pp. 44–49. See also *Det norske Veritas 1864–1914*.

[49] L. Pettersen *et al.*, *The Sound of Grieg* (Bergen, 1984), pp. 2–4.

[50] H. Sande, "Norsk skipsmeglervirksomhet gjennom tidene og i dag," in *Skipsnytt* (1980), 3, no. 2 6–11. See also Pettersen, *Sound of Grieg*, p. 7.

in the maritime web, sitting on the most valuable of all commodities in a business based on rapid change: information.[51]

The telegraph became crucial to the rapidly-growing international brokerage trade. With the North Sea and transatlantic cables it became possible to work as quick as the trade's inherent characteristics demanded.[52] When the Association of Norwegian Shipbrokers was established in 1919, 93 brokerage firms were represented and one year later there were 126 member firms.[53] Some of these firms probably had a rather short life, due to the boom in shipping during and immediately after the First World War. But a substantial number survived into the 1920s.

There were two main reasons for the international breakthrough of the Norwegian ship brokers between 1870 and 1920. The first stemmed from the particular trades the Norwegians were serving. As a tramp fleet sailing between third countries, carrying mainly bulk goods, the importance of arranging freight contracts became crucial. The shift from leaving this task to the captain because the owners were not in contact with the ship during the sailing season to arrange it through cabled orders made it possible to ensure a better profit. As the Norwegian owners were unable to enter the liner trade, brokerage services became even more important. The shipping lines had set up their own organizations for handling goods, either internally or through agents. Brokers came to be associated with individual and casual cargoes or charter parties for shorter or longer periods. The solution was the same as that on the north-west coast yards 100 years later: to organize the service outside the firms themselves. A typical case was the development of the timber trade where the brokers established close contacts between Norwegian owners and shippers all around the Baltic.[54] With the arrival of the steamship era, the need for continuous sailing throughout the year increased, strengthening the brokers' position. The cause was, of course, the much larger capital investment tied up in each ship. So while the transition to steam was catastrophic for the old yards, and difficult for the owners, it proved advantageous on balance to the brokers.

In the years following the abolition of brokerage regulation law in 1869, it was not uncommon to mix shipbrokerage and shipping. This was the case for Fearnley & Eger, Chr. Th. Boe & Son and for Winge & Co. The mixture of shipping and brokerage naturally had advantages since both activities demanded a good knowledge of markets and shippers. These characteristics were reflected in shipbrokers' training. Young men were sent abroad to Hamburg and London

[51] Baalsrud and Gram, *R. S. Platou A. S. 1936–1986*; Kr. A. Olsen, *Mitt skip er lastet av Winge & Co gjennom 100 år. 1855–1955* (Olso, 1954); J. D. Knudsen, *Chr. Th. Boe & Søn A/S. Skipsmeglere Arendal 150 år 1836–1986* (Arendal, 1985); G. C. Wasberg and K. Pettersen, *Fearnley & Eger 1869–1969* (Oslo, 1971); Pettersen, *Sound of Grieg*; Sande, "Norsk skipsmeglervirksomhet gjennom tidene og i dag."

[52] Wasberg and Pettersen, *Fearnley & Eger*, pp. 260–262.

[53] Sande, "Norsk skipsmeglervirksomhet gjennom tidene og i dag," p. 7.

[54] Olsen, *Mitt skip er lastet av Winge & Co gjennom 100 år. 1855–1955*; and Knudsen, *Chr. Th. Boe & Søn A/S. Skipsmeglere Arendal 150 år 1836–1986.*

to "learn the trade" and not least to establish and maintain good contacts in the major international centers. One such contact was the British firm of H. Clarkson in England which took on as many as 130 Norwegian trainees between 1860 and 1960.[55] Though the tendency in this century has been towards the professionalization of brokerage, we can also observe spin offs in the other direction: brokers who became owners and started their own firms, a phenomenon which was particularly evident during the tanker boom of the 1930s.[56]

In summary, then, the Norwegian maritime complex had acquired a tripartite structure during the second half of the nineteenth century. At the local level it consisted of integrated communities making their living from building and sailing ships. The ownership of the fleet was widely diffused, including farmers, craftsmen, skilled sailors and merchants as well as public servants in higher positions. When specialized shipping companies began to emerge the majority remained rather small, with only a few vessels each. They concentrated mainly on transport of bulk goods outside the large shipping lines' domain. At a collective level, institutions were developed to service the common interests of this large and fragmented fleet, first in insurance and then in classification and certification. The latter developed into a kind of common pool of technological expertise, open to all the members. It thus came to compensate for the lack of technological knowledge among the small shipping companies and shareowners. Brokers, finally, started to perform services outside their local communities, expanding their field of activity from the arrangement of freight shipments to and from their home town into the international negotiation of freight agreements and ship sales. As classification had become the knowledge base for technology, brokerage became the base for business and markets.

The decline of sections of this complex around the turn of the century was mainly due to developments in international shipping. First and foremost, larger and more capital-intensive steam and steel ships became increasingly competitive with sailing vessels for all types of cargo. For a long time the Norwegian owners tried to pursue a labor-intensive strategy, using cheap second-hand sailing ships, rigging them down to reduce the size of the crew, and competing in low-value trades where speed was less important. But Norwegian decline was not total, as much of the network of brokerage, insurance and classification activities survived.[57] The distinctive organization of the maritime complex, which had been necessitated by the lack of indigenous resources, thus proved fruitful in the sense that the network of supporting institutions surmounted the crisis, while the frontline firms in shipping and most notably in shipbuilding did not.

But the new organizational form combined with the new technology rendered the labor-intensive strategy obsolete, and the Norwegian fleet slowly became a convoy of old hulks with an accident rate triple the world average. What in 1875 looked like a modern fleet was reduced by 1910 to a tragedy of backward

[55] *Baalsrud and Gram, R. S. Platou A.S. 1936–1986*, p. 10.
[56] MA theses under preparation at the Department of History, University of Oslo.
[57] O. Gjølberg, *Skipsfart, økonomi og hitorie, 1866–1914* (Bergen: NHH, 1980). See also Andersen and Collett, *Anchor and Balance*.

technology, even if it still made money and had already begun to transform itself. For the south coast communities, this tragedy had been apparent since the late 1880s when new construction of sailing ships had come to a total halt. The center of the new steamship business moved to the larger cities, Oslo and above all Bergen, where the new and higher capital requirements could be met. Even some of the south coast owners who managed the shift from sail to steam moved their main offices to these cities. But many were forced to leave the business. As in the case of the modern yards of the north-west coast discussed earlier, the change in frontline firms also brought a change in their geographical localization.

During this period, too, shipbuilding and sailing became very different tasks, both in terms of skills and location. The skills and requirements involved in building wooden sailing ships and steel steamships were not at all compatible. The result was that shipbuilding moved away from the south coast to the engineering workshops in the larger cities. On the coast this resulted in tragedies which can perhaps best be read in the statistics for emigration to the United States. In the cities of Oslo, Bergen and Trondheim the engineering workshops developed into large shipyards with the mechanical trades as the core of the knowledge base. Even in these large-scale yards work remained craft-like, based on teams and shops for different trades, but the social relations in the large urban workshops were very different from those of the south coast. At the same time, shipowning companies developed into multi-vessel limited companies of a more standard type, even if most of them still remained rather small. The novel development, however, was that they now bought their ships, new or old, from abroad.

Developments during the second half of the nineteenth century thus represent an example of differentiation of what had previously been considered a single business. First came the establishment of collective support institutions in classification, insurance, capital formation and technology, then the parallel emergence of a separate institution for market information through the brokerage firms and finally the separation of shipbuilding from shipping. This development was carried through not as a second-best solution, but rather as the best solution given the constraints under which the Norwegian maritime community was working. In other countries such as Britain, the large shipping houses developed most of these services internally, and where they did not, as in the case of classification, their use of the external institution was quite different.[58]

Steel and steam: the British ideal

The first half of the twentieth century brought deep structural changes to both shipbuilding and shipping. Steel was different than wood as a construction material, requiring different trades and different types of equipment. Oil was a different type of bulk commodity than coal or timber, requiring different ships and organizational patterns. In this section we will analyze these changes and the

[58] G. Blake, *Lloyd's Register of Shipping 1760–1960* (London, n.d.); P. Le Conte, *Le Bureau Veritas 1824–1924* (Paris, 1928).

strategies selected both by owners and yards in adapting to the new environment. We will begin with the organization and technology of steel shipbuilding, before moving on to examine the shipping firms and the market strategies which carried them into the oil trade. This last theme will bring us back to shipbuilding, since the largest tanker-building nation would not be Norway but instead neighboring Sweden.

Norwegian steel shipbuilding never regained the position it had held in the age of sail. But it is noteworthy that hardly any of the former wooden shipbuilding companies managed to change over to steel construction. Instead it was the engineering workshops in the larger cities along the coast which took over the shipbuilding business. The most serious threat to the Norwegian shipbuilding industry came from Great Britain which gained a position among Norwegian owners which they maintained until the early 1950s. The yards along the Clyde, Tyne and Wear rivers became the dominant suppliers of ships throughout the world. To a considerable extent, the Norwegian yards saw them as an ideal or model, trying to emulate their business strategies and production methods.[59] Many of the older shops were established in close contact with the British tradition of engineering and shipbuilding.[60] But it proved difficult to copy the British yards, mainly because their regional concentration facilitated a superior supply of skilled labor for the hull-construction trades. The best of the Norwegian yards succeeded fairly well in following British layouts, and during the years before the sharp post-war recession of 1920 they both modernized and experimented with new construction methods.[61]

The organization of production inside the Norwegian yards can best be described as craft-orientated teamwork based on collaboration between separate trades and a low degree of planning. Most of the construction process was devolved to the foremen of the different groups (often corresponding to trades and workshops) and to the senior skilled workers who served as leaders of their squads. The construction process was linear in the sense that it centered on the berths, slowly building up the ship, plate after plate, profile after profile. The surrounding workshops of the different trades were subordinated to the construction on the berths, providing service functions in line with the needs of each step of the construction process. Internal transport remained undeveloped, because each trade controlled its own business.[62]

[59] S. Pollard, and P. Robertson, *The British Shipbuilding Industry 1870–1914* (Cambridge MA: Harvard University Press, 1979); see also Jonathan Zeitlin's article in this book. For Norwegian yards and workshops inspired by the British model, see H. Aagotnes, *Skipsbygging, jernarbeiderar og fagleg kamp* (Bergen: Notat nr. 2 i skriftserien fra hum. orient. arbeidslivshistorie); K. Kjelstadli, *Jerntid*, University of Oslo, 1987. See also Andersen, *Fra det britiske til det amerikanske produksjonsideal.*

[60] K. Bruland, "Norsk mekanisk verkstedindustri og teknologioverføring 1840–1900," in Lange (ed.), *Teknologi i virksomhet.*

[61] Aagotnes, *Skipsbygging, jernarbeiderar og fagleg kamp.*

[62] Pollard and Robertson, *British Shipbuilding Industry*; Andersen, *Fra det britiske til det amerikanske produksjonsideal.*

The engineering workshops adapted easily to the shipbuilding business, even if they often started as outsiders to the maritime complex. The technological changes which undermined wooden shipyards of the south coast at the same time favored the engineering workshops, just as they did the brokers and probably also the insurers and the classification society. Hence the engineering workshops brought to the maritime complex a combination of their own traditions and what they regarded as the most modern technology of the day: the British way of building ships.

The classification society, Det norske Veritas experienced severe difficulties during the later years of the sailing ship as the deteriorating standards of the fleet brought political problems and problems of legitimacy and independence. But in 1909, the board appointed as their new managing director a Danish senior engineer from the British classification society Lloyd's. His experience with the steam and steel technology in Britain was considered very important in the struggle to transform the Norwegian shipping community. This director put all his energy into the business of developing high standards for the steam fleet, setting new rules that were even stricter than those of Lloyd's.[63] It was, however, the First World War that finally brought an end to the old tradition. Norway lost a tonnage equivalent to 50 percent of the 1914 fleet, a development that was tragic for the seamen, but enabled the owners to rebuild a newer and more modern fleet thanks to war insurance and the like.[64]

In addition to the old yards, Norwegian shipowners also had great problems shifting to steel and steam. Several succeeded, but their very success spelled farewell to the old forms of life on the south coast. The need for large credits and capital favored the shipping circles in Oslo and Bergen. Some moved their business from the south coast to the capital, while others quit altogether. But as a result of the business knowledge of the brokers and the technological competence of the classification companies allied with the new yards and the British tradition, it proved possible to obtain both capital and cargoes for the new and much more capital-intensive shipping business.

In other words, the technological shift from sail to steam only affected the most market-exposed sections of the maritime complex, the builders and the owners. For the supporting institutions or infrastructure, the advantages of this change perhaps outweighed the disadvantages insofar as shipping became more capital-intensive and thus a more secure foundation for brokerage (more reliable arrival and departure times) and insurance (a declining accident rate). The classification society could also aspire to catch up with British Lloyd's and the Bureau Veritas.

The shipping companies remained rather small, and even though some firms did manage to enter the liner trade despite the conferences and oligopolistic competition, most continued to specialize in single cargoes or short charter parties. And they could do so precisely because of the very strong infrastructure

[63] Andersen and Collett, *Anchor and Balance*, pp. 66–69.

[64] J. Schreiner, *Norges skipsfart under krig og høykonjunktur 1914–1920* (Oslo: Cappelen, 1963). See also the annual statistics from *Det norske Veritas*.

on which they could rely. It was not a second-rate strategy, just a different strategy. To understand this we need to look more closely at the changing organization of the bulk cargo trades. The size of a shipowning company is often measured in gross registered tons (grt) or deadweight tons (dwt). Usually this is misleading. A better concept would be to measure the size in terms of the number of ships each capable of carrying freight. But the big difference is that between companies providing liner services and freight companies specializing in short or long charter parties. In the first case it would take a company several years of losses to build up a line and a comparatively large organization to keep it going with agents and services to the more or less frequent users. For a ship operating under charter-party conditions on the other hand, a very modest organization could suffice. Each agreement is a single charter party, which can keep the ship occupied for years in some cases. Thus the company's only obligations are maintenance and operations, both of which are often taken care of by the staff on board. The need for formal in-house expertise and administration at the head office is small, so long as business, market and technical services can be obtained through some sort of common infrastructure.

Paradoxically we can say that even if the Norwegian fleet was one of the most competitive in the world – at least in the charter market – it was also extremely flexible since the central administration was small and the supporting infrastructure well developed. The result was that Norwegian owners often pioneered new and rapidly growing cargo trades which were later taken over by liner companies or the shippers themselves. A typical case in point was the banana traffic between the West Indies and North America initially developed by the Grieg brokers and Norwegian owners, which was eventually taken over by United Fruit itself,[65] and many similar examples can also be adduced.[66]

The most extreme story of this type was that of oil transport, particularly after the mid-1920s. Its trajectory was the reverse of the United Fruit case as the oil companies began to charter ships instead of owning the whole fleet themselves. The sale of several old Anglo-Saxon (Royal Dutch Shell) steam tankers proved particularly important to the Norwegian owners for this business. The ships were old, but rather inexpensive, and most significantly, they came with fixed-price charters to transport oil, often for as long as eight years. This became the opportunity for many former south coast owners and others who wanted to enter the oil trade. The more established firms were reluctant to take this chance, mainly for two reasons. First, they considered it very dangerous to be on the margin, transporting the last 20–30 percent of an oil company's requirements and to rely totally on a single customer. Second, they reckoned the maintenance expenditures to be larger than included in the time-charter rates. Thus the field was open for new entrants with nothing to lose. Credits were obtainable since the

[65] Pettersen, *Sound of Grieg*; Baalsrud and Gram, *R. S. Platou A.S. 1936–1986*. See also Worm-Müller, *Den norske sjøfartens historie*.

[66] Other trades pioneered by Norwegian shipowners included peanuts, grain, locomotives (to India), sheep and later cars.

ships had long time-charters with the oil companies, and the risks were taken mainly by the new owners.[67]

The oil transport business proved successful and after a short time the steam-tanker owners started building large, modern motor ships for this trade using new charter agreements with the oil companies as security. This way of doing business turned out to be profitable mostly because of the enormous expansion in international oil consumption and the transport strategy of the oil companies. As soon as these conditions changed, as in the mid-1970s, the result would be catastrophic, with one Norwegian ship among every four laid up in the world. But in the intervening period, the Norwegian deep sea merchant fleet became above all a tanker fleet, with the proportion reaching 50 percent of grt between 1950 and 1970.[68] As the business diversified several shipping firms changed over to single cargo terms, sometimes combined with time-chamber arrangements. But the central characteristics remained the same: companies which remained small in terms of numbers of ships and internal organization, but which grew larger in terms of tonnage, turnover and investments. Transport of other bulk goods such as iron ore, wheat and bauxite/alumina also increased in volume from the mid-1950s, so by 1971 83 percent of the Norwegian-owned tonnage was classified as bulk and oil carriers.[69]

One might have expected the turn towards tankers to be important for the Norwegian shipbuilding industry, particularly as many British yards were reluctant to enter this market. But here again the impact of market and technological trends had different effects on different sections of the maritime complex. The Norwegian shipyards were also reluctant to start building tankers. This appears surprising since the interwar years were difficult ones for both British and Norwegian shipyards, as orders and output fluctuated dramatically against a background of general depression until the onset of rearmament after 1936. In Norway, if anything, the situation was even worse as the government tried to revalue the Norwegian crown during the mid-1920s.

Why Norwegian shipbuilders did not switch to tankers is a question which has been debated by a number of historians. The explanations offered include insufficient investment (Nordvik), inadequately-developed technology (Hanisch) and business strategies rooted in their technological and organizational base (Andersen).[70] The true explanation may of course be a combination of all

[67] L. Nørgård, *Tankfartens etablerings- og introduksjonsperiode i norsk skipsfart 1912–1913 og 1927–1930* (Bergen, NHH, 1961). See also K. Olsson, *Från panserbåtvarv til Tankfartygsvarv* (Göteborg: Svenska Varv, 1983); H. W. Andersen, B. Stråth, and T. Svensson (eds.), *Olje, verft og redere – et seminar om norsk og svensk skipsbyggingsindustri og norske redere i vårt århundrede* (Trondheim: University of Trondheim, 1980).

[68] *Maritime Transport* (1954, 1958); *Lloyd's Register of Shipping, Statistical Tables* (1967–72).

[69] *Ibid.*

[70] H. W. Nordvik, "The Norwegian Shipbuilding Industry. The Transition from Wood to Steel, 1880–1980," in F. M. Walker and A. Slaven (eds.), *European Shipbuilding, One Hundred Years of Change* (London, 1983); T. J. Hanisch, in T. Berg *et al.* (eds.) *Norge fra u-land til i-land* (Oslo, 1983); T. J. Hanisch and E. Lange, *Vitenskap for industrien: NTH – en høyskole i utvikling gjennom 75 år* (Oslo, 1985); Andersen, *Fra det britiske til det amerikanske produksjonsideal.*

three. The British shipbuilding tradition of flexible workshops discouraged specialization on a narrow range of products. Firms chose between products according to their relative profitability at each moment, all the more so since their fixed capital was relatively low compared to wages. Thus the British industry adopted a hire-and-fire strategy, with employment closely linked to business fluctuations, and the Norwegian yards followed suit. In this tradition shipbuilding did not always mean building new ships. The same staff and equipment were also used for repairs, overhauls and industrial production. Firms had a wide range of commercial opportunities depending on the market situation in different fields. The various shops could deliver whatever kind of service was in demand. But the tradition of mixing new construction and repair also had certain distinct drawbacks. First of all, it meant a prolongation of the building time. Second, it meant the yards lost their competitive edge on new construction when they were doing too much repair work. Following Nelson and Winter, we may say that the routines of new construction were forgotten by the firms.[71] When the system worked well, it meant that the yards could keep all trades employed independent of the stage the construction had reached. This was also achieved having several new ships in the berths at the same time, each at a different stage of completion.

During the interwar years, the flow of orders did not permit firms to build several new ships at the same time. Thus the yards opted to seek profits in repairs. The market for repair services was much less affected by the low demand for new ships for simple geographical reasons, and the fact that time laid up was the most important element in the owners' calculations. Paradoxically, we may say that the reason the Norwegian yards could survive in this way was the existence of a large Norwegian fleet and the shipping infrastructure. From a business perspective, it was much easier to obtain profits from repairs than from new construction where the international prices were very low.

The new motor tankers therefore came to be built mainly in another country, with different industrial traditions: Sweden. The linkages between Norwegian tanker owners and Swedish shipyards in Gothenburg and Malmö became very strong. In Sweden the modernization of the yards was an industrial strategy more than a shipping strategy. The yards were also linked with financial institutions which made it possible to advance long credits to the buyers as part of the contracts. The Swedish yards became specialized factories for new ship construction to a degree previously unknown in peacetime. In a number of areas, they followed a typical Fordist strategy, emphasizing planning, simplification and long series of standard ships. All this was sharply opposed to the dominant British tradition which stressed tailor-made products, designed and produced in a craftsmanlike way.[72]

[71] R. R. Nelson, and S. G. Winter, *An Evolutionary Theory of Economic Change* (Cambridge MA: Harvard University Press, 1982).

[72] There is a series of books from a large research project on the development of the Swedish shipbuilding experience. For an updated overview see B. Stråth, "Les chantiers navals suédois et leurs ouvriers de 1890 à 1990," *Le Mouvement Social*, no. 156 (1991). Important works for this study include: Olsson, *Från panserbåtvarv til Tankfartygsvarv*; T. Svensson, *Från ackord till månadisløn* (Göteborg: Svenska Varv, 1983); and J. Bohlin, *Tre perioder i den*

There were several reasons why it proved possible [illegible] Swedes to enter shipbuilding in this way. One was of course the produ[illegible] tankers, which were usually large ships, design to travel at a moderate a[illegible]mic speed. This meant that the hull designs were wide and some would [illegible]y, as opposed to the liners with nice lines and high speeds. The aesthet[illegible] versus Fordist production were important in established shipping ci[illegible]n though economics won out in the end. Secondly, the major clients of [illegible] Swedish yards, the Norwegian owners, were not orientated towards prestigious and beautiful ships, but rather towards profits from a rather weak financial base. This meant that they were willing to accept standard ships so long as the price was low and the credit high. The success of the Swedes' Fordist strategy thus rested on a fruitful alliance between outsiders in both shipping and shipbuilding. In the Norwegian yards, however, no one wanted to take the chance of becoming mass producers of hulls so long as it was possible to survive on repair and occasional new construction, even if they were regarded as backward and placed under political pressure to modernize.[73] Probably this was the same for the British yards. To specialize on new construction was regarded as dangerous, precisely because of the lack of flexibility and the potential problems of market adjustment, a fate that would be tragically experienced by the Fordist yards during the late 1970s. The Swedes also pioneered several new product and production technologies. The combination of welding and prefabrication was perhaps the most important which gave them an advantage later on, as did their emphasis on diesel motors. In addition came organizational changes which broke down the shop-orientation of the traditional yards by developing the planning and preparation of work together with internal transport. In a number of ways, the Swedes thus anticipated and predated American mass production of ships during the Second World War.

The "British" period reshaped the Norwegian maritime complex in a number of ways. It strengthened a new type of yard, situated in different regions (the major coastal cities) and destroyed the older way of life based on the integration of building and sailing. Most of the new yards maintained a tradition of multipurpose shops, neglecting new construction when other sectors of the market such as repair or industrial products offered better returns. The shipowners continued in the tramp and bulk cargo business, but with oil transport as one of the few growing markets (another was whaling) this took a new form, which gave them a more secure position at the same time as it tied them more closely to a single customer. At the same time, however, the shippers continued to rely on the national support network. The brokers became crucial both in finding and negotiating contracts for cargoes and ships. Veritas strengthened its position as a collective knowledge base for ship technology and insurance clubs developed into more conventional commercial firms. The Swedish yards developed into ship suppliers for Norwegian owners from a position as outsiders. As the Norwegian owners took advantage of the oil trade, so too the Swedes took

svenska varvsindustriens lönsamhet och finansiering under 1900–talet (Göteborg: Ekonomisk historiska institutionen, Göteborgs Universitet, 1987).

[73] Stortingsmelding no. 27, 1936: *Om fremme av skipsbygningsindustrien* (parliamentary paper).

advantage of the Norwegian owners' need for large motor tankers. Taken as a whole, the maritime complex lost its integrated and regionalized characteristics during this period, but continued to develop both market flexibility and the collective networks that made its strategies work.

The American ideal of production: towards Fordism

Whatever the merits of the British ideal, the period from the Second World War to the 1970s was marked by a typical Fordist strategy based mainly on larger yards with international ambitions. Not only the yards themselves, but also the ships and their owners became much more specialized. In this section we will examine what from the perspective of the 1980s had become the declining section of the industry, but now we will analyze prospectively the roots of that decline. First and foremost, this is a story of an ever-expanding market which came to an "incredible" halt in the mid-1970s. Once again we shall see that the supporting institutions did not undergo the same changes as the frontline firms. Instead Veritas, for example, had to keep abreast of all types of ship technology, as did the brokers in the case of the shipping market, and naturally the insurers and consultants as well. Investments in shipping and shipbuilding technology were highly specialized, while the supporting institutions continued to serve as a general information base for the frontline firms. Nor were all sections of the industry drawn into this web: as we saw earlier, the north-west coast yards followed a different strategy which would later become both more important and more visible.

Although the Swedish yards had set the stage for specialized series production of ships, it was the Second World War that defined a new ideal for international shipbuilding.[74] The American achievement in which ships spent as little as three weeks on the berths made a serious impression everywhere. Welding, standardization, planning and internal transport all came to express the feeling of a "modern" way to build ships. Enormous plants with a wide network of suppliers of standardized parts all under a central command became the post-war ideal. But what the early observers failed to notice was that this system was designed to produce an enormous quantity of ships quickly, not for the market but rather for a war economy. This meant, of course, an economy freed from normal commercial constraints, which was a crucial condition for such large-scale

[74] During the First World War, the American yards tried to develop a sort of mass production. The most extreme case was Henry Ford's attempt to produce a patrol ship, the *Eagle*: see D. A. Hounshell, "Ford Eagle Boats and Mass Production during World War I" in M. R. Smith, *Military Enterprise and Technological Change. Perspectives on the American Experience* (Cambridge MA: MIT Press, 1985). Other yards tried with merchant ships, but did not quite succeed, or at least did not become a symbol of how ships should be produced during peacetime. See F. G. Fassett, jr. (ed.), *The Shipbuilding Business in the United States of America*, 2 vols. (New York, 1948); D. Levine and S. A. Platt, "The Contribution of U.S. Shipbuilding and the Merchant Marine to the Second World War," in R. A. Kilmax (ed.), *America's Maritime Legacy* (Boulder CO, 1979).

production. The result was impressive in quantity, but not in quality or the use for resources. Swedish motor tankers were produced at lower cost than even the final ships in the long series of US T2 tankers, and their quality was also much higher. The necessary condition for the success of American war production was an abundant demand for ships at almost any price. Under normal market conditions, it would have been very difficult to sell these ships and still recover the production costs.[75]

For Norwegian owners the war was a difficult period, but for the seamen a deep tragedy. As in the First World War, Norway lost half her merchant fleet through military actions. For many a seaman, the ice-cold North Atlantic became a watery grave. But the organization of the fleet during the war once more relied on the infrastructure of the maritime complex. As much as 90 percent of the fleet escaped German control as it was sailing all over the world. As a result of the Norwegian surrender to Germany in 1940, one might have expected the Allies to have confiscated the whole fleet as enemy property. As one of the largest merchant fleets in the world, it was of the greatest importance for the allied nations to put it to work for them. Quick initiatives by the government, the owners and the brokers resulted in a publicly-owned company encompassing nearly all Norwegian ships outside German control. This company, Nortraship, became the largest shipowning company in the world. It was staffed by owners and brokers living abroad in close cooperation with the government-in-exile. There were multiple advantages of this system for all groups. The government-in-exile received revenues which financed their expenditures, the owners retained some control over the ships and the brokers could continue their businesses in other forms. It was also possible to save some of the profits and build up funds for maintenance and reinvestment when shipbuilding would become possible once again under normal market conditions. Of particular importance was the supply of new ships in compensation for wartime losses. As a governmental-controlled company, Nortraship could negotiate with the allied administration for quotas. During the later stages of the war, the Norwegians collectively negotiated new construction contracts with the Swedes and they also were in an excellent position to buy up American standard ships as soon as peace was declared.[76]

The war pushed the Norwegian shipyards further into repair, maintenance and industrial production. But except for a few yards that were partially destroyed in military actions, their activities and knowledge were preserved. During the war, some shipyards were involved in secret planning for post-war reconstruction. In a secret report produced by the SBL (the Association of

[75] See Olsson, *Från panserbåtvarv til Tankfartysgsvarv*, for productivity at Swedish yards; Andersen, *Fra det britiske til det amerikanske produksjonsideal*, for the Norwegian yards; and the chapters by H. G. Smith and L. C. Brown in Fassett, *Shipbuilding Business in the United States of America*, vol. I.

[76] The history of NORTRASHIP is now being written and will eventually be published in 5 volumes. The information presented here draws on drafts and articles by two of the authors, A. Towsen and B. Basberg, but responsibility for the interpretation is mine alone.

Shipbuilders) in 1944, it was argued forcefully that the yards had to modernize and take up the new methods developed by the Americans (standardization, welding, prefabrication, internal transport, planning, subcontracting) which had made a deep impression. Some of the yards also developed plans for this, although not all of them were realized.[77] All of these plans were too strongly oriented towards production efficiency regardless of market conditions, and in this sense tended to be a bit naive.

The adoption of the new production paradigm was long and gradual rather than quick and abrupt. During the war the Norwegian yards had been able to reduce their indebtedness, and with a market that wanted new ships above all the prospects seemed very good. But the availability of raw materials proved problematic, particularly in the case of steel, and the yards developed a cooperative organization for its supply. Without a domestic steel industry, it was difficult to obtain the necessary quantities for new construction. But by coordinating their orders and using their international network and supporting institutions, it was possible to overcome this problem.[78] The period from the late 1940s to the first severe shipbuilding recession in 1957–58 was a transitional period during which the larger Norwegian yards changed their ideal from British multi-shop craft-orientated production towards one based on American mass production. Strictly speaking, the model was more Swedish, but the differences were not great, and American industrial practice became the ideal vision of how a modern Norwegian shipbuilding industry should be organized. We may hesitate to call this ideal "Fordist," since it was obvious that the market never would absorb mass production of ships on a scale comparable to that of industries like automobiles. But the movement towards Fordist strategies was very clear. Fordism became synonymous with modernity and with the Social-Democratic regime which wanted to modernize Norway.[79]

The most extreme Scandinavian example of the new strategies was the Swedish Arendalsvarvet, a greenfield yard erected outside Gothenburg during the late 1950s. This yard was constructed as a single production line, a couple of kilometers long, with steel coming in at one end and a large tanker coming out at the other every three months. The layout of the yard was totally different than that of the British tradition with its many berths, quays and shops scattered around along a relatively long coastline. At Arendalsvarvet there were only two building docks, and a very short coastline; the whole construction focussed on the inland production line, along which larger and larger sections moved before being assembled in the docks. The alternative for the older yards was to redesign

[77] G. Vedeler, "Orientering i forbindelse med den påtenkte storskipsbyggingen i Norge," unpublished manuscript, later published in a slightly different version in "Skipsbyggesaken," *Teknisk ukeblad*, no. 11, 1946. For plans at other yards, see Andersen, *Fra det britiske til det amerikanske produksjonsideal*, pp. 208–224.

[78] Minutes and memoranda from the archive of *Skipsbyggerienes Landsforening* (SBL). See also Dokument 1, 1947 in *Stortingsforhandlingene* (Parliamentary Papers).

[79] See for example *Retningslinjer for gjennomfoiJring av arbeidsstudier*, Oslo, 1947; and Andersen, *Fra det britiske til det amerikanske produksjonsideal*, chs. 13–14.

their plant for flow production. The main physical change was the creation of welding sheds and areas for building up large prefabricated sections. One solution for urban yards which had extra space was to close down some berths, use the area for prefabrication and arrange the production line in an S-shaped curve. This was the case at the large Aker yard in Oslo which closed down two of four berths and one of two dry docks to obtain space for sectional construction,[80] and similar processes occurred at other yards.[81]

The road towards Fordism was long and complex, and different yards followed different paths. But there were some common themes. Several yards took up construction of large tankers, which became prototypical of "modern" shipbuilding. This meant a sharp change from the multiple product strategy of the British ideal. It was possible to earn money on new construction, as demand was booming and berths were scarce, while the repair business stagnated as the proportion of ships sailing to Norway declined, and tankers in particular rarely cast anchor in domestic harbours. This shift also altered the relations between yards and owners. At the Aker yard, for example, there was a complete turnover of customers as the firm moved from producing liners and cargo ships for the larger well-established shipping firms to producing tankers for a new and hitherto unknown clientele. While the earlier relationship with the elite liner firms was close and intimate, a new and much more businesslike attitude emerged. The changes in the firm's structure meant that brokers and consultant firms played a larger role in contract negotiation contracts and design as opposed to the liner firms who had their own in-house capacities. Other yards, such as that of Rosenberg in Stavanger, were taken over by one of the largest tanker-owners and began to produce for this firm.[82]

The introduction of planning departments and other administrative functions at the yards also meant a tremendous change from the old way of organizing work. The many small empires of the foremen and supervisors disappeared as the personnel and planning departments took control of the yards. Labor policies also shifted dramatically. At the Aker yard in Oslo, the wage system was totally altered in 1957 and all workers were employed on a salaried basis with promotion opportunities dependent on skills and seniority.[83] Four distinct elements underlay this fundamental policy shift. First was the tight labor market in the Oslo region: the local yards in the area had experienced turnover rates of up to 50 percent a year, and the scarcity of labor appeared permanent. Secondly, the war experience – when output remained steady despite orders to work as slowly as possible – clearly showed that it was very difficult for skilled craftsmen to reduce the pace of production.[84] Thirdly and most important to the shipyard directors

[80] *Ibid.*, p. 189.

[81] *Frederikstad mek. Verksted*, Fredrikstad, 1970; "Utbyggingen ved Rosenberg 1945–1955," *Rosenberg*, 10/11 (Stavanger, 1954); H. Myran and K. Fasting, *Herfra går skibe BmV 1855–1955* (Bergen, 1955); T. Stamsø and K. Fasting, *Skipsbygging på Horten gjennom 150år, 1818–1968* (Horten, 1968). [82] "Utbyggingen av Rosenberg."

[83] H. W. Andersen, "Kapitalisme med et menneskelig ansikt?", *Tidskrift for arbeiderbevegelsens historie*, 2 (1981). [84] Author's interview with senior engineer J. Sundnes, 1979.

was the fact that the old piece-pay system made it impossible to introduce proper detailed planning. The piece rate system stemmed from the British tradition of large subcontract jobs involving whole squads of workers, often lasting for a week or more. This made the pay system inflexible, and with a tight labor market it was difficult to confront the very strong union at the yard.[85]

The fourth element of the situation proved important in resolving this problem. The chairman of the union at the yard became after some conflicts rather friendly with the managing director. They came from very different traditions, but in a way shared the same aim. The chairman of the local union was a veteran of the Furubotn group, a dissident faction excluded from the Communist Party in 1949, which had its stronghold in the unions. On the ideological side this group thought the way to Communism was to develop the nation's productive forces.[86] Together with the rather young managing director and his struggle to introduce modern Fordish strategies with regard to planning, production flow, etc., this meant a possible alliance based on the modernization of the yard and the introduction of salaried status. It thus became a kind of gentlemen's agreement that productivity would not fall so long as the wages were kept at a good level.[87] To the young director this meant that he was free to introduce "scientific management" as he called it, while the union was satisfied that the yard should modernize and that they would have the same wage conditions as the staff.

The labor market situation also led the yard to expand outside Oslo, establishing a new yard for large hulls in an agricultural and fishing community at Stord on the west coast, as well as to acquire other yards.[88] In 1960 the Aker group consisted of five large yards and a number of other firms. Its organization was based on the strategy of producing large tankers which had been formulated in 1956–57. As the size of these ships made them impossible to produce at the two Oslo yards, the firm intended to produce hulls in the regions, and then to tow them to Oslo for outfitting and motor installation. This division of labor was extended as the tankers grew larger to include the production of center tanks at one yard, the surrounding hull at another, outfitting at a third and motor production at a fourth. Other yards developed different strategies, buying in hulls from external firms, etc.

85 Andersen, "Kapitalisme med et menneskelig ansikt?"
86 M. Thorsen, *Ragnar Kalheim – et liv i strid* (Oslo: Tiden, 1977); Interview with Peder Furubotn, *Telemark Arbeiderblad*, July 19, 1952.
87 Andersen, "Kapitalisme med et menneskelig ansikt?"
88 During the 1930s there were discussions about establishing a large new publicly owned yard in Norway. It was then forcefully argued that such a yard would have to be located in Oslo in order to obtain an adequate supply of skilled workers in the shipbuilding trades. In 1956 a yard similar to the proposed yard of the 1930s was established by a private firm at Stord, far outside any major city. The change of technology and production regime from the British to the American ideal meant that the social qualifications of the workforce became much more important than the actual skills. See T. C. Hageman, *Norsk skipsbygging kan brøfø 50,000 mennesker* (Oslo: DnAs forlag 1934); and Andersen, *Fra det britiske til det amerikanske produksjonsideal.*

This extreme product specialization came to an end as the market deteriorated in the early 1960s. Other types of ships had to be produced to fill all the companies' berths and docks. After some years they also diversified into other business areas. Bulk-ships became important in the 1960s and were often a more suitable size for the Norwegian yards. The economic results of the new strategies were rather good for the firms. At the Aker group, for example, the number of hours needed to build a ship of a particular size was roughly halved between 1953 and 1960, though they never reached the low level which the Swedish yards had achieved years earlier.[89]

Well into the 1960s both the Norwegian and the Swedish yards had reaped the easy profits of reorganization. From then on heavy investments and large credits were needed to compete for orders. Reinvestment proved the critical step in the downfall of the firms which had chosen the Fordist strategy. The tanker market boomed during the late 1960s and early 1970s, and investment followed suit. When the oil price shock sharply reduced transport demand, the tanker yards experienced severe difficulties arising from the cancellation of orders. Despite workforce reductions, it became impossible to pay off their heavy fixed investments. There was no way back to the British model of a multi-product, labor-intensive strategy. Shipowners' demand for large credits also worked to undermine the yards' financial situation as the burden of interest charges combined with that of investment debt could not be sustained for long once the market for their product disappeared.[90]

The crisis in shipbuilding rapidly became a political problem as it threatened the jobs of thousands of workers. All over Europe various schemes for subsidizing the yards were established. The Swedish yards were taken over by the state and organized into a single company, Svenska varv. The Norwegian yards shifted into the offshore business, building structures and semi-submersibles for the North Sea. But this solution posed its own problems, and only the two largest yards, at Stavanger and Stord, succeeded in specializing on large integrated deck structures which were sheltered from international competition by the simple fact that it was impossible to transport them over large distances. In any case, many of these contracts were allocated on political criteria aimed at averting potential unemployment.

The Fordist strategy in shipbuilding thus proved to be a short-term success while the market was good, but long-term tragedy when the market suddenly disappeared. It was not a failure in the technical sense, but was ill-adapted to changing market conditions. The attempted solutions internationally were mainly financial, with a whole array of governmental measures including public subsidies and credit facilities.[91] The British model of diversified products and

[89] My own calculations based on material from the firm's archives. See Andersen, *Fra det britiske til det amerikanske produksjonsideal*, appendix 5, for details.

[90] Bohlin, *Tre perioder i den svenska varvsindustriens lönsamhet och finansiering under 1900-talet.*

[91] D. Stokland, Økonomisk politikk og politisk økonomi: statlige støtteordninger i internasjonal skipsbyggingsindustri 1955–1983 (Oslo: UiO, 1986).

workshops with little fixed capital actually proved much more robust between the wars than the Fordist strategy after 1945. During the long period of depression from the mid-1970s to the late 1980s, most of the thirty larger shipbuilding yards in Norway were closed down. Some did survive, however, following strategies that seemed more akin to the old British model. But only public subsidies, orders and other such arrangements made this possible. It was in the shadow of this decline that the north-west coast yards quietly expanded, as we saw in the first section of this chapter.

The maritime infrastructure grew and developed regardless of the production strategies of the yards, with some important exceptions. Concentration on tankers and bulk ships meant that the structure of the shipowning companies remained small and undeveloped. A 1964 study found that of 335 Norwegian shipowning firms, 78 percent had five or fewer ships.[92] The structure of the shipping industry thus enabled the brokers and consultants to flourish by providing the knowledge which the small firms could not generate in-house. As we noted earlier, Oslo became a major international brokerage center by concentrating on the needs of Norwegian shipping firms. The resulting expertise also proved useful to owners from other countries entering similar markets during the post-war period, thereby increasing the brokers' independence and reducing their vulnerability to the changing fortunes of any single group of customers.

Veritas used the booming shipping sector to build up its research department and to provide consultancy services to firms. During the 1950s, in particular, the aspiration to a leading position in all fields of ship technology became crucial to its commercial success. But Veritas worked closely not only with the shipowners but also with the Norwegian and Swedish yards through a joint body called the Scandinavian Technical Committee, thereby managing to integrate both groups into its organization very successfully.[93] On the other hand, however, Veritas was not dependent on production technology, but rather on the product. When the offshore oil business began to flourish, they tried to incorporate it, but ran into problems both with the public authorities' safety requirements, and with the very different operational style of the large multinationals which were the main customers for such equipment. But Veritas did succeed in winning a larger share of the international fleet and in this way they too became less dependent on the Norwegian owners.[94]

Both for Veritas and for the brokers this independence was important in enabling them to survive when the owners and yards went bankrupt or closed down. Another reason for their survival was of course the simple fact that ships

[92] NTNF, *Skipsfart og skipsbygging* (Oslo 1964). The list includes members of the Shipowners Association, Summer 1963.

[93] The aim of the Technical Committee was to "gain a certain goodwill with the technical directors of shipyards and engineering companies throughout the Scandinavian countries," Minutes of the Annual General Meeting at *Det norske Veritas* (1953), p. 25.

[94] Andersen and Collett, *Anchor and Balance*, chs. 11 and 15.

need cargoes even if rates are low and they need classification and certificates whatever the fluctuations in the freight market. The crisis of the 1970s was above all a crisis for large-scale transport of oil and other bulk commodities. Both the brokers and Veritas had to turn away from new construction contracts towards freight and shipping operations. This meant reduced profits, but as typical labor-intensive service industries a contraction of this type was easier to absorb than for the frontline firms which were burdened with heavy fixed investments. The decline of the bulk transport business also spurred the supporting institutions to reorientate themselves towards new products such as supply ships for the oil industry and towards new partners such as the north-west coast yards with which we began.

Conclusions

In a sense, the Norwegian fleet operated at the margin of the world fleet for 140 years – if it is possible to call a fleet that ranked between the third and the fifth-largest in the world marginal. The concept of marginality does not therefore refer to size, but to their orientation towards the most turbulent sectors of the shipping market. Their strategies and forms of organization proved successful as a whole insofar as the Norwegian fleet became one of the major fleets in the world. But such success was always precarious. We can identify at least four separate crises, each of which threatened to destroy Norwegian shipping once and for all: the transition from sail to steam, the two world wars in which around half the fleet was lost and the crisis of the 1970s. In shipbuilding, too, we find at least two severe crashes that sent most of the established yards into bankruptcy: the crisis of wooden shipbuilding on the south coast during the late nineteenth century and crisis of the 1970s that closed nearly all of the larger steel construction yards. But each time, whether in shipping or shipbuilding, the industry revitalized itself albeit in other forms. This raises an important question: how was it possible to revive such apparently broken-backed activities?

The key to explaining both success and revitalization can be found, in my view, in the distinction between what I have called the frontline firms and the supporting institutions. This distinction originated precisely in characteristic Norwegian mode of operation – on the margin. Hence what elsewhere were considered internal functions such as technical and market knowledge came to be organized externally, in separate institutions. In this way it became easier to cope with the risks of frontline operations, enabling the owners and yards to become much more flexible and take advantage of new market opportunities. Typical examples are the development of mutual insurance clubs and later the classification society. The lack of resources and the high fees charged both by Hamburg insurers and Bureau Veritas fostered collective action among the owners and builders both in establishing mutual clubs and a classification society. As these institutions acquired an independent existence, they became crucial in the revitalization of the shipping and shipbuilding firms after the different crises. Not all of these supporting institutions, however, were intended to be collective. Brokerage grew out of the owners' market strategy but remained commercial and individual. It came nonetheless to perform similar functions to those of the more collective institutions in providing a thorough knowledge of markets and business opportunities. Hence a basic understanding of the Norwegian maritime complex must depend heavily on its infrastructure of supporting institutions, which facilitates new ventures by making available technical, market, insurance and financial knowledge and contacts. If the frontline firms were dying newcomers could thus replace them by drawing on the knowledge contained in the supporting infrastructure.

The larger foreign liner companies, by contrast, tried to build this knowledge in-house. British Lloyd's, for example, never secured the same position vis-à-vis the British fleet as Det norske Veritas won among the Norwegian. In shipbuilding, too, this distinction can be seen. Why didn't the British shipbuilding industry prosper through the 1960s as did its Norwegian or Swedish counterparts, and why did Swedish shipbuilding disappear after the crisis of Fordism in the 1970s? These questions are too large to answer here, but some suggestions can nonetheless be advanced. For the British yards there were at least two important factors. If the north-west coast yards look similar to the British in their methods, they differ in at least two respects: collaborative rather than adversarial industrial relations, and lesser reliance on vertical integration. The Swedish yards, on the other hand, came as close as possible to mass production in shipbuilding. This strategy, precisely the opposite to that of the British yards, had its triumphs in the 1950s and 1960s, but its heavy fixed investments could never pay off in times of crisis like the 1970s. In the absence of a support structure like the Norwegian, the result was the permanent closure of Swedish merchant shipbuilding.

Our main objective, however, has not been to account for the trajectory of international shipbuilding or shipping as a whole, but rather to explain how it was possible for a small country to build and support one of the world's largest merchant fleets for more than a century during which it was among the most exposed to the fluctuations of the international market. In this story, small firms, regional clustering and diversification of tasks proved crucial for the survival and renewal of the fleet and the maritime complex, blending fierce competition and close collaboration in ways appropriate to the maneuvering of the owners and builders. Hence in an age of large-scale integrated companies, this maritime complex simultaneously provided both the strength and the resources to take advantage of market opportunities. As we have clearly seen, this mode of operation was no transient way-station along the road towards large-scale firms. On the contrary, it was perhaps one of the best strategies for survival in some of the roughest seas of international competition.

Index